WITHDRAWN
WRIGHT STATE UNIVERSITY LIBRARIES

Computer Modeling in Bioengineering

Computer Modeling in Bioengineering

Theoretical Background, Examples and Software

Miloš Kojić
Harvard School of Public Health, USA
University of Kragujevac, Serbia
University of Texas Health Science Center at Houston, USA

Nenad Filipović
Faculty of Mechanical Engineering, University of Kragujevac, Serbia

Boban Stojanović
Center for Scientific Research of Serbian Academy of Sciences and Arts
and University of Kragujevac, Serbia

Nikola Kojić
Harvard-MIT Division of Health Sciences and Technology, USA
Harvard Medical School, USA

John Wiley & Sons, Ltd

Copyright © 2008 John Wiley & Sons Ltd, The Atrium, Southern Gate, Chichester,
West Sussex PO19 8SQ, England

Telephone (+44) 1243 779777

Email (for orders and customer service enquiries): cs-books@wiley.co.uk
Visit our Home Page on www.wiley.com

All Rights Reserved. No part of this publication may be reproduced, stored in a retrieval system or transmitted in any form or by any means, electronic, mechanical, photocopying, recording, scanning or otherwise, except under the terms of the Copyright, Designs and Patents Act 1988 or under the terms of a licence issued by the Copyright Licensing Agency Ltd, 90 Tottenham Court Road, London W1T 4LP, UK, without the permission in writing of the Publisher. Requests to the Publisher should be addressed to the Permissions Department, John Wiley & Sons Ltd, The Atrium, Southern Gate, Chichester, West Sussex PO19 8SQ, England, or emailed to permreq@wiley.co.uk, or faxed to (+44) 1243 770620.

Designations used by companies to distinguish their products are often claimed as trademarks. All brand names and product names used in this book are trade names, service marks, trademarks or registered trademarks of their respective owners. The Publisher is not associated with any product or vendor mentioned in this book.

This publication is designed to provide accurate and authoritative information in regard to the subject matter covered. It is sold on the understanding that the Publisher is not engaged in rendering professional services. If professional advice or other expert assistance is required, the services of a competent professional should be sought.

Other Wiley Editorial Offices

John Wiley & Sons Inc., 111 River Street, Hoboken, NJ 07030, USA

Jossey-Bass, 989 Market Street, San Francisco, CA 94103-1741, USA

Wiley-VCH Verlag GmbH, Boschstr. 12, D-69469 Weinheim, Germany

John Wiley & Sons Australia Ltd, 42 McDougall Street, Milton, Queensland 4064, Australia

John Wiley & Sons (Asia) Pte Ltd, 2 Clementi Loop #02-01, Jin Xing Distripark, Singapore 129809

John Wiley & Sons Canada Ltd, 6045 Freemont Blvd, Mississauga, ONT, L5R 4J3

Wiley also publishes its books in a variety of electronic formats. Some content that appears in print may not be available in electronic books.

Library of Congress Cataloging in Publication Data

Computer modeling in bioengineering : theoretical background, examples, and
  software / Miloš Kojić . . . [et al.].
    p. ; cm.
  Includes bibliographical references and index.
  ISBN 978-0-470-06035-3 (cloth : alk. paper)
  1. Biomedical engineering—Computer simulation. I. Kojic, Milos, 1941–
  [DNLM: 1. Biomedical Engineering.   2. Biomedical Technology.
  3. Computer Simulation.   4. Software. QT 36 C7375 2008]
  R858.C6415 2008
  610.28—dc22
                                                             2008002698

British Library Cataloguing in Publication Data

A catalogue record for this book is available from the British Library

ISBN 978-0-470-06035-3 (H/B)

Typeset in 10/12pt Times by Integra Software Services Pvt. Ltd, Pondicherry, India
Printed and bound in Great Britain by Antony Rowe Ltd, Chippenham, Wiltshire

About this Book

This book is comprised of the following entities:

A. The Main Text

Presented in three parts: I II and III
Organized in chapters and sections with reference to other entities on the web.

B. Theory – available at www.wiley.com/go/kojic

Additional details are provided which complement the main text and extend it to give the more complete presentation of the text in the Entity A.

C. Examples – available at www.wiley.com/go/kojic

Additional details are provided for examples from Entity A, and some additional examples are given.

D. Software – the link for the software can be accessed via www.wiley.com/go/kojic

User can run examples by the Software with menu for each example. The menu contains example parameters, options for execution and results display, and Tutorial. Examples are organized according to chapters and sections of the main text (Entity A). Some examples are from the main text and a number of additional examples are included.

The Software will continuously be updated by the authors.

To access parts B, C and D on the web, the user will need to use the password: nanotechnology

Contents

Contributors — xv

Preface — xvii

Part I Theoretical Background of Computational Methods — 1

1 **Notation – Matrices and Tensors** — 3
   1.1 Matrix representation of mathematical objects — 3
   1.2 Basic relations in matrix algebra — 4
   1.3 Definition of tensors and some basic tensorial relations — 6
   1.4 Vector and tensor differential operations and integral theorems — 8
   1.5 Examples — 11

2 **Fundamentals of Continuum Mechanics** — 15
   2.1 Definitions of stress and strain — 15
      2.1.1 Stress — 15
      2.1.2 Strain and strain rate — 19
      2.1.3 Examples — 21
   2.2 Linear elastic and viscoelastic constitutive relations — 26
      2.2.1 Linear elastic constitutive law — 26
      2.2.2 Viscoelasticity — 29
      2.2.3 Transformation of constitutive relations — 30
      2.2.4 Examples — 30
   2.3 Principle of virtual work — 37
      2.3.1 Formulation of the principle of virtual work — 37
      2.3.2 Examples — 38
   2.4 Nonlinear continuum mechanics — 40
      2.4.1 Deformation gradient and the measures of strain and stress — 41
      2.4.2 Nonlinear elastic constitutive relations — 45
      2.4.3 Examples — 47

3 **Heat Transfer, Diffusion, Fluid Mechanics, and Fluid Flow through Porous Deformable Media** — 51
   3.1 Heat conduction — 51
      3.1.1 Governing relations — 52
      3.1.2 Examples — 53

	3.2	Diffusion	55
		3.2.1 Differential equations of diffusion	55
		3.2.2 Examples	57
	3.3	Fluid flow of incompressible viscous fluid with heat and mass transfer	58
		3.3.1 Governing equations of fluid flow and of heat and mass transfer	59
		3.3.2 Examples	60
	3.4	Fluid flow through porous deformable media	63
		3.4.1 The governing equations	64
		3.4.2 Examples	66

Part II Fundamentals of Computational Methods   69

4 Isoparametric Formulation of Finite Elements   71
 4.1 Introduction to the finite element method   71
 4.2 Formulation of 1D finite elements and equilibrium equations   73
  4.2.1 Truss finite element   73
  4.2.2 Equilibrium equations of the FE assemblage and boundary conditions   78
  4.2.3 Examples   80
 4.3 Three-dimensional (3D) isoparametric finite element   81
  4.3.1 Element formulation   81
  4.3.2 Examples   84
 4.4 Two-dimensional (2D) isoparametric finite elements   85
  4.4.1 Formulation of the elements   86
  4.4.2 Examples   89
 4.5 Isoparametric shell finite element for general 3D analysis   91
  4.5.1 Basic assumptions about shell deformation   92
  4.5.2 Formulation of a four-node shell element   94
  4.5.3 Examples   95

5 Dynamic Finite Element Analysis   99
 5.1 Introduction to dynamics of structures   99
 5.2 Differential equations of motion   100
 5.3 Integration of differential equations of motion   101
 5.4 System frequencies and modal shapes   103
 5.5 Examples   104

6 Introduction to Nonlinear Finite Element Analysis   109
 6.1 Introduction   109
 6.2 Principle of virtual work and equilibrium equations in nonlinear incremental analysis   113
  6.2.1 Discrete system   113
  6.2.2 Principle of virtual work for a continuum   114
  6.2.3 Finite element model   115
  6.2.4 Finite element model with logarithmic strains   117
 6.3 Examples   118

CONTENTS ix

7 Finite Element Modeling of Field Problems — 121
7.1 Introduction — 121
7.1.1 General considerations — 122
7.1.2 The Galerkin method — 122
7.2 Heat conduction — 124
7.2.1 The finite element equations — 124
7.2.2 Examples — 125
7.3 Diffusion — 127
7.3.1 The finite element equations — 127
7.3.2 Examples — 128
7.4 Fluid flow with heat and mass transfer — 129
7.4.1 The finite element equations — 129
7.4.2 Examples — 133
7.5 FE equations for modeling large change of fluid domain – Arbitrary Lagrangian–Eulerian (ALE) formulation — 135
7.5.1 The ALE formulation — 135
7.5.2 Examples — 138
7.6 Solid–fluid interaction — 139
7.6.1 Loose coupling method — 140
7.6.2 Examples — 141
7.7 Fluid flow through porous deformable media — 143
7.7.1 Finite element balance equations — 143
7.7.2 Examples — 145

8 Discrete Particle Methods for Modeling of Solids and Fluids — 147
8.1 Molecular dynamics — 147
8.1.1 Introduction — 147
8.1.2 Differential equations of motion and boundary conditions — 148
8.1.3 Examples — 150
8.2 Dissipative Particle Dynamics (DPD) method — 151
8.2.1 Introduction to mesoscale DPD modeling — 151
8.2.2 Basic DPD equations — 152
8.2.3 Examples — 154
8.3 Multiscale modeling, coupling DPD-FE for fluid flow — 155
8.3.1 Introduction to multiscale modeling — 155
8.3.2 Basic equations and boundary conditions — 156
8.3.3 Examples — 160
8.4 Smoothed Particle Hydrodynamics (SPH) — 161
8.4.1 Introduction — 161
8.4.2 The basic equations of the SPH method — 161
8.4.3 Examples — 164
8.5 Element-Free Galerkin (EFG) method — 164
8.5.1 Introduction — 164
8.5.2 Formulation of the EFG method — 165
8.5.3 Examples — 168

Part III Computational Methods in Bioengineering — 171

9 Introduction to Bioengineering — 173
9.1 The subject and scope of bioengineering — 173
9.2 The role of computer modeling in bioengineering — 175
    9.2.1 Computational models — 175
    9.2.2 Future advances in computer modeling — 177

10 Bone Modeling — 181
10.1 The structure and forms of bones — 181
    10.1.1 The structure of bone tissue — 181
    10.1.2 The form of bones — 183
    10.1.3 Osteoporosis and bone density — 184
10.2 The mechanical properties of bone and FE modeling — 185
10.3 Bone fracture – medical treatment and computer modeling — 187
    10.3.1 General considerations — 187
    10.3.2 Fracture treatment — 188
    10.3.3 FE modeling of femur comminuted fracture — 190
10.4 Internal fixation of hip fracture – two solutions and computer models — 194
    10.4.1 Solutions by parallel screws and by dynamic hip implant — 194
    10.4.2 Finite element models of intracapsular fractures of the femoral neck — 195

11 Biological Soft Tissue — 201
11.1 Introduction to mechanics of biological tissue — 201
    11.1.1 Structure and function of biological tissue — 201
    11.1.2 Basic experiments and mechanical models — 203
11.2 Modeling methods for isotropic tissue — 207
    11.2.1 General concept of computational procedures — 207
    11.2.2 Biaxial models of membranes, hardening and hysteretic behavior, action of surfactant — 209
    11.2.3 Use of strain energy functions — 215
11.3 Examples — 217

12 Skeletal Muscles — 227
12.1 Introduction — 227
    12.1.1 Basic physiology of muscle mechanics — 227
    12.1.2 Basics of muscle finite element modeling — 231
12.2 Muscle modeling — 234
    12.2.1 Hill's phenomenological model — 234
    12.2.2 Determination of stresses within muscle fiber — 235
    12.2.3 Hill's model which includes fatigue — 239
    12.2.4 An extension of Hill's model to include different fiber types — 242
12.3 Examples — 245

13 Blood Flow and Blood Vessels — 249
13.1 Introduction to the cardiovascular system — 249
    13.1.1 The circulatory system — 249

13.1.2 Blood	251
13.1.3 Blood vessels	255
13.2 Methods of modeling blood flow and blood vessels	256
13.2.1 Introduction	256
13.2.2 Methods of blood flow modeling in large blood vessels	257
13.2.3 Modeling the deformation of blood vessels	260
13.2.4 Blood–blood vessel interaction	261
13.3 Human aorta	262
13.3.1 Introduction	262
13.3.2 Finite element model of the aorta	263
13.3.3 Results and discussion	264
13.4 Abdominal Aortic Aneurysm (AAA)	265
13.4.1 Introduction	265
13.4.2 Modeling of blood flow within the AAA	267
13.4.3 Results	268
13.5 Blood flow through the carotid artery bifurcation	270
13.5.1 Introduction	270
13.5.2 Finite element model of the carotid artery bifurcation	271
13.5.3 Example solutions	273
13.6 Femoral artery with stent	276
13.6.1 Femoral artery anatomical and physiological considerations and endovascular solutions	276
13.6.2 Analysis of the combined effects of the surrounding muscle tissue and inner blood pressure to the arterial wall with implanted stent	278
13.7 Blood flow in venous system	282
13.7.1 Introduction	282
13.7.2 Modeling blood flow through the veins	283
13.8 Heart model	286
13.8.1 Description of heart functioning	286
13.8.2 Computational model	289
14 Modeling Mass Transport and Thrombosis in Arteries	**295**
14.1 Introduction	295
14.2 Modeling mass transport in arteries by continuum-based methods	297
14.2.1 The basic relations for mass transport in arteries	297
14.2.2 Finite element modeling of diffusion–transport equations	298
14.2.3 Examples	299
14.3 Modeling thrombosis by continuum-based methods	302
14.3.1 Model description	302
14.3.2 Examples	304
14.4 Modeling of thrombosis by DPD	306
14.4.1 General considerations	306
14.4.2 Examples	308
15 Cartilage Mechanics	**313**
15.1 Introduction	313
15.2 Differential equations of balance in cartilage mechanics	316

15.2.1 Basic physical quantities, swelling pressure and electrokinetic coupling	316
15.2.2 Equations of balance	318
15.3 Finite element modeling of cartilage deformation	320
15.3.1 Finite element balance equations	320
15.4 Examples	322

16 Cell Mechanics   331
16.1 Introduction to mechanics of cells	331
16.2 Cell mechanical models	334
16.2.1 Stabilizing influence of CSK prestress – cellular tensegrity model	334
16.2.2 Mathematical model of a six-strut tensegrity structure	336
16.2.3 Biphasic models	339
16.3 Examples: modeling of cell in various mechanical conditions	340

17 Extracellular Mechanotransduction: Modeling Ligand Concentration Dynamics in the Lateral Intercellular Space of Compressed Airway Epithelial Cells   349
17.1 Autocrine signaling in airway epithelial cells	350
17.1.1 Introduction	350
17.1.2 The EGF–receptor autocrine loop in the LIS	351
17.1.3 Modeling the effects of compressive stress on epithelial cells *in vitro*	352
17.2 The dynamic diffusion model	356
17.2.1 Introduction	356
17.2.2 Finite element model of dynamic diffusion	357
17.2.3 Exploring the parameter space of the diffusion equation	359
17.3 The dynamic diffusion and convection model	362
17.3.1 Introduction	362
17.3.2 Finite element model of coupled diffusion and convection	363
17.3.3 Exploring the parameter space of the governing equations	366
17.3.4 Rate sensitivity of extracellular mechanotransduction	368
17.3.5 HB-EGF vs. TGF-alpha concentration dynamics	372
17.3.6 Discussion	375

18 Spider Silk: Modeling Solvent Removal during Synthetic and *Nephila clavipes* Fiber Spinning   379
18.1 Determination of the solvent diffusion coefficient in a concentrated polymer solution	380
18.1.1 Introduction	380
18.1.2 Numerical procedure	381
18.1.3 Example	386
18.2 Modeling solvent removal during synthetic fiber spinning	388
18.2.1 Introduction	388
18.2.2 Governing process during synthetic solvent removal	390
18.2.3 Numerical modeling of synthetic internal solvent diffusion	392
18.2.4 Example: Synthetic fiber spinning	394

18.3	Modeling solvent removal during *Nephila clavipes* fiber spinning	397
	18.3.1 Introduction	397
	18.3.2 *Nephila* water diffusion coefficient	398
	18.3.3 Modeling of internal water diffusion	400
	18.3.4 Example: The *Nephila* spinning canal	402

19 Modeling in Cancer Nanotechnology — 407

19.1	Introduction	407
19.2	The transport of particulates in capillaries	409
19.3	The mathematical model	414
	19.3.1 The governing equations	415
	19.3.2 The initial and boundary conditions	416
	19.3.3 Solution for K_0 and f_0	417
	19.3.4 Solution for K_1 and f_1	418
	19.3.5 Solution for K_2	419
	19.3.6 The velocity distribution (effect of boundary depletion of the solvent)	420
19.4	The concentration profile	422
	19.4.1 The mean dimensionless concentration Ψ_m	423
	19.4.2 The local dimensionless concentration Ψ	424
19.5	Comments and discussions of the analytical models and solutions	427
19.6	Numerical modeling of particle motion within capillary	428
	19.6.1 Computational procedure	428
	19.6.2 Example – trajectories of spherical and elliptical particles	429

Index — **433**

Contributors

Paolo Decuzzi
University of Magna Graecia, Italy
The University of Texas Health Science Center, USA

Mauro Ferrari
The University of Texas Health Science Center, USA
Rice University, USA

Francesco Gentile
University of Magna Graecia, Italy

Nenad Grujovic
University of Kragujevac, Serbia

Velibor Isailović
University of Kragujevac, Serbia

Miloš Ivanović
University of Kragujevac, Serbia

Nikola Jagić
University of Kragujevac, Serbia

Gareth McKinley
Massachusetts Institute of Technology, USA

Srboljub Mijailović
Harvard School of Public Health, USA

Vladimir Miloradović
University of Kragujevac, Serbia

Božidar Novaković
University of Kragujevac, Serbia

Vladimir Ranković
University of Kragujevac, Serbia

Branko Ristić
University of Kragujevac, Serbia

Mirko Rosić
University of Kragujevac, Serbia

Radovan Slavković
University of Kragujevac, Serbia

Dimitrije Stamenović
Boston University, USA

Daniel Tschumperlin
Harvard School of Public Health, USA

Akira Tsuda
Harvard School of Public Health, USA

Ivo Vlastelica
University of Kragujevac, Serbia

Miroslav Živković
University of Kragujevac, Serbia

Preface

Bioengineering in recent years has become one of the most attractive fields of research and development in industry, education and medicine. The development mainly relies on experimental investigations, which have increasingly been coupled to computer modeling.

The aim of this book is to provide basic information about the methods used in computer modeling and simulation of biological systems and processes and to present typical results of modeling. The book is accompanied by software on the world wide web for studying the representative biomechanical problems in more detail. The primary goal of the book is to serve as a textbook in various bioengineering university courses, as well as a support for basic and clinical research.

The presented text, results and software rely on the work of the authors over a number of years, together with collaborators and contributors from the University of Kragujevac, Serbia and other universities (Harvard University, Boston University, University of Texas Health Science Center at Houston, The Hong Kong Polytechnic University). Most of the topics are presented as an introduction, referring to our modeling results, rather than as an extensive overview of various approaches. On the other hand, we have given in-depth analyses where we considered useful to further elucidate the biomedical problem.

The book is divided into three parts: I Theoretical Background of Computational Methods, II Fundamentals of Computational Methods, and III Computational Methods in Bioengineering. In the first part, Chapters 1–3, the basic relations are summarized for the ease of the presentations in the subsequent chapters and for overall completeness. This summary is accompanied by a rather small number of solved examples to illustrate applications of the theoretical considerations.

Part II, Chapters 4–8, covers computational methods that are subsequently implemented in modeling of bioengineering problems. The basis is the Finite Element (FE) method which is commonly used in engineering, science and medicine. Here, we give the essence of the method, for solids, general field problems and coupled physical fields, in linear and nonlinear domains. Also, we present in Chapter 8 the fundamentals of more recent methods, such as Dissipative Particle Dynamics (DPD), Smoothed Particle Hydrodynamics (SPH), and Element-Free Galerkin (EFG) method, as well as coupling of these methods to the FE method, and a multiscale approach. The last methods are especially well suited for bioengineering applications and are also implemented in the subsequent chapters. The most representative example solutions are shown at the end of sections, and for most of them the software on the web provides a more detailed analysis.

In the last part, Chapters 9–19, the computational methods of Part II are applied to various bioengineering problems. Each chapter contains the following: physiological background and significance, description of the computational methods used for modeling of the considered problems, and example solutions. The examples are selected to be representative and the most important results are shown in the form of graphs or fields of the considered physical quantity. The solutions are mainly obtained using the current stage of the computer package PAK (University of Kragujevac, Serbia).

On the web, accompanying the book, the following sections are provided: Theory, Examples, and Software. In Theory, we provide additions to the text in the book in order to make some of the topics more complete and with more detail of certain derivations. In Examples we give further details for some of the examples from the book, or present more example solutions. The Software contains most of examples from the book, with extra examples given. For each of the examples a specific interface is developed so that the model can be generated by the menu with the suitable model parameters. These parameters can be changed within a given range in order to elucidate the effects of various conditions for the considered problem. Post-processing of results can be selected by the menu and results can be displayed by the selected option. Also, a Tutorial is provided for each example where the example is described in detail and guidance through the solution approach is suggested. The Software relies on the current stage of the software package PAK and the size of each example, as well as the range of the model parameters, are limited. Questions about the use of the Software and solutions of problems which are more general or over the prescribed limits, can be sent to the authors: *http://www.wiley.com/go/kojic*.

The book is prepared mainly as a textbook for upper-level undergraduate or graduate courses in bioengineering, engineering and applied sciences in general, and medicine. For the courses where the computational methods are important, the chapters within Part I and Part II should be used. There, a selection can be made with respect to problems of solids or field problems. In the case of the emphases on solids, Chapters 2, 4, 5, 6 and 8 provide the theoretical background and basics of the computational methods. When the emphasis is on the field problems, then Chapters 2 (with selection of sections), 3, 7 and 8 can be used. After these theoretical background chapters a number of chapters within Part III can be examined. In the case where the fundamentals of the computational methods are not essential, an overview of topics within Part I and Part II can be made, or both parts can be skipped; followed by a selection of Chapters in Part III. We have organized chapters in Part III to be self-sufficient in a way that each chapter has physiological considerations, a presentation of computational methods, and example solutions. Here, also, some of the computational methods description can be skipped without losing the essence of the computer modeling goals and purpose.

We consider that support by the Software on the web should be of great help for lecturers when organizing classes. The theoretical presentations (either from the theoretical background or from bioengineering applications) can be accompanied by use of the Software with menu-driven modeling and solution display. Use of the Software can also aid students when studying various theoretical or bioengineering problems.

The book is also prepared to be useful for researchers in various fields related to bioengineering as well as other scientific fields, including medical applications. The book provides basic information about how a bioengineering (or medical) problem can be modeled, which computational models can be used, and the background of the applied computer models. Each of the bioengineering problems treated in this book has been analyzed elsewhere from different aspects, with more detail and particular theoretical considerations. We have referred to these analyses to a certain extent, but these referrals are far from being complete, since the field of computer modeling in bioengineering is vast and consistently expanding.

PREFACE

We are indebted to all contributors who, as experts in their fields, added to the book's overall depth and scope: P. Decuzzi – Chapter 19; M. Ferrari – 9,19; F. Gentile – 19; N. Grujović – 4–7, Development of PAK program; V. Isailović – 16; M. Ivanović – 8,14; N. Jagić – 13; G. McKinley – 18; S. Mijailović – 12,15; V. Miloradović – 13; B. Novaković – 13; V. Ranković – 10, 11,13; B. Ristić – 10; M. Rosić – 12, 13; R. Slavković – 4–7, Development of PAK program; D. Stamenović – 9, 16; D. Tschumperlin – 17; A. Tsuda – 11, 14; I. Vlastelica – 8, 11, Development of PAK program; M. Živković – 4–7, Development of PAK program.

The authors are grateful to a number of institutions and individuals who supported this challenging project. Among the institutions, we are grateful to: Serbian Academy of Sciences and Arts (Center for Scientific Research of Serbian Academy of Sciences and Arts and University of Kragujevac); University of Kragujevac (including Faculty of Mechanical Engineering and Faculty of Medicine); Harvard School of Public Health; University of Texas Health Science Center at Houston, School of Health Information Sciences, and Alliance for NanoHealth Texas; The Hong Kong Polytechnic University; University of Vienna (Austria); Boston University. The research leading to the results presented in the book has been supported by a number of institutions: Serbian Ministry of Science, City of Kragujevac (Serbia), NIH (USA), NASA (USA), Institute for Water Resources 'Jaroslav Černi' (Belgrade, Serbia).

The authors express their deep gratitude to the contributors (the list is given in the second page) and to other collaborators who put enormous effort in preparing the examples, running the software, and drawing the figures, among whom are: Vladimir Ranković, Velibor Isailović, Mileta Nedeljković, Danko Milašinović, Radun Vulović, Dejan Veljković and Dejan Petrović, all from Center for Scientific Research of Serbian Academy of Sciences and Arts and University of Kragujevac. The authors thank Dr. Jeffrey Drazen (Harvard Medical School), Prof. Bora Mikić (MIT) and Prof. David Kaplan (Tufts University) for their advice and support.

We are all deeply grateful to our families who have supported us over many years in this work.

Miloš Kojić, Nenad Filipović,
Boban Stojanović, Nikola Kojić

Part I

Theoretical Background of Computational Methods

1

Notation – Matrices and Tensors

In this chapter, we give the definition of matrices and tensors for the purpose of notation used in the book. We summarize the basic relations that are useful for reading the text, without any proofs or in-depth presentation. More details are given on the web (Theory, Chapter 1, Examples, Chapter 1). For further reading the reader should consult books specialized to this topic (e.g. Fung 1965, Malvern 1969, Mase & Mase 1999).

1.1 Matrix representation of mathematical objects

For some physical quantities, a single number is sufficient to define the quantity. For example, we use T to denote temperature at a given material point, and associate a certain value with T (e.g. $T = 20\,°C$). The quantity specified by a single number is called a *scalar*. We will denote scalars by italic letters.

However, many quantities need more than one number to be completely defined. For example, in order to define velocity of a particle, we need to know not only the magnitude of the velocity, but also its direction and orientation in space. The spatial direction, magnitude and orientation are defined by, say, three velocity components in a Cartesian coordinate system: v_1, v_2, v_3. We denote the velocity by a bold letter **v**, associating with it three scalars v_1, v_2, v_3. In general, we use a bold lower case letter for a *vector* **b**, or a column *matrix* of the order $1 \times n$, defined as

$$\mathbf{b} = \begin{bmatrix} b_1 \\ b_2 \\ \cdot \\ \cdot \\ b_{n-1} \\ b_n \end{bmatrix} \qquad (1.1.1)$$

Computer Modeling in Bioengineering Edited by M. Kojić, N. Filipović, B. Stojanović, N. Kojić
© 2008 John Wiley & Sons, Ltd

We will be using notation of a transpose of a vector, $\mathbf{b}^T$, which assumes an interchange of the rows and columns, i.e.,

$$\mathbf{b}^T = [b_1 \; b_2 \; .. \; b_{n-1} \; b_n] \qquad (1.1.2)$$

For some physical quantities we need more complex representation than a vector. For example, the state of stress at a material point is represented by values of forces per unit area at three orthogonal planes (details are shown on the web – Theory, Chapter 1). Hence we need nine scalars (three for each force), which we order in a two-dimensional matrix form as σ_{11}, σ_{12}, σ_{13}; . . . . . ; σ_{31}, σ_{32}, σ_{33}. In general, a *two-dimensional matrix* $\mathbf{B}$ (capital bold letter is used for a matrix) of order $m \times n$ is defined as a mathematical object with terms B_{ij} in which the first index denotes the row number, and the second index represents the column number,

$$\mathbf{B} = \begin{bmatrix} B_{11} & B_{12} & . & . & B_{1n-1} & B_{1n} \\ B_{21} & B_{22} & . & . & B_{2n-1} & B_{2n} \\ . & . & . & . & . & . \\ . & . & . & . & . & . \\ B_{m-1,1} & B_{m-1,2} & . & . & B_{m-1,n-1} & B_{m-1,n} \\ B_{m1} & B_{m2} & . & . & B_{m,n-1} & B_{m,n} \end{bmatrix} \qquad (1.1.3)$$

We will be mainly using *square matrices*, where $m = n$. If the columns and rows are interchanged, then we have the transposed matrix,

$$\left(\mathbf{B}^T\right)_{ij} = B_{ji} \qquad (1.1.4)$$

Following this definition of matrix, it is possible to extend the matrix to have more than two dimensions. However, in our presentation throughout the book, by a matrix we assume a *two-dimensional square matrix*, unless otherwise stated. A square matrix is *symmetric* when

$$B_{ji} = B_{ij} \quad i, j = 1, 2, \ldots, n \qquad (1.1.5)$$

1.2 Basic relations in matrix algebra

We list here some of the basic matrix algebra relationships that are used in this book.

The *addition* of vectors $\mathbf{a}$ and $\mathbf{b}$ is expressed by

$$\mathbf{c} = \mathbf{a} + \mathbf{b} \quad \text{or} \quad c_i = a_i + b_i \quad i = 1, 2, \ldots, n \qquad (1.2.1)$$

resulting in the vector $\mathbf{c}$ with components c_i. In the case of subtraction we have a 'minus' instead of 'plus' sign. Summation of matrices $\mathbf{A}$ and $\mathbf{B}$ assumes the same order of these matrices, say $m \times n$, and is given as

$$\mathbf{C} = \mathbf{A} + \mathbf{B} \quad \text{or} \quad C_{ij} = A_{ij} + B_{ij} \quad i = 1, 2, \ldots, m; \; j = 1, 2, \ldots, n \qquad (1.2.2)$$

The resulting matrix $\mathbf{C}$ is also of the order $m \times n$, with terms C_{ij}.

NOTATION – MATRICES AND TENSORS

The *scalar (matrix) multiplication* of vectors **a** and **b** results in the scalar c according to the relation

$$c = \mathbf{a}^T \mathbf{b} = \mathbf{b}^T \mathbf{a} = \sum_{s=1}^{n} a_s b_s = a_s b_s \qquad (1.2.3)$$

We will generally omit the summation sign by using the convention that the summation is carried over repeated indices (here the index 's' is also called the dummy index). This is known as the *Einstein summation convention*. On the other hand, a dyadic multiplication $\mathbf{ab}^T$ gives the matrix **C**,

$$\mathbf{C} = \mathbf{ab}^T \quad \text{or} \quad C_{ij} = a_i b_j \qquad (1.2.4)$$

with the terms C_{ij}.

The *matrix multiplication* between a matrix and a vector, or between two matrices, is defined as follows:

$$\mathbf{c} = \mathbf{Ab}, \quad c_i = A_{ik} b_k \qquad (1.2.5)$$

$$\mathbf{C} = \mathbf{AB}, \quad C_{ij} = A_{ik} B_{kj} \qquad (1.2.6)$$

Note that the following relation can be proved

$$\mathbf{C}^T = (\mathbf{AB})^T = \mathbf{B}^T \mathbf{A}^T, \quad \left(\mathbf{C}^T\right)_{ij} = B_{ki} A_{jk} \qquad (1.2.7)$$

The scalar multiplication of two matrices is given as

$$c = \mathbf{A} \cdot \mathbf{B} = A_{ij} B_{ij} \qquad (1.2.8)$$

The inverse of a matrix **A**, denoted as $\mathbf{A}^{-1}$, is the matrix which satisfies the relation

$$\mathbf{AA}^{-1} = \mathbf{I}, \quad A_{ik} A_{kj}^{-1} = \delta_{ij} \qquad (1.2.9)$$

where **I** is the identity matrix (diagonal terms equal to one, all other terms equal to zero), and δ_{ij} are the Kronecker delta symbols: $\delta_{ij} = 1$ for $i = j$; $\delta_{ij} = 0$ for $i \neq j$.

The determinant of a matrix **A**, denoted as $det\mathbf{A}$ or $\|\mathbf{A}\|$, is defined as

$$\det \mathbf{A} \equiv \|\mathbf{A}\| = e_{ijk} A_{1i} A_{2j} A_{3k}, \quad i,j,k = 1,2,3 \qquad (1.2.10)$$

where e_{ijk} is the permutation symbol, with values: $e_{ijk} = 0$ for $i = j$ or $j = k$ or $i = k$ or $i = j = k$; $e_{ijk} = 1$ for even permutation of 1,2,3; $e_{ijk} = -1$ for odd permutation of 1,2,3. Calculation of the matrix determinant is needed for the matrix inversion, therefore the inverse matrix exists if the determinant of the matrix is not equal to zero.

Two matrices **A** and **B** are orthogonal if the following relationship is satisfied:

$$\mathbf{AB} = \mathbf{I}, \quad \text{or} \quad A_{ik} B_{kj} = \delta_{ij}, \quad i,j = 1,2,\ldots,m; \; k = 1,2,\ldots,n \qquad (1.2.11)$$

Note that the identity matrix is of dimension $m \times m$. The matrix **A** is orthogonal if

$$\mathbf{A}^T \mathbf{A} = \mathbf{I}, \quad \text{or} \quad A_{ki} A_{kj} = \delta_{ij}, \quad i,j = 1,2,\ldots,m; \; k = 1,2,\ldots,n \qquad (1.2.12)$$

1.3 Definition of tensors and some basic tensorial relations

Tensors are mathematical objects defined by components associated with a coordinate system. These components change when the coordinate system is changed, according to certain rules (tensorial transformation rules). Note, however, that tensors (as well as vectors) do not change with the change of the coordinate system, only their components change. We will use tensors to represent physical quantities for which the transformation rules have the physical background. Throughout the book we will be using the Cartesian coordinate system with unit vectors (triad) $\mathbf{i}_1$, $\mathbf{i}_2$, $\mathbf{i}_3$, shown in Fig. 1.3.1.

A first-order tensor $\mathbf{b}$ is represented in two coordinate systems (with unit vectors $\mathbf{i}_k$ and $\bar{\mathbf{i}}_k$) as

$$\mathbf{b} = b_k \mathbf{i}_k = b_1 \mathbf{i}_1 + b_2 \mathbf{i}_2 + b_3 \mathbf{i}_3 = \bar{b}_k \bar{\mathbf{i}}_k = \bar{b}_1 \bar{\mathbf{i}}_1 + \bar{b}_2 \bar{\mathbf{i}}_2 + \bar{b}_3 \bar{\mathbf{i}}_3 \qquad (1.3.1)$$

where b_k and $\bar{b}_k$ are the vector components in the two coordinate systems. The relationships between the components in the two systems are given by

$$\bar{b}_j = T_{jk} b_k \qquad (1.3.2)$$

where T_{jk} are the cosines of angles between the unit vectors $\bar{\mathbf{i}}_j$ and $\mathbf{i}_k$ of the two coordinate systems, $T_{jk} = \cos(\bar{\mathbf{i}}_j, \mathbf{i}_k)$. Note that this equation represents the matrix multiplication of the form (1.2.5), involving the 3×3 transformation matrix $\mathbf{T}$ and the 3×1 vector of the form (1.1.1).

A second-order tensor $\mathbf{B}$ is defined as

$$\begin{aligned}\mathbf{B} &= B_{jk} \mathbf{i}_j \mathbf{i}_k = B_{11} \mathbf{i}_1 \mathbf{i}_1 + B_{12} \mathbf{i}_1 \mathbf{i}_2 + B_{13} \mathbf{i}_1 \mathbf{i}_3 + \ldots + B_{31} \mathbf{i}_3 \mathbf{i}_1 + B_{32} \mathbf{i}_3 \mathbf{i}_2 + B_{33} \mathbf{i}_3 \mathbf{i}_3 = \\ &= \bar{B}_{jk} \bar{\mathbf{i}}_j \bar{\mathbf{i}}_k = \bar{B}_{11} \bar{\mathbf{i}}_1 \bar{\mathbf{i}}_1 + \bar{B}_{12} \bar{\mathbf{i}}_1 \bar{\mathbf{i}}_2 + \bar{B}_{13} \bar{\mathbf{i}}_1 \bar{\mathbf{i}}_3 + \ldots + \bar{B}_{31} \bar{\mathbf{i}}_3 \bar{\mathbf{i}}_1 + \bar{B}_{32} \bar{\mathbf{i}}_3 \bar{\mathbf{i}}_2 + \bar{B}_{33} \bar{\mathbf{i}}_3 \bar{\mathbf{i}}_3\end{aligned} \qquad (1.3.3)$$

with components B_{jk} and $\bar{B}_{jk}$ in the coordinate systems $\mathbf{i}_k$ and $\bar{\mathbf{i}}_k$, respectively. These components can be represented in the matrix form (1.1.3). The transformation of the tensorial components due to change of the coordinate system is

$$\bar{B}_{jm} = T_{jk} B_{ks} T_{ms} \qquad (1.3.4a)$$

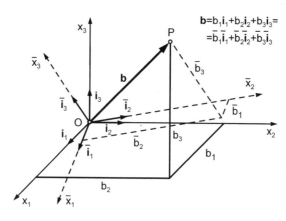

Fig. 1.3.1 Graphical representation of a vector $\mathbf{b}$ in two Cartesian systems

NOTATION – MATRICES AND TENSORS

which corresponds to the matrix multiplication (1.2.6),

$$\overline{\mathbf{B}} = \mathbf{TBT}^T \tag{1.3.4b}$$

Tensors of higher order can be defined, following (1.3.3), but we will use the second-order tensors and will call them tensors.

We further cite the tensorial relations used in the book. The *dot product (multiplication)* of two vectors, tensor and vector, and two tensors, are consecutively defined as follows:

$$c = \mathbf{a} \cdot \mathbf{b} = a_k \mathbf{i}_k \cdot b_m \mathbf{i}_m = a_k b_k \tag{1.3.5}$$

$$\mathbf{c} = \mathbf{Ab} = A_{jk} \mathbf{i}_j \mathbf{i}_k \cdot b_m \mathbf{i}_m = A_{jk} b_m \mathbf{i}_j \mathbf{i}_k \cdot \mathbf{i}_m = A_{jk} b_m \mathbf{i}_j \delta_{km} = A_{jk} b_k \mathbf{i}_j \tag{1.3.6}$$

$$c_j = A_{jk} b_k$$

$$\mathbf{C} = \mathbf{AB} = A_{jk} \mathbf{i}_j \mathbf{i}_k \cdot B_{ms} \mathbf{i}_m \mathbf{i}_s = A_{jk} B_{ms} \mathbf{i}_j \mathbf{i}_k \cdot \mathbf{i}_m \mathbf{i}_s = A_{jk} B_{ms} \delta_{km} \mathbf{i}_j \mathbf{i}_s = A_{jk} B_{ks} \mathbf{i}_j \mathbf{i}_s \tag{1.3.7}$$

$$C_{js} = A_{jk} B_{ks}$$

Here we have employed the orthogonality of the unit vectors $\mathbf{i}_k$ and $\mathbf{i}_m$, i.e., $\mathbf{i}_k \cdot \mathbf{i}_m = \delta_{km}$. The dot product of two vectors is also called the scalar product. It can be seen that the dot product of two vectors gives a scalar, the dot product of tensor and vector gives vector, and the dot product of two tensors gives a tensor.

We will also use the *cross-product* of two vectors defined as

$$\mathbf{c} = \mathbf{a} \times \mathbf{b}, \quad \text{or} \quad c_i = e_{ijk} a_j b_k \tag{1.3.8}$$

The *scalar product* of two tensors gives a scalar, and is defined as

$$c = \mathbf{A} \cdot \mathbf{B} = A_{ij} B_{ij} \tag{1.3.9}$$

The *Euclidean norms* of a vector and a tensor are

$$\|\mathbf{b}\| = (b_j b_j)^{1/2}, \quad \|\mathbf{A}\|_2 = (A_{ij} A_{ij})^{1/2} \tag{1.3.10}$$

The *rotation tensor* $\mathbf{R}$ corresponding to two coordinate systems with unit vectors $\mathbf{i}_k$ and $\overline{\mathbf{i}}_k$ is defined by the components

$$R_{km} = \cos(\mathbf{i}_k, \overline{\mathbf{i}}_m) \tag{1.3.11}$$

It can be shown that the following relationship holds (see web – Theory, Chapter 1)

$$\overline{\mathbf{i}}_m = \mathbf{R} \mathbf{i}_m \tag{1.3.12}$$

leading to a rotation of vector $\mathbf{i}_m$. Multiplication of any vector $\mathbf{b}$ by the rotation tensor $\mathbf{R}$ rotates this vector as it rotates the vector $\mathbf{i}_m$. The following relation is valid $\mathbf{R} = \mathbf{T}^T$. It is important to emphasize that multiplication of a vector $\mathbf{b}$ by the transformation matrix $\mathbf{T}$ gives the vector components in a rotated coordinate system of the same vector (see (1.3.2) and Fig. 1.3.1; also see web – Theory, Chapter 1). On the other hand, multiplication of a vector **b** by the rotation tensor **R** produces another vector $\bar{\mathbf{b}}$, rotated with respect to **b** (see web – Theory, Chapter 1).

Finally, we define the principal values and principal directions of a tensor. In order to introduce these quantities, consider the following equation

$$\mathbf{Ap} = \lambda \mathbf{p} \quad \text{or} \quad (\mathbf{A} - \lambda \mathbf{I})\mathbf{p} = \mathbf{0} \qquad (1.3.13)$$

where λ is a scalar, and **p** is a unit vector. The equation has a nontrivial solution ($\mathbf{p} \neq \mathbf{0}$) if determinant of the system matrix $\mathbf{A} - \lambda \mathbf{I}$ is equal to zero,

$$\det(\mathbf{A} - \lambda \mathbf{I}) = 0, \quad \text{or} \quad \lambda^3 - I_1 \lambda^2 + I_2 \lambda - I_3 = 0 \qquad (1.3.14)$$

where I_1, I_2, I_3 are the first, second and third invariants of the tensor **A**,

$$I_1 = tr\mathbf{A} = A_{ii} = A_{11} + A_{22} + A_{33}, \quad I_2 = \frac{1}{2}\left(A_{ii}A_{jj} - A_{ij}A_{ji}\right),$$
$$I_3 = \det \mathbf{A} = e_{ijk}A_{1i}A_{2j}A_{3k}, \quad i,j,k = 1,2,3 \qquad (1.3.15)$$

If the matrix **A** is symmetric and its terms are real numbers, then there are three real solutions λ_1, λ_2, λ_3 which are the *principal values*, or *eigenvalues* of matrix **A**. To each principal value λ_k there corresponds the *principal vector*, or *eigenvector* $\mathbf{p}_k$. It can be shown that principal vectors $\mathbf{p}_k$ are orthogonal (or orthonormal if eigenvectors are unit vectors), forming the *principal basis* of the tensor **A** (see Example 1.5-4). Therefore, the tensor **A** in the principal basis, written in a matrix and tensorial form, is:

$$\mathbf{A} = \begin{bmatrix} \lambda_1 & 0 & 0 \\ 0 & \lambda_2 & 0 \\ 0 & 0 & \lambda_2 \end{bmatrix}, \quad \text{or} \quad \mathbf{A} = \lambda_1 \mathbf{p}_1 \mathbf{p}_1 + \lambda_2 \mathbf{p}_2 \mathbf{p}_2 + \lambda_3 \mathbf{p}_3 \mathbf{p}_3 \qquad (1.3.16)$$

In practical calculations of the principal vectors, we use (for a given principal value λ_k) two of the equations (1.3.13) and one representing the condition that $\mathbf{p}_k$ is a unit vector. Therefore the following system of equations is solved:

$$(A_{11} - \lambda_k)p_{(k)1} + A_{12}p_{(k)2} + A_{13}p_{(k)3} = 0$$
$$A_{21}p_{(k)1} + (A_{22} - \lambda_k)p_{(k)2} + A_{23}p_{(k)3} = 0, \quad \text{no sum on } k \qquad (1.3.17)$$
$$(p_{(k)1})^2 + (p_{(k)2})^2 + (p_{(k)3})^2 = 1$$

Further details about the eigenvalue problem can be seen in Examples 1.5-4; 2.1-4 and 2.1-6 on the web (Examples, Chapter 2).

1.4 Vector and tensor differential operations and integral theorems

Here, the vector and tensor differential operations and integral theorems used in this book are summarized. Throughout the text, we mainly refer to the Cartesian coordinate system.

NOTATION – MATRICES AND TENSORS

Differential Operations

We start with the *differential operator* 'nabla' or 'del', defined as

$$\nabla = \frac{\partial}{\partial x_1}\mathbf{i}_1 + \frac{\partial}{\partial x_2}\mathbf{i}_2 + \frac{\partial}{\partial x_3}\mathbf{i}_3 \equiv \frac{\partial}{\partial x_k}\mathbf{i}_k \tag{1.4.1}$$

which is a vector-operator with components $\partial/\partial x_k$, $k = 1, 2, 3$. When applied to a scalar function $\phi(x_1, x_2, x_3)$, the *gradient of the function* ϕ is obtained,

$$\nabla\phi = \frac{\partial\phi}{\partial x_1}\mathbf{i}_1 + \frac{\partial\phi}{\partial x_2}\mathbf{i}_2 + \frac{\partial\phi}{\partial x_3}\mathbf{i}_3 \equiv \frac{\partial\phi}{\partial x_k}\mathbf{i}_k \tag{1.4.2}$$

as a vector, with the components $\partial\phi/\partial x_k$. The *gradient of a vector field* $\mathbf{b}(x_1, x_2, x_3)$ is

$$\nabla\mathbf{b} = \frac{\partial}{\partial x_k}(b_j\mathbf{i}_j)\mathbf{i}_k = \frac{\partial b_j}{\partial x_k}\mathbf{i}_j\mathbf{i}_k \tag{1.4.3}$$

and represents the second-order tensor with components $(\nabla\mathbf{b})_{kj} = \partial b_j/\partial x_k$. The $\nabla\mathbf{b}$ is called the *dyadic product* of the vectors ∇ and $\mathbf{b}$.

The *divergence of a vector field* $\mathbf{b}(x_1, x_2, x_3)$ is defined as

$$\nabla\cdot\mathbf{b} \equiv div\mathbf{b} = \frac{\partial}{\partial x_k}(b_j\mathbf{i}_j)\cdot\mathbf{i}_k = \frac{\partial b_j}{\partial x_k}\mathbf{i}_j\cdot\mathbf{i}_k = \frac{\partial b_j}{\partial x_k}\delta_{jk} = \frac{\partial b_k}{\partial x_k} \tag{1.4.4}$$

and represents a scalar. The *divergence of a tensor field* $\mathbf{A}(x_1, x_2, x_3)$ is

$$\nabla\cdot\mathbf{A} \equiv div\mathbf{A} = \mathbf{i}_k\frac{\partial}{\partial x_k}\cdot(A_{jm}\mathbf{i}_j\mathbf{i}_m) = \frac{\partial A_{jm}}{\partial x_k}\mathbf{i}_k\cdot\mathbf{i}_j\mathbf{i}_m = \frac{\partial A_{km}}{\partial x_k}\mathbf{i}_m \tag{1.4.5}$$

Therefore, $\nabla\cdot\mathbf{A}$ is a vector with components $(\nabla\cdot\mathbf{A})_m = \partial A_{km}/\partial x_k$.

The *curl of a vector field* is (see the definition of cross-product (1.3.8)) is

$$\nabla\times\mathbf{b} = e_{mjk}\frac{\partial b_k}{\partial x_j}\mathbf{i}_m = \left(\frac{\partial b_3}{\partial x_2} - \frac{\partial b_2}{\partial x_3}\right)\mathbf{i}_1 + \left(\frac{\partial b_1}{\partial x_3} - \frac{\partial b_3}{\partial x_1}\right)\mathbf{i}_2 + \left(\frac{\partial b_2}{\partial x_1} - \frac{\partial b_1}{\partial x_2}\right)\mathbf{i}_3 \tag{1.4.6}$$

representing a vector with components shown here.

The *Laplacian operator* is defined as

$$\nabla\cdot\nabla \equiv \Delta = \mathbf{i}_k\frac{\partial}{\partial x_k}\cdot\mathbf{i}_m\frac{\partial}{\partial x_m} = \frac{\partial^2}{\partial x_k\partial x_k} \equiv \frac{\partial^2}{(\partial x_1)^2} + \frac{\partial^2}{(\partial x_2)^2} + \frac{\partial^2}{(\partial x_3)^2} \tag{1.4.7}$$

which is a scalar differential operator. Hence, the *Laplacian of a scalar field* is the scalar

$$\nabla\cdot\nabla\phi \equiv \Delta\phi = \frac{\partial^2\phi}{\partial x_k\partial x_k} \equiv \frac{\partial^2\phi}{(\partial x_1)^2} + \frac{\partial^2\phi}{(\partial x_2)^2} + \frac{\partial^2\phi}{(\partial x_3)^2} \tag{1.4.8}$$

while the *Laplacian of a vector field* is the vector

$$\nabla\cdot\nabla\mathbf{b} \equiv \Delta\mathbf{b} = \frac{\partial^2 b_j}{\partial x_k\partial x_k}\mathbf{i}_j = (\Delta b_j)\mathbf{i}_j \tag{1.4.9}$$

with components Δb_j.

Integral Theorems

We list here the integral theorems that are used subsequently. The mostly used is the *Gauss Theorem* (Fung 1965, Bird et al. 2002). If a closed volume V in space is bounded by a surface S, then for a vector field we have

$$\int_V \nabla \cdot \mathbf{b}\, dV = \int_S \mathbf{n} \cdot \mathbf{b}\, dS, \quad \text{or} \quad \int_V \frac{\partial b_k}{\partial x_k} dV = \int_S b_j n_j\, dS \qquad (1.4.10)$$

where $\mathbf{n}$ is the unit normal of the surface element dS. This relation is also known as the Gauss–Ostrogradskii or the Divergence Theorem. In the case of a scalar field $\phi(x_k)$ and a tensor field $\mathbf{A}(x_k)$, the Gauss theorem is

$$\int_V \nabla \phi\, dV = \int_S \phi \mathbf{n}\, dS, \quad \text{or} \quad \int_V \frac{\partial \phi}{\partial x_i} dV = \int_S \phi n_i\, dS \qquad (1.4.11)$$

and

$$\int_V \nabla \cdot \mathbf{A}\, dV = \int_S \mathbf{n} \cdot \mathbf{A}\, dS, \quad \text{or} \quad \int_V \frac{\partial A_{ki}}{\partial x_k} dV = \int_S n_k A_{ki}\, dS \qquad (1.4.12)$$

Note that in case of a two-dimensional domain, the volume V becomes the surface S and the surface S in the above integrals becomes the two-dimensional domain contour line L. Proof of the Gauss theorem and additional details are given on the web – Theory, Chapter 1.

Next, we give the expression for the so-called *material derivative of volume integral*. Assume that a continuum is moving in space. Considering the continuum as a set of material particles, we have that physical quantities, such as mass density, temperature, velocity, stresses, are associated with each material particle, changing over time while particles move. Let ψ_V be the total value of a quantity ψ (it is a scalar: temperature, density, ..., or a component of a vector or tensor) over all material particles occupying currently a fixed closed space volume V_{space}, i.e.

$$\psi_V = \int_{V_{space}} \psi\, dV \qquad (1.4.13)$$

If we want to find the rate of change of ψ_V, we obtain

$$\frac{D\psi_V}{Dt} = \int_{V_{space}} \frac{\partial \psi}{\partial t} dV + \int_{S_{space}} \psi \mathbf{v} \cdot \mathbf{n}\, dS =$$

$$= \int_{V_{space}} \left(\frac{\partial \psi}{\partial t} + \frac{\partial \psi}{\partial x_k} v_k + \psi \frac{\partial v_k}{\partial x_k} \right) dV = \int_{V_{space}} \left(\frac{D\psi}{Dt} + \psi \frac{\partial v_k}{\partial x_k} \right) dV \qquad (1.4.14)$$

The integral over the surface S_{space}, which encloses the volume V_{space}, represents the flux of ψ through the surface. We have used the Gauss theorem (1.4.11) to transform the surface integral to the volume integral. Also we use the notation $D\psi_V/Dt$ to indicate that the time derivative is evaluated assuming the same material particles. Consequently, the derivative

$$\frac{D\psi}{Dt} \equiv \frac{\partial \psi}{\partial t} + \frac{\partial \psi}{\partial x_k} v_k \qquad (1.4.15)$$

is called the *material (or substantial) derivative* of ψ. If a spatial field of a physical quantity ψ which changes with time is defined, $\psi(x_k, t)$, then the derivative $\partial \psi / \partial t$ is the *local derivative* assuming constant spatial coordinates x_k; while the term $(\partial \psi / \partial x_k) v_k$ is the *convective derivative* which takes into account motion of the material particle. Therefore, for a given spatial field ψ, the sum of these last two derivatives gives the rate of change $D\psi/Dt$ for a material particle (material point) at a given space position. The material derivatives are used when transport phenomena are studied (e.g. mass and heat transport).

Additional details about the relations presented in this section are given on the web – Theory, Chapter 1.

1.5 Examples

Example 1.5-1. Prove the e-δ identity
The following relations between the permutation symbol e_{ijk} and the Kronecker-delta symbol δ_{ij} can be proved:

$$e_{ijk} e_{imn} = \delta_{jm} \delta_{kn} - \delta_{jn} \delta_{km} \quad \text{(E1.5-1.1)}$$

Details of the poof of these relations are given on the web – Examples, Section 1.5.

Example 1.5-2. Derive the procedure for calculation of the inverse matrix
For simplicity, consider a 3×3 matrix $\mathbf{A}$. The matrix $\mathbf{A}^{-1}$ is the matrix which satisfies the relation (1.2.9). We write the matrix $\mathbf{A}^{-1}$ as

$$\mathbf{A}^{-1} = \left[A_{ij}^{-1} \right] = \left[\mathbf{x}_{(1)} \mathbf{x}_{(2)} \mathbf{x}_{(3)} \right] \quad \text{(E1.5-2.1)}$$

where the vector $\mathbf{x}_{(i)}$ is the i-th column of the matrix $\mathbf{A}^{-1}$, i.e. we have that $x_{(i)j} = A_{ji}^{-1}$. The equations (1.2.9) can be written as a system of three equations:

$$\mathbf{A} \mathbf{x}_{(i)} = \boldsymbol{\delta}_{(i)}, \quad i = 1, 2, 3 \quad \text{(E1.5-2.2)}$$

where the vectors $\boldsymbol{\delta}_{(1)}, \boldsymbol{\delta}_{(2)}, \boldsymbol{\delta}_{(3)}$ have the components $\delta_{(i)j} = \delta_{ij}$.

By solving this system of equations, we obtain that the terms A_{ij}^{-1} of the matrix $\mathbf{A}^{-1}$ are:

$$A_{ij}^{-1} = \frac{1}{D} D^{ji}, \quad \text{or} \quad \mathbf{A}^{-1} = \frac{1}{D} \mathbf{D}^T \quad \text{(E1.5-2.3)}$$

where $\mathbf{D} = \left[D^{ij} \right]$ is the matrix of cofactors of the matrix $\mathbf{A}$. Details of this derivation and problems for exercise are given on the web – Examples, Section 1.5.

Example 1.5-3. Determine the rotation tensor using the relation (1.3.12)
The component form of this relation is

$$\bar{i}_{(\alpha)k} = R_{kj} i_{(\alpha)j} \quad \text{(E1.5-3.1)}$$

We multiply this equation by $i_{(\alpha)s}$ and sum such three equations on α to obtain

$$\sum_{\alpha=1}^{3} \bar{i}_{(\alpha)k} i_{(\alpha)s} = \sum_{\alpha=1}^{3} R_{kj} i_{(\alpha)j} i_{(\alpha)s} = R_{kj} \sum_{\alpha=1}^{3} i_{(\alpha)j} i_{(\alpha)s} = R_{kj} \delta_{js} = R_{ks} \quad \text{(E1.5-3.2)}$$

We have used here the orthogonality property of the base vectors $\mathbf{i}_{(\alpha)}$. Namely, we have that $i_{(\alpha)k} = i_{(k)\alpha}$ and then

$$\sum_{\alpha=1}^{3} i_{(j)\alpha} i_{(s)\alpha} = \mathbf{i}_{(j)} \cdot \mathbf{i}_{(s)} = \delta_{js} \quad \text{(E1.5-3.3)}$$

The relation (E1.5-3.1) can be written in a dyadic (direct notation) as

$$\mathbf{R} = \sum_{\alpha=1}^{3} \bar{\mathbf{i}}_{(\alpha)} \mathbf{i}_{(\alpha)} = \bar{\mathbf{i}}_{(1)} \mathbf{i}_{(1)} + \bar{\mathbf{i}}_{(2)} \mathbf{i}_{(2)} + \bar{\mathbf{i}}_{(3)} \mathbf{i}_{(3)} \quad \text{(E1.5-3.4)}$$

The tensor $\mathbf{R}$ written in a matrix form is

$$\mathbf{R} = \begin{array}{c|ccc} & \bar{x}_1 & \bar{x}_2 & \bar{x}_3 \\ \hline x_1 & l_1 & l_2 & l_3 \\ x_2 & m_1 & m_2 & m_3 \\ x_3 & n_1 & n_2 & n_3 \end{array} \quad \text{(E1.5-3.5)}$$

where the coefficients l_i, m_i, n_i are the cosines of the angles between, respectively, the axes x_1, x_2, x_3 and $\bar{x}_1, \bar{x}_2, \bar{x}_3$. Note that the transformation matrix $\mathbf{T}$ in (1.3.2) is

$$\mathbf{T} = \mathbf{R}^T \begin{array}{c|ccc} & x_1 & x_2 & x_3 \\ \hline \bar{x}_1 & l_1 & m_1 & n_1 \\ \bar{x}_2 & l_2 & m_2 & n_2 \\ \bar{x}_3 & l_3 & m_3 & n_3 \end{array} \quad \text{(E1.5-3.6)}$$

EXERCISE
(a) Determine the rotation tensor $\mathbf{R}$ using the relation (1.3.12) and two orthogonal bases $\mathbf{p}_{(\alpha)}$ and $\bar{\mathbf{p}}_{(\alpha)}$ which do not coincide. The solution is (see web – Examples, Section 1.5):

$$\mathbf{R} = \sum_{\alpha}^{3} \bar{\mathbf{p}}_{(\alpha)} \mathbf{p}_{(\alpha)} \quad \text{(E1.5-3.7)}$$

(b) Write the rotation tensor in the bases $\mathbf{p}_{(\alpha)}$ and $\bar{\mathbf{p}}_{(\alpha)}$.

(c) Prove that the rotation tensor is orthogonal, i.e. $\mathbf{R}^T \mathbf{R} = \mathbf{I}$.

Example 1.5-4. Show that that the principal vectors are orthogonal and that the symmetric tensor in the principal directions is diagonal

Consider two principal vectors $\mathbf{p}_m$ and $\mathbf{p}_n$ of the symmetric matrix $\mathbf{A}$, corresponding to the eigenvalues $\lambda_m \neq \lambda_n$. Then, according to (1.3.13) we have

$$\begin{aligned} \mathbf{A}\mathbf{p}_m &= \lambda_m \mathbf{p}_m \\ \mathbf{A}\mathbf{p}_n &= \lambda_n \mathbf{p}_n, \quad \text{no sum on } m \text{ and } n \end{aligned} \quad \text{(E1.5-4.1)}$$

NOTATION – MATRICES AND TENSORS

If we multiply the first equation by $\mathbf{p}_n^T$ and the second equation by $\mathbf{p}_m^T$ and subtract the resulting scalar equations, we obtain

$$\mathbf{p}_n^T \mathbf{A} \mathbf{p}_m - \mathbf{p}_m^T \mathbf{A} \mathbf{p}_n = 0 = (\lambda_m - \lambda_n) \mathbf{p}_n^T \mathbf{p}_m \tag{E1.5-4.2}$$

since the matrix $\mathbf{A}$ is symmetric, and $\mathbf{p}_n^T \mathbf{p}_m = \mathbf{p}_m^T \mathbf{p}_n$. The vectors $\mathbf{p}_m$ and $\mathbf{p}_n$ are the unit vectors, therefore we have that

$$\mathbf{p}_n^T \mathbf{p}_m = \delta_{mn} \tag{E1.5-4.3}$$

which shows that the principal vectors are orthogonal (or orthonormal). The orthogonality of the principal vectors is also applicable in the case when some principal values are equal. Details of the proof for this case can be found elsewhere, e.g. in Bathe (1996).

To prove that $\mathbf{A}$ is diagonal in the coordinate base $\mathbf{p}_m$, we write the system of equations (1.3.13) in a matrix form as

$$\mathbf{A}\mathbf{P} = \mathbf{P}\Lambda \tag{E1.5-4.4}$$

where the matrices $\mathbf{P}$ and Λ are

$$\mathbf{P} = [\mathbf{p}_1 \mathbf{p}_2 \mathbf{p}_3], \quad \Lambda = \begin{bmatrix} \lambda_1 & 0 & 0 \\ 0 & \lambda_2 & 0 \\ 0 & 0 & \lambda_3 \end{bmatrix} \tag{E1.5-4.5}$$

Note that $\mathbf{P}^T \mathbf{P} = \mathbf{I}$ due to the othogonality condition (E1.5-4.3). From multiplication of (E1.5-4.4) from the left by $\mathbf{P}^T$ it follows

$$\mathbf{P}^T \mathbf{A} \mathbf{P} = \mathbf{P}^T \mathbf{P} \Lambda = \Lambda, \quad \text{hence} \quad \overline{\mathbf{A}} = \Lambda \tag{E1.5-4.6}$$

where $\overline{\mathbf{A}}$ is the matrix in the coordinate system with the base vectors $\mathbf{p}_m$ (see also the transformation rule (1.3.4a,b)).

EXERCISE
Express the matrix $\mathbf{A}$ in terms of matrices $\mathbf{P}$ and Λ (spectral decomposition of A) and determine $\mathbf{A}^{-1}$. Solution is given on the web – Examples, Section 1.5.

Example 1.5-5. Determine the differential operator 'nabla' in cylindrical coordinate system

The relations between Cartesian x, y, z and cylindrical system r, θ, z are (see Fig. E1.5-5, where P is a point in space and P' is its projection onto $x - y$ or $r - x$ plane),

$$x = r\cos\theta, \quad y = r\sin\theta, \quad z = z \tag{E1.5-5.1}$$

The relations between the unit vectors $\mathbf{i}_x, \mathbf{i}_y, \mathbf{i}_z$ and $\mathbf{r}_0, \mathbf{c}_0, \mathbf{i}_z$ are

$$\mathbf{r}_0 = \cos\theta \mathbf{i}_x + \sin\theta \mathbf{i}_y, \quad \mathbf{c}_0 = -\sin\theta \mathbf{i}_x + \cos\theta \mathbf{i}_y, \quad \mathbf{i}_z = \mathbf{i}_z \tag{E1.5-5.2}$$

Relations between the partial derivatives in the two coordinate systems are

$$\frac{\partial}{\partial x} = \cos\theta \frac{\partial}{\partial r} - \frac{1}{r}\sin\theta \frac{\partial}{\partial \theta}, \quad \frac{\partial}{\partial y} = \sin\theta \frac{\partial}{\partial r} + \frac{1}{r}\cos\theta \frac{\partial}{\partial \theta} \tag{E1.5-5.3}$$

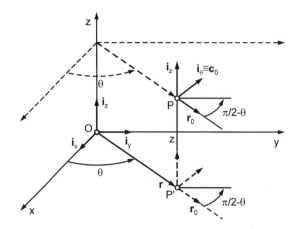

Fig. E1.5-5 Cartesian and cylindrical coordinate systems

Using the relations (E1.5-5.2) and (E1.5-5.3) we obtain that the operator ∇ in the cylindrical coordinate system is

$$\nabla = \mathbf{r}_0 \frac{\partial}{\partial r} + \mathbf{c}_0 \frac{1}{r}\frac{\partial}{\partial \theta} + \mathbf{i}_z \frac{\partial}{\partial z} \qquad (E1.5\text{-}5.4)$$

Note that the following relations are valid:

$$\begin{aligned}\frac{\partial \mathbf{r}_0}{\partial \theta} &= -\sin\theta\, \mathbf{i}_x + \cos\theta\, \mathbf{i}_y = \mathbf{c}_0 \\ \frac{\partial \mathbf{c}_0}{\partial \theta} &= -\cos\theta\, \mathbf{i}_x - \sin\theta\, \mathbf{i}_y = -\mathbf{r}_0\end{aligned} \qquad (E1.5\text{-}5.5)$$

which are obtained from (E1.5-5.2). Derivatives of $\mathbf{r}_0, \mathbf{c}_0, \mathbf{i}_z$ with respect to r and z are equal to zero.

Detailed derivations of the above relations and problems for exercise are given on the web – Examples, Section 1.5.

References

Bathe, K.J. (1996). *Finite Element Procedures*, Prentice-Hall, Englewood Cliffs, NJ.
Bird, R.B., Stewart, W.E. & Lightfoot, E.N. (2002). *Transport Phenomena*, 2nd edition, John Wiley & Sons, Ltd., Chichester, England.
Fung, Y.C. (1965). *Foundations of Solid Mechanics*, Prentice-Hall, Englewood Cliffs, NJ.
Malvern, L.E. (1969). *Introduction to the Mechanics of a Continuous Medium*, Prentice-Hall, Englewood Cliffs, NJ.
Mase, G.T. & Mase, G.E. (1999). *Continuum Mechanics for Engineers*, CRC Press, Boca Raton, FL.

2

Fundamentals of Continuum Mechanics

In this chapter we briefly summarize the fundamentals of continuum mechanics of solids. We first define the basic concepts in continuum mechanics, such as stress and strain and then present the constitutive relations that are used in modeling biosolids. Further, the principle of virtual work is given in a simple form for linear problems. This is the most fundamental principle on which rely almost all numerical methods in the book. Finally, the basic relations of nonlinear continuum mechanics are presented. Each section is followed by typical examples. Additional details and examples are given on the web (Theory, Chapter 2; Examples, Chapter 2). A number of books cover in-depth the mechanics of continuous media (e.g. Fung 1965, Malvern 1969, Mase & Mase 1999).

2.1 Definitions of stress and strain

We define stress and strain as the basic mechanical quantities. The definitions assume small deformation of a continuous medium. It is assumed that the continuum represents a deformable solid which moves in space and deforms under mechanical action. The stresses and strains are defined at a material point of the body. During body deformation, the stresses and strains change and the aim of continuum mechanics is to determine these changes within the entire body, i.e. for all material points. This approach, where we follow the history of changes of quantities (such as stresses and strains) at material points, is called the *Lagrangian description* of continuous media.

2.1.1 Stress

In order to define stress, consider a material body $\mathcal{B}$ which deforms under mechanical action, as schematically shown in Fig. 2.1.1a. Internal forces are generated within the body

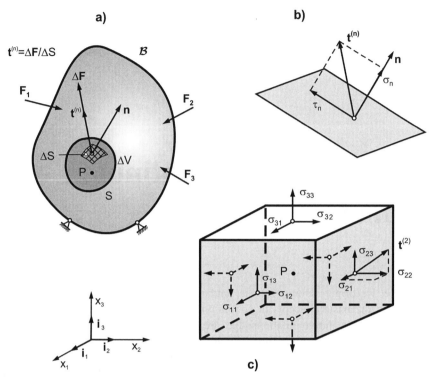

Fig. 2.1.1 Stresses within a material body produced by mechanical action. (a) Elementary material volume ΔV around point P, force $\Delta \mathbf{F}$ and stress vector $\mathbf{t}^{(n)}$ on the surface ΔS with normal $\mathbf{n}$; (b) Normal σ_n and tangential τ_n components of the stress vector $\mathbf{t}^{(n)}$; (c) Components σ_{ij} of the stress tensor $\boldsymbol{\sigma}$ at the material point P

due to this deformation. These internal mechanical forces can be specified if we consider a small volume ΔV of material around a *material point* P in $\mathcal{B}$. The mechanical action of the surrounding material on the material within the elementary volume ΔV occurs over surface ΔS and the force per unit area $\mathbf{t}^{(n)}$ is referred to as the stress vector, where the superscript n indicates an outer unit normal $\mathbf{n}$ to ΔS. Thus,

$$\mathbf{t}^{(n)} = \lim_{\Delta S \to 0} \frac{\Delta \mathbf{F}}{\Delta S} = \frac{d\mathbf{F}}{dS} \tag{2.1.1}$$

where $\Delta \mathbf{F}$ is the force acting on ΔS. The stress vector $\mathbf{t}^{(n)}$ can be decomposed into its normal and tangential components – the normal and tangential stresses, σ_n and τ_n, respectively, where σ_n is in the direction of $\mathbf{n}$ and τ_n lies in the plane with normal $\mathbf{n}$,

$$\mathbf{t}^{(n)} = \boldsymbol{\sigma}_n + \boldsymbol{\tau}_n \tag{2.1.2}$$

as shown in Fig. 2.1.1b.

Consider an elementary volume around material point P bounded by surfaces parallel to the coordinate planes of the Cartesian system x_1, x_2, x_3 (Fig. 2.1.1c). We can specify the stress vectors on each of the elementary surfaces with the normals $\mathbf{i}_1, \mathbf{i}_2, \mathbf{i}_3$ and get normal

FUNDAMENTALS OF CONTINUUM MECHANICS

and tangential components of these vectors, denoted as $\sigma_{11}, \sigma_{12}, \sigma_{13}, \ldots, \sigma_{31}, \sigma_{32}, \sigma_{33}$, with the directions considered positive in continuum mechanics; components $\sigma_{11}, \sigma_{22}, \sigma_{33}$ are the normal stress components, while $\sigma_{ij}, i \neq j$, are the shear stress components. The stress components can be written in a matrix form as follows

$$\boldsymbol{\sigma} = \begin{bmatrix} \sigma_{11} & \sigma_{12} & \sigma_{13} \\ \sigma_{21} & \sigma_{22} & \sigma_{23} \\ \sigma_{31} & \sigma_{32} & \sigma_{33} \end{bmatrix} \qquad (2.1.3)$$

It can be shown that the stress components σ_{ij} transforms according to the tensorial transformation rule (1.3.4),

$$\overline{\boldsymbol{\sigma}} = \mathbf{T}\boldsymbol{\sigma}\mathbf{T}^T \qquad (2.1.4)$$

therefore the stress at a material point, expressed in terms of the components σ_{ij}, represents the *stress tensor*,

$$\boldsymbol{\sigma} = \sigma_{km}\mathbf{i}_k\mathbf{i}_m \qquad (2.1.5)$$

It also can be shown that the stress tensor is symmetric. The proofs that the stress components σ_{ij} are symmetric and of a tensorial character rely on the physical condition of balance of forces acting on a material element (see web – Theory, Chapter 2).

Stress vector $\mathbf{t}^{(n)}$ on a plane with the normal $\mathbf{n}$ can be represented as follows

$$\mathbf{t}^{(n)} = \boldsymbol{\sigma}\mathbf{n}, \qquad t_i^{(n)} = \sigma_{ij}n_j \qquad (2.1.6)$$

This relation is known as the Cauchy representation theorem (or Cauchy formula) and $\boldsymbol{\sigma}$ is the Cauchy stress tensor. Therefore, if the stress tensor is known at a material point, we can calculate stresses (vector $\mathbf{t}^{(n)}$), as well as the normal and shear stresses, on any plane passing through that point (details are given on the web – Theory, Chapter 2; Examples, Chapter 2).

The principal values of stresses $\sigma_1, \sigma_2, \sigma_3$ acting on planes with unit normals $\mathbf{p}_1, \mathbf{p}_2, \mathbf{p}_3$ can be obtained by performing the eigenanalysis (see Section 1.3 and web – Theory, Chapter 2). Then, the stress matrix has a diagonal form,

$$\boldsymbol{\sigma} = \begin{bmatrix} \sigma_1 & 0 & 0 \\ 0 & \sigma_2 & 0 \\ 0 & 0 & \sigma_3 \end{bmatrix} \qquad (2.1.7)$$

and the stress tensor is

$$\boldsymbol{\sigma} = \sigma_1\mathbf{p}_1\mathbf{p}_1 + \sigma_2\mathbf{p}_2\mathbf{p}_2 + \sigma_3\mathbf{p}_3\mathbf{p}_3 \qquad (2.1.8)$$

Note that the shear stresses on the principal planes are equal to zero.

The stress state at a material point can be represented by Mohr's circles (see web – Theory, Chapter 2), with a use of the so-called pole of the Mohr circle. From this representation, it can be found that maximum shear stresses are acting on the planes rotated for $\pi/4$ with respect to the principal axes.

In many applications, it is necessary to represent the stress state by the mean stress σ_m,

$$\sigma_m = \frac{1}{3}(\sigma_{11} + \sigma_{22} + \sigma_{33}) = \frac{1}{3}(\sigma_1 + \sigma_2 + \sigma_3) \quad (2.1.9)$$

and by the deviatoric stress $\boldsymbol{\sigma}'$, with the components

$$\sigma'_{ij} = \sigma_{ij} - \sigma_m \delta_{ij} \quad (2.1.10)$$

Obviously, the deviatoric stress $\boldsymbol{\sigma}'$ differs from the stress $\boldsymbol{\sigma}$ in the normal components only. According to equation (2.1.9), the mean stress is one-third of the trace of $\boldsymbol{\sigma}$ and thus it has the same value in any coordinate system (i.e. it is invariant with respect to the coordinate transformation).

It is of fundamental importance that the stress components within a continuum are such that the equilibrium equations (balance of linear momentum) must be satisfied (see web – Theory and Examples, Chapter 2),

$$\frac{\partial \sigma_{ik}}{\partial x_k} + f_i^V = 0 \quad i,k = 1,2,3; \quad \text{or} \quad (\nabla^T \boldsymbol{\sigma})^T + \mathbf{f}^V = \mathbf{0} \quad (2.1.11)$$

where f_i^V (N/m^3) are components of the body force per unit volume $\mathbf{f}^V$; ∇ is the 'nabla' differential operator defined in (1.4.1); and $\nabla^T \boldsymbol{\sigma}$ is the matrix multiplication (1.2.5) between vector ∇ and matrix $\boldsymbol{\sigma}$. If inertial forces are taken into account, we obtain (from the balance of linear momentum) the dynamic equations of motion

$$-\rho \ddot{u}_i + \frac{\partial \sigma_{ik}}{\partial x_k} + f_i^V = 0, \quad i,k = 1,2,3 \quad (2.1.12)$$

where ρ is the mass density and $\ddot{u}_i$ are the components of acceleration of the material point.

Finally, we give a representation of stress in the form of a one-dimensional matrix (whose size is 6×1), since due to the symmetry of stresses there are only six different stress components. We can define the stress as

$$[\boldsymbol{\sigma}]^T = [\sigma_{11} \; \sigma_{22} \; \sigma_{33} \; \sigma_{12} \; \sigma_{23} \; \sigma_{13}] \quad (2.1.13)$$

where, obviously, $\sigma_1 \equiv \sigma_{11}, \sigma_2 \equiv \sigma_{22}, \ldots, \sigma_6 \equiv \sigma_{13}$. Then, the transformation (2.1.4) can be written in a form

$$[\bar{\boldsymbol{\sigma}}] = \mathbf{T}^\sigma [\boldsymbol{\sigma}] \quad (2.1.14)$$

where the matrix $\mathbf{T}^\sigma$ contains cosines of angles between the axes $\bar{x}_i$ and x_i (see Example 2.2-2). Note that the equilibrium equations (2.1.11) can be written using the one-index stress representation (2.1.13), as

$$\mathbf{L}\boldsymbol{\sigma} + \mathbf{f}^V = \mathbf{0}, \quad \text{or} \quad L_{ij}\sigma_j + f_i^V = 0 \quad (2.1.15)$$

where operator $\mathbf{L}$ is

$$\mathbf{L} = \begin{bmatrix} \partial/\partial x_1 & 0 & 0 & \partial/\partial x_2 & 0 & \partial/\partial x_3 \\ 0 & \partial/\partial x_2 & 0 & \partial/\partial x_1 & \partial/\partial x_3 & 0 \\ 0 & 0 & \partial/\partial x_3 & 0 & \partial/\partial x_2 & \partial/\partial x_1 \end{bmatrix} \quad (2.1.16)$$

It is possible to write three stress invariants, scalar functions of the stresses, which do not change when the stress components are changed via coordinate transformations (see web – Theory, Chapter 2).

FUNDAMENTALS OF CONTINUUM MECHANICS

The equilibrium equations in the cylindrical coordinate system r, θ, z are

$$\frac{\partial \sigma_{rr}}{\partial r} + \frac{1}{r}\frac{\partial \sigma_{r\theta}}{\partial \theta} + \frac{\partial \sigma_{rz}}{\partial z} + \frac{\sigma_{rr} - \sigma_{\theta\theta}}{r} + f_r^V = 0$$

$$\frac{\partial \sigma_{r\theta}}{\partial r} + \frac{1}{r}\frac{\partial \sigma_{\theta\theta}}{\partial \theta} + \frac{\partial \sigma_{\theta z}}{\partial z} + \frac{2\sigma_{r\theta}}{r} + f_\theta^V = 0 \quad (2.1.17)$$

$$\frac{\partial \sigma_{rz}}{\partial r} + \frac{1}{r}\frac{\partial \sigma_{\theta z}}{\partial \theta} + \frac{\partial \sigma_{zz}}{\partial z} + \frac{\sigma_{rz}}{r} + f_z^V = 0$$

Details of the derivation of these equations are given on the web – Theory, Chapter 2.

2.1.2 Strain and strain rate

Strain is a mechanical quantity used to specify a measure of material deformation. It is commonly said that strain is used to define kinematics of material deformation. We here assume that the deformation is small, i.e. that displacements of material points due to deformation are small.

Consider first the strain e of a material element of the elementary length ds (see Fig. 2.1.2a), as

$$e = \frac{d(ds)}{ds} \quad (2.1.18)$$

giving the change of unit length. Following this definition, we can define the strains of material elements in the directions of a coordinate system x, y, z (we also use notation x, y, z instead of x_1, x_2, x_3 for ease of presentation)

$$e_{xx} = \frac{d(dx)}{dx}, \quad e_{yy} = \frac{d(dy)}{dy}, \quad e_{zz} = \frac{d(dz)}{dz}, \quad \text{or}$$

$$e_{ii} = \frac{d(dx_i)}{dx_i}, \quad i = 1, 2, 3 \text{ (no sum on } i\text{)} \quad (2.1.19)$$

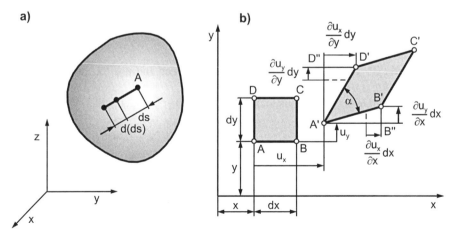

Fig. 2.1.2 Definition of small strains. (a) Extension of a line element; (b) Deformation of a rectangular material element

However, in order to describe distortion of the material, it is necessary to introduce shear strains as measures of shape change. These strains are defined as change of the angle between any two initially orthogonal line segments (Fig. 2.1.2b). Hence, for a coordinate system x, y, z we have

$$\gamma_{xy} = \frac{\pi}{2} - \alpha_{xy}, \quad \gamma_{yz} = \frac{\pi}{2} - \alpha_{yz}, \quad \gamma_{xz} = \frac{\pi}{2} - \alpha_{xz}, \quad \text{or}$$

$$\gamma_{ij} = \frac{\pi}{2} - \alpha_{ij}, \quad i \neq j, \quad i, j = 1, 2, 3$$

(2.1.20)

where α_{ij} is the angle between the line segments dx_i and dx_j after deformation. Clearly, the shear strains are symmetric, i.e. $\gamma_{ji} = \gamma_{ij}$. The shear strains γ_{ij} are the *engineering shear strains* traditionally used in engineering literature. On the other hand, the *tensorial shear strains* are used in continuum mechanics, defined as

$$e_{ij} = \frac{1}{2}\gamma_{ij}, \quad i \neq j, \quad i, j = 1, 2, 3 \qquad (2.1.21)$$

It can be shown that the strain components e_{ij}, with the shear strains (2.1.21), transform according to the tensorial rule (1.3.4). This property of e_{ij} follows from kinematics of deformation (see web – Theory, Chapter 2). Therefore, from the components e_{ij} we can obtain the strains $\bar{e}_{ij}$ in a rotated coordinate system $\bar{x}_i$ and determine change in lengths and distortions associated with other directions (see web – Theory, Chapter 2). Strain tensor, $\mathbf{e}$, can be expressed in a matrix form, or in a tensorial form (see (2.1.3) and (2.1.5) for stress tensor) as

$$\mathbf{e} = e_{km}\mathbf{i}_k\mathbf{i}_m \qquad (2.1.22)$$

This strain tensor $\mathbf{e}$ is called the *small (linear) strain tensor*.

The strain tensor can be represented in the principal directions or by use of Mohr circles. Also, we will use the strain represented by a 6×1 matrix,

$$[\mathbf{e}]^T = [e_{11} \; e_{22} \; e_{33} \; \gamma_{12} \; \gamma_{23} \; \gamma_{13}] \qquad (2.1.23)$$

Then, the transformation (2.1.4) can be written as

$$[\bar{\mathbf{e}}] = \mathbf{T}^e [\mathbf{e}] \qquad (2.1.24)$$

with the matrix $\mathbf{T}^e$ containing cosines of angles between the axes $\bar{x}_i$ and x_i (see Example 2.2-2).

The strain components can be calculated from the displacements u_i,

$$e_{ij} = \frac{1}{2}\left(\frac{\partial u_i}{\partial x_j} + \frac{\partial u_j}{\partial x_i}\right), \quad \text{or} \quad \mathbf{e} = \frac{1}{2}\left(\nabla \mathbf{u} + (\nabla \mathbf{u})^T\right) \qquad (2.1.25)$$

and therefore the strain field can be obtained from the displacement field. In the cylindrical coordinate system the strains are

$$e_{rr} = \frac{\partial u_r}{\partial r}, \quad e_{\theta\theta} = \frac{1}{r}\frac{\partial u_\theta}{\partial \theta} + \frac{u_r}{r}, \quad e_{zz} = \frac{\partial u_z}{\partial z}$$

$$e_{r\theta} = \frac{1}{2}\left(\frac{1}{r}\frac{\partial u_r}{\partial \theta} + \frac{\partial u_\theta}{\partial r} - \frac{u_\theta}{r}\right), \quad e_{\theta z} = \frac{1}{2}\left(\frac{\partial u_\theta}{\partial z} + \frac{1}{r}\frac{\partial u_z}{\partial \theta}\right), \quad e_{zr} = \frac{1}{2}\left(\frac{\partial u_r}{\partial z} + \frac{\partial u_z}{\partial r}\right)$$

(2.1.26)

FUNDAMENTALS OF CONTINUUM MECHANICS

where u_r, u_θ, u_z are radial, circumferential and axial components of the displacement vector **u**, respectively. Details of derivation of the expressions (2.1.26) are given on the web – Theory, Chapter 2.

Finally, we introduce the rate of strain (or strain rate, or rate of deformation) tensor **D** and the spin tensor **W**. The strain rate tensor is defined as

$$D_{ij} = \dot{e}_{ij} = \frac{1}{2}\left(\frac{\partial \dot{u}_i}{\partial x_j} + \frac{\partial \dot{u}_j}{\partial x_i}\right), \quad \text{or} \quad \dot{\mathbf{e}} = \frac{1}{2}\left(\nabla \dot{\mathbf{u}} + (\nabla \dot{\mathbf{u}})^T\right) \quad (2.1.27)$$

and the spin tensor is

$$W_{ij} = \frac{1}{2}\left(\frac{\partial \dot{u}_i}{\partial x_j} - \frac{\partial \dot{u}_j}{\partial x_i}\right), \quad \text{or} \quad \mathbf{W} = \frac{1}{2}\left(\nabla \dot{\mathbf{u}} - (\nabla \dot{\mathbf{u}})^T\right) \quad (2.1.28)$$

where $\nabla \dot{\mathbf{u}}$ is defined according to (1.4.3). Note that the spin tensor is skew-symmetric, i.e. $W_{ii} = 0$; $W_{ji} = -W_{ij}$. Therefore, we have in general three nonzero components W_{12}, W_{23}, W_{31} and can associate a rotation vector **w** with components $w_1 = -W_{31}$, $w_2 = -W_{23}$, $w_3 = -W_{12}$. The strain rate tensor gives the rate of deformation, while the spin tensor (or the rotation vector) gives the rate of material rotation.

As in the case of the stress tensor, the strain tensor invariants also can be defined (see web – Theory, Chapter 2).

2.1.3 Examples

Example 2.1-1. Representation of stress state for two-dimensional problems by the Mohr circle and with use of the pole of Mohr circle

Let $x-y$ be a plane in which we perform the analysis for a two-dimensional problem (plane strain, plane stress, or axisymmetric conditions; see Section 2.2). Then, the stress tensor is

$$\sigma = \begin{bmatrix} \sigma_{xx} & \sigma_{xy} & 0 \\ \sigma_{yx} & \sigma_{yy} & 0 \\ 0 & 0 & \sigma_{zz} \end{bmatrix} \quad \text{(E2.1-1.1)}$$

To this stress tensor, there correspond two principal stresses in the $x-y$ plane, σ_1 and σ_2. The given stresses (E2.1-1.1) can be expressed in terms of the principal stresses and angle α between the first principal direction and the x-axis (see (2.1.4) and web – Theory, Chapter 2),

$$\sigma_{xx} = \frac{1}{2}(\sigma_1 + \sigma_2) + \frac{1}{2}(\sigma_1 - \sigma_2)\cos 2\alpha, \quad \sigma_{yy} = \frac{1}{2}(\sigma_1 + \sigma_2) - \frac{1}{2}(\sigma_1 - \sigma_2)\cos 2\alpha$$

$$\sigma_{xy} = \frac{1}{2}(\sigma_1 - \sigma_2)\sin 2\alpha$$

(E2.1-1.2)

By eliminating $\cos 2\alpha$ and $\sin 2\alpha$ from the first and last equation of the system (E2.1-1.2), we obtain that

$$\left(\sigma_{xx} - \frac{\sigma_1 + \sigma_2}{2}\right)^2 + \sigma_{xy}^2 = \left(\frac{\sigma_1 - \sigma_2}{2}\right)^2 \quad \text{(E2.1-1.3)}$$

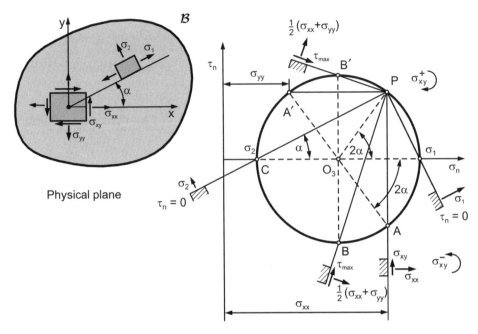

Fig. E2.1-1 Mohr circle for a two-dimensional problem and pole of the Mohr circle

This equation represents a Mohr circle in the $\sigma_n-\tau_n$ plane, as shown in Fig. E2.1-1. Details of derivation of (E2.1-1.2) are given on the web – Theory, Chapter 2.

A graphical method for determination of stresses for two-dimensional problems can be formulated by use of the Mohr circle pole. To outline this method, we first emphasize that the stresses considered here are acting on planes in which lies the z-axis. Each plane intersects the $x-y$ plane along a straight line; for example, the plane on which act the stresses σ_{xx} and σ_{xy} intersects the $x-y$ plane along a straight line parallel to y-axis. The method is described as follows. We first determine a point P, called the pole of the Mohr circle. Assuming that the stresses $\sigma_{xx}, \sigma_{yy}, \sigma_{xy}$ are given, we draw a circle passing through points $A(\sigma_{xx}, \sigma_{xy})$ and $A'(\sigma_{yy}, \sigma_{yx})$, with the center $O_3((\sigma_{xx}+\sigma_{yy})/2, 0)$, Fig. E2.1-1. A positive direction for the shear stress is shown in the figure; it is the opposite with respect to the direction used in calculations. Then, we draw a vertical line through point A, representing the direction of the plane on which the stresses σ_{xx} and σ_{xy} are acting and obtain the point P – *the pole of Mohr circle* – at the intersection with the circle. Note that we can draw a horizontal line through point A' and obtain the same point P. The pole has the following characteristic: by drawing the direction of a plane through P we obtain the intersection with the circle which gives the stresses σ_n and τ_n on that plane. For example, by drawing the line PC we obtain the direction of the principal plane on which the stress σ_2 is acting; the plane is inclined by the angle α with respect to the x-axis. Or, planes PB and PB' are the planes with the maximum shear stress $\tau_{\max}$. Note that the relations (E2.1-1.2) also follow from the graphical representation in Fig. E2.1-1. For a given state of stress there corresponds one pole P. Hence, the pole can be considered as the characteristic point in the $\sigma_n-\tau_n$ plane representing the state of stress at a material point. This unique mapping between the state of stress and the pole in the $\sigma_n-\tau_n$ plane has been used to graphically determine stress fields in plasticity (e.g. Prager 1959, Kojić & Cheatham 1974).

FUNDAMENTALS OF CONTINUUM MECHANICS

EXERCISE
Select a direction at an angle of $\pi/4$ with respect to the x-axis and find normal and shear stresses on that plane using the pole of Mohr circle in Fig. E2.1-1.

Example 2.1-2. Homogenous deformation
A *homogenous deformation* is a deformation of a material body where the displacement field is given as (Mase & Mase 1999)

$$u_i = a_{ij}{}^0 x_j \qquad (\text{E2.1-2.1})$$

where a_{ij} are constants and ${}^0 x_j$ are the initial coordinates of material points. For this displacement field we obtain that the small (linear) strains are (see (2.1.25)),

$$e_{ij} = \frac{1}{2}\left(\frac{\partial u_i}{\partial^0 x_j} + \frac{\partial u_j}{\partial^0 x_i}\right) = \frac{1}{2}(a_{ij} + a_{ji}) = const. \qquad (\text{E2.1-2.2})$$

Therefore, the strains are independent of material coordinates. A number of interesting cases and results can be analyzed for the homogenous deformation. We here consider several examples.

(a) *Uniaxial deformation.* If $a_{11} = a_0$, and all other coefficients a_{ij} are equal to zero, we have (the upper left index t is used to denote a deformed configuration)

$$\begin{aligned} u_1 &= a_0{}^0 x_1, & {}^t x_1 &= (1+a_0)\,{}^0 x_1; & e_{11} &= a_0 \\ u_2 &= u_3 = 0, & {}^t x_2 &= {}^0 x_2,\ {}^t x_3 = {}^0 x_3; & e_{ij} &= 0\,(\text{except } e_{11}) \end{aligned} \qquad (\text{E2.1-2.3})$$

Note that if the displacement vector **u** is along a direction $\bar{x}$ with the unit vector **n** (n_1, n_2, n_3) (see Fig. E2.1-2Aa), so that

$$\bar{u} = a_0{}^0\bar{x}, \quad \text{and} \quad u_i = a_0{}^0\bar{x} n_i = a_0\left({}^0 x_1 n_1 + {}^0 x_2 n_2 + {}^0 x_3 n_3\right) n_i \qquad (\text{E2.1-2.4})$$

where ${}^0 x_1, {}^0 x_2, {}^0 x_3$ are coordinates of the vector ${}^0\bar{x}\mathbf{n}$, we further obtain the strain components as

$$e_{ij} = a_0 n_i n_j \qquad (\text{E2.1-2.5})$$

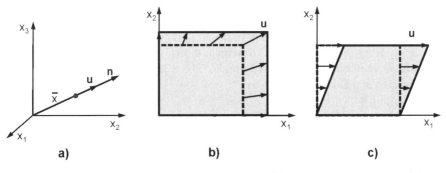

Fig. E2.1-2A Homogenous deformation of material. (a) Uniaxial deformation; (b) Expansion in plane x_1-x_2; (c) Pure shear in plane x_1-x_2

By applying the transformation (2.1.24), we obtain $\bar{e}_{11} = a_0$, whereas the other strain components $\bar{e}_{ij} = 0$.

(b) *Expansion in plane* $x_1 - x_2$. If $a_{11} \neq 0$ and $a_{22} \neq 0$ and all other coefficients $a_{ij} = 0$, we obtain that

$$u_1 = a_{11}{}^0x_1, \quad {}^tx_1 = (1+a_{11}){}^0x_1; \quad e_{11} = a_{11}$$
$$u_2 = a_{22}{}^0x_2, \quad {}^tx_2 = (1+a_{22}){}^0x_2; \quad e_{22} = a_{22} \quad \text{other } e_{ij} = 0$$
(E2.1-2.6)

The deformed material element is shown in Fig. E2.1-2Ab.

(c) *Pure shear in plane* $x_1 - x_2$. If only $a_{12} \neq 0$ and all other coefficients $a_{ij} = 0$, we have that the displacement field and the strain field are

$$u_1 = a_{12}{}^0x_2, \quad {}^tx_1 = (1+a_{12}){}^0x_2; \quad e_{12} = \frac{\gamma_{12}}{2} = \frac{1}{2}a_{12}; \quad \text{other } e_{ij} = 0 \quad \text{(E2.1-2.7)}$$

The displacement field is shown in Fig. E2.1-2Ac.

(d) *Material plane remains plane* under homogenous deformation. Consider a material plane in the undeformed configuration defined by

$$b_1{}^0x_1 + b_2{}^0x_2 + b_3{}^0x_3 + b_0 = 0 \quad \text{(E2.1-2.8)}$$

where b_j are constants. After deformation (E2.1-2.1), the coordinates of material points become

$$^tx_i = (\delta_{ik} + a_{ik}){}^0x_k \quad \text{or} \quad {}^t\mathbf{x} = (\mathbf{I}+\mathbf{A}){}^0\mathbf{x} \quad \text{(E2.1-2.9)}$$

where $\mathbf{A}$ is the matrix of the coefficients a_{ij} and $\mathbf{I}$ is the identity matrix. By solving (E2.1-2.9) for $^0\mathbf{x}$, we obtain that

$$^0x_k = c_{ki}{}^tx_i, \quad \text{where} \quad [c_{ij}] = [\mathbf{I}+\mathbf{A}]^{-1} \quad \text{(E2.1-2.10)}$$

providing that the matrix $(\mathbf{I}+\mathbf{A})$ is nonsingular. Substituting (E2.1-2.10) into (E2.1-2.8), we obtain that

$$d_1{}^tx_1 + d_2{}^tx_2 + d_3{}^tx_3 + b_0 = 0 \quad \text{(E2.1-2.11)}$$

where $d_k = b_s c_{sk}$ are constants. Equation (E2.1-2.11) represents a plane, hence the material points initially lying in a plane (E2.1-2.8) remain in that plane after deformation.

(e) *Straight material line remains straight*. The equation for a straight line passing through points A and B can be written as

$$^0x_1 = k_{13}{}^0x_3 + k_{10}, \quad {}^0x_2 = k_{23}{}^0x_3 + k_{20} \quad \text{(E2.1-2.12)}$$

where $k_{13} = {}^0(AB)_1/{}^0(AB)_3$, $k_{10} = -k_{13}{}^0x_3^A + {}^0x_1^A$, $k_{23} = {}^0(AB)_2/{}^0(AB)_3$, $k_{20} = -k_{23}{}^0x_3^A + {}^0x_2^A$. After deformation (E2.1-2.1), we obtain for the coordinate $^tx_3$:

$$^tx_3 = a_{31}\left(k_{13}{}^0x_3 + k_{10}\right) + a_{32}\left(k_{23}{}^0x_3 + k_{20}\right) + (1+a_{33}){}^0x_3 \quad \text{(E2.1-2.13)}$$

FUNDAMENTALS OF CONTINUUM MECHANICS

From this equation, we express $^0x_3$ in terms of $^tx_3$ and substitute into (E2.1-2.12). Hence, we have

$$^0x_3 = d_{33}{}^tx_3 + d_{30}, \quad {}^tx_1 = d_{13}{}^tx_3 + d_{10}, \quad {}^tx_2 = d_{23}{}^tx_3 + d_{20} \qquad (\text{E2.1-2.14})$$

The expressions for $^tx_1$ and $^tx_2$ show that the points lay on the straight line after deformation. The coefficients d_{ij} follow from (E2.1-2.13) and (E2.1-2.14).

(f) *A sphere deforms into ellipsoid*. Consider a sphere

$$\left(^0x_1 - {}^0x_{O1}\right)^2 + \left(^0x_2 - {}^0x_{O2}\right)^2 + \left(^0x_3 - {}^0x_{O3}\right)^2 = R^2 \qquad (\text{E2.1-2.15})$$

where $^0x_{Oi}$ are coordinates of the center of the sphere O. By substituting coordinates ^0x_i from (E2.1-2.10), which correspond to the deformation (E2.1-2.1), into (E2.1-2.15), the following equation is obtained

$$B_{ij}{}^tx_i{}^tx_i + C_i^t x_i + D = 0 \qquad (\text{E2.1-2.16})$$

where B_{ij}, C_i and D are the coefficients from this coordinate substitution. The surface (E2.1-2.16) represents an ellipsoid. Change of the sphere into ellipsoid is graphically represented in Fig. E2.1-2B, where the material element aligned with the principal axes of deformation, $\mathbf{p}_1, \mathbf{p}_2, \mathbf{p}_3$, remain orthogonal and lie along a rotated axes, $\bar{\mathbf{p}}_1, \bar{\mathbf{p}}_2, \bar{\mathbf{p}}_3$. The principal

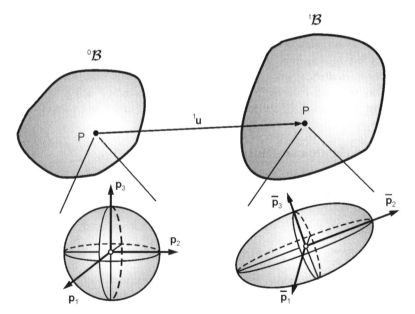

Fig. E2.1-2B Material points lying on the sphere surrounding a material point P at initial configuration $^0\mathcal{B}$ (left panel) form an ellipsoid after deformation at the deformed configuration $^t\mathcal{B}$ (right panel). The displacement vector of point P is $^t\mathbf{u}$. The infinitesimal material lines along the principal deformation directions, $\mathbf{p}_1, \mathbf{p}_2, \mathbf{p}_3$, remain orthogonal and lie on the directions, $\bar{\mathbf{p}}_1, \bar{\mathbf{p}}_2, \bar{\mathbf{p}}_3$

directions after deformation also remain the principal directions of the strain ellipsoid (see Example 2.1.5 on the web). Note that in the case when the deformation is given by the displacement field

$$u_i = a_0{}^0 x_i \qquad \text{(E2.1-2.17)}$$

the sphere (E2.1-2.15) becomes

$$({}^t x_1 - {}^t x_{O1})^2 + ({}^t x_2 - {}^t x_{O2})^2 + ({}^t x_3 - {}^t x_{O3})^2 = ((1 + a_0) R)^2 \qquad \text{(E2.1-2.18)}$$

which is the sphere with an increased (decreased) radius if $a_0 > 0 (< 0)$. Here, ${}^t x_{Oi} = (1 + a_0)^0 x_{Oi}$ are the coordinates of the center of the sphere after deformation. Therefore, the material expands uniformly, with the expansion strain $e_{11} = e_{22} = e_{33} = a_0 = e_V / 3$.

EXERCISE

Assume that the nonzero constants in (E2.1-2.1) are: $a_{11} = 0.01$, $a_{12} = a_{21} = 0.05$, $a_{22} = 0.02$. Draw the material elements after deformation if they initially lie in the $x - y$ plane and have the shape of: (a) a square with the side length of 2 m and (b) a sphere with the radius of 2 m.

2.2 Linear elastic and viscoelastic constitutive relations

Under a given stress field within a continuum the extent of material deformation measured by strains depends on material characteristics of the continuum. These characteristics are represented by the relationships between stresses and strains, called *constitutive relations*. They are phenomenological laws, established by experimental observations and are also termed here as *material models*. We present the most common and simple constitutive relations (constitutive laws): linear elastic, nonlinear elastic and linear viscoelastic.

2.2.1 Linear elastic constitutive law

In presenting this simplest material model, we will use the matrix representation of stress and strain (2.1.13) and (2.1.23), with the 'vector' terms

$$\begin{aligned} \sigma_1 &\equiv \sigma_{11}, & \sigma_2 &\equiv \sigma_{22}, & \sigma_3 &\equiv \sigma_{33}, & \sigma_4 &\equiv \sigma_{12}, & \sigma_5 &\equiv \sigma_{23}, & \sigma_6 &\equiv \sigma_{13} \\ e_1 &\equiv e_{11}, & e_2 &\equiv e_{22}, & e_3 &\equiv e_{33}, & e_4 &\equiv \gamma_{12}, & e_5 &\equiv \gamma_{23}, & e_6 &\equiv \gamma_{13} \end{aligned} \qquad (2.2.1)$$

Note that stresses $\sigma_1, \sigma_2, \sigma_3$ and e_1, e_2, e_3 should not be confused with the principal stresses and strains (see (2.1.7)). The linear constitutive relations for an *isotropic material* (Hooke's law) can be expressed by

$$\boldsymbol{\sigma} = \mathbf{C}\mathbf{e}, \quad \text{or} \quad \sigma_i = C_{ik} e_k, \quad i, k = 1, 2, \ldots, 6 \qquad (2.2.2)$$

where C_{ik} are elastic constants of the *elastic constitutive matrix*. Using material isotropy conditions, it can be shown (Wang 1953, Fung 1965) that the elastic constants can be

FUNDAMENTALS OF CONTINUUM MECHANICS

expressed in terms of two distinct material constants. For example, the *elastic modulus*, or *Young's modulus*, E; and the *Poisson ratio* ν. The Young modulus represents the slope of the linear stress–strain relationship σ-e in case of uniaxial loading of material,

$$\sigma = Ee \qquad (2.2.3)$$

while the Poison ratio is the ratio between the magnitude of lateral strain e_{yy} and longitudinal strain e_{xx}, when the material is loaded in the x-direction, i.e.

$$\nu = \left|\frac{e_{yy}}{e_{xx}}\right|_{\sigma=\sigma_{xx}}. \qquad (2.2.4)$$

(see web – Theory, Chapter 2 for details).

Below are given different forms of the elastic matrix $\mathbf{C}$ in terms of the elastic constants E and ν, corresponding to several physical conditions. Details about these conditions are given on the web – Theory, Chapter 2. For a general three-dimensional deformation, $\mathbf{C}$ is

$$\mathbf{C} = \frac{E(1-\nu)}{(1+\nu)(1-2\nu)} \begin{bmatrix} 1 & \frac{\nu}{(1-\nu)} & \frac{\nu}{(1-\nu)} & 0 & 0 & 0 \\ \frac{\nu}{1-\nu} & 1 & \frac{\nu}{(1-\nu)} & 0 & 0 & 0 \\ \frac{\nu}{1-\nu} & \frac{\nu}{1-\nu} & 1 & 0 & 0 & 0 \\ 0 & 0 & 0 & \frac{1-2\nu}{2(1-\nu)} & 0 & 0 \\ 0 & 0 & 0 & 0 & \frac{1-2\nu}{2(1-\nu)} & 0 \\ 0 & 0 & 0 & 0 & 0 & \frac{1-2\nu}{2(1-\nu)} \end{bmatrix} \qquad (2.2.5)$$

In the case of axisymmetric problems (see Fig. 2.2.1), the nonzero strains (shear strain γ_{xy} and normal strains – radial e_{xx}, axial e_{yy}, and circular e_{zz}) will be ordered in the book as $e_1 = e_{xx}, e_2 = e_{yy}, e_3 = \gamma_{xy}, e_4 = e_{zz}$. Then, the 4×4 constitutive matrix $\mathbf{C}$ is

$$\mathbf{C} = \frac{E(1-\nu)}{(1+\nu)(1-2\nu)} \begin{bmatrix} 1 & \frac{\nu}{1-\nu} & 0 & \frac{\nu}{1-\nu} \\ \frac{\nu}{1-\nu} & 1 & 0 & \frac{\nu}{1-\nu} \\ 0 & 0 & \frac{1-2\nu}{2(1-\nu)} & 0 \\ \frac{\nu}{1-\nu} & \frac{\nu}{1-\nu} & 0 & 1 \end{bmatrix} \qquad (2.2.6)$$

For plane strain conditions, $e_{zz} = 0$ and the 3×3 constitutive matrix $\mathbf{C}$ is of the form (2.2.6) with deleted fourth row and column. In the case of plane stress (membrane) conditions, the 3×3 constitutive matrix $\mathbf{C}$ is obtained from (2.2.6) by imposing the condition $\sigma_{zz} = 0$. The procedure of imposing this condition is called static condensation. The resulting matrix is (see Example 2.2-1)

$$\mathbf{C} = \frac{E}{1-\nu^2}\begin{bmatrix} 1 & \nu & 0 \\ \nu & 1 & 0 \\ 0 & 0 & \frac{1-\nu}{2} \end{bmatrix} \qquad (2.2.7)$$

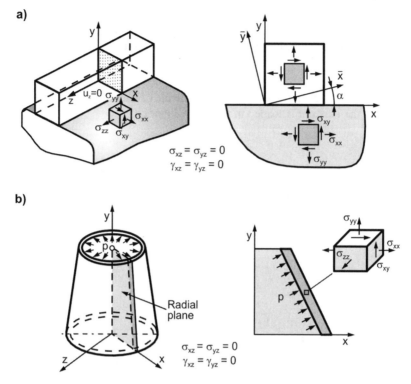

Fig. 2.2.1 Stresses and strains for two-dimensional conditions, physical space and stress representation in $x-y$ plane. (a) Plane strain; (b) Axial symmetry

This matrix is also applicable to the tangential plane of curved space membranes (see Fig. E2.2-1b). When transversal shear stresses and strains are taken into account for a membrane (also with bending), we have a shell structure, and the constitutive matrix is

$$\mathbf{C} = \frac{E}{1-\nu^2} \begin{bmatrix} 1 & \nu & 0 & 0 & 0 & 0 \\ \nu & 1 & 0 & 0 & 0 & 0 \\ 0 & 0 & 0 & 0 & 0 & 0 \\ 0 & 0 & 0 & \frac{1-\nu}{2} & 0 & 0 \\ 0 & 0 & 0 & 0 & \frac{1-\nu}{2} & 0 \\ 0 & 0 & 0 & 0 & 0 & \frac{1-\nu}{2} \end{bmatrix} \quad (2.2.8)$$

We have written a 6×6 matrix with zero third row and third column, corresponding to normal stress and strain in the normal shell direction, as it is usually used in numerical methods. Note that the nonzero normal strain e_{zz} for the plane stress conditions (membrane or shell) is related to the in-plane stresses and strains as

$$e_{zz} = -\frac{\nu}{E}\left(\sigma_{xx} + \sigma_{yy}\right) = -\frac{\nu}{1-\nu}(e_{xx} + e_{yy}) \quad (2.2.9)$$

This expression for e_{zz} will be used in the numerical methods presented later in the book.

FUNDAMENTALS OF CONTINUUM MECHANICS

The elastic constitutive relations between the deviatoric stresses σ'_{ij} (2.1.10) and deviatoric strains $e'_{ij} = e_{ij} - e_m$, where $e_m = (e_{11} + e_{22} + e_{33})/3$ is the mean strain, are

$$\sigma'_{ij} = 2G e'_{ij} \tag{2.2.10}$$

Here $G = E/(2(1+\nu))$ is the shear modulus. The relation between the mean stress σ_m and the volumetric strain $e_V = 3e_m$ is

$$\sigma_m = K e_V \tag{2.2.11}$$

where $K = E/(3(1-2\nu))$ is the bulk modulus.

We will also use the inverse relationship with respect to (2.2.2), i.e. the relationship between strain and stress tensors as

$$\mathbf{e} = \mathbf{C}^{-1}\boldsymbol{\sigma}, \quad \text{or} \quad e_i = C_{ik}^{-1}\sigma_k, \quad i,k = 1,\ldots,6 \tag{2.2.12}$$

or

$$\mathbf{e} = \begin{Bmatrix} e_{xx} \\ e_{yy} \\ e_{zz} \\ \gamma_{xy} \\ \gamma_{yz} \\ \gamma_{xz} \end{Bmatrix} = \begin{bmatrix} 1/E & -\nu/E & -\nu/E & 0 & 0 & 0 \\ -\nu/E & 1/E & -\nu/E & 0 & 0 & 0 \\ -\nu/E & -\nu/E & 1/E & 0 & 0 & 0 \\ 0 & 0 & 0 & 1/G & 0 & 0 \\ 0 & 0 & 0 & 0 & 1/G & 0 \\ 0 & 0 & 0 & 0 & 0 & 1/G \end{bmatrix} \begin{Bmatrix} \sigma_{xx} \\ \sigma_{yy} \\ \sigma_{zz} \\ \sigma_{xy} \\ \sigma_{yz} \\ \sigma_{xz} \end{Bmatrix} \tag{2.2.13}$$

where $\mathbf{C}^{-1}$ is the compliance matrix.

Constitutive relations for orthotropic elasticity, and isotropic and anisotropic thermoelasticity are given on the web – Theory, Chapter 2.

2.2.2 Viscoelasticity

The material constitutive behavior described above shows that the current stress state depends on the current state of deformation, therefore the current material response does not depend on the history (or time) of deformation. Experiments show that there are materials in which the current state depends not only on the current state of deformation, but also on the history of deformation (materials with 'memory' of the deformed configurations which occurred prior to the current one). These materials with memory are called *viscoelastic*. Biological materials can have these constitutive characteristics, displaying damping resistance, i.e. the resistance dependent on the rate of deformation (see description of muscle models in Chapter 12).

We consider here *linear viscoelasticity*, where the constitutive law for uniaxial loading can be written as (Fung 1965, Malvern 1969),

$$^t\sigma = E^t e + \int_0^t G(t-\tau) \frac{de}{dt} d\tau \tag{2.2.14}$$

where $^t\sigma$ and $^t e$ are the current stress and strain, E is Young's modulus, and $G(t)$ is called the *relaxation function*. This viscoelastic constitutive law is linear since the addition to the elastic part of the stress is proportional to the relaxation function. This constitutive law relies on a generalization of the Maxwell and Kelvin–Voigt models, consisting of an elastic spring and a dashpot, which represent the elastic and viscous characteristics of the models (see Example 2.2-4 and Software on the web).

The one-dimensionl relationship (2.2.14) can be generalized to tensorial viscoelastic relationships of the form (Malvern 1969),

$$^t\boldsymbol{\sigma}' = 2G\left[\mathbf{e}' - \int_0^t G_S(t-\tau)\frac{d\mathbf{e}'}{dt}d\tau\right], \quad ^t\sigma_m = K\left[^te_V - \int_0^t G_V(t-\tau)\frac{de_V}{dt}d\tau\right] \quad (2.2.15)$$

where $G_S(t)$ and $G_V(t)$ are the relaxation functions corresponding to deviatoric and volumetric stresses and strains.

Finally, we give a form of three-dimensional viscoelastic constitutive relationships for a linear constitutive law (Evans & Hochmuth 1976, With 2006). This law represents a generalization of the Kelvin model (see Example 2.2-4):

$$\sigma_{ij} = \sigma_{ij}^E + \sigma_{ij}^v,$$
$$\sigma_{ij}^v = \lambda^v \dot{e}_V \delta_{ij} + 2\mu^v \dot{e}_{ij} \quad (2.2.16)$$

where σ_{ij}^E and σ_{ij}^v are elastic and viscous stresses; $\dot{e}_V$ is volumetric strain rate, and $\dot{e}_{ij}$ are components of the strain rate; and λ^v and μ^v are material constants for the viscous constitutive law.

2.2.3 Transformation of constitutive relations

The transformation of constitutive relations due to change of coordinate systems is very important in practical applications. We here give the relations to which we will refer in the subsequent chapters.

By substituting (2.1.14) and (2.1.24) into the constitutive relations (2.2.2), we obtain the constitutive matrix $\overline{\mathbf{C}}$ in the rotated coordinate system as (details are given on web – Theory, Chapter 2)

$$\overline{\mathbf{C}} = \mathbf{T}^\sigma \mathbf{C} (\mathbf{T}^\sigma)^T \quad (2.2.17)$$

For a general three-dimensional deformation we have that $\overline{\mathbf{C}} = \mathbf{C}$ since the material is isotropic. However, the transformation (2.2.17) of the membrane or shell matrices given in (2.2.7) and (2.2.8) preserve the isotropy only for coordinate systems obtained by rotation around the z-axis normal to the membrane or shell surface (see web – Examples, Chapter 2).

The transformation (2.2.17) is also applicable to constitutive matrices of nonlinear material models, and will be employed in the subsequent text.

2.2.4 Examples

Example 2.2-1. Derivation of the constitutive matrix for plane stress and shell conditions
In the case of plane stress (membrane) in the x–y plane (see Fig. E2.2-1a) we have the condition

$$\sigma_{zz} = 0 \quad (\text{E2.2-1.1})$$

FUNDAMENTALS OF CONTINUUM MECHANICS

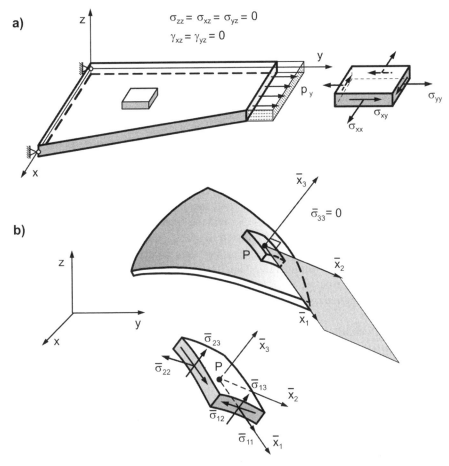

Fig. E2.2-1 Plane stress and shell conditions for stresses. (a) Plane stress (plate); (b) Shell

Also, the following shear strains and stresses are equal to zero

$$\gamma_{xz} = \gamma_{yz} = 0, \quad \sigma_{xz} = \sigma_{yz} = 0 \quad \text{(E2.2-1.2)}$$

With use of (2.2.2) and (2.2.5), the condition (E2.2-1.1) can be written as

$$\sigma_{zz} = 0 = \frac{E(1-2\nu)}{(1+\nu)(1-2\nu)} \left[\frac{\nu}{(1-\nu)}(e_{xx} + e_{yy}) + e_{zz} \right] \quad \text{(E2.2-1.3)}$$

from which we obtain the normal strain through the thickness (see (2.2.9))

$$e_{zz} = -\frac{\nu}{(1-\nu)}(e_{xx} + e_{yy}) \quad \text{(E2.2-1.4)}$$

Now we substitute this expression for e_{zz} into the first two relations (2.2.2) and obtain

$$\sigma_{xx} = \frac{E(1-2v)}{(1+v)(1-2v)}\left[e_{xx} + \frac{v}{1-v}e_{yy} - \frac{v^2}{(1-v)^2}(e_{xx}+e_{yy})\right] = \frac{E}{1-v^2}(e_{xx}+ve_{yy})$$

$$\sigma_{yy} = \frac{E(1-2v)}{(1+v)(1-2v)}\left[\frac{v}{1-v}e_{xx} + e_{yy} - \frac{v^2}{(1-v)^2}(e_{xx}+e_{yy})\right] = \frac{E}{1-v^2}(ve_{xx}+e_{yy})$$
(E2.2-1.5)

Therefore, the constitutive matrix is as given in (2.2.7). The procedure of correction of the constitutive matrix is called *static condensation*.

We apply the same static condensation procedure for the shell, using the shell local coordinate system $\bar{x}_1, \bar{x}_2, \bar{x}_3$, shown in Fig. E2.2-1b. The resulting constitutive matrix in the local shell system is as given in (2.2.8). The shell constitutive matrix is isotropic with respect to coordinate systems obtained by rotation around the shell normal $\bar{x}_3$, but not isotropic with respect to coordinate systems arbitrary oriented in space (see web – Example 2.2-6).

Finally, note that we can further perform the static condensation of the matrix (2.2.7) to obtain the uniaxial stress–strain law. Namely, assuming that $\sigma_{yy} = 0$, from the second of equations (E2.2-1.5) follows $e_{yy} = -ve_{xx}$, and than from the first equation we obtain $\sigma_{xx} = Ee_{xx}$ which is the uniaxial Hooke's law (2.2.3).

Example 2.2-2. Derivation of transformation matrices for transformation of stresses and strains written as one-dimensional arrays

We apply the transformation (2.1.4) to the stress components and then write the stress as a one-dimensional array (2.1.13). Further, the relation (2.1.4) can be written in the form (2.1.14),

$$\bar{\boldsymbol{\sigma}} = \mathbf{T}^\sigma \boldsymbol{\sigma} \qquad (E2.2\text{-}2.1)$$

where the stress components σ_k and $\bar{\sigma}_k$, $k = 1, 2, \ldots, 6$ correspond to the coordinate systems x_i and $\bar{x}_i$, respectively. Matrix $\mathbf{T}^\sigma$ is

$$\mathbf{T}^\sigma = \begin{bmatrix} l_1^2 & m_1^2 & n_1^2 & 2l_1m_1 & 2m_1n_1 & 2n_1l_1 \\ l_2^2 & m_2^2 & n_2^2 & 2l_2m_2 & 2m_2n_2 & 2n_2l_2 \\ l_3^2 & m_3^2 & n_3^2 & 2l_3m_3 & 2m_3n_3 & 2n_3l_3 \\ l_1l_2 & m_1m_2 & n_1n_2 & l_1m_2+m_1l_2 & m_1n_2+n_1m_2 & n_1l_2+l_1n_2 \\ l_2l_3 & m_2m_3 & n_2n_3 & l_2m_3+m_2l_3 & m_2n_3+n_2m_3 & n_2l_3+l_2n_3 \\ l_3l_1 & m_3m_1 & n_3n_1 & l_3m_1+m_3l_1 & m_3n_1+n_3m_1 & n_3l_1+l_3n_1 \end{bmatrix} \qquad (E2.2\text{-}2.2)$$

The coefficients l_i, m_i, n_i are the cosines of angles between the axes of the two coordinate systems and together they define the transformation matrix $\mathbf{T}$ in Example 1.5-3,

$$\mathbf{T} = \begin{array}{c|ccc} & x_1 & x_2 & x_3 \\ \hline \bar{x}_1 & l_1 & m_1 & n_1 \\ \bar{x}_2 & l_2 & m_2 & n_2 \\ \bar{x}_3 & l_3 & m_3 & n_3 \end{array} \qquad (E2.2\text{-}2.3)$$

FUNDAMENTALS OF CONTINUUM MECHANICS

The transformation of the strain tensor is given by (2.1.4) as in the case of the stress tensor, and the transformation can be written as (equation (2.1.24))

$$\bar{\mathbf{e}} = \mathbf{T}^e \mathbf{e} \tag{E2.2-2.4}$$

The matrix $\mathbf{T}^e$ is

$$\mathbf{T}^e = \begin{bmatrix} l_1^2 & m_1^2 & n_1^2 & l_1 m_1 & m_1 n_1 & n_1 l_1 \\ l_2^2 & m_2^2 & n_2^2 & l_2 m_2 & m_2 n_2 & n_2 l_2 \\ l_3^2 & m_3^2 & n_3^2 & l_3 m_3 & m_3 n_3 & n_3 l_3 \\ 2l_1 l_2 & 2m_1 m_2 & 2n_1 n_2 & l_1 m_2 + m_1 l_2 & m_1 n_2 + n_1 m_2 & n_1 l_2 + l_1 n_2 \\ 2l_2 l_3 & 2m_2 m_3 & 2n_2 n_3 & l_2 m_3 + m_2 l_3 & m_2 n_3 + n_2 m_3 & n_2 l_3 + l_2 n_3 \\ 2l_3 l_1 & 2m_3 m_1 & 2n_3 n_1 & l_3 m_1 + m_3 l_1 & m_3 n_1 + n_3 m_1 & n_3 l_1 + l_3 n_1 \end{bmatrix} \tag{E2.2-2.5}$$

The inverse relations of (E2.2-2.1) and (E2.2-2.4) are

$$\boldsymbol{\sigma} = \overline{\mathbf{T}}^\sigma \bar{\boldsymbol{\sigma}} \quad \text{and} \quad \mathbf{e} = \overline{\mathbf{T}}^e \bar{\mathbf{e}} \tag{E2.2-2.6}$$

It can be shown that the following relations hold:

$$\overline{\mathbf{T}}^\sigma = \mathbf{T}^{eT} \quad \text{and} \quad \overline{\mathbf{T}}^e = \mathbf{T}^{\sigma T} \tag{E2.2-2.7}$$

Also, the following orthogonality relation, that can be verified, holds

$$\mathbf{T}^{\sigma T} \mathbf{T}^e = \mathbf{I}_6 \tag{E2.2-2.8}$$

where $\mathbf{I}_6$ is a 6×6 identity matrix.

Example 2.2-3. Determine the reactions of a restrained pipe under internal pressure

A straight pipe is restrained from axial deformation as shown in Fig. E2.2-3a. The pipe is subjected to loading by internal pressure p. In this analysis we assume that the pipe wall thickness δ is small compared to the pipe radius R. Also, it is assumed that the pipe supports allow change of the pipe radius. Under these assumptions it can be considered that the stress–strain state within the entire pipe is uniform.

The nonzero stresses within the pipe wall are the axial stress σ_{aa} and circular stress σ_{cc}. The stress through the pipe wall thickness can be neglected with respect to σ_{aa} and σ_{cc} because the thickness of the wall is small when compared to the radius. Therefore, we have the plane stress conditions for the pipe material. On the other hand, if we consider a cross-section shown in Fig. E2.2-3a, the circular stress can be determined from the equilibrium of vertical forces due to pressure and due to circular stress. Thus, we have

$$F_p = 2R_i p, \; F_c = 2\sigma_{cc} \delta; \quad F_p = F_c \Rightarrow \sigma_{cc} = \frac{R_i}{\delta} p \tag{E2.2-3.1}$$

Here F_p and F_c are vertical forces per unit axial length due to pressure and due to circular stress.

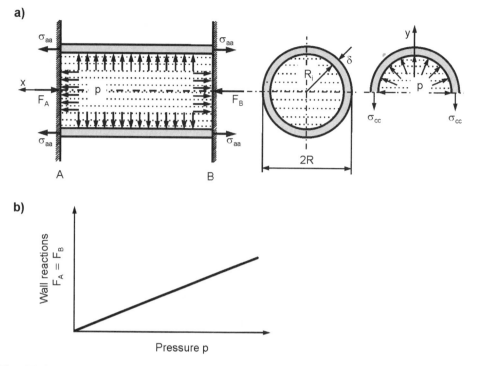

Fig. E2.2-3 Restrained pipe under internal pressure. Wall reactions F_A and F_B include action of pressure and axial stress σ_{aa} in the pipe wall. (a) Geometry of the structure, stresses and forces; (b) Wall reactions in terms of pressure p

Due to axial restrain by the supports A and B, the axial strain in the pipe wall is $e_{aa} = 0$. Thus, from (2.2.2) and the constitutive matrix (2.2.7) we obtain that

$$\sigma_{aa} = \frac{\nu E}{1-\nu^2} e_{cc} \qquad (E2.2\text{-}3.2)$$

From (2.2.13) we obtain the hoop strain e_{cc} as

$$e_{cc} = \frac{1}{E}(\sigma_{cc} - \nu \sigma_{aa}) \qquad (E2.2\text{-}3.3)$$

We now substitute e_{cc} into (E2.2-3.2) and use σ_{cc} to obtain the relationship between the axial stress σ_{aa} and the pressure p,

$$\sigma_{aa} = \nu \frac{R_i}{\delta} p \qquad (E2.2\text{-}3.4)$$

where $R_i = R - \delta$ is the internal pipe radius.

The reactions $F_A = F_B$ are due to the direct loading by pressure and due to stress σ_{aa} in the pipe wall. Using the directions of the reactions as shown in Fig. E2.2-3a, we obtain

$$F_A = F_B = \pi R_i^2 p - \pi D_m \delta \sigma_{aa} = \pi R_i [(1-2\nu) R_i - \nu \delta] p \qquad (E2.2\text{-}3.5)$$

FUNDAMENTALS OF CONTINUUM MECHANICS

where $D_m = 2R_i + \delta$ is the mean pipe radius. Here, we used the expression (E2.2-3.4) for the stress σ_{aa}. It can be seen that the reactions increase linearly with increasing pressure p, as schematically shown in Fig. E2.2-3b.

Example 2.2-4. Determine the force–elongation relations for Maxwell, Voigt and Kelvin bodies

The most commonly used viscoelastic models are the Maxwell, Voigt and Kelvin models (Fung 1993) shown schematically in Fig. E2.2-4A. They are the one-dimensional mechanical analogs consisting of elastic (spring) and viscous (dashpot) elements to represent the viscoelastic behavior of material. The constitutive relation is here expressed as the force–elongation relationship and for each model it is obtained as follows.

In the case of the Maxwell model we have

$$F_k = ku_k, \quad F_\mu = \mu \dot{u}_\mu \qquad (E2.2\text{-}4.1)$$

where F_k and F_μ are the elastic and viscous forces, u_k is spring elongation, $\dot{u}_\mu$ is the dashpot elongation velocity; and k and μ are the spring constant and viscous damping coefficient, respectively. Since $F_k = F_\mu = F$ and $u = u_k + u_\mu$ (Fig. E2.2-4Aa), the constitutive relation is obtained as

$$\dot{F}/k + F/\mu = \dot{u} \qquad (E2.2\text{-}4.2)$$

Therefore, the governing constitutive equation of the Maxwell model is represented by a first-order linear differential equation.

For the Voigt model we have (Fig. E2.2-4Ab)

$$F = F_k + F_\mu, \quad u_k = u_\mu = u \qquad (E2.2\text{-}4.3)$$

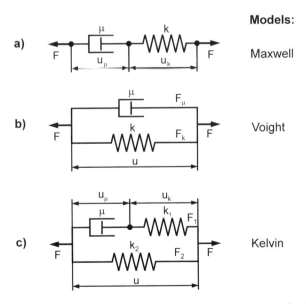

Fig. E2.2-4A Schematic representations of three linear viscoelastic models

and the constitutive relation (also known as equation of motion) is

$$F = ku + \mu \dot{u} \tag{E2.2-4.4}$$

Finally, for the Kelvin model the governing equation of motion is (Fig. E2.2-4Ac)

$$F = F_1 + F_2 = k_1 u_k + k_2 u = k_1 (u - u_\mu) + k_2 u = (k_1 + k_2) u - k_1 u_\mu \tag{E2.2-4.5}$$

where the geometric relation $u_k = u - u_\mu$ has been used. If we take the time derivative of this equation and use that $\mu \dot{u}_\mu = k_1 u_k$ (the force in the dashpot and spring with the spring constant k_1 is equal to F_1), then, combining with (E2.2-4.5), the constitutive equation is obtained as

$$F + \frac{\mu}{k_1} \dot{F} = k_2 u + \mu \left(1 + \frac{k_2}{k_1}\right) \dot{u} \tag{E2.2-4.6}$$

The above relationships (E2.2-4.2), (E2.2-4.4) and (E2.2-4.6) can be generalized and used by relating the stress to the strains and strain rates.

If the material is subjected to the unit step force, the material response $u(t)$, known as creep, for these three models is as shown in Fig. E2.2-4B. A notable difference in the creep responses is obtained form these three models. Material response for other loading conditions can be obtained by using the software (see web – Software, Chapter 2). Also, in the Software is an example with a model defined in (2.2.15) and variable loading.

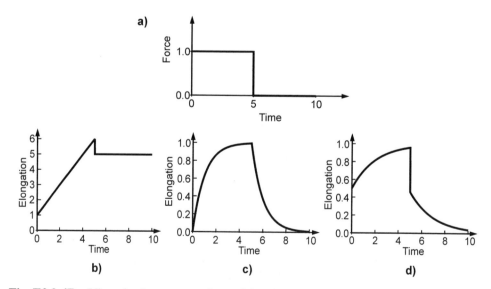

Fig. E2.2-4B Viscoelastic response of material under unit step force. All material coefficients are set to unity. (a) Force function; material elongation with time (creep) for: (b) Maxwell; (c) Voigt; (d) Kelvin models

FUNDAMENTALS OF CONTINUUM MECHANICS

2.3 Principle of virtual work

The principle of virtual work is one of the most fundamental principles in mechanics. It is used in many numerical methods as a basis for the development of necessary relations. Here we derive this principle for linear problems: linear material model and small strains.

2.3.1 Formulation of the principle of virtual work

Consider a deformable body in equilibrium, shown in Fig. 2.3.1, subjected to external loadings and with given boundary conditions. Let us assume that a field of virtual displacements $\delta \mathbf{u}$ is imposed, keeping the loading (and stresses) unaltered. Those displacements are infinitesimally small and satisfy the displacement boundary conditions. Virtual strains corresponding to the virtual displacements are

$$\delta e_{ks} = \frac{1}{2}\left[\delta\left(\frac{\partial u_k}{\partial x_s}\right) + \delta\left(\frac{\partial u_s}{\partial x_k}\right)\right] \quad (2.3.1)$$

We note here that there are two types of *boundary conditions*: (a) stress (loading) and (b) displacement boundary conditions. In case (a) the stresses can be zero (free surface), or can be given, as in the case of pressure loading shown in Fig. 2.3.1. Part of the surface where the loading is prescribed is denoted by S_σ. Displacement boundary conditions mean that displacements are prescribed at some points, as the zero displacements at the supports A and B, or for part of the surface, such as $\mathbf{u}_n = 0$ shown in the figure. Concentrated forces might act on the body, such as forces $\mathbf{F}_1$ and $\mathbf{F}_2$ shown in Fig. 2.3.1. It can be proved that in the case of a linear elastic material and small displacements, the solution for a displacement field within the body is *unique* (uniqueness theorem) for given boundary and loading conditions.

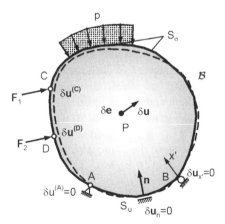

Fig. 2.3.1 Schematics of deformable body used for the derivation of the principle of virtual work. Virtual displacements and virtual strain at a material point P are $\delta\mathbf{u}$ and $\delta\mathbf{e}$. Virtual displacements at points of action of forces are $\delta\mathbf{u}^{(C)}$ and $\delta\mathbf{u}^{(D)}$, while the virtual displacements at the supports are restrained. Virtual displacements correspond to the equilibrium state of the body under given loads. Parts of the surface where stresses and displacements are prescribed are S_σ and S_u, respectively

Starting from the equilibrium equations and using the boundary conditions, we finally obtain the following result (detailed derivation is given on web – Theory, Chapter 2)

$$\delta W_{int} = \delta W_{ext} \qquad (2.3.2)$$

i.e. *the virtual work of internal forces δW_{int} and virtual work of external forces δW_{ext} are equal.* The internal and external virtual works are

$$\delta W_{int} = \int_V \sigma_{ij} \delta e_{ij} dV = \int_V \delta \mathbf{e}^T \boldsymbol{\sigma} dV \qquad (2.3.3)$$

and

$$\delta W_{ext} = \int_V F_k^V \delta u_k dV + \int_{S_\sigma} f_k^S \delta u_k^S dV + \sum_i F_k^{(i)} \delta u_k^{(i)}; \quad \text{sum on } k, k = 1, 2, 3 \qquad (2.3.4)$$

Here f_k^S and δu_k^S are the distributed surface forces and virtual displacements at the surface S_σ. Also, $F_k^{(i)}$ and $\delta u_k^{(i)}$ are the components of the concentrated force 'i' and virtual displacement of the material point where this force is acting on the body. Note that the matrix form of virtual work in (2.3.3) assumes the stress and strain vectors (2.1.13) and (2.1.23).

2.3.2 Examples

Example 2.3-1. Determine the deformation of an artery bifurcation

Consider a simplified model of artery bifurcation, shown in Fig. E2.3-1. The bifurcation consists of the main artery and two daughter branches. The artery vessels are represented by pipes of circular cross-sections and uniform thickness. The geometrical parameters are shown in the figure. Assuming that the arteries are restrained from axial elongation, determine the displacement of the branching cross-section B under the blood pressure p.

When the artery structure is loaded by pressure, the material deforms and the point B moves along the x-axis since the structure and loading are symmetric with respect to the x-axis. We assume that the wall thickness of each artery is small with respect to the cross-sectional radius, so that the stresses and strains are uniform within the wall. Also, the stress and strains are uniform along the length of the arteries.

In order to determine displacement of the point B, we apply the principle of virtual work (2.3.2). Here, the external forces $\mathbf{F}_1$, $\mathbf{F}_2$, $\mathbf{F}_a$ are the axial forces due to pressure, as shown in Fig. E2.3-1, where $F_1 = F_2$ because the branches are the same. The internal virtual work corresponds to the axial virtual strain in each of the three components of the bifurcation. Therefore, we have

$$(F_a - 2F_1 \cos \alpha) \delta u_B = A \int_L \sigma_{aa} \delta e_{aa} + 2A_1 \int_{L_1} \sigma_{aa1} \delta e_{aa1} \qquad (E2.3-1.1)$$

where A and A_1 are the cross-sectional areas equal to $\pi \left(R^2 - (R-\delta)^2 \right)$ and $\pi \left(R_1^2 - (R_1-\delta_1)^2 \right)$ for the main artery and the branches, respectively; L and L_1 are the lengths; σ_{aa} and σ_{aa1} are the axial stresses, and δe_{aa} and δe_{aa1} are the virtual axial strains in the main artery and in the branches, respectively. Here, it is assumed that the virtual displacement δu_B

FUNDAMENTALS OF CONTINUUM MECHANICS

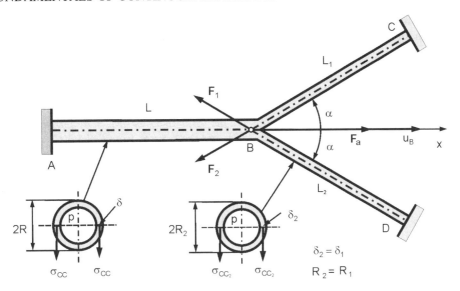

Fig. E2.3-1 Geometry of artery bifurcation simplified model. The branched arteries have the same geometrical data (cross-section areas and lengths). The pressure inside the arteries is p; the axial forces generated by the pressure are F_a and $F_1 = F_2$; and the hoop stresses due to pressure are σ_{cc} and $\sigma_{cc1} = \sigma_{cc2}$. The axial displacements of the artery ends are restrained

of point B is in the positive x-direction (it is equally correct to assume δu_B in the negative x-direction, since it is kinematically admissible).

The axial forces due to pressure are:

$$F_a = A_{int} p, \quad F_1 = F_2 = A_{1int} p \tag{E2.3-1.2}$$

where $A_{int} = \pi (R - \delta)^2$, $A_{1int} = \pi (R_1 - \delta_1)^2$ are the internal surface areas of the artery cross-sections (also $R_1 = R_2$, $\delta_1 = \delta_2$). The axial strains in the arteries can be expressed in terms of the displacement u_B, as follows:

$$e_{aa} = u_B/L, \quad e_{aa1} = -u_B \cos \alpha / L_1 \tag{E2.3-1.3}$$

Determination of the axial stresses requires the following considerations. As shown in Example 2.2.3, the internal pressure within straight pipes produces the hoop stress given in (E2.2-3.1), hence in this example we have

$$\sigma_{cc} = \left(\frac{R}{\delta} - 1\right) p, \quad \sigma_{cc1} = \left(\frac{R_1}{\delta_1} - 1\right) p \tag{E2.3-1.4}$$

Therefore, the hoop stresses do not depend on the axial deformation of arteries and are considered known quantities for a given pressure p. Furthermore, for the main artery, the following relations can be obtained from Hooke's law (2.2.13) and (2.2.2):

$$e_{cc} = \frac{1}{E} (\sigma_{cc} - \nu \sigma_{aa}), \quad \sigma_{aa} = \frac{E}{1 - \nu^2} (e_{aa} + \nu e_{cc}) \tag{E2.3-1.5}$$

where E and ν are Young's modulus and Poisson's ratio of tissue, respectively. We have used the elastic matrix (2.2.7) because the material of the artery wall deforms under plane stress conditions (stress in the normal direction to the wall surface can be considered equal to zero for small ratio δ/R). From these two equations it follows that

$$\sigma_{aa} = E e_{aa} + \nu \sigma_{cc}, \quad \sigma_{aa1} = E e_{aa1} + \nu \sigma_{cc1} \tag{E2.3-1.6}$$

Here, the expression for the stress σ_{aa1} is also given and it follows from the analogy with the derivation for the main artery.

Finally, we substitute the expressions for the virtual strains $\delta e_{aa} = \delta u_B/L$ and $\delta e_{aa1} = -\delta u_B \cos\alpha/L_1$, which follow from (E2.3-1.3), and the stresses from (E2.3-1.6), into (E2.3-1.1) and obtain

$$(F_a - 2F_1 \cos\alpha)\,\delta u_B$$
$$= \left[\frac{A}{L} \int_L (E e_{aa} + \nu \sigma_{cc})\, dL - 2\frac{A_1}{L_1} \cos\alpha \int_{L_1} (E e_{aa1} + \nu \sigma_{cc1})\, dL_1 \right] \delta u_B \tag{E2.3-1.7}$$

This equation must be satisfied for any value of δu_B, and the terms multiplying δu_B on the left- and right-hand side must be equal. Substituting further e_{aa} and e_{aa1} from (E2.3-1.3), and taking into account that all terms within the integrals are independent of the axial coordinate, we obtain the solution for the displacement u_B as

$$u_B = \frac{F_a - 2F_1 \cos\alpha - \nu(A\sigma_{cc} - 2A_1\sigma_{cc1}\cos\alpha)}{E(A/L + 2A_1\cos^2\alpha/L_1)} \tag{E2.3-1.8}$$

From this solution, we can compute axial stresses and strains within each artery, as well as the reactions at the end points A, C and D (see Example 2.2.3).

EXERCISE
Determine displacement u_B and reactions at the end of arteries using the following data:
$E = 0.5\,N/mm^2$, $\nu = 0.5$, $L = L_1 = L_2 = 50\,mm$, $\delta_1 = \delta_2 = \delta = 1.5\,mm$, $\alpha = 30°$, $p = 1.5\,N/mm^2$

(a) Assume that the pressure is constant over time
(b) Assume that the pressure varies over time as

$$p = p_0 \sin\left(2\pi \frac{t}{T}\right), \quad \text{where } p_0 = 2\,N/mm^2,\ T = 1s$$

In the case of variable pressure with time, draw the dependence on time of: displacement u_B, reactions at the arterial ends, and axial and hoop stresses and strains.

2.4 Nonlinear continuum mechanics

There are basically two types of nonlinearities in continuum mechanics: (a) material and (b) geometrical. In case (a) we have small strains, but the constitutive law is nonlinear; in case (b) the strains are large and the constitutive law can be either linear – then the problem is geometrically nonlinear only, or nonlinear – the problem is geometrically and materially nonlinear. We here give the basic kinematic quantities used in describing geometric nonlinearity of deformation.

FUNDAMENTALS OF CONTINUUM MECHANICS

2.4.1 Deformation gradient and the measures of strain and stress

Basically all large strain kinematic quantities rely on the deformation gradient. We start by introducing a *configuration* $^0\mathcal{B}$ of a continuum defined by coordinates of all material points $^0\mathbf{x}$ corresponding to initial time $(t=0)$, i.e.

$$^0\mathbf{x} = \mathbf{x}(t=0) \tag{2.4.1}$$

The time parameter t can be real time or a parameter for defining the configuration. Usually, the configuration $^0\mathcal{B}$ corresponds to the *stress-free (undeformed) reference configuration*. Under loading the material deforms and at a 'time' t (load level t) we have the *current configuration* $^t\mathcal{B}$ when the material coordinates are $^t\mathbf{x}$ (Fig. 2.4.1). Note that a Cartesian coordinate system can be used to follow the motion of material points in space (Fig. 2.4.1a), or the finite element description can be employed where the natural coordinates of material points within each element remain constant (in Fig. 2.4.1b the coordinates r and s remain

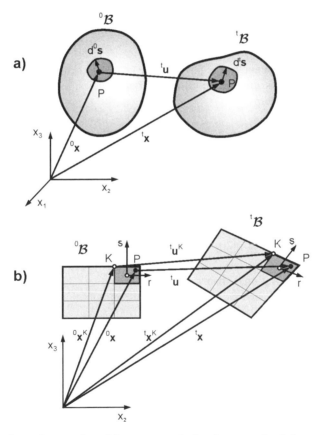

Fig. 2.4.1 Configurations and position vectors during large strain deformation of material body. (a) General representation; (b) Use of finite element modeling, with material coordinates r, s within the finite elements (Reproduced with permission from Kojić & Bathe: *Inelastic Analysis of Solids and Structures*, Springer-Verlag, Berlin, 2005)

unchanged, see Sections 4.1 and 4.2). *Deformation gradient* ${}_0^t\mathbf{F}$, with components ${}_0^t F_{ij}$, is defined as (Kojić & Bathe 2005)

$${}_0^t\mathbf{F} = \frac{\partial {}^t\mathbf{x}}{\partial {}^0\mathbf{x}}, \quad {}_0^t F_{ij} = \frac{\partial {}^t x_i}{\partial {}^0 x_j} \qquad (2.4.2)$$

The deformation gradient can be expressed in terms of displacements ${}^t\mathbf{u}$ as

$${}_0^t\mathbf{F} = \mathbf{I} + \frac{\partial {}^t\mathbf{u}}{\partial {}^0\mathbf{x}}, \quad {}_0^t F_{ij} = \delta_{ij} + \frac{\partial {}^t u_i}{\partial {}^0 x_j} \qquad (2.4.3)$$

The inverse deformation gradient is

$${}_t^0\mathbf{F} \equiv {}_0^t\mathbf{F}^{-1} = \frac{\partial {}^0\mathbf{x}}{\partial {}^t\mathbf{x}}, \quad {}_t^0 F_{ij} = \frac{\partial {}^0 x_i}{\partial {}^t x_j} \qquad (2.4.4)$$

since $\partial {}^t x_i / \partial {}^t x_j = \delta_{ij} = \left(\partial {}^t x_i / \partial {}^0 x_k \right)\left(\partial {}^0 x_k / \partial {}^t x_j \right)$. From (2.4.3) and (2.4.4) it follows that

$${}_t^0\mathbf{F} = \mathbf{I} - \frac{\partial {}^t\mathbf{u}}{\partial {}^t\mathbf{x}}, \quad {}_t^0 F_{ij} = \delta_{ij} - \frac{\partial {}^t u_i}{\partial {}^t x_j} \qquad (2.4.5)$$

The deformation gradient has a physical meaning. It relates the differential material vectors $d\,{}^t\mathbf{s}$ and $d\,{}^0\mathbf{s}$, consisting of the same material particles, in two configurations ${}^t\mathcal{B}$ and ${}^0\mathcal{B}$, through the relationship

$$d\,{}^t\mathbf{s} = {}_0^t\mathbf{F}\, d\,{}^0\mathbf{s}, \quad d\,{}^t x_i = \frac{\partial {}^t x_i}{\partial {}^0 x_k} d\,{}^0 x_k = {}_0^t F_{ik}\, d\,{}^0 x_k \qquad (2.4.6)$$

where $d\,{}^t x_k$ and $d\,{}^0 x_k$ are the components of $d\,{}^t\mathbf{s}$ and $d\,{}^0\mathbf{s}$. Also, the relationship between elementary material volumes $d\,{}^t V$ and $d\,{}^0 V$ in two configurations (Bathe 1996) is

$$d\,{}^t V = \det\left({}_0^t\mathbf{F}\right) d\,{}^0 V \qquad (2.4.7)$$

Proof of this relation is given on the web – Theory, Chapter 2 (Section 2.4e).

Cauchy–Green Deformation Tensors

The squared lengths of a material element at initial configuration ${}^0\mathcal{B}$ and a current configuration ${}^t\mathcal{B}$ can be related as

$$(d\,{}^t s)^2 = \frac{\partial {}^t x_k}{\partial {}^0 x_m}\frac{\partial {}^t x_k}{\partial {}^0 x_s} d\,{}^0 x_m d\,{}^0 x_s, \quad (d\,{}^t s)^2 = d\,{}^0\mathbf{s}^T {}_0^t\mathbf{C}\, d\,{}^0\mathbf{s} \qquad (2.4.8)$$

or

$$(d\,{}^0 s)^2 = \frac{\partial {}^0 x_k}{\partial {}^t x_m}\frac{\partial {}^0 x_k}{\partial {}^t x_s} d\,{}^t x_m d\,{}^t x_s, \quad (d\,{}^0 s)^2 = d\,{}^t\mathbf{s}^T {}_t^0\mathbf{B}\, d\,{}^t\mathbf{s} \qquad (2.4.9)$$

where ${}_0^t\mathbf{C}$ is the *right Cauchy–Green deformation tensor*,

$${}_0^t\mathbf{C} = {}_0^t\mathbf{F}^T\, {}_0^t\mathbf{F}, \quad \text{or} \quad {}_0^t C_{ms} = \frac{\partial {}^t x_k}{\partial {}^0 x_m}\frac{\partial {}^t x_k}{\partial {}^0 x_s} \qquad (2.4.10)$$

FUNDAMENTALS OF CONTINUUM MECHANICS

and the *Finger deformation tensor*,

$$^0_t\mathbf{B} = {}^0_t\mathbf{F}^T\,{}^0_t\mathbf{F}, \quad \text{or} \quad {}^0_t B_{ms} = \frac{\partial^0 x_k}{\partial {}^t x_m}\frac{\partial^0 x_k}{\partial {}^t x_s} \qquad (2.4.11)$$

We will use in the numerical methods the *left Cauchy–Green deformation tensor* ${}^t_0\mathbf{B}$, as the inverse of ${}^0_t\mathbf{B}$,

$$^t_0\mathbf{B} = {}^t_0\mathbf{F}\,{}^t_0\mathbf{F}^T, \quad {}^t_0 B_{ms} = \frac{\partial {}^t x_m}{\partial {}^0 x_k}\frac{\partial {}^t x_s}{\partial {}^0 x_k} \qquad (2.4.12)$$

It can be shown (see web – Theory, Chapter 2, Section 2.4) that the principal values of the tensors ${}^t_0\mathbf{C}$ and ${}^t_0\mathbf{B}$ are the squares of the principal stretches ${}^t_0\lambda_i^2$ in the directions of the principal vectors of the right basis ${}^t\mathbf{p}_i$ and the left basis ${}^t\bar{\mathbf{p}}_i$, see Fig. 2.4.2. Hence, we can write

$$^t_0\mathbf{C} = \sum_{i=1}^{3} {}^t_0\lambda_i^2\,{}^t\mathbf{p}_i\,{}^t\mathbf{p}_i, \quad \text{and} \quad {}^t_0\mathbf{B} = \sum_{i=1}^{3} {}^t_0\lambda_i^2\,{}^t\bar{\mathbf{p}}_i\,{}^t\bar{\mathbf{p}}_i \qquad (2.4.13)$$

Figure 2.4.2 illustrates the polar decomposition theorem, according to which the deformation at a material point can be decomposed into pure stretch in the principal directions and material rotation. In accordance with this theorem, the deformation gradient can be expressed as:

$$^t_0\mathbf{F} = {}^t_0\mathbf{R}\,{}^t_0\mathbf{U}, \quad \text{and} \quad {}^t_0\mathbf{F} = {}^t_0\mathbf{V}\,{}^t_0\mathbf{R} \qquad (2.4.14)$$

where ${}^t_0\mathbf{R}$ is the rotation tensor, and ${}^t_0\mathbf{U}$ and ${}^t_0\mathbf{V}$ are the right and left stretch tensors. Proof of the relations is given on the web – Theory, Chapter 2 (see also Example 2.4-1).

Strain Measures

Various strain measures have been introduced by employing the polar decomposition theorem. We list some of them that are commonly used, and also give their representation in the principal directions:

Green-Lagrange strain, $\quad {}^t_0\mathbf{E}^{GL} = \dfrac{1}{2}\left({}^t_0\mathbf{C} - \mathbf{I}\right)\;$ or

$$^t_0\mathbf{E}^{GL} = \frac{1}{2}\sum_{i=1}^{3}\left({}^t_0\lambda_i^2 - 1\right){}^t\mathbf{p}_i\,{}^t\mathbf{p}_i \qquad (2.4.15)$$

Almansi strain, $\quad {}^t_t\mathbf{E}^A = \dfrac{1}{2}\left(\mathbf{I} - {}^0_t\mathbf{B}\right)\;$ or

$$^t_t\mathbf{E}^A = \frac{1}{2}\sum_{i=1}^{3}\left(1 - {}^t_0\lambda_i^{-2}\right){}^t\bar{\mathbf{p}}_i\,{}^t\bar{\mathbf{p}}_i \qquad (2.4.16)$$

Logarithmic (or Hencky) strain, $\quad {}^t_0\mathbf{E}^H = \sum_{i=1}^{3}\ln\left({}^t_0\lambda_i\right){}^t\bar{\mathbf{p}}_i\,{}^t\bar{\mathbf{p}}_i$

$$^t_0 E^H_{mn} = \sum_{i=1}^{3}\ln\left({}^t_0\lambda_i\right)\left({}^t\bar{\mathbf{p}}_i\right)_m\left({}^t\bar{\mathbf{p}}_i\right)_n \qquad (2.4.17)$$

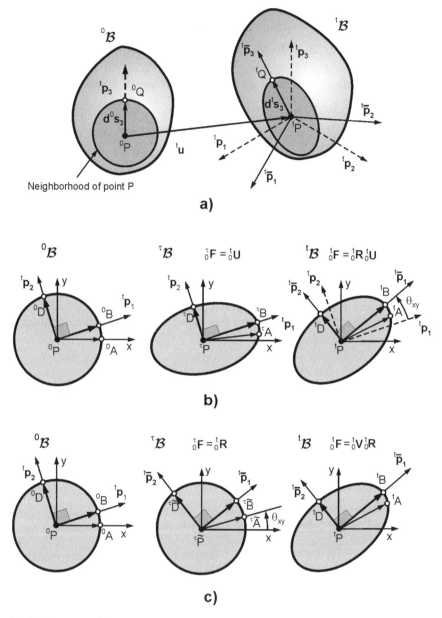

Fig. 2.4.2 Decomposition of deformation into stretch and rotation: ${}_0^t\mathbf{F} = {}_0^t\mathbf{R}\,{}_0^t\mathbf{U} = {}_0^t\mathbf{V}\,{}_0^t\mathbf{R}$. (a) A sphere is deformed into an ellipsoid, with the principal stretch directions $\mathbf{p}_i$ (right basis) and $\bar{\mathbf{p}}_i$ (left basis); (b) Material is first stretched and then rotated; (c) Rotation followed by stretch (Reproduced with permission from Kojić & Bathe 2005: *Inelastic Analysis of Solids and Structures*, Springer-Verlag, Berlin)

Here ${}_0^t E_{mn}^H$ are the components the Hencky strains in a coordinate system x_i. Note that for stretches close to one, all large strains reduce to small strains (2.1.25) (see Examples, Chapter 2, on the web). Also, the shear components in the principal bases are equal to zero.

FUNDAMENTALS OF CONTINUUM MECHANICS

Stress Measures

The mechanical power per unit volume ${}^t\dot{W}$ is given by

$${}^t\dot{W} = {}^t\sigma_{ij}\,{}^tD_{ij} \tag{2.4.18}$$

where ${}^t\sigma_{ij}$ are the *Cauchy (true) stresses* (forces per unit current area); and ${}^tD_{ij}$ are the strain rates (2.1.27). This mechanical power must be equal to the mechanical power calculated as the product of the rate of a given strain and the corresponding stress measure. From this equality of the mechanical power, we find the relationship between a stress measure, work-conjugate to a given strain measure, and the Cauchy stresses ${}^t\sigma_{ij}$. It can be shown (Bathe 1996, Kojić & Bathe 2005) that the Cauchy stress can be considered conjugate to the Almansi and logarithmic strains. The conjugate stress to the Green–Lagrange strain is the second Piola–Kirchhoff stress ${}^t_0\mathbf{S}$, related to the Cauchy stress as,

$${}^t_0\mathbf{S} = \frac{{}^0\rho}{{}^t\rho}\,{}^0_t\mathbf{F}\,{}^t\boldsymbol{\sigma}\,{}^0_t\mathbf{F}^T, \quad {}^t_0S_{ij} = \frac{{}^0\rho}{{}^t\rho}\frac{\partial\,{}^0x_i}{\partial\,{}^tx_k}\frac{\partial\,{}^0x_j}{\partial\,{}^tx_m}\,{}^t\sigma_{km} \tag{2.4.19}$$

where ${}^0\rho$ and ${}^t\rho$ are mass densities at the initial and current configurations, respectively. Note that the ratio ${}^0\rho/{}^t\rho$ can be expressed as ${}^0\rho/{}^t\rho = \det\left({}^t_0\mathbf{F}\right)$, see web – Theory, Chapter 2, equation (T2.4-33).

2.4.2 Nonlinear elastic constitutive relations

The stress–strain relationships for many materials, and biological materials in particular, are nonlinear. Then the constitutive matrix depends on the strain level and changes during the course of material deformation. The material is considered elastic if it returns to the initial zero-stress and zero-strain state after it is unloaded.

There are various nonlinear elastic material models (Desai & Siriwardane 1984, Fung 1965, 1993, Ogden 1997). Here we consider the *hyperelastic* models. These materials are characterized by the existence of a strain energy function (see web – Theory, Chapter 2, Section 2.2d). This function of strains is such that its derivatives with respect to strains give the stresses.

The strain energy function for large strains is usually expressed in terms of the Green–Lagrange strains, $W = W\left(E^{GL}_{ij}\right)$. Then the derivatives with respect to strains give the conjugate second Piola–Kirchhoff stresses,

$${}^t_0S_{ij} = \frac{\partial\,{}^tW}{\partial\,{}^tE^{GL}_{ij}} \tag{2.4.20}$$

Commonly used expressions of the strain energy function have exponential form (Fung *et al.* 1979, Fung 1993, Holzapfel *et al.* 2000). This form can in general be written as (Humphrey 1995)

$$W = \frac{1}{2}c\left[\exp(Q) - 1\right] \tag{2.4.21}$$

where Q is a quadratic form of the Green–Lagrange strain components. Nine coefficients and the modified Green–Lagrange strains (which preserve material incompressibility) are used. In Fig. 2.4.3 we give a graphical representation of a two-dimensional form of this function (Fung *et al.* 1979, Takamizawa & Hayashi 1987) and the stress–stretch relationships for

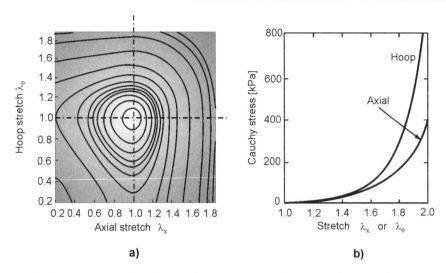

Fig. 2.4.3 Strain energy function in the $\lambda_x - \lambda_\theta$ coordinates and uniaxial stress–stretch curves for blood vessel wall (Takamizawa & Hayashi 1987, Holzapfel *et al.* 2000). (a) Strain potential showing convexity; (b) Uniaxial stress–axial strain constitutive relationships. Data: $Q = a_1 \left(E^{GL}_{\theta\theta}\right)^2 + a_2 \left(E^{GL}_{xx}\right)^2 + 2a_4 E^{GL}_{\theta\theta} E^{GL}_{xx}$; $c = 28.58\,kPa$; dimensionless coefficients: $a_1 = 0.8329$, $a_2 = 0.6004$, $a_4 = 0.0169$ (Reproduced with permission from Takamizawa, K. & Hayashi, K. 1987: Strain energy density function and uniform strain hypothesis for arterial mechanics, *J. Biomechanics*, 20, 7–17)

uniaxial material loadings. Details about this function and other forms of the strain energy function are given on the web – Theory, Chapter 2.

Finally, we show nonlinear elastic constitutive behavior for biological membranes. The constitutive relations are represented by the uniaxial and biaxial stress–stretch curves. These

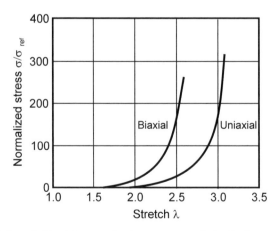

Fig. 2.4.4 Uniaxial and biaxial constitutive curves of dog's lung tissue slab (Lee & Hoppin 1972). Stress is normalized with respect to a reference stress σ_{ref} (Reproduced with permission from Lee, C.G. & Hoppin, F.G. Jr.: Lung elasticity. In Y.C. Fung, N. Perrone & M. Anliker (eds.), *Biomechanics – Its Foundations and Objectives* (pp. 317–35), Prentice-Hall, Englewood Cliffs, NJ, 1972)

FUNDAMENTALS OF CONTINUUM MECHANICS

curves are shown in Fig. 2.4.4 for a lung tissue (Lee & Hoppin 1972). The main property shown in these curves is the material nonlinear hardening: for small stretches, the stresses are small, while the material becomes very stiff in the domain of large stretches. More details about mechanical behavior of biological membranes are given in Chapter 11.

2.4.3 Examples

Here, we present one example. Additional examples are given on the web – Examples, Section 2.4.

Example 2.4-1. Determine principal stretches and principal directions for pure shear
Deformation of the material element in the case of pure shear is shown in Fig. E2.4-1. It is assumed that all material points move in the $x-y$ plane and straight-line elements remain straight, with no elongational (normal) strains in the x and y directions. The displacement field is defined by

$$^t u_x \equiv {}^t u = y(V/H)t = y\dot{\gamma}t = y{}^t\gamma, \quad {}^t\gamma = \dot{\gamma}t \quad (E2.4\text{-}1.1)$$

where V is constant velocity of the line element at the top surface, for $y = H$; and $\dot{\gamma} = V/H = const$. Note that for small strains ${}^t\gamma = \tan{}^t\gamma \approx {}^t\gamma$ is the shear strain (see (2.1.20)). In this analysis, we assume large displacements $u(y, t)$.

The x, y components of deformation gradient follow from (2.4.3) and (E2.4-1.1),

$${}^t_0\mathbf{F} \equiv \mathbf{F} = \begin{bmatrix} 1 & \gamma \\ 0 & 1 \end{bmatrix} \quad (E2.4\text{-}1.2)$$

For simplicity, we omit the left indices 't' and '0' in the text below, but assume that the reference configuration corresponds to the undeformed configuration at $t = 0$. Thus, the deformation tensors $\mathbf{C}$ and $\mathbf{B}$, given in (2.4.10) and (2.4.12), are

$$\mathbf{C} = \mathbf{F}^T\mathbf{F} = \begin{bmatrix} 1 & \gamma \\ \gamma & 1+\gamma^2 \end{bmatrix}, \quad \mathbf{B} = \mathbf{F}\mathbf{F}^T = \begin{bmatrix} 1+\gamma^2 & \gamma \\ \gamma & 1 \end{bmatrix} \quad (E2.4\text{-}1.3)$$

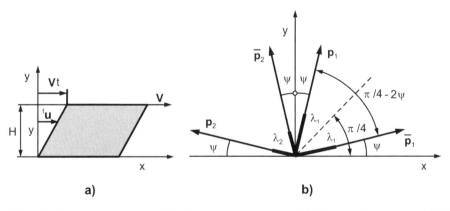

Fig. E2.4-1 Pure shear of material element, large strains. (a) Deformed element; (b) Principal directions of the right $(\mathbf{p}_1, \mathbf{p}_2)$ and left $(\bar{\mathbf{p}}_1, \bar{\mathbf{p}}_2)$ bases. While material deforms, the principal directions of the two bases rotate in the opposite directions (according to Kojić & Bathe 2005)

The characteristic equation (1.3.14) for both tensors **C** and **B** is

$$\left\|\begin{matrix} 1-\lambda^2 & \gamma \\ \gamma & 1+\gamma^2-\lambda^2 \end{matrix}\right\| = \left\|\begin{matrix} 1+\gamma^2-\lambda^2 & \gamma \\ \gamma & 1-\lambda^2 \end{matrix}\right\| = \lambda^4 - \lambda^2(2+\gamma^2) + 1 = 0 \tag{E2.4-1.4}$$

where λ^2 is the principal value. The solutions of this equation are

$$\lambda_1 = \cot\psi, \quad \lambda_2 = \tan\psi \tag{E2.4-1.5}$$

where

$$\psi = \frac{1}{2}\tan^{-1}(2/\gamma) \tag{E2.4-1.6}$$

We now determine the principal vectors of the right and left bases. The first equation of the system (1.3.17) for the vector $\mathbf{p}_1$ is

$$(1-\cot^2\psi)p_{1x} + \gamma p_{1y} = 0, \quad \tan\alpha_1 = \frac{p_{1y}}{p_{1x}} = \frac{1-\cot^2\psi}{\gamma} = \frac{1-\cot^2\psi}{2\cot 2\psi} = \cot\psi \tag{E2.4-1.7}$$

Therefore, the angle α_1 between the vector $\mathbf{p}_1$ is $\alpha_1 = \pi/2 - \psi$ as shown in Fig. E2.4-1b. Note that the second equation of the system (1.3.17) for the vector $\bar{\mathbf{p}}_1$ is the same as in (E2.4-1.7). Thus we have that

$$\gamma\bar{p}_{1x} + (1-\cot^2\psi)\bar{p}_{1y} = 0, \quad \tan\bar{\alpha}_1 = \frac{\bar{p}_{1y}}{\bar{p}_{1x}} = -\frac{\gamma}{1-\cot^2\psi} = -\frac{2\cot 2\psi}{1-\cot^2\psi} = \tan\psi \tag{E2.4-1.8}$$

and $\bar{\alpha}_1 = \psi$. The vectors of the right and the left bases and angles ψ, α_1 and $\bar{\alpha}_1$ are shown in the figure. Note that at the initial configuration ($t=0$), we have that $^0\psi = {}^0\alpha_1 = {}^0\bar{\alpha}_1 = \pi/4$, the right and left bases coincide, and the right basis rotates counterclockwise, while the left basis rotates clockwise (see figure). As time increases, the material element approaches to the x-axis, therefore the first principal direction $\bar{\mathbf{p}}_1$ approaches the x-axis, with large elongation stretch λ_1, while the vector $\bar{\mathbf{p}}_2$ approaches to the y-axis with large compressive stretch λ_2.

The rotation tensor **R** can be obtained from (E1.5-3.7),

$$\mathbf{R} = \bar{\mathbf{P}}_{(1)}\mathbf{P}_{(1)} + \bar{\mathbf{P}}_{(2)}\mathbf{P}_{(2)} = \sin 2\psi \mathbf{i}_1\mathbf{i}_1 - \cos 2\psi \mathbf{i}_1\mathbf{i}_2 + \cos 2\psi \mathbf{i}_2\mathbf{i}_1 + \sin 2\psi \mathbf{i}_2\mathbf{i}_2 \tag{E2.4-1.9}$$

In the matrix form, **R** is given as

$$\mathbf{R} = \begin{bmatrix} \sin 2\psi & \cos 2\psi \\ -\cos 2\psi & \sin 2\psi \end{bmatrix} \tag{E2.4-1.10}$$

Note that the orthogonality condition (1.2.12) $\mathbf{R}^T\mathbf{R} = \mathbf{R}\mathbf{R}^T = \mathbf{I}$ is satisfied.

The right and left stretch tensors **U** and **V** follow from (2.4.14),

$$\mathbf{U} = \mathbf{R}^T\mathbf{F} = \begin{bmatrix} \sin 2\psi & \cos 2\psi \\ \cos 2\psi & \gamma\cos 2\psi + \sin 2\psi \end{bmatrix}$$

$$\mathbf{V} = \mathbf{F}\mathbf{R}^T = \begin{bmatrix} \gamma\cos 2\psi + \sin 2\psi & \cos 2\psi \\ \cos 2\psi & \sin 2\psi \end{bmatrix} \tag{E2.4-1.11}$$

FUNDAMENTALS OF CONTINUUM MECHANICS

The rate of deformation tensor and the spin tensor, given in (2.1.27) and (2.1.28), are

$$\mathbf{D} = \frac{V}{H}\begin{bmatrix} 0 & 1 \\ 1 & 0 \end{bmatrix}, \quad \mathbf{W} = \frac{V}{H}\begin{bmatrix} 0 & 1 \\ -1 & 0 \end{bmatrix} \qquad \text{(E2.4-1.12)}$$

We note that the above analysis of principal stretches and principal directions can be obtained graphically using the Mohr circles (see web – Examples, Sections 2.1 and 2.4).

Various stress and strain measures are calculated in Example 2.4.2, see web – Examples, Section 2.4.

EXERCISE

Draw the Mohr circles for the deformation tensors **C** and **B** and graphically determine principal stretches and principal stretch directions. Determine values of the maximum shear and directions of the maximum shear. Show how the principal stretches and the maximum shear change over time, as well as their directions (see web – Examples, Section 2.4).

References

Bathe, K.J. (1996). *Finite Element Procedures*, Prentice-Hall, Englewood Cliffs, NJ.
Desai, C.S. & Siriwardane, H.J. (1984). *Constitutive Laws for Engineering Materials*, Prentice-Hall, Englewood Cliffs, NJ.
Evans, E.A. & Hochmuth, R.M. (1976). Membrane viscoelasticity, *Biophysical J.*, **16**(1), 1–11.
Fung, Y.C. (1965). *Foundations of Solid Mechanics*, Prentice-Hall, Englewood Cliffs, NJ.
Fung, Y.C. (1993). *Biomechanics: Mechanical Properties of Living Tissue*, 2nd edition, Springer-Verlag, New York.
Fung, Y.C., Fronek, K. & Patitucci, P. (1979). Pseudoelasticity of arteries and the choice of its mathematical expression, *Am. J. Physiol.*, **237**, 620–31.
Holzapfel, G.A., Gasser, C.T. & Ogden, R.W. (2000). A new constitutive framework for arterial wall mechanics and comparative study of material models, *J. Elasticity*, **61**, 1–48.
Humphrey, J.D. (1995). Mechanics of arterial wall: review and directions, *Crit. Rev. Biomed. Eng.*, **23**, 1–162.
Kojić, M. & Bathe, K.J. (2005). *Inelastic Analysis of Solids and Structures*, Springer-Verlag, Berlin.
Kojić, M. & Cheatham, J.B. Jr. (1974). Theory of plasticity of fluid flow through porous media with fluid flow, *Soc. Petrol. Eng. J.*, **14**(3), 263–70.
Lee, C.G. & Hoppin, F.G. Jr. (1972). Lung elasticity. In Y.C. Fung, N. Perrone & M. Anliker (eds.), *Biomechanics – Its Foundations and Objectives* (pp. 317–35), Prentice-Hall, Englewood Cliffs, NJ.
Malvern, L.E. (1969). *Introduction to the Mechanics of a Continuous Medium*, Prentice-Hall, Englewood Cliffs, NJ.
Mase, G.T. & Mase, G.E. (1999). *Continuum Mechanics for Engineers*, CRC Press, Boca Raton, FL.
Ogden, R.W. (1997). *Nonlinear Elastic Deformations*, Dover, New York.
Prager, W. (1959). *Introduction to Plasticity*, Addison-Wesley, Reading, MA.
Takamizawa, K. & Hayashi, K. (1987). Strain energy density function and uniform strain hypothesis for arterial mechanics, *J. Biomechanics*, **20**, 7–17.
Wang, C.T. (1953). *Applied Elasticity*, McGraw-Hill, New York.
With de, G. (2006). *Structure, Deformation, and Integrity of Materials I, II*, Wiley-VCH Verlag GmbH & Co, KGaA, Weinheim, Germany.

3

Heat Transfer, Diffusion, Fluid Mechanics, and Fluid Flow through Porous Deformable Media

The basic relations in heat conduction, diffusion, fluid mechanics, and flow through porous deformable media are summarized in this chapter. Also, heat transfer and diffusion within the flowing fluid is considered. The presented relations are used in further developments of numerical methods. We first present heat transfer, then diffusion, followed by fluid mechanics and fluid flow through porous deformable media. Many textbooks are available for these topics (e.g. Mills 1999, 2001, Incropera & DeWitt 1996, Munson *et al*. 1998, Lewis & Schrefler 1987).

3.1 Heat conduction

Heat energy propagates through a continuous medium when there is temperature difference (temperature gradient) within the continuum. In the case of a solid, the heat energy propagation is called *heat conduction*, while heat propagation within a moving fluid is called *heat transfer*. We here derive the differential equations of heat conduction as the basic ones, while the heat transfer equations will be derived in Section 3.3.

3.1.1 Governing relations

The differential equation of heat conduction relies on the balance of internal energy within an elementary material volume dV and unit time of the continuum,

$$\frac{dU}{dt} = \frac{dQ}{dt} \tag{3.1.1}$$

where dU/dt is the rate of change of internal energy, and dQ/dt is transfer rate of energy as heat. Rate of change of the internal energy can be expressed in terms of rate of temperature change, dT/dt,

$$\frac{dU}{dt} = \rho c \frac{dT}{dt} dV \tag{3.1.2}$$

where ρ is mass density (kg/m^3), c is specific heat (J/kgK), and $T(K)$ is temperature. We further introduce the surface heat flux $\mathbf{q}$ vector, with the components q_i (or q_x, q_y, q_z in the x, y, z coordinate system, Fig. 3.1.1) (W/m^2), and use the *Fourier law* of heat conduction,

$$\mathbf{q} = -\mathbf{k}\nabla T, \qquad q_i = -k_{ik}\frac{\partial T}{\partial x_k} \tag{3.1.3}$$

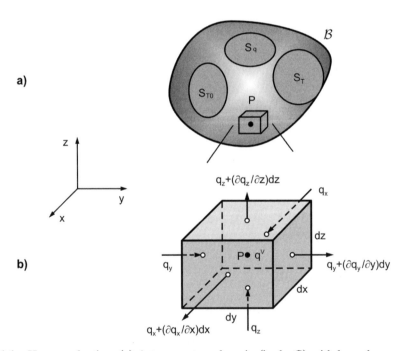

Fig. 3.1.1 Heat conduction. (a) A temperature domain (body $\mathcal{B}$) with boundary conditions on surfaces: S_T – surface with prescribed temperature, S_{T0} – surface with given surrounding temperature (boundary condition expressed as $q_{conv} = h(T_S - T_0)$, where q_{conv} is convective flux, T_S is temperature at the surface, T_0 is the surrounding temperature, and h coefficient of convection (W/m^2K)), S_q – given flux through the surface; (b) Elementary material volume with heat fluxes through the surfaces and heat source $q^V(W/m^3)$

HEAT TRANSFER, DIFFUSION, FLUID MECHANICS, AND FLUID FLOW

where **k** is the heat conduction matrix,

$$\mathbf{k} = \begin{bmatrix} k_x & 0 & 0 \\ 0 & k_y & 0 \\ 0 & 0 & k_x \end{bmatrix}, \quad k_{ik} = k_i \delta_{ik}, \quad i, k = 1, 2, 3 \tag{3.1.4}$$

where k_i (or k_x, k_y, k_z) (W/m K) are the heat conduction coefficients corresponding to an orthotropic medium; for isotropic medium $k_i = k$.

Equation (3.1.1) finally becomes (see detailed derivations on web – Theory, Chapter 3)

$$-\rho c \frac{\partial T}{\partial t} + \nabla^T(\mathbf{k}\nabla T) + q^V = 0, \quad \text{or} \quad -\rho c \frac{\partial T}{\partial t} + \frac{\partial}{\partial x_i}\left(k_i \frac{\partial T}{\partial x_i}\right) + q^V = 0, \quad i = 1, 2, 3 \tag{3.1.5}$$

where we have assumed that the heat conduction coefficients k_i depend on temperature. When k_i are temperature independent, the term with derivatives with respect to x_i have the form $k_i \partial^2 T/\partial^2 x_i$, with summation over the dummy index i.

3.1.2 Examples

Example 3.1-1. Steady heat conduction laterally through long column
Uniform temperatures T_1 and T_2 are given along all column sides, Fig. E3.1-1a. There is no heat flow in the axial column direction, hence the problem is two-dimensional, with identical temperature field in $x-y$ planes. Also, there is a symmetry plane due to symmetry of the boundary conditions, shown in the figure.

For $T_2 = 0$, the analytical steady state solution can be obtained (Polivka & Wilson 1976) in a form of infinite series:

$$T(x,y) = \frac{4T_1}{\pi} \sum_{n=0}^{\infty} \frac{1}{(2n+1)} \sin \frac{(2n+1)\pi y}{a} \sinh \frac{(2n+1)(a-x)\pi}{a} \operatorname{cosech}(2n+1)\pi \tag{E3.1-1.1}$$

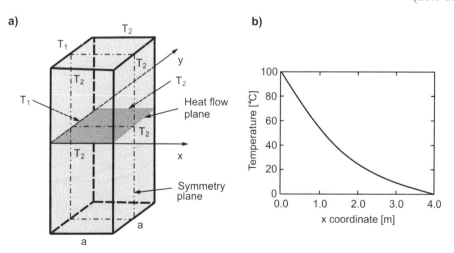

Fig. E3.1-1 Steady lateral heat conduction through a column with given temperatures along the column sides. (a) Geometry and boundary conditions; (b) Temperature distribution along symmetry plane for: $k = 1 \ W/mK$, $a = 4m$, $T_1 = 100°C$, $T_2 = 0°C$

Temperature distribution along the symmetry plane is shown in Fig. E3.1-1b. The finite element solution for this example is presented in Section 7.2.2 (see also web – Software).

Example 3.1-2. Unsteady heat conduction through a semi-infinite medium
A constant flux is given on the surface bounding the semi-infinite solid shown schematically in Fig. E3.1-2a. Initially, temperature of the solid is equal to zero.

It can be considered that the heat flow in the solid occurs in the direction normal to the surface, shown in the figure as the x-axis. Then, the analytical solution can be obtained (Polivka & Wilson 1976),

$$T(x,t) = \frac{2}{k}\left[\sqrt{\frac{kt}{\pi}}\exp(-x^2/(4kt)) - \frac{x}{2}\mathrm{erfc}\left(x/\left(2\sqrt{kt}\right)\right)\right] \qquad (E3.1\text{-}2.1)$$

where

$$\mathrm{erfc}(x) = \frac{2}{\pi}\int_x^\alpha \exp(-u^2)\,du \qquad (E3.1\text{-}2.2)$$

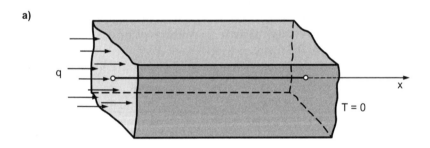

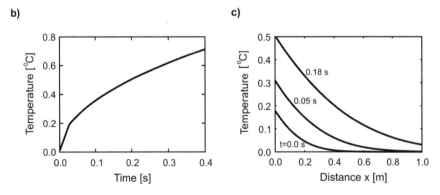

Fig. E3.1-2 Unsteady heat conduction through semi-infinite solid. (a) Schematic representation of the solid with 1D heat conduction in the x-direction; (b) Increase of temperature with time at the surface where constant flux is given; (c) Temperature profiles along the distance x from the surface with flux. Data: flux $q = 1\,W/m^2$, heat conduction coefficient $k = 1\,W/m\,K$, mass density $\rho = 1\,kg/m^3$, specific heat $c = 1\,J/kgK$, initial temperature $T_{initial} = 0\,^\circ C$

Temperature at the boundary and distribution along the x-axis are shown in Figs. E3.1-2b,c. The temperature increases with time, on the surface with flux as well as within the body. The temperature profiles show that the temperature decreases with x. The finite element solutions are given in Example 7.2.2 (see also web – Software).

3.2 Diffusion

Diffusion is a process of mass transport within a mixture of several constituents. We will consider mass transport of constituents within a liquid solution. There are two distinct cases which we are going to consider: diffusion of dilute substances, and diffusion in concentrated solutions. We first introduce the basic terms and Fick's law, which is the fundamental diffusion law and is analogous to Fourier's law in heat conduction.

3.2.1 Differential equations of diffusion

Definitions of concentration

There are various definitions of concentration of a constituent within a mixture (Mills 2001). We will use two of them: mass concentration and volumetric concentration. *Mass concentration* c_{mj} and partial density ρ_j (kg/m^3) of species 'j', and the density of mixture (total density) ρ, are defined as (see Fig. 3.2.1)

$$c_{mj} = \frac{\rho_j}{\rho}, \quad \rho_j = \frac{\Delta m_j}{\Delta V}, \quad \rho = \frac{\Delta m}{\Delta V}, \tag{3.2.1}$$

where Δm_j and Δm are, respectively, elementary masses of the species and mixture within an elementary mixture volume ΔV. The relations following from these definitions are

$$\rho = \sum_j \rho_j, \quad \sum_j c_{mj} = 1 \tag{3.2.2}$$

Note that the mass density can also be defined as the mass per mol (kg/mol). We further introduce the *substance density* $\bar{\rho}_j$ which occupies a volume ΔV_j within the volume ΔV and consequently the *volumetric concentration* c_{Vj},

$$\bar{\rho}_j = \frac{\Delta m_j}{\Delta V_j} = (const)_j, \quad c_{Vj} = \frac{\Delta V_j}{\Delta V} = \frac{\rho_j}{\bar{\rho}_j}, \quad \sum_j c_{Vj} = 1 \tag{3.2.3}$$

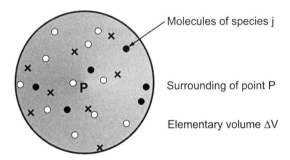

Fig. 3.2.1 Elementary volume of a mixture

where it is indicated that the substance densities are material properties and are constant, and the sum of c_{Vj} is equal to 1.

Fick's Law

The Fick law is a phenomenological law which states that the mass flux of a species 'j', $\mathbf{q}_{mj}$ ($kg/m^2 s$), is in the negative direction of the spatial gradient of the species concentration, ∇c_{mj}, expressed in a mathematical form as

$$\mathbf{q}_{mj} = -\rho D_j \nabla c_{mj} \quad \text{no sum on } j \tag{3.2.4}$$

Here D_j (m^2/s) is the *diffusion coefficient*. If the volumetric concentration is used, the Fick law is

$$\mathbf{q}_{mj} = -\rho \bar{\rho}_j D_j \nabla (c_{Vj}/\rho) \quad \text{no sum on } j \tag{3.2.5}$$

When the total density ρ is constant, the Fick law becomes

$$\mathbf{q}_{mj} = -\bar{\rho}_j D_j \nabla c_{Vj} \quad \text{no sum on } j \tag{3.2.6}$$

Governing Equations of Diffusion

From the equation of balance of mass for a species 'j', in analogy with the derivation of equation (3.1.5) for heat conduction (see Fig. 3.1.1, where the heat flux should be replaced by the mass flux, and the heat source term q^V by the mass source term q_{mj}^V ($kg/m^3 s$)), replacing the Fourier law (3.1.3) by the Fick law (3.2.4), we obtain

$$-\frac{\partial \rho_j}{\partial t} + \nabla^T (\rho D_j \nabla c_{mj}) + q_{mj}^V = 0, \quad \text{or}$$

$$-\frac{\partial \rho_j}{\partial t} + \frac{\partial}{\partial x_i}\left(\rho D_j \frac{\partial c_{mj}}{\partial x_i}\right) + q_{mj}^V = 0, \quad \text{no sum on } j, \text{ sum on } i: \ i = 1, 2, 3 \tag{3.2.7}$$

Therefore, we have the equation of mass balance for each species of the mixture. If the number of constituents is M, the number of equations to be solved is $M - 1$ because the sum of concentrations is equal to one (see (3.2.2) and (3.2.3)). These equations correspond to the *concentrated solution*, where change of the mass concentrations of species causes change of the mixture density.

For a *dilute solution*, we have a solvent (incompressible fluid) as the dominant medium in mass, which also occupies the space, with a small mass (and volumetric) part belonging to species. Then it can be considered that the mixture density $\rho = const.$, and the equations of mass balance (3.2.7) can be written as

$$-\frac{\partial c_{Vj}}{\partial t} + \nabla^T (D_j \nabla c_{Vj}) + q_{mj}^V/\bar{\rho}_j = 0, \quad \text{or}$$

$$-\frac{\partial c_{Vj}}{\partial t} + \frac{\partial}{\partial x_i}\left(D_j \frac{\partial c_{Vj}}{\partial x_i}\right) + q_{mj}^V/\bar{\rho}_j = 0, \quad \text{no sum on } j, \text{ sum on } i: \ i = 1, 2, 3 \tag{3.2.8}$$

These equations have the same form as the heat balance equation (3.1.5).

HEAT TRANSFER, DIFFUSION, FLUID MECHANICS, AND FLUID FLOW

3.2.2 Examples

Example 3.2-1. One-dimensional unsteady diffusion of a dilute solution

In the case of one-dimensional diffusion without the mass source, and for one spices ($j = 1$), the diffusion equation (3.2.8) reduces to

$$-\frac{\partial c}{\partial t} + \frac{\partial}{\partial x}\left(D\frac{\partial c}{\partial x}\right) = 0 \qquad \text{(E3.2-1.1)}$$

Consider the solution of this equation within the region $-x_0 \leq x \leq x_0$ with the initial condition:

$$c(x, t_0) = \exp\left(\frac{-x^2}{4Dt_0}\right) \qquad \text{(E3.2-1.2)}$$

where $t_0 > 0$ is the initial time; and $D = \text{const}$. The spatial boundary conditions are assumed as

$$c(\pm x_0, t) = \sqrt{\frac{t_0}{t}} \exp\left(\frac{-x_0^2}{4Dt}\right) \qquad \text{(E3.2-1.3)}$$

The analytical solution of this problem is (Crank 1979)

$$c(\pm x, t) = \sqrt{\frac{t_0}{t}} \exp\left(\frac{-x^2}{4Dt}\right) \qquad \text{(E3.2-1.4)}$$

Graphical representation of the initial concentration and at $t = 5\,s$ are shown in Fig. E3.2-1. This example is also solved using the finite element method, see Example 7.3-1.

EXERCISE

Compute and plot concentration distribution for different diffusion coefficient D and different initial time t_0 (see web – Software).

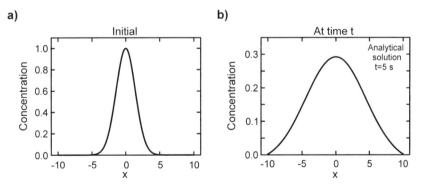

Fig. E3.2-1 Unsteady one-dimensional diffusion through infinite medium. (a) Initial concentration (eq. (E3.2-1.2)); (b) Analytical solution for concentration distribution along the distance x for $t = 5\,s$. Data: diffusion coefficient $D = 1.0\,m^2/s$, $t_0 = 1.0\,s$, mass density $\rho = 1\,kg/m^3$

3.3 Fluid flow of incompressible viscous fluid with heat and mass transfer

Biological fluids can be considered incompressible and viscous and we present here the governing equations for these fluids: the equation of balance of mass (continuity equation) and equations of balance of linear momentum. These equations are used in the numerical applications. Also, we present the equations of heat transfer (heat conduction within the fluid, with convective effects) and mass transfer (diffusion with convection).

In deriving the governing equations we adopt a *control volume*, which represents an elementary volume fixed in space, with dimensions dx, dy and dz (Fig. 3.3.1b), through which the fluid is flowing. The control volume surrounds the point in space, with coordinates x, y, z (Fig. 3.3.1a). The goal of fluid mechanics is to determine the *spatial fields* of fluid variables, such as pressure and velocity. This approach in studying a continuous medium is called the *Eulerian (or spatial) description*. We pointed out in Section 2.1 that the *Lagrangian (or material) description* is used in solid mechanics.

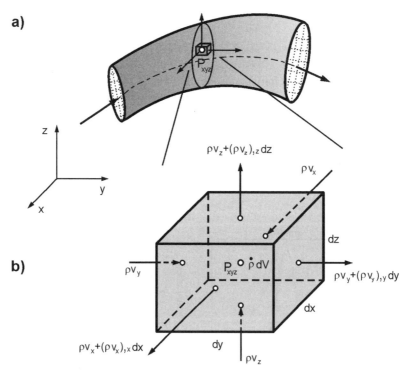

Fig. 3.3.1 Control volume of a fluid, surrounding spatial point P_{xyz} (the position of P_{xyz} is fixed in space), through which the fluid is flowing. (a) Control volume within the fluid domain; (b) Enlarged control volume with mass fluxes through elementary surfaces parallel to stationary coordinate planes, and rate of change of mass within the volume $\dot{\rho}dV$ due to fluid density change

HEAT TRANSFER, DIFFUSION, FLUID MECHANICS, AND FLUID FLOW

3.3.1 Governing equations of fluid flow and of heat and mass transfer

Continuity Equation

Figure 3.3.1a shows a control volume within a fluid flow domain, while in Fig. 3.3.1b this control volume is enlarged – with the mass fluxes through the elementary surfaces. The fluid density is ρ and velocity components are v_x, v_y, v_z. The equation of balance of mass within the control volume leads to the continuity equation (detailed derivation is given on the web – Theory, Chapter 3),

$$\frac{\partial \rho}{\partial t} + \nabla \cdot (\rho \mathbf{v}) = 0, \quad \text{or} \quad \frac{\partial \rho}{\partial t} + \frac{\partial (\rho v_i)}{\partial x_i} = 0; \quad \text{or} \quad \frac{D\rho}{Dt} + \rho \nabla \cdot \mathbf{v} = 0 \qquad (3.3.1)$$

We have used here $D\rho/Dt = \partial \rho/\partial t + (\partial \rho/\partial x_i) v_i$ to denote the so-called *total (material) derivative* of the fluid density ρ (see (1.4.15)). The material derivative $D(\bullet)/Dt$ of any fluid quantity is used in the Eulerian description, which has a general form,

$$\frac{D(\bullet)}{Dt} = \frac{\partial(\bullet)}{\partial t} + \nabla(\bullet) \cdot \mathbf{v}, \quad \text{or} \quad \frac{D(\bullet)}{Dt} = \frac{\partial(\bullet)}{\partial t} + \frac{\partial(\bullet)}{\partial x_i} v_i \qquad (3.3.2)$$

The derivative $\partial(\bullet)/\partial t$ is the local derivative at the spatial point (assuming no motion of the fluid); and the term $[\partial(\bullet)/\partial x_i] v_i$ represents the convective part of the material derivative, which takes into account motion of the fluid (see detailed description on the web – Theory, Chapter 3). In case of incompressible fluid we have $\rho = const.$, and

$$\nabla \cdot (\rho \mathbf{v}) = 0, \quad \text{or} \quad \frac{\partial v_i}{\partial x_i} \equiv \frac{\partial v_x}{\partial x} + \frac{\partial v_y}{\partial y} + \frac{\partial v_z}{\partial z} = 0 \qquad (3.3.3)$$

The scalar product of the operator $\nabla \equiv \partial/\partial x_k \mathbf{i}_k$ and a vector is called the divergence of the vector (see (1.4.4)), therefore the continuity equation for incompressible fluid is expressed by the condition that the *divergence of velocity is equal to zero* at each point of the fluid domain.

Constitutive Relations

The stresses σ_{ij} acting on a fluid element can be decomposed into pressure p and viscous stresses τ_{ij}, hence we have

$$\sigma_{ij} = -p\delta_{ij} + \tau_{ij} \qquad (3.3.4)$$

where δ_{ij} is the Kronecker delta symbol ($= 1$ for $i = j$; $= 0$ for $i \neq j$). Obviously, it is assumed, as in solids (see Section 2.1.1), that the tensile stress is positive. The viscous stress components are proportional to the strain rates $\dot{e}_{ij}$ defined in (2.1.27),

$$\begin{aligned} \tau_{ij} &= 2\mu \dot{e}_{ij} = \mu \left(v_{i,j} + v_{j,i} \right), \quad \text{or} \\ \tau_{xx} &= 2\mu v_{x,x}, \quad \tau_{yy} = 2\mu v_{y,y}, \quad \tau_{zz} = 2\mu v_{z,z}, \\ \tau_{xy} &= \mu \left(v_{x,y} + v_{y,x} \right), \quad \tau_{yz} = \mu \left(v_{y,z} + v_{z,y} \right), \quad \tau_{xz} = \mu \left(v_{x,z} + v_{z,x} \right) \end{aligned} \qquad (3.3.5)$$

where μ (Pa s) is the *viscosity coefficient* (or *dynamic viscosity*).

The Navier–Stokes Equations

These equations represent the equations of balance of linear momentum, as we have for a solid – equations (2.1.12). However, since here the Eulerian description of the fluid motion is used, it is necessary to employ the material derivative for evaluating the inertial term, and the equations of motion (for $\rho = const.$) are

$$\rho \left(\frac{\partial v_i}{\partial t} + \frac{\partial v_i}{\partial x_k} v_k \right) = -\frac{\partial p}{\partial x_i} + \frac{\partial \tau_{ik}}{\partial x_k} + f_i^V \quad i = 1, 2, 3; \quad \text{sum on } k: \ k = 1, 2, 3 \quad (3.3.6)$$

where f_i^V are the volumetric force components as in the case of a solid. Substituting now the constitutive equations (3.3.5) for the viscous stresses, we obtain

$$\rho \left(\frac{\partial v_i}{\partial t} + \frac{\partial v_i}{\partial x_k} v_k \right) = -\frac{\partial p}{\partial x_i} + \mu \frac{\partial^2 v_i}{\partial x_k \partial x_k} + f_i^V \quad i = 1, 2, 3; \quad \text{sum on } k: \ k = 1, 2, 3 \quad (3.3.7)$$

Here, the incompressibility condition (3.3.3) is taken into account. Detailed derivation of (3.3.3) and (3.3.7) is given on the web – Theory, Chapter 3.

Heat Transfer with Convection

In order to obtain the equation for heat conduction in flowing fluid (usually called heat transfer) we simply replace the time derivative $\partial T/\partial t$ in (3.1.5) by the material derivative DT/Dt according to (3.3.2) and obtain

$$-\rho c \left(\frac{\partial T}{\partial t} + \frac{\partial T}{\partial x_i} v_i \right) + \frac{\partial}{\partial x_i} \left(k_i \frac{\partial T}{\partial x_i} \right) + q_V = 0, \quad \text{sum on } i: \ i = 1, 2, 3 \quad (3.3.8)$$

Diffusion with Convection

As for heat transfer, we now substitute the time derivative $\partial c_{Vj}/\partial t$ in (3.2.8) by the material derivative Dc_{Vj}/Dt (following (3.3.2)), and obtain the diffusion equations for each diluted species,

$$-\left(\frac{\partial c_{Vj}}{\partial t} + \frac{\partial c_{Vj}}{\partial x_i} v_i \right) + \frac{\partial}{\partial x_i} \left(D_j \frac{\partial c_{Vj}}{\partial x_i} \right) + q_{mj}^V / \bar{\rho}_j = 0, \quad (3.3.9)$$

no sum on $j, j = 1, 2, \ldots, M;$ sum on $i: \ i = 1, 2, 3$

Details on the derivations of (3.3.8) and (3.3.9) are given on the web – Theory, Chapter 3.

In practical applications we first solve for the fluid velocities and then use the velocity field to calculate temperatures or concentrations from (3.3.8) or (3.3.9), respectively, with the corresponding initial and boundary conditions for temperatures and concentrations.

3.3.2 Examples

We provide a few examples for which the finite element solutions are presented in Section 7.4. Other examples can be found on the web – Examples and Software.

Example 3.3-1. Two-dimensional steady flow between two parallel walls
Viscous fluid is flowing between two parallel walls, which are long in the z-direction (Fig. E3.3-1a). Under these conditions it can be considered that fluid flow is two-dimensional,

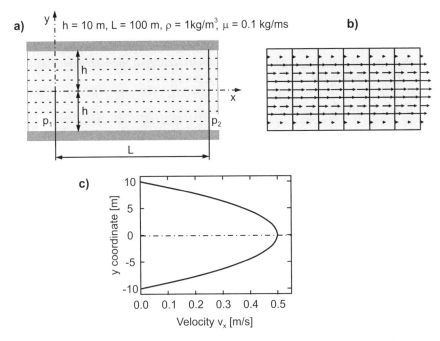

Fig. E3.3-1 Two-dimensional steady fluid flow between parallel walls. (a) Geometrical and material data; (b) Velocity vector field; (c) Velocity profile $v(y) = v_x(y)$

with velocities in the x-direction only. Therefore, we have one of the equations (3.3.7), with zero external force. Geometrical and material data, and boundary conditions are given in the figure. It is assumed that the flow is stationary. Boundary conditions consist of given pressures p_1 and p_2 at the entrance and at the outlet, respectively; and zero velocities at the walls. This fluid flow is known as the Poiseuille flow.

The differential equation of motion is

$$-\frac{\partial p}{\partial x} + \mu \frac{\partial^2 v}{\partial y^2} = 0 \qquad (E3.3\text{-}1.1)$$

and it can be satisfied only if each of the terms is equal to the same constant. Therefore, we have that $\partial p/\partial x = const.$, or $\partial p/\partial x = -(p_1 - p_2)/L$. Substituting this value for the pressure gradient into the above equation follows the dependence $v(y)$ which satisfies the boundary conditions, as

$$v = \frac{1}{2\mu} \frac{p_1 - p_2}{L} (h^2 - y^2) \qquad (E3.3\text{-}1.2)$$

The velocity vector field and parabolic-shape curve representing the dependence $v(y)$, for the given data are shown in Fig. E3.3-1b,c. Comparison of the finite element solution with this analytical solution is given in Example 7.4.1, with the software on the web (see web – Software) to obtain solutions for various data.

EXERCISE
(a) Compute and plot the velocity distribution when the viscosity is changed as $\mu = k\mu_0$, where μ_0 is the referent value given in Fig. E3.3-1, and k is a positive scalar. (b) Determine the shear stress τ and plot the graphs $\tau(y)$ for various values of the viscosity μ. (c) Find the rate of energy dissipation per unit axial length of the channel.

Example 3.3-2. Unsteady fluid flow on a plate

The motion of a fluid, which lies on a plate, is caused by plate motion with constant velocity v_0. The motion of the plate starts at $t = 0$, when the fluid is at rest. The fluid domain is considered to be semi-infinite in the y-direction (Fig. E3.3-2a) and the velocity of fluid tends to zero when y tends to $-\infty$ for any time $t > 0$. Influence of gravity is neglected. Under the assumption that flow is parallel to the plate, that convective term is neglected, and with constant pressure in time and space, the Navies–Stokes equations (3.3.7) reduce to the wave equation

$$\frac{\partial v}{\partial t} = \nu \frac{\partial^2 v}{\partial y^2} \qquad (E3.3\text{-}2.1)$$

where $\nu = \mu/\rho$ (m^2/s) is kinematic viscosity. The analytical solution of this equation (in the domain L, near the plate) is (Batchelor 1967)

$$v(y, t) = v_0 + \frac{4v_0}{\pi} \sum_{n=1}^{\infty} \frac{(-1)^n}{2n-1} e^{-(2n-1)^2 \pi^2 \nu t / 4L^2} \cos \frac{(2n-1)\pi y}{2L} \qquad (E3.3\text{-}2.2)$$

The velocity distribution in the domain L and for several times is shown in Fig. E3.3-2b. It can be seen that the velocity tends to the plate velocity v_0 in the entire domain as time increases. Finite element solutions are shown in Example 7.4-2.

EXERCISE
Calculate shear stress distribution along the y-axis, in the vicinity of the moving plate, for times used in Fig. E3.3-2b. Use $n = 5$ in (E3.3-2.2).

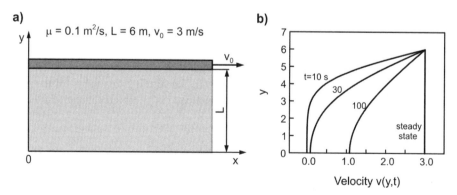

Fig. E3.3-2 Unsteady fluid flow (seminfinite domain in the y-direction) caused by the moving plate on the top of the fluid domain. (a) Geometrical and material data; (b) Velocity distribution in the domain L near the moving plate for several times

HEAT TRANSFER, DIFFUSION, FLUID MECHANICS, AND FLUID FLOW

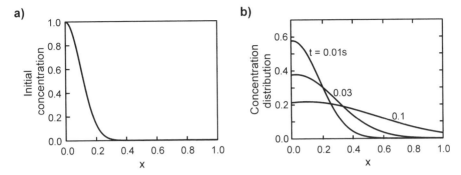

Fig. E3.3-3 Linear convection–diffusion equation (Burger's viscous equation). (a) Initial conditions according to (E3.3-3.2), constant $B = 50$; (b) Solutions for several times

Example 3.3-3. Linear convection–diffusion equation (Burger's viscous equation)
If in the equation (3.3.8) for heat transfer with convection, and in the equation (3.3.9) for diffusion with convection – all material parameters are taken as unity (ρ, c and k in (3.3.8), and D in (3.3.9)), then the one-dimensional problems for heat conduction and diffusion can be represented by the same equation (known as Burger's viscous equation)

$$\frac{\partial u}{\partial t} + \frac{\partial u}{\partial x} = \frac{\partial^2 u}{\partial x^2} \qquad (E3.3\text{-}3.1)$$

where u is either temperature T or concentration c_V. Also, it is assumed that the fluid velocity is constant and equal to unity, and that there are no source terms. We use the initial condition for this equation as (Chawla et al. 2000):

$$u(x, 0) = \exp(-Bx^2) \qquad (E3.3\text{-}3.2)$$

graphically shown in Fig. E3.3-3a; where B is a constant.

The analytical solution of (E3.3-3.1) which satisfies the initial conditions (E3.3-3.2) is

$$u(x, t) = \frac{1}{\sqrt{s}} \exp\left(-B\frac{(x-t)^2}{s}\right), \quad s = 1 + 200t \qquad (E3.3\text{-}3.3)$$

and is shown in Fig. E3.3-3b. Finite element solutions are given in Example 7.4-3 (see also web – Software, Chapter 7).

EXERCISE
Compute and plot temperature/concentration distribution for different values of constant B and compare the solutions for the same times t_k. For selected values of time t and coordinate x (i.e. $t = t_k$, $x = x_p$) find the change of u among the solutions; can this change be computed analytically (and compared with the values obtained from graphs)?

3.4 Fluid flow through porous deformable media

In this section we present the governing equations for flow of fluid through porous deformable media. It is assumed that the fluid is compressible, the solid medium is elastic and strains within the solid are small (Lewis & Schrefler 1987, Kojić et al. 1998, 2001).

3.4.1 The governing equations

Consider a solid whose material is porous, with pores filled by a fluid. It is assumed that the solid material deforms, while the fluid moves relative to the solid matrix. Let the current deformed configuration at time t be ${}^t\mathcal{B}$ as shown in Fig. 3.4.1. The current position of a material point P within the solid–fluid mixture – considered as a continuum – is ${}^t\mathbf{r}$. The physical quantities at the material point, used further to describe deformation of the solid and the fluid flow, are: displacement of solid $\mathbf{u}$, relative fluid velocity with respect to the solid (Darcy's velocity) $\mathbf{q}$, and fluid pressure p. Darcy's velocity is the volumetric flux (volume per unit time) of fluid through a unit area of the mixture (units are *m/s*) and represents the mean relative velocity of the fluid with respect to the moving solid.

Here we present the governing equations for the coupled problem described above. First, consider the equilibrium equation of the solid,

$$(1-n)\nabla^T \boldsymbol{\sigma}_s + (1-n)\rho_s \mathbf{b} + \mathbf{k}^{-1} n \mathbf{q} - (1-n)\rho_s \ddot{\mathbf{u}} = \mathbf{0} \tag{3.4.1}$$

where $\boldsymbol{\sigma}_s$ is stress in the solid phase, n is porosity, $\mathbf{k}$ is permeability matrix, ρ_s is density of solid, $\mathbf{b}$ is body force per unit mass, and $\ddot{\mathbf{u}}$ is acceleration of the solid material. If the one-index notation is employed, then the differentiation operator $\mathbf{L}^T$ must be used, see (2.1.16) and web – Theory, Chapter 2. Equation (3.4.1) and the others that follow correspond to the current configuration ${}^t\mathcal{B}$ and time 't', but we omit the left upper index 't' in order to simplify the notation.

The equilibrium equation of the fluid phase is

$$n\nabla p + n\rho_f \mathbf{b} - \mathbf{k}^{-1} n \mathbf{q} - n\rho_f \dot{\mathbf{v}}_f = \mathbf{0} \tag{3.4.2}$$

where p is pore fluid pressure, ρ_f is fluid density and $\dot{\mathbf{v}}_f$ is fluid acceleration. This equation is also known as the generalized Darcy's law. Using the relation between Darcy's velocity $\mathbf{q}$ and the fluid velocity $\mathbf{v}_f$,

$$\mathbf{q} = n\left(\mathbf{v}_f - \dot{\mathbf{u}}\right) \tag{3.4.3}$$

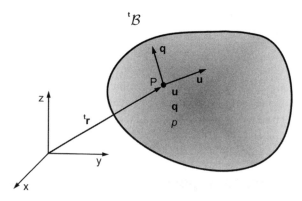

Fig. 3.4.1 Configuration ${}^t\mathcal{B}$ at time t and variables at a material point P of the mixture: displacement of solid $\mathbf{u}$, relative fluid velocity with respect to the solid (Darcy's velocity) $\mathbf{q}$, and fluid pressure p

HEAT TRANSFER, DIFFUSION, FLUID MECHANICS, AND FLUID FLOW

we transform (3.4.2) into

$$-\nabla p + \rho_f \mathbf{b} - \mathbf{k}^{-1}\mathbf{q} - \rho_f \ddot{\mathbf{u}} - \frac{\rho_f}{n}\dot{\mathbf{q}} = \mathbf{0} \tag{3.4.4}$$

Note that the above equilibrium equations are written per unit volume of the mixture.
If we multiply (3.4.4) by n and add to (3.4.1) we obtain

$$\nabla^T \boldsymbol{\sigma} + \rho \mathbf{b} - \rho \ddot{\mathbf{u}} - \rho_f \dot{\mathbf{q}} = \mathbf{0} \tag{3.4.5}$$

where $\boldsymbol{\sigma}$ is the total stress which can be expressed in terms of $\boldsymbol{\sigma}_s$ and p, as

$$\boldsymbol{\sigma} = (1-n)\boldsymbol{\sigma}_s - n\mathbf{m}p \tag{3.4.6}$$

and $\rho = (1-n)\rho_s + n\rho_f$ is the mixture density. Here $\mathbf{m}$ is a constant vector defined as $\mathbf{m}^T = [1\ 1\ 1\ 0\ 0\ 0]$ to indicate that the pressure contributes to the normal stresses only. It is assumed that the pressure has a positive sign in compression, while tensional stresses are considered positive. The one-index notation for stress is used (see Section 2.2), and it will be further used for stresses and strains to simplify the presentation.

The next fundamental equation is the constitutive relation for the solid,

$$\boldsymbol{\sigma}' = \mathbf{C}^E (\mathbf{e} - \mathbf{e}^P) \tag{3.4.7}$$

where $\mathbf{C}^E$ is the elastic constitutive matrix of the solid skeleton, $\mathbf{e}$ is the total strain, and $\mathbf{e}^P$ is the strain in the solid due to pressure (Lewis & Schrefler 1987)

$$\mathbf{e}^P = -\frac{\mathbf{m}}{3K_s}p \tag{3.4.8}$$

where K_s is the bulk modulus (see (2.2.11)) of the solid material – solid grains. Obviously, the tensional strains are considered positive. Also, $\boldsymbol{\sigma}'$ is the so-called effective stress defined as (according to Terzaghi 1936)

$$\boldsymbol{\sigma}' = \boldsymbol{\sigma} + \mathbf{m}p \tag{3.4.9}$$

Further, the fluid continuity equation can be written as

$$\nabla^T \mathbf{q} + \left(\mathbf{m}^T - \frac{\mathbf{m}^T \mathbf{C}^E}{3K_s}\right)\dot{\mathbf{e}} + \left(\frac{1-n}{K_s} + \frac{n}{K_f} - \frac{\mathbf{m}^T \mathbf{C}^E \mathbf{m}}{9K_s^2}\right)\dot{p} = 0 \tag{3.4.10}$$

Details about derivation of this equation are given on the web – Theory, Chapter 3.

The above continuity equation is written for the current configuration and the current porosity n. However, in the incremental (finite element) analysis change of porosity can be taken into account, particularly when the mixture experiences significant porosity change. In order to find change of porosity, we write the continuity equation for the fluid as

$$\nabla^T (\rho_f \mathbf{q}) + \frac{\partial(\rho_f n)}{\partial t} = 0 \tag{3.4.11}$$

With use of the constitutive relation for fluids: $\partial \rho_f / \partial t = -(\partial p / \partial t) \rho_f / K_f$, where K_f is the compressibility modulus of a fluid, this equation becomes

$$\nabla^T (\rho_f \mathbf{q}) + \frac{n \rho_f}{K_f} \frac{\partial p}{\partial t} + \rho_f \frac{\partial n}{\partial t} = 0 \qquad (3.4.12)$$

This equation relates the changes of Darcy's velocity, pressure and porosity.

Finally, we note that various stress and strain measures can be employed in this analysis. In the case of small strains, the Cauchy stress and small strains are employed and the above relations are applicable. In the case of large strains, logarithmic strains and Cauchy stresses may be used (see Section 2.4), with the proper modifications of the above relations. Details about these modifications are given on the web – Theory, Chapter 3.

3.4.2 Examples

Example 3.4-1. One-dimensional creep response of human spinal motion segment (SMS)

The model of an SMS is based on the consolidation of a one-dimensional column of poroelastic material. The cylindrical column lies on the rigid foundation and is loaded at the top by pressure. The column is constrained laterally, while the fluid is not allowed to flow through the column boundary and the bottom of the cylinder. A step load p_0 is applied and free drainage is allowed at the top surface. Geometrical and material data are given in Fig. E3.4-1.

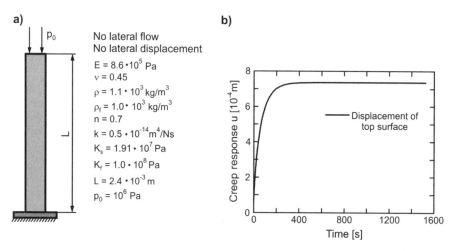

Fig. E3.4-1 One-dimensional creep response of an SMS. (a) Geometrical and material data (notation: E is Young's modulus, v is Poisson's coefficient, ρ is mixture density, ρ_f is fluid density, n is porosity, k is permeability coefficient, K_s is compressibility modulus of solid, K_f is compressibility modulus of fluid, L is column length and p_0 is applied pressure); (b) Displacement of the column top surface during creep

The analytical solution for the column settlement (displacement of the top surface), u_c, can be obtained in the form of a series (Biot 1941)

$$u_c(t) = p_0 L a - \frac{8(a-a_i)p_0 L}{\pi^2} \sum_{n=0}^{\infty} \frac{\exp\left\{-\left[\frac{(2n+1)\pi}{2L}\right]ct\right\}}{(2n+1)^2} \quad \text{(E3.4-1.1)}$$

where a, a_i and c are constants given in terms of the poroelastic material properties (Simon et al. 1985):

$$a = \frac{(1-2\nu)(1+\nu)}{E(1-\nu)}, \quad a_i = \frac{a}{1+a\alpha^2 Q}, \quad c = \frac{k}{a\alpha^2 + Q^{-1}},$$

$$Q^{-1} = \frac{n}{K_f} + \frac{\alpha - n}{K_s}, \quad \alpha = 1 - \frac{K_D}{K_s}, \quad K_D = \frac{E}{3(1-2\nu)} \quad \text{(E3.4-1.2)}$$

This example is also solved by the finite element method, see Example 7.7.1 and Software on the web.

References

Batchelor, G.K. (1967). *An Introduction to Fluid Dynamics*, Cambridge University Press, Cambridge, England.

Biot, M.A. (1941). General theory of three-dimensional consolidation, *J. Appl. Physics*, **12**, 155–64.

Chawla, M.M., Al-Zanaidi, M.A. & Al-Aslab, M.G. (2000). On approximate solutions of the diffusion and convection-diffusion equations, *Comp. Math. Applic.*, **39**, 71–87.

Crank, J. (1979). *The Mathematics of Diffusion*, Oxford University Press, Oxford, England.

Incropera, F.P. & DeWitt, D.P. (1996). *Fundamentals of Heat and Mass Transfer*, John Wiley & Sons, Inc., New York.

Kojić, M., Filipović, N. & Mijailović, S. (2001). A large strain finite element analysis of cartilage deformation with electrokinetic coupling, *Comp. Meth. Appl. Mech. Eng.*, **190**, 2447–64.

Kojić, M., Filipović, N., Vulovic, S. & Mijailović, S. (1998). A finite element solution procedure for porous medium with fluid flow and electromechanical coupling, *Commun. Numer. Meth. Eng.*, **14**, 381–92.

Lewis, R.W. & Schrefler, B.A. (1987). *The Finite Element Method in the Deformation and Consolidation of Porous Media*, John Wiley & Sons, Ltd, Chichester, England.

Mills, A.F. (1999). *Heat Transfer*, Prentice-Hall, Englewood Cliffs, NJ.

Mills, A.F. (2001). *Mass Transfer*, Prentice-Hall, Englewood Cliffs, NJ.

Munson, B.R., Young, D.F. & Okiishi, T.H. (1998). *Fundamentals of Fluid Mechanics*, John Wiley & Sons, Inc., New York.

Polivka, R.M. & Wilson, E.L. (1976). Finite element analysis of nonlinear heat transfer problems. Technical Report UC SESM 76-2, Berkeley, CA.

Simon, B.R., Wu, J.S.S., Carlton, M.W., Evans, J.H. & Kazarian, L.E. (1985). Structural models for human spinal motion segments based on a poroelastic view of the intervertebral disc, *J. Biomech. Eng.*, **107**, 327–35.

Terzaghi, K. (1936). The shearing resistance of saturated soil and the angle between the planes of shear, *Proc. 1st Int. SMFE Conference*, **1**, 54–6, Harvard, Cambridge, MA.

Part II

Fundamentals of Computational Methods

4

Isoparametric Formulation of Finite Elements

This is the introductory chapter for the finite element method (FEM). The FEM is the dominant method for general computer modeling in engineering and science, as well as in bioengineering. We only present the isoparametric formulation as the most general that is employed in subsequent chapters. This formulation assumes that the same interpolation functions are used to interpolate domain geometry and the fields of physical quantities which are to be obtained by the FEM. Other formulations and many theoretical and practical aspects of the FEM, which are not essential for this book, are not discussed; they can be found in a large body of literature (e.g. Huebner 1975, Sekulovic 1984, Hughes 1987, Crisfield 1991, Bathe 1996, Kojić *et al.* 1998, Kojić & Bathe 2005).

After the introduction, we present the isoparametric formulation starting from a simple one-dimensional finite element, and then proceed to more general, three-dimensional, two-dimensional and shell finite element formulations. A number of solved examples demonstrate applicability of the FEM. These and additional examples are supported by the Software on the web.

4.1 Introduction to the finite element method

The finite element method was introduced in the 1960s as a generalization of the matrix methods of structural analysis. Since then the FEM has become the most general method in computational mechanics and other scientific and engineering fields. The FE methodology advanced from modeling linear problems in solid mechanics to nonlinear problems in any physical field, including biology and medicine. The enormous advances in the FEM are based on a large number of researchers in this field, developments in computer technology, and practical benefits in industrial and scientific applications. Many journals are almost entirely devoted to the FEM, such as the *International Journal for Numerical Methods in Engineering*, and *Computer Methods in Applied Mechanics and Engineering*. A large number of industrial and scientific software packages based on the FEM have been developed and are in use in everyday industrial and scientific applications. In recent years, the FEM is becoming more attractive in bioengineering and medicine, as a tool for computer modeling.

The basic idea of the FEM is that any physical field can be discretized into a finite number of *subdomains* called *finite elements*. This discretization is schematically shown in Fig. 4.1.1 (see color plate) for a displacement field due to deformation of a blood vessel wall and for the field of blood velocity. Here, we consider deformation of solids, while the FE modeling of fluid flow will be presented in Chapter 7.

To introduce the FEM we assume static deformation of a solid. The displacement field $\mathbf{u}(x, y, z)$ caused by a mechanical action is approximated within each finite element by the displacement vector fields $\mathbf{u}^e(r, s, t)$ (we further use $\mathbf{u}(r, s, t)$ to simplify notation), where r, s, t are the local coordinates of the finite element ('t' will generally be used for 'time', but in the text the specific meaning of 't' will always be defined if it does not represent time). The approximate displacement is expressed in terms of the displacement vector $\mathbf{U}^e$ (further written as $\mathbf{U}$) of the finite element nodes, as

$$\mathbf{u} = \mathbf{NU}, \quad u_i = \sum_{K=1}^{N} N_K U_i^K \equiv N_K U_i^K, \quad i = 1, 2, 3; \ K = 1, 2, \ldots, N \qquad (4.1.1)$$

where N_K are the interpolation functions, U_i^K are the components (x, y, z) of the displacement vector of the node K, and N is the number of the finite element nodes. By a proper assemblage of the equilibrium equations of the finite elements, we obtain the equilibrium equation for the entire domain (body) in the form

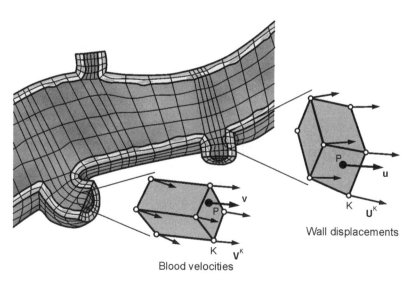

Fig. 4.1.1 Discretization into finite elements of the blood velocity field and displacement field of blood vessel deformation. The velocity $\mathbf{v}$ and displacement $\mathbf{u}$ at a material point within a finite element are obtained, respectively, by interpolations from the nodal points vectors $\mathbf{V}^K$ and $\mathbf{U}^K$ (see Plate 1)

ISOPARAMETRIC FORMULATION OF FINITE ELEMENTS

$$\mathbf{K}_{sys}\mathbf{U}_{sys} = \mathbf{F}_{sys}^{ext}, \quad \text{or}$$
$$(K_{sys})_{IJ}(U_{sys})_J = (F_{sys})_I^{ext}, \quad I, J = 1, 2, \ldots, N_{total}, \quad \text{sum on } J \quad (4.1.2)$$

Here $\mathbf{U}_{sys}$ is the vector of nodal displacements of all nodes; $\mathbf{K}_{sys}$ is the stiffness matrix of the entire system; $\mathbf{F}_{sys}^{ext}$ is the vector of external forces, represented by forces acting on the finite element nodes; and N_{total} is the total number of degrees of freedom of the system ($N_{total} \leq 3N_{nodes}$ for three-dimensional finite element discretization shown in Fig. 4.1.1, where N_{nodes} is the total number of FE nodes). We further use *capital letters* for *node numbers*. By solving for the displacement vector $\mathbf{U}_{sys}$, displacements $\mathbf{u}$ within each finite element can be further calculated, and also the strains (see (2.1.25)) and stresses (see (2.2.2)).

The presented basic idea is applicable to more complex problems, such as dynamic and nonlinear problems of solids, as well as to general field problems. In this and subsequent chapters we give details (to a certain extent) of how the basic finite element relations are formulated and how more complex problems can be modeled by the FEM.

4.2 Formulation of 1D finite elements and equilibrium equations

We introduce the basic relations of the isoparametric FE formulation through a simple one-dimensional (1D) structural example and then describe the assemblage procedure for the system of finite elements. The final relations, written in general form, are subsequently used for other types of finite elements.

4.2.1 Truss finite element

Let us consider a simple bar structure consisting of two straight bars and subjected to a force **F** (Fig. 4.2.1a). A schematic representation by line elements is shown in Fig. 4.2.1b. In this scheme we have specified two elements and '1' and '2' and three nodes. The bars

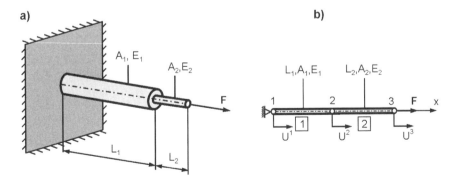

Fig. 4.2.1 A simple bar structure loaded by axial force **F**. (a) Structure geometry – lengths L_1 and L_2, cross-sectional areas A_1 and A_2, Young's moduli E_1 and E_2; (b) Schematic representation by two truss finite elements, with FE nodes 1, 2 and 3, and nodal displacements U_1, U_2 and U_3

are called *truss finite elements*, and the end points of the bars are called the finite element *nodal points* (1, 2 and 3 in the figure). Now we present the isoparametric formulation of a truss finite element.

Interpolation Functions

Consider a truss finite element shown in Fig. 4.2.2a. The forces F^1 and F^2 are called the *nodal forces*. As shown in the figure, the nodal force for element 1 at node 1 comes from the support, while at node 2 it comes from element 2. On the other hand, for element 2 the nodal force at node 2 comes from element 1, while the nodal force at node 3 is the external force with respect to the structure (external body action).

Due to element deformation we have a displacement u of the cross-section at a coordinate x (Fig. 4.2.2b), while the deformation e is

$$e = \frac{du}{dx} = \frac{\sigma}{E} = \frac{F}{AE} \tag{4.2.1}$$

Hooke's law (2.2.3) and the value for stress $\sigma = F/A$ are used here. Since the axial force is the same along the element, the displacement along the element can be obtained by integration of this equation,

$$u = \frac{F}{AE}x + U^1 \tag{4.2.2}$$

where the condition: $u = U^1$ for $x = 0$ is used. From (4.2.2) it follows that $U^2 = (F/AE)L + U^1$, where L is the element length, so that the displacement $u(x)$ can be written as

$$u = \left(1 - \frac{x}{L}\right)U^1 + \frac{x}{L}U^2 \tag{4.2.3}$$

Instead of the coordinate x, we introduce the *natural coordinate*

$$r = -1 + 2\frac{x}{L} \tag{4.2.4}$$

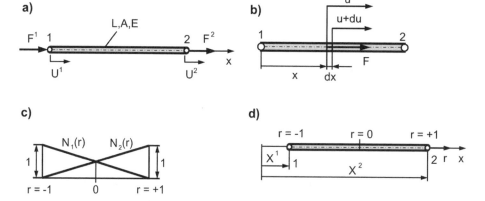

Fig. 4.2.2 Truss finite element. (a) Nodal displacements and forces; (b) Displacement of a cross-section; (c) Isoparametric functions $N_1(r)$ and $N_2(r)$; (d) Interpolation of geometry

ISOPARAMETRIC FORMULATION OF FINITE ELEMENTS

Note that the value of the natural coordinate changes from -1 to 1, for any element length, which is of fundamental importance in the isoparametric FE formulations. Using (4.2.4) we substitute x by r in (4.2.3) and obtain

$$u = \mathbf{NU}, \quad \text{or} \quad u(r) = N_1(r)U^1 + N_2(r)U^2 \qquad (4.2.5)$$

where $\mathbf{N}$ is a 1×2 *interpolation matrix* with the terms $N_1(r)$ and $N_2(r)$ called the *interpolation functions*; and $\mathbf{U}$ is the *nodal point displacement vector* with components U^1 and U^2. The interpolation functions are

$$N_1(r) = \frac{1}{2}(1-r), \quad N_2(r) = \frac{1}{2}(1+r) \qquad (4.2.6)$$

A graphical representation of the interpolation functions is shown in Fig. 4.2.2c. Note the following relation is applicable

$$x = \mathbf{NX}, \quad \text{or} \quad x(r) = N_1(r)X^1 + N_2(r)X^2 \qquad (4.2.7)$$

and is in agreement with (4.2.3). Here $\mathbf{X}$ is the vector of nodal coordinates of the element (note that the origin for coordinate x in Fig. 4.2.2d is not at node 1 as in Fig. 4.2.2b). The term *isoparametric formulation* comes from the fact that the same interpolation matrix is used for displacements and coordinates (element geometry). Note that the linear interpolation functions $N_1(r)$ and $N_2(r)$ are introduced using Hooke's law (4.2.1), and the relation (4.2.5) gives the exact displacement field (4.2.2).

Stiffness Matrix and Equilibrium Equation

Using (4.2.1) for the strain and (4.2.5) for the displacement, we first express the strain e as

$$e = \mathbf{BU}, \quad \text{or} \quad e = B_1 U^1 + B_2 U^2 \qquad (4.2.8)$$

where the matrix $\mathbf{B}$ is

$$\mathbf{B} = [B_1 \ B_2] = [N_{1,x} \ N_{2,x}] \qquad (4.2.9)$$

The derivatives $N_{1,x} \equiv dN_1/dx$ and $N_{2,x} \equiv dN_2/dx$ can be obtained with use of (4.2.4) as

$$\mathbf{B} = J^{-1}[N_{1,r} \ N_{2,r}] \qquad (4.2.10)$$

where J^{-1} is the inverse of the Jacobian of the transformation between the Cartesian and natural coordinate system,

$$J = \frac{dx}{dr} \qquad (4.2.11)$$

The relations (4.2.8)–(4.2.11) have a general form employed in the isoparametric finite element formulation. In our case we have that

$$\mathbf{B} = \begin{bmatrix} -\frac{1}{L} & \frac{1}{L} \end{bmatrix}, \quad J = \frac{L}{2}, \quad J^{-1} = \frac{2}{L}, \quad e = \frac{U^2 - U^1}{L} \qquad (4.2.12)$$

To obtain the element equilibrium equation we apply the principle of virtual work (2.3.2). First, the virtual strain δe can be expressed using (4.2.8) as

$$\delta e = \delta \mathbf{U}^T \mathbf{B}^T \tag{4.2.13}$$

Then, with Hooke's law (2.2.3) and expression (4.2.8) for the strain e, the internal virtual work δW^{int} for the finite element can be written in the form (see (2.3.3))

$$\delta W^{int} = \int_V \delta e \sigma dV = \int_V \delta e E e dV = \delta \mathbf{U}^T \mathbf{K} \mathbf{U} = \delta \mathbf{U}^T \mathbf{F}^{int} \tag{4.2.14}$$

Here, V is the element volume, and

$$\mathbf{K} = \int_V \mathbf{B}^T E \mathbf{B} dV \tag{4.2.15}$$

is the *element stiffness matrix*. We have introduced the *internal nodal force* vector $\mathbf{F}^{int}$ (with components F_1^{int}, F_2^{int}) of the element, which produces the same work on the nodal virtual displacement $\delta \mathbf{U}$, as does the stress σ on virtual strain δe over the element volume. Physically, the force $\mathbf{F}^{int}$ is the *element resistance force* by which the element reacts to the external loading exerted on the element. From (4.2.14) it follows that

$$\mathbf{F}^{int} = \mathbf{K} \mathbf{U} \tag{4.2.16}$$

The expressions (4.2.15) and (4.2.16) have a general form which is also used later. Note that the *stiffness matrix is symmetric*. The symmetry property is valid in general for all finite element types. In our case of truss elements, the stiffness matrix is

$$\mathbf{K} = \int_{-1}^{1} \begin{bmatrix} -1/L \\ 1/L \end{bmatrix} E \begin{bmatrix} 1/L & 1/L \end{bmatrix} A \frac{L}{2} dr = \frac{EA}{L} \begin{bmatrix} 1 & -1 \\ -1 & 1 \end{bmatrix} \tag{4.2.17}$$

The relation

$$dV = A dx = A J dr = A \frac{L}{2} dr \tag{4.2.18}$$

is used here, which also has a general form within the isoparametric formulation. Note that the stiffness matrix in (4.2.17) is given in the closed (analytical) form since the terms within the integral are constant. However, in the case of nonlinear material behavior, Young's modulus E in the constitutive law (4.2.1) depends on the strain e and the integration might be performed numerically. Details about numerical integration are given on the web – Theory, Chapter 4.

The external virtual work results from the virtual work of the forces F^1 and F^2 on the virtual displacements δU^1 and δU^2, hence

$$\delta W^{ext} = \delta \mathbf{U}^T \mathbf{F}^{ext}, \quad \text{or} \quad \delta W^{ext} = \delta U^1 F^1 + \delta U^2 F^2 \tag{4.2.19}$$

From the principle of virtual work (2.3.2), and expressions (4.2.14) and (4.2.19), follows

$$\delta \mathbf{U}^T \mathbf{K} \mathbf{U} = \delta \mathbf{U}^T \mathbf{F}^{ext}, \quad \text{or} \quad \delta \mathbf{U}^T (\mathbf{K} \mathbf{U} - \mathbf{F}^{ext}) = 0 \tag{4.2.20}$$

ISOPARAMETRIC FORMULATION OF FINITE ELEMENTS

Since the virtual displacements are arbitrary (hence, nonzero), from (4.2.20) follows the *equilibrium equation of the finite element*,

$$\mathbf{KU} = \mathbf{F}^{ext}, \quad \text{or} \quad K_{IJ}U_J = F_I^{ext}, \quad I, J = 1, 2 \qquad (4.2.21)$$

where the element stiffness matrix components K_{ij} are given in (4.2.17). The displacement vector components U_j are $U_1 = U^1, U_2 = U^2$ (Fig.4.2.2a). The external force components F_i^{ext} are $F_1^{ext} = F^1, F_2^{ext} = F^2$ and in general include the action of the neighboring elements and/or forces produced by other external mechanical actions.

The equilibrium equation (4.2.21) has a general form applicable in the finite element method. In the subsequent sections several isoparametric elements will be described, where we will only give specifics for determination of the stiffness matrix and nodal forces of these elements.

Transformation of the Stiffness Matrix

The derived truss stiffness matrix corresponds to the coordinate system in which one coordinate axis (in our derivation it is the x-axis) is along the truss axis. That coordinate system is usually called the *local system*. In applications, however, it is necessary to use a coordinate system in which the truss element is not aligned to any of the coordinate axes (Fig. 4.2.3).

Let the cosines of angles between the axis $\bar{x}$ along the element (with unit vector $\mathbf{\bar{i}}$) and coordinate axes x, y, z be $l = \cos\alpha, m = \cos\beta, n = \cos\gamma$, and the displacement vector along the element be $\mathbf{\bar{U}}\left(\bar{U}_1 = \bar{U}^1, \bar{U}_2 = \bar{U}^2\right)$. The displacement vector of the element in the coordinate system x, y, z is defined as $\mathbf{U}\left(U_1^1 U_2^1 U_3^1 U_1^2 U_2^2 U_3^2\right)$, with components $U_1 = U_1^1 \equiv U_x^1, U_2 = U_2^1 \equiv U_y^1, \ldots, U_6 = U_3^2 \equiv U_z^2$. Then, the relationships between the components of the element displacement vector in the two systems can be written as

$$\mathbf{\bar{U}} = \mathbf{TU}, \quad \text{or} \quad \bar{U}_i = T_{ik}U_k, \quad i = 1, 2; \quad k = 1, 2, \ldots, 6 \qquad (4.2.22)$$

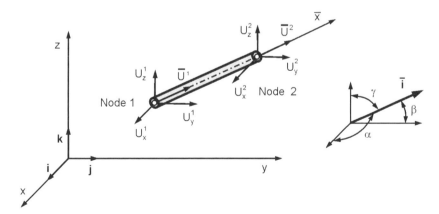

Fig. 4.2.3 Truss finite element inclined to the coordinate axes, with nodal displacement components with respect to the global x, y, z coordinate system and to local element axis $\bar{x}$. The direction of the unit vector $\mathbf{\bar{i}}$ is defined by the angles α, β, γ

where the 2×6 transformation matrix $\mathbf{T}$ is

$$\mathbf{T} = \begin{bmatrix} l & m & n & 0 & 0 & 0 \\ 0 & 0 & 0 & l & m & n \end{bmatrix} \quad (4.2.23)$$

Note that the transformation matrix satisfies the orthogonality condition (1.2.12), i.e.

$$\mathbf{T}^T \mathbf{T} = \mathbf{I}, \quad \text{or} \quad T_{ki} T_{kj} = \delta_{ij}, \quad i,j = 1, 2, \ldots, 6; \quad k = 1, 2 \quad (4.2.24)$$

We now impose the condition that the internal virtual work expressed in the two systems, $\delta \overline{W}^{\text{int}}$ and δW^{int}, must be equal, i.e. $\delta \overline{W}^{\text{int}} = \delta W^{\text{int}}$. Then from (4.2.14) and (4.2.22) follows

$$\delta \overline{\mathbf{U}}^T \overline{\mathbf{K}} \overline{\mathbf{U}} = \delta \mathbf{U}^T \mathbf{K} \mathbf{U}; \quad \delta \mathbf{U}^T \mathbf{T}^T \overline{\mathbf{K}} \mathbf{T} \mathbf{U} = \delta \mathbf{U}^T \mathbf{K} \mathbf{U} \quad (4.2.25)$$

and consequently,

$$\mathbf{K} = \mathbf{T}^T \overline{\mathbf{K}} \mathbf{T} \quad (4.2.26)$$

This relation expresses the transformation rule for the stiffness matrix when the coordinate system is changed from the element local to the global coordinate system. The element stiffness matrix $\mathbf{K}$ has the dimension 6×6, it is symmetric, and the rows and columns correspond to the displacements $U_x^1, U_y^1, \ldots, U_z^2$.

Note that the relationships between the components of the nodal force in the two coordinate systems have the form (4.2.22),

$$\overline{\mathbf{F}} = \mathbf{T} \mathbf{F}, \quad \text{or} \quad \overline{F}_i = T_{ik} F_k, \quad i = 1, 2; k = 1, 2, \ldots, 6 \quad (4.2.27)$$

where the element force vectors $\overline{\mathbf{F}}$ and $\mathbf{F}$ are: $\overline{\mathbf{F}} \left(\overline{F}_1 = F^1, \overline{F}_2 = F^2 \right)$ and $\mathbf{F} \left(F_1 = F_1^1, F_2 = F_2^1, \ldots, F_6 = F_3^2 \right)$.

4.2.2 Equilibrium equations of the FE assemblage and boundary conditions

In order to derive the equilibrium equations we consider our simple structure in Fig. 4.2.1. Using displacements shown in the figure, the equilibrium equations (4.2.21) for the two elements can be obtained as:

$$\begin{aligned} K_{11}^1 U^1 + K_{12}^1 U^2 &= F^{\text{sup}} \\ K_{21}^1 U^1 + K_{22}^1 U^2 &= F_2^1 \\ K_{11}^2 U^2 + K_{12}^2 U^3 &= F_2^2 \\ K_{21}^2 U^2 + K_{22}^2 U^3 &= F \end{aligned} \quad (4.2.28)$$

Here F^{sup} is the support reaction force; F_2^1 is the force of element 1 due to the action of element 2; F_2^2 is the force of element 2 due to the action of element 1. The stiffness matrix terms of elements 1 and 2 are K_{11}^1, K_{12}^1 and K_{21}^2, K_{22}^2, respectively. The second and the third

ISOPARAMETRIC FORMULATION OF FINITE ELEMENTS

equations represent the equations of balance corresponding to the same displacement U^2. Therefore, by summing these two equations, a system of three equations is obtained:

$$K_{11}^1 U^1 + K_{12}^1 U^2 = F^{\text{sup}}$$
$$K_{21}^1 U^1 + \left(K_{22}^1 + K_{11}^2\right) U^2 + K_{12}^2 U^3 = 0 \tag{4.2.29}$$
$$K_{21}^2 U^2 + K_{22}^2 U^3 = F$$

where the relation $F_2^2 = -F_2^1$ is used (Newton third law of action and reaction). Assuming that the displacement U^1 is nonzero, we have three equations with three unknown displacements.

Analyzing the system matrix in (4.2.29) we note that the term corresponding to the displacement U^2, common to both finite elements, represents the sum of the terms K_{22}^1 and K_{11}^2. This is an important result in forming the stiffness matrix $\mathbf{K}_{sys}$ (see (4.1.2)) of the FE assemblage. Also, on the right-hand side we only have the external structural load vector $\mathbf{F}_{sys}^{ext}[F^{\text{sup}}F]$, since the interaction forces between the elements cancel (internal structural nodal forces). Therefore, the equilibrium equation of the finite element assemblage (4.1.2) is obtained.

Generalizing the above result, we have that the *structural stiffness matrix*, or the *system stiffness matrix*, is

$$\mathbf{K}_{sys} = \sum_{e=1}^{N_E} \mathbf{K}^e \tag{4.2.30}$$

where N_E is the number of finite elements. The summation assumes that the element stiffness matrix terms (corresponding to the common displacement) are added. Note that the system stiffness matrix is symmetric. Also, the assemblage (4.2.30) is applicable in any finite element analysis.

Boundary Conditions

Assuming that the displacement U^1 is nonzero and unknown, we find that, with use of (4.2.17), the system stiffness matrix in (4.2.29) is *singular*, i.e. $\det \mathbf{K} = 0$. This singularity reflects the physical condition that the structure is free to move in space; it has a *rigid body displacement*, or mode. Therefore, we have to eliminate this singularity by imposing boundary conditions that prevent rigid body motion. In our case, the structure is attached to the rigid foundation (Fig. 4.2.1), hence

$$U^1 = 0 \tag{4.2.31}$$

Then, we remove the first equation from (4.2.28) and substitute $U^1 = 0$ into the last two equations. The nonsingular system is

$$\left(K_{22}^1 + K_{11}^2\right) U^2 + K_{12}^2 U^3 = 0$$
$$K_{21}^2 U^2 + K_{22}^2 U^3 = F \tag{4.2.32}$$

which can be solved for the unknowns U^2 and U^3.

With these solutions for displacements, the support reaction F^{sup} can be obtained from the first of the equations (4.2.29). Note that the condition (4.2.31) is imposed by deleting the row and the column from the system of equations (4.2.29) corresponding to the zero displacement U^I. This procedure of *deleting the rows and columns* from the system of equilibrium equations is generally applicable in finite element analysis. A similar procedure can be applied when some nonzero displacements are prescribed (see web – Software).

4.2.3 Examples

Example 4.2-1. Analyze truss structure under static loading
The truss structure consisting of four trusses is attached to the rigid foundation at points 1, 2, 3, 4 (Fig. E4.2-1). The trusses are connected at a joint A (point 5) and all connections of trusses are pin-joints. The structure is loaded by a vertical force **F**.

The deformed structure and the deflection u_A are shown in the figure. Suggestions for exercise and detailed analysis of this example are given within the Software on the web (Example 4.2-1).

Example 4.2-2. Analyze dependence of the size of the stiffness matrix on element numbering
A truss structure consisting of 17 trusses represents a typical structure used in the design of bridges, Fig. E4.2-2. The structure is lying in the $x-y$ plane and all displacements under the loading occur in this plane.

The system matrices for the two node numbering ((a) and (b)) are shown in the figure. A significant difference in the number of matrix terms can be observed. Detailed analysis of the example can be performed using the software (Example 4.2-2), where options for exercise are suggested. Numbering of displacements follows the node numbers. The term $(K_{sys})_{IJ} \neq 0$ when the displacements U_I and U_J belong to the same finite element.

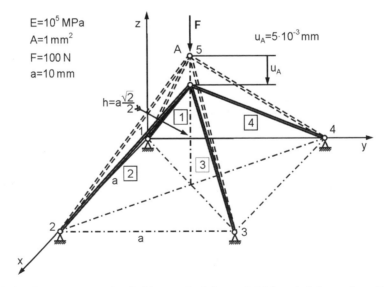

Fig. E4.2-1 Truss structure loaded by vertical force. Initial and deformed configurations. Due to symmetry of the structure and the loading, the point A displaces vertically under the action of force **F**

ISOPARAMETRIC FORMULATION OF FINITE ELEMENTS

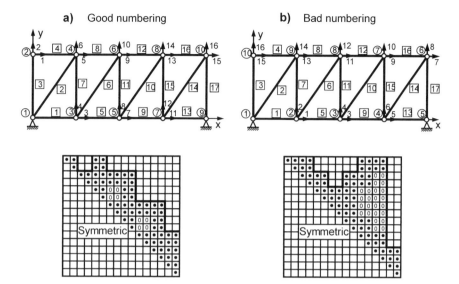

Fig. E4.2-2 Dependence of the matrix size on the node numbering. Matrices are shown for two numberings, (a) and (b), demonstrating that the numbering (a) gives smaller matrix than the numbering (b). The symmetric part of the matrix, used in the calculation, occupies space between the skyline and the main diagonal. Numbers in circles are node numbers, numbers in squares are element numbers, and other numbers represent the displacements

4.3 Three-dimensional (3D) isoparametric finite element

Following the above basic ideas about the isoparametric finite element formulation, we extend these ideas to a general 3D finite element formulation. We first consider 3D problems followed by 2D conditions, because 2D finite elements have specific subtypes and their presentation is more complex.

4.3.1 Element formulation

Interpolation of Geometry and Displacements
A 3D finite element is shown in Fig. 4.3.1a. We consider a simple element bounded by six plane surfaces, and with eight nodes numbered by $1, 2, \ldots, 8$. We here describe the eight-node element, and a generalization to 3D elements with larger number of nodes and curved surfaces (higher order elements) is straightforward (see web – Theory, Chapter 4).

The element geometry and the displacement field are interpolated by the relationships:

$$\mathbf{x} = \begin{Bmatrix} x_1 \\ x_2 \\ x_3 \end{Bmatrix} \equiv \begin{Bmatrix} x \\ y \\ z \end{Bmatrix} = \mathbf{NX}, \quad \text{or} \quad x_i = \sum_{K=1}^{N} N_K X_i^K, \quad i = 1, 2, 3 \quad (4.3.1)$$

and

$$\mathbf{u} = \begin{Bmatrix} u_1 \\ u_2 \\ u_3 \end{Bmatrix} \equiv \begin{Bmatrix} u_x \\ u_y \\ u_z \end{Bmatrix} = \mathbf{NU}, \quad \text{or} \quad u_i = \sum_{K=1}^{N} N_K U_i^K, \quad i = 1, 2, 3 \quad (4.3.2)$$

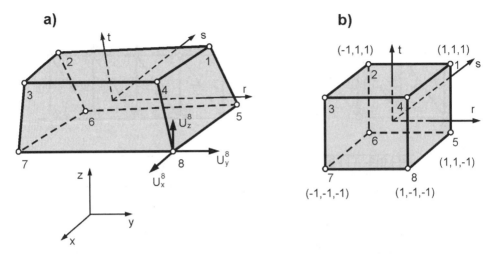

Fig. 4.3.1 Three-dimensional eight-node finite element. (a) Physical space (nodal displacement components only shown for node 8); (b) Element mapped to natural coordinate space (numbers in parentheses for some nodes are the values of natural coordinates at these nodes)

where **x** is the position vector of a material point within the element (we use both notations x_1, x_2, x_3 and x, y, z for coordinates of a point); and **u** is the displacement vector of a point, with components u_1, u_2, u_3 (or u_x, u_y, u_z). The vector **X** of nodal coordinates and the nodal point displacement vector **U** are defined as

$$\mathbf{X}^T = \begin{bmatrix} X_1^1 X_2^1 X_3^1 \ldots \ldots X_1^N X_2^N X_3^N \end{bmatrix} \tag{4.3.3}$$

and

$$\mathbf{U}^T = \begin{bmatrix} U_1^1 U_2^1 U_3^1 \ldots \ldots U_1^N U_2^N U_3^N \end{bmatrix} \tag{4.3.4}$$

where X_i^1 ($i = 1, 2, 3$) and U_i^1 ($i = 1, 2, 3$) are coordinates and displacements of the first node, ..., X_i^N and U_i^N are coordinates and displacements of the node N, and N is number of nodes, in our case $N = 8$. The interpolation matrix **N** is

$$\mathbf{N} = \begin{bmatrix} N_1 & 0 & 0 & N_2 & 0 & 0 & . & . & N_N & 0 & 0 \\ 0 & N_1 & 0 & 0 & N_2 & 0 & . & . & 0 & N_N & 0 \\ 0 & 0 & N_1 & 0 & 0 & N_2 & . & . & 0 & 0 & N_N \end{bmatrix} \tag{4.3.5}$$

where $N_K(r, s, t)$, $K = 1, 2, \ldots, N$ are the interpolation functions of the natural coordinates r, s, t. The interpolation functions can be written in the form

$$N_K(r, s, t) = \frac{1}{8}(1 + r_K r)(1 + s_K s)(1 + t_K t), \quad K = 1, 2, \ldots, 8 \tag{4.3.6}$$

where the natural nodal coordinates r_K, s_K, t_K for some nodes are given in Fig. 4.3.1b. Note that the natural coordinates of the nodes are equal either to 1 or -1. Also, the interpolation functions have the property that $N_K = 1$ at the node 'K', while $N_K = 0$ at all other nodes (see Fig. 4.2.2c for the 1D element). The eight-node element is called *linear* because all interpolation functions are linear with respect to each natural coordinate.

ISOPARAMETRIC FORMULATION OF FINITE ELEMENTS

Strains

To obtain the expression for the strain components we use the matrix representation of strains (2.1.23), the expressions (2.1.25) and interpolation of displacements (4.3.2). Then we have

$$\mathbf{e} = \begin{Bmatrix} e_{xx} \\ e_{yy} \\ e_{zz} \\ \gamma_{xy} \\ \gamma_{yz} \\ \gamma_{zx} \end{Bmatrix} = \begin{Bmatrix} u_{1,1} \\ u_{2,2} \\ u_{3,3} \\ u_{1,2}+u_{2,1} \\ u_{2,3}+u_{3,2} \\ u_{1,3}+u_{3,1} \end{Bmatrix} = \begin{bmatrix} N_{1,1} & 0 & 0 & \cdots & N_{N,1} & 0 & 0 \\ 0 & N_{1,2} & 0 & \cdots & 0 & N_{N,2} & 0 \\ 0 & 0 & N_{1,3} & \cdots & 0 & 0 & N_{N,3} \\ N_{1,2} & N_{1,1} & 0 & \cdots & N_{N,2} & N_{N,1} & 0 \\ 0 & N_{1,3} & N_{1,2} & \cdots & 0 & N_{N,3} & N_{N,2} \\ N_{1,3} & 0 & N_{1,1} & \cdots & N_{N,3} & 0 & N_{N,1} \end{bmatrix} \begin{Bmatrix} U_1^1 \\ U_2^1 \\ U_3^1 \\ \vdots \\ U_1^N \\ U_2^N \\ U_3^N \end{Bmatrix} = \mathbf{BU}$$

(4.3.7)

where the derivatives of the interpolation functions are denoted as $N_{K,i} \equiv \partial N_K / \partial x_i$, and the *strain-displacement relation matrix* **B** is defined by these derivatives. Since the interpolation functions are defined in terms of the natural coordinates, with a use of the chain rule for the derivatives $\partial N_K / \partial x_i$, it follows that

$$N_{K,j} = \frac{\partial N_K}{\partial r} \frac{\partial r}{\partial x_j} + \frac{\partial N_K}{\partial s} \frac{\partial s}{\partial x_j} + \frac{\partial N_K}{\partial t} \frac{\partial t}{\partial x_j} \qquad (4.3.8)$$

We next introduce the *Jacobian of transformation* (see also (4.2.11)),

$$\mathbf{J} = \left[\frac{\partial \mathbf{x}}{\partial \mathbf{r}} \right] = \begin{bmatrix} \frac{\partial x}{\partial r} & \frac{\partial y}{\partial r} & \frac{\partial z}{\partial r} \\ \frac{\partial x}{\partial s} & \frac{\partial y}{\partial s} & \frac{\partial z}{\partial s} \\ \frac{\partial x}{\partial t} & \frac{\partial y}{\partial t} & \frac{\partial z}{\partial t} \end{bmatrix} \qquad (4.3.9)$$

where **r** stands for the natural coordinate position vector ($r_1 \equiv r, r_2 \equiv s, r_3 \equiv t$), and its inverse $\mathbf{J}^{-1}$,

$$\mathbf{J}^{-1} = \left[\frac{\partial \mathbf{r}}{\partial \mathbf{x}} \right] = \begin{bmatrix} \frac{\partial r}{\partial x} & \frac{\partial s}{\partial x} & \frac{\partial t}{\partial x} \\ \frac{\partial r}{\partial y} & \frac{\partial s}{\partial y} & \frac{\partial t}{\partial y} \\ \frac{\partial r}{\partial z} & \frac{\partial s}{\partial z} & \frac{\partial t}{\partial z} \end{bmatrix} \qquad (4.3.10)$$

Note that the derivatives in the Jacobian (4.3.9) can be computed using (4.3.1) as

$$J_{mn} = \sum_{K=1}^{N} \frac{\partial N_K}{\partial r_m} X_n^K \qquad (4.3.11)$$

In general, the relations between the derivatives can be written as

$$\frac{\partial (\bullet)}{\partial \mathbf{x}} = \mathbf{J}^{-1} \frac{\partial (\bullet)}{\partial \mathbf{r}}, \quad \text{and} \quad \frac{\partial (\bullet)}{\partial \mathbf{r}} = \mathbf{J} \frac{\partial (\bullet)}{\partial \mathbf{x}} \qquad (4.3.12)$$

Therefore, the derivatives of the interpolation functions in (4.3.8) can be written as

$$\frac{\partial N_K}{\partial \mathbf{x}} = \mathbf{J}^{-1} \frac{\partial N_K}{\partial \mathbf{r}}, \quad \text{or}$$
$$\frac{\partial N_K}{\partial x_i} = J_{ij}^{-1} \frac{\partial N_K}{\partial r_j}, \quad K = 1, 2, \ldots, N; \, i, j = 1, 2, 3 \tag{4.3.13}$$

In practical applications, the strains at a material point within the element (at given coordinates r, s, t) are determined by calculating: the Jacobian $\mathbf{J}$ and its inverse $\mathbf{J}^{-1}$; the derivatives $\partial N_K / \partial x_i$ according to (4.3.13); the matrix $\mathbf{B}$ in (4.3.7); and finally the strains from (4.3.7).

Stiffness Matrix and Nodal Forces
The internal virtual work can be expressed as (see (4.2.14))

$$\delta W^{int} = \int_V \delta \mathbf{e}^T \boldsymbol{\sigma} dV = \delta \mathbf{U}^T \int_V \mathbf{B}^T \mathbf{C} \mathbf{B} dV \mathbf{U} = \delta \mathbf{U}^T \mathbf{K} \mathbf{U} \tag{4.3.14}$$

where we have employed the relation (4.3.7) from which $\delta \mathbf{e}^T = \delta \mathbf{U}^T \mathbf{B}^T$, and the constitutive relationship (2.2.2) $\boldsymbol{\sigma} = \mathbf{C} \mathbf{e}$. Clearly, the *stiffness matrix* $\mathbf{K}$ is

$$\mathbf{K} = \int_V \mathbf{B}^T \mathbf{C} \mathbf{B} dV \tag{4.3.15}$$

and the element internal force $\mathbf{F}^{int}$ is given by the expression (4.2.16). The stiffness matrix is symmetric and has dimensions $3N \times 3N$ (in our case 24×24) and the force vector $\mathbf{F}^{int}$ is of size $3N$, $\mathbf{F}^{int}\left(F_x^{(int)1} \, F_y^{(int)1} \, F_z^{(int)1} \, \ldots \, F_x^{(int)N} \, F_y^{(int)N} \, F_z^{(int)N}\right)$.

In the case when body forces are present, the corresponding nodal forces are calculated from the equality of virtual work:

$$\int_V \delta \mathbf{u}^T \mathbf{f}^V dV = \delta \mathbf{U}^T \int_V \mathbf{N}^T \mathbf{f}^V dV = \delta \mathbf{U}^T \mathbf{F}^V \Rightarrow \mathbf{F}^V = \int_V \mathbf{N}^T \mathbf{f}^V dV \tag{4.3.16}$$

where $\mathbf{f}^V$ is the force per unit volume, and $\mathbf{F}^V$ is the vector of equivalent volumetric nodal forces. Here, the displacement interpolation (4.3.2) has been used.

The external nodal forces resulting from the pressure on an element surface are calculated by employing again the equivalence of virtual work. A simple approximation for the eight-node element is to calculate the total force as $F_p = pA$ (where p is the mean pressure and A is the area of the element side) and use $F_p/4$ at each node in the mean normal surface direction. Details about calculation of the nodal pressure forces are given on the web – Theory, Chapter 4.

Calculation of the above volumetric integrals must be performed numerically. Details about numerical integration are given on the web – Theory, Chapter 4.

4.3.2 Examples

We here give one example only. Other examples are described on the web (Software).

ISOPARAMETRIC FORMULATION OF FINITE ELEMENTS

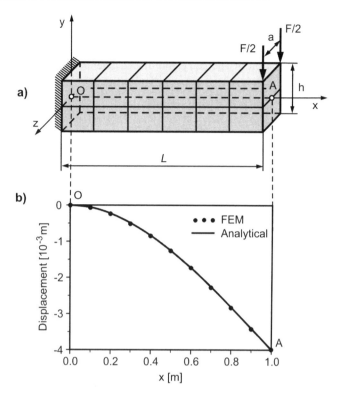

Fig. E4.3-1 Bending of cantilever modeled by 3D finite elements. (a) Schematics of the FE mesh and loading; (b) Deflection of the mid-plane $x-y$ (FEM and analytical solutions). Data: Lengths (m) $L=1$, $h=0.1$, $a=0.05$; $E=2\times 10^5\,MPa$, $v=0.3$, $F=0.01\,MN$

Example 4.3-1. Bending of cantilever modeled by 3D finite elements
A cantilever structure is subjected to loading by a force. The cantilever is modeled by 3D finite elements, as shown in Fig. E4.3-1a.
Distribution of deflection of the cantilever mid-plane (material line OA) for data given in the figure is shown in Fig. E4.3-1b. It can be seen that the agreement between the FE and analytical solutions is very good. Detailed analysis of beam deformation and options of solving this example are given on the web (see Software).

4.4 Two-dimensional (2D) isoparametric finite elements

In the case of 2D problems, deformation is described by the displacement field in a plane. We will use the coordinate system x_1, x_2, or x, y in this plane. Consequently, 2D isoparametric elements can be considered as special cases of 3D elements described in Section 4.3. The expressions for the element stiffness matrix and nodal forces have the same form as for 3D elements, as well as the integration scheme over the element volume. Some particular details are given on the web – Theory, Chapter 4.

4.4.1 Formulation of the elements

The four-node and nine-node elements are shown in Fig. 4.4.1. The natural coordinates are now r and s. The interpolation of geometry and displacements follows directly from (4.3.1) and (4.3.2),

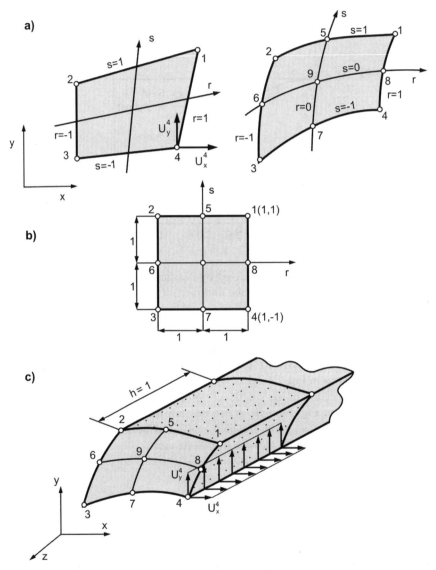

Fig. 4.4.1 Two-dimensional four-node and nine-node finite elements. (a) Physical space – linear element (left) and parabolic element (right), components of the nodal displacements are only shown for node 4; (b) Elements mapped to natural coordinate space; (c) Axonometric view of the element with displacements at node 4, the element thickness is h

ISOPARAMETRIC FORMULATION OF FINITE ELEMENTS 87

$$\mathbf{x} = \begin{Bmatrix} x_1 \\ x_2 \end{Bmatrix} \equiv \begin{Bmatrix} x \\ y \end{Bmatrix} = \mathbf{NX}, \quad \text{or} \quad x_i = \sum_{K=1}^{N} N_K X_i^K, \quad i = 1, 2 \tag{4.4.1}$$

and

$$\mathbf{u} = \begin{Bmatrix} u_1 \\ u_2 \end{Bmatrix} \equiv \begin{Bmatrix} u_x \\ u_y \end{Bmatrix} = \mathbf{NU}, \quad \text{or} \quad u_i = \sum_{K=1}^{N} N_K U_i^K, \quad i = 1, 2 \tag{4.4.2}$$

where N is number of element nodes; and the vectors $\mathbf{X}$ and $\mathbf{U}$ are $\mathbf{X}\left(X_1^1 \ X_2^1 \ldots \ldots X_1^N \ X_2^N\right)$ and $\mathbf{U}\left(U_1^1 \ U_2^1 \ldots \ldots U_1^N \ U_2^N\right)$. In the case of a four-node element, the interpolation functions can be expressed in the form (4.3.6), with the third coordinate $t = 0$, and with the coefficient 1/4 instead of 1/8. We will mainly use linear elements, with $N = 4$. Details about higher order elements (parabolic) are given on the web – Theory, Chapter 4.

In order to determine strains and stresses within a 2D element we have to distinguish three types of these elements: plane strain, plane stress and axisymmetric elements, shown schematically in Fig. 4.4.2.

Plane Strain Element
In the case of plane strain deformation, displacements and strains in the z-direction are equal to zero (Fig.4.4.2a). Deformation of material is independent of the z-coordinate and

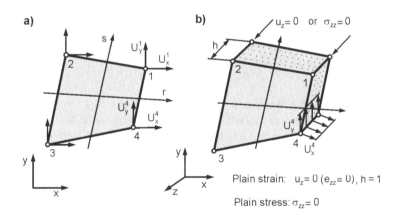

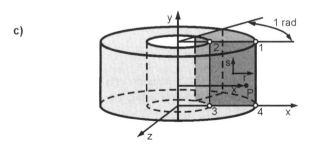

Fig. 4.4.2 Two-dimensional four-node finite elements. Plane strain and plane stress elements (a) In plane; (b) Axonometric representations; (c) Axisymmetric element – volume corresponding to one radian is used

the displacement $u_z = 0$. The nonzero strains are $e_{xx}, e_{yy}, \gamma_{xy}$ and they follow from the relations (4.3.7),

$$\mathbf{e} = \begin{Bmatrix} e_{xx} \\ e_{yy} \\ \gamma_{xy} \end{Bmatrix} = \begin{bmatrix} u_{1,1} \\ u_{2,2} \\ u_{1,2} + u_{2,1} \end{bmatrix} \begin{bmatrix} N_{1,1} & 0 & N_{2,1} & 0 & \cdots & N_{N,1} & 0 \\ 0 & N_{1,2} & 0 & N_{2,2} & \cdots & 0 & N_{N,2} \\ N_{1,2} & N_{1,1} & N_{2,2} & N_{2,1} & \cdots & N_{N,2} & N_{N,1} \end{bmatrix} \begin{Bmatrix} U_1^1 \\ U_2^1 \\ \cdot \\ \cdot \\ \cdot \\ U_1^N \\ U_2^N \end{Bmatrix} = \mathbf{B}\mathbf{U}$$

(4.4.3)

The stresses are obtained using the elastic constitutive relationship (see (2.2.2)) and the constitutive matrix (2.2.6) – without fourth row and column,

$$\boldsymbol{\sigma} = \mathbf{C}\mathbf{e}, \quad \text{or}$$

$$\boldsymbol{\sigma} = \begin{Bmatrix} \sigma_{xx} \\ \sigma_{yy} \\ \sigma_{xy} \end{Bmatrix} = \frac{E(1-\nu)}{(1+\nu)(1-2\nu)} \begin{bmatrix} 1 & \dfrac{\nu}{1-\nu} & 0 \\ \dfrac{\nu}{1-\nu} & 1 & 0 \\ 0 & 0 & \dfrac{1-2\nu}{2(1-\nu)} \end{bmatrix} \begin{Bmatrix} e_{xx} \\ e_{yy} \\ \gamma_{xy} \end{Bmatrix} \quad (4.4.4)$$

Besides these stress components, the nonzero stress is σ_{zz} which can be obtained from the expression for e_{zz} in (2.2.13) and the condition $e_{zz} = 0$:

$$\sigma_{zz} = \nu\left(\sigma_{xx} + \sigma_{yy}\right) \quad (4.4.5)$$

The stiffness matrix is of the dimension 3×3 and has the form (4.3.15).

Note that the integration over the volume to obtain the element stiffness matrix assumes integration over the element area and unit thickness $h = 1$ (see Fig. 4.4.2b) in the z-direction. Also, the nodal forces represent the forces due to stresses or due to surface forces evaluated over the unit element thickness. Equilibrium equations assume the unit element thickness, and therefore the nodal forces are in *N/m*.

Plane Stress Element

The physical condition for a plane stress (membrane) element is that the normal stress to the plane $x - y$ is equal to zero,

$$\sigma_{zz} = 0 \quad (4.4.6)$$

Now we use the constitutive matrix (2.2.7) to obtain stresses in (4.4.4). The strain through the element thickness is obtained from the stresses using the third row in equation (2.2.13) and the condition (4.4.6) (see also (2.2.9)),

$$e_{zz} = -\frac{\nu}{E}\left(\sigma_{xx} + \sigma_{yy}\right) = -\frac{\nu}{1-\nu}(e_{xx} + e_{yy}) \quad (4.4.7)$$

The element stiffness matrix is calculated by integration over the element surface with multiplication by the element thickness h,

ISOPARAMETRIC FORMULATION OF FINITE ELEMENTS 89

$$\mathbf{K} = h \int_A \mathbf{B}^T \mathbf{C} \mathbf{B} \, dA \tag{4.4.8}$$

The nodal forces correspond to the thickness h (see Fig. 4.4.2b) and are in N.

Axisymmetric Element
This element is used for modeling the problems with axial symmetry in geometry, loading and boundary conditions. Then, we model one radial plane (see Fig. 4.4.2c) by 2D axially symmetric elements. Displacements occur in the radial planes, but now the circumferential (hoop) strain is generated, which is related to the radial displacement as

$$e_{zz} = \frac{u_r}{r} = \frac{u_x}{x} \tag{4.4.9}$$

where $r = x$ is the radial distance of a material point, and $u_r = u_x$ is the radial displacement. In the strain–displacement relationship (4.4.3) we add the relation (4.4.9) and obtain

$$\mathbf{e} = \begin{Bmatrix} e_{xx} \\ e_{yy} \\ \gamma_{xy} \\ e_{zz} \end{Bmatrix} = \begin{bmatrix} u_{1,1} \\ u_{2,2} \\ u_{1,2} + u_{2,1} \\ \dfrac{u_1}{x} \end{bmatrix} \begin{bmatrix} N_{1,1} & 0 & N_{2,1} & 0 & \cdots & N_{N,1} & 0 \\ 0 & N_{1,2} & 0 & N_{2,2} & \cdots & 0 & N_{N,2} \\ N_{1,2} & N_{1,1} & N_{2,2} & N_{2,1} & \cdots & N_{N,2} & N_{N,1} \\ \dfrac{N_1}{x} & 0 & \dfrac{N_2}{x} & 0 & \cdots & \dfrac{N_N}{x} & 0 \end{bmatrix} \begin{Bmatrix} U_1^1 \\ U_2^1 \\ \cdot \\ \cdot \\ \cdot \\ U_1^N \\ U_2^N \end{Bmatrix} = \mathbf{BU}$$

(4.4.10)

For a point at the axis of symmetry $(x = 0)$ we use in practical applications $e_{zz} = e_{xx}$.

The integration performed for evaluation of the element matrices and vectors assumes, respectively, the integration over the element volume and area corresponding to one radian in the circumferential direction, see Fig. 4.4.2c. The nodal forces represent the forces for the one-radian volume or the one-radian surface. Details are given on the web – Theory, Chapter 4.

4.4.2 Examples

Example 4.4-1. Deformation of a thick-walled cylinder subjected to internal and external pressures
A thick-walled cylinder, shown in Fig. E4.4-1a, is loaded by internal and external pressures p_i and p_e. In the axial direction the cylinder can be free to deform or it can be restrained. Axisymmetric finite elements are used and the FE mesh in a radial plane is shown in Fig. E4.4-1b.

Radial distribution of stresses along the cylinder radius are shown in Figs. E4.4-1c,d for data given in the figure. The solutions compare well with the analytical solutions.

Suggestions for detailed analysis and solutions using the Software are given on the web.

Example 4.4-2. Deformation of a plate with a hole subjected to uniaxial tension
A plate with a hole, shown in Fig. E4.4-2a, is subjected to uniaxial loading (extension or compression). It is assumed that the plate has unit thickness. Data are given in the figure.

90 COMPUTER MODELING IN BIOENGINEERING

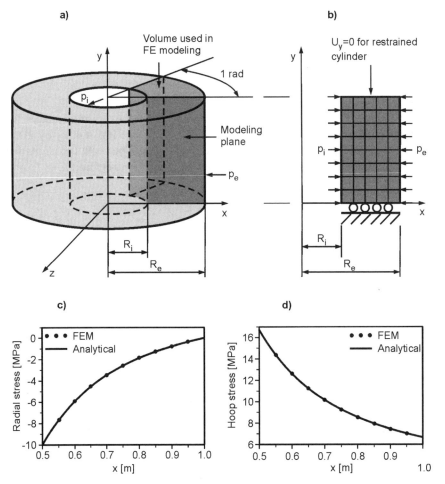

Fig. E4.4-1 Cylinder loaded by internal and external pressures, free to deform or restrained axially. (a) Three-dimensional representation of the cylinder, with the section modeled by finite elements; (b) Finite element model in a radial plane $x-y$; (c) Radial stress distribution; (d) Hoop stress distribution. Data: $R_i = 0.5\,m$, $R_e = 1.0\,m$, $E = 2 \times 10^5\,MPa$, $v = 0.3$, $p_i = 10\,MPa$, $p_e = 0$

One-quarter of the plate is modeled due to symmetry conditions. Deformed configuration, with the field of the extensional stress σ_{xx}, is shown in Fig. E4.4-2b. The hole in the plate causes the increase of stress near the hole (stress concentration), as can be seen from the graph in Fig. E4.4-2c. The analytical solution for small d/b (if d/b is less then 1/4 the error is less then 6%, Timoshenko & Goodier 1951) is

$$\sigma_{xx} = \frac{p}{2}\left[2+\left(\frac{d}{2y}\right)^2+3\left(\frac{d}{2y}\right)^4\right] \quad \text{(E4.4-2.1)}$$

Solutions for various geometrical data and loading can be obtained using the Software on the web.

ISOPARAMETRIC FORMULATION OF FINITE ELEMENTS

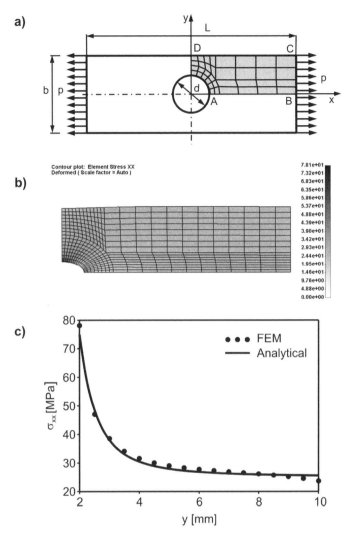

Fig. E4.4-2 A plate with a hole subjected to uniaxial loading. (a) Geometry of the plate, with the section ABCD modeled by finite elements; (b) Field of extensional stress σ_{xx} and FE mesh – deformed configuration; (c) Distribution of extensional stress along the line $x = 0$ (analytical and numerical solutions), with notable stress concentration. Data: Lengths (mm) $L = 56$, $b = 20$, $d = 4$; $E = 7 \times 10^4$ MPa, $v = 0.25$, $p = 25$ MPa

4.5 Isoparametric shell finite element for general 3D analysis

Thin-walled structures in which structural surface dominates are called shells. These structures are broadly used in technical practice. Examples of shell structures are a car body, an airplane structure, domes, cooling towers, etc. The efficiency of the material in carrying the loads (defined as the ratio between the load value and the mass of the structure) is very high for shells. Consequently, shells have been used in design since ancient times and will likely be used in future for numerous applications.

The mechanical behavior of shells is complex and various shell theories have been developed. Also, shell finite element analysis has been the challenging topic in the FE research and a number of approaches have been introduced. The results of these efforts are different formulations of shell finite elements for general and specific conditions (see e.g. Hughes 1987, Bathe 1996, Chapelle & Bathe 2003). We present a simple four-node isoparametric shell element for general 3D analysis which provides reliable and accurate results for the problems of interest in bioengineering applications.

4.5.1 Basic assumptions about shell deformation

The basic shell geometry is shown in Fig. 4.5.1 (Kojić et al. 1998). The shell is defined by its *mid-surface* mathematically expressed by a function

$$f(x, y, z) = 0 \qquad (4.5.1)$$

and by its thickness measured along the normals to the mid-surface. The *unit normal* **n** is determined by

$$\mathbf{n} = \nabla f / \|\nabla f\| \quad \text{or} \quad n_i = \frac{\partial f / \partial x_i}{\|\nabla f\|} \qquad (4.5.2)$$

where ∇f is the gradient to the mid-surface, with components $\partial f/\partial x_i$, $i = 1, 2, 3$. The position vector **x** of a material point P can be expressed as the sum of the position vector $\mathbf{x}_0$ of a point P_0 in the mid-surface and the relative position vector along the shell normal,

$$\mathbf{x} = \mathbf{x}_0 + \bar{z}\mathbf{n}, \quad \text{or} \quad x_i = x_{0i} + \bar{z}n_i \qquad (4.5.3)$$

where $\bar{z}$ is the coordinate in the direction of the normal ($-h \leq \bar{z} \leq h$, where h is the shell thickness).

The physical assumptions for the kinematics of deformation rely on Mindlin's plate theory. These assumptions can be summarized as follows: Displacement **u** of a material point P due to shell deformation can be expressed as the sum of the displacement $\mathbf{u}_0$ of the point P_0 and the rotation displacement $\mathbf{u}^{rot}$,

$$\mathbf{u} = \mathbf{u}_0 + \mathbf{u}^{rot}, \quad \text{i.e.} \quad \mathbf{u} = {}^t\mathbf{x} - {}^0\mathbf{x} = {}^t\mathbf{x}_0 - {}^0\mathbf{x}_0 + \bar{z}\left({}^t\mathbf{n} - {}^0\mathbf{n}\right), \qquad (4.5.4)$$

Here, as shown in Figs. 4.5.1a,b, the vectors ${}^0\mathbf{x}$, ${}^0\mathbf{x}_0$ and ${}^t\mathbf{x}$, ${}^t\mathbf{x}_0$ are the position vectors of the points P and P_0 before and after deformation, respectively. Also, ${}^0\mathbf{n}$ and ${}^t\mathbf{n}$ are the normals before and after deformation. This assumption about the displacements of material points implies that the *straight material elements*, originally in the normal direction to the shell surface, *remain straight*; this is expressed by the term $\bar{z}\left({}^t\mathbf{n} - {}^0\mathbf{n}\right)$ in equation (4.5.4). Change of the normal is due to the rotation $\boldsymbol{\varphi}$ which as the vector lies in the shell *tangential plane*. Then, change of the normal can be expressed as

ISOPARAMETRIC FORMULATION OF FINITE ELEMENTS

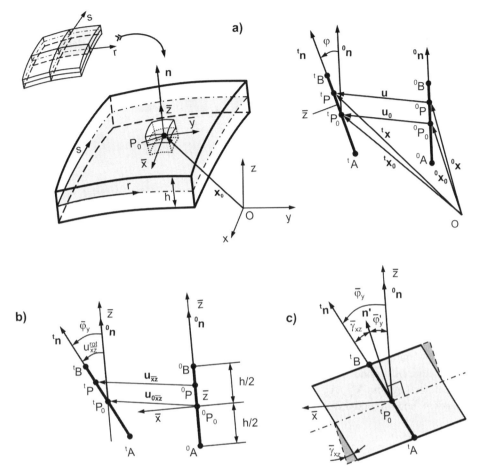

Fig. 4.5.1 Geometry and deformation of shell. (a) Geometry and position vectors of a material element P_0P before and after deformation; (b) Displacements in the local shell plane $\bar{x}-\bar{z}$; (c) Rotation of the material element P_0P in the $\bar{x}-\bar{z}$ plane which includes rotation and shear of the cross-section

$$^t\mathbf{n} - {}^0\mathbf{n} = \boldsymbol{\varphi} \times {}^0\mathbf{n}, \quad \text{or} \quad ({}^t\mathbf{n} - {}^0\mathbf{n})_i = e_{ijk}\varphi_j\,{}^0n_k, \quad \text{or}$$
$$({}^t\mathbf{n} - {}^0\mathbf{n})_{\bar{x}} = \overline{\varphi}_y, \quad ({}^t\mathbf{n} - {}^0\mathbf{n})_{\bar{y}} = -\overline{\varphi}_x \tag{4.5.5}$$

where e_{ijk} is the permutation symbol (see (1.3.8)).

The components of change of the normal are written in the local shell system $\bar{x}, \bar{y}, \bar{z}$ to emphasize that the rotation vector lies in the tangential plane. This fact is illustrated in Fig. 4.5.1c (in plane $\bar{x}-\bar{z}$). We have graphically shown a very important physical assumption about the shell deformation: rotation of the shell normal consists of the part $\boldsymbol{\varphi}'$ which is due to bending, and a part due to transversal shear. In Fig. 4.5.1c we have $\overline{\varphi}_y = \overline{\varphi}'_y + \overline{\gamma}_{xz}$, where $\overline{\varphi}'_y$ is the rotation due to bending moment around the axis $\bar{y}$, and $\overline{\gamma}_{xz}$ is the shear strain in plane $\bar{x}-\bar{z}$ due to transversal loading in plane $\bar{x}-\bar{z}$.

4.5.2 Formulation of a four-node shell element

The shell element described here relies on the above assumptions about the deformation. We present a four-node element (Dvorkin & Bathe 1984). The natural coordinate system has the origin in the mid-surface, with the r, s coordinates representing the position of a material point in the mid-surface, and t the position along the shell normal. The coordinates of a material point in the global coordinate system x_i ($x_1 \equiv x, x_2 \equiv y, x_3 \equiv z$), can be expressed as

$$x_i = \sum_{K=1}^{N} N_K X_i^K + \frac{t}{2} \sum_{K=1}^{N} N_K h_K V_{ni}^K, \quad i=1,2,3 \tag{4.5.6}$$

where $N = 4$ is the number of nodes; $N_K(r, s)$ are the interpolation functions; h_K are the thicknesses at shell nodes; and $V_{ni}^K \equiv n_i^K$ are the components of the normals at nodal points.

Using (4.5.4) and (4.5.5) the displacements $u_i(r, s, t)$ can be written in the form (see Fig. 4.5.2)

$$u_i = \sum_{K=1}^{N} N_K U_i^K + \frac{t}{2} \sum_{K=1}^{N} h_K N_K \left(-V_{2i}^K \alpha_K + V_{1i}^K \beta_K \right), \quad i=1,2,3 \tag{4.5.7}$$

Here, U_i^K are the nodal point displacements; V_{1i}^K and V_{2i}^K are the components of the vectors $\mathbf{V}_1^K$ and $\mathbf{V}_2^K$ which lie in the tangential plane at nodes (which can be suitably selected, for example by multiplying the unit vector of one coordinate axis and normal $\mathbf{n}$); and α_K and β_K are the rotations around $\mathbf{V}_1^K$ and $\mathbf{V}_2^K$, as the components of the nodal rotation vector $\boldsymbol{\varphi}^K$ in the local shell system $\mathbf{V}_1^K, \mathbf{V}_2^K, \mathbf{V}_n^K$. From (4.5.7) it follows that each node has *five degrees of freedom*, and the nodal displacement vector for the node K is

$$\left(\mathbf{U}^K\right)^T = \begin{bmatrix} U_x^K & U_y^K & U_z^K & \alpha_K & \beta_K \end{bmatrix} \tag{4.5.8}$$

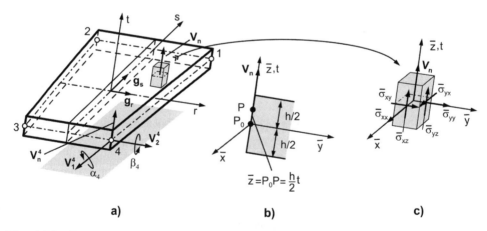

Fig. 4.5.2 Four-node shell finite element. (a) Element geometry, the natural coordinate system r, s, t and base vectors $\mathbf{g}_r, \mathbf{g}_s$ in the tangential plane; (b) Position of a material point P with respect to the mid-surface point P_0 and the local shell coordinate system $\bar{x}, \bar{y}, \bar{z}$ at point P_0; (c) Nonzero stresses in the local shell coordinate system

ISOPARAMETRIC FORMULATION OF FINITE ELEMENTS

The strains can be obtained by applying the relationship (4.3.7) to the displacement field (4.5.7):

$$\mathbf{e} = [\mathbf{B}^1 \mathbf{B}^2 \ldots \mathbf{B}^N] \begin{Bmatrix} \mathbf{U}^1 \\ \mathbf{U}^2 \\ \cdot \\ \cdot \\ \mathbf{U}^N \end{Bmatrix} \quad (4.5.9)$$

The submatrix $\mathbf{B}^K$ is

$$\mathbf{B}^K(r,s,t) = \begin{bmatrix} N_{K,1} & 0 & 0 & tg_{11}^K N_{K,1} & tg_{21}^K N_{K,1} \\ 0 & N_{K,2} & 0 & tg_{12}^K N_{K,2} & tg_{22}^K N_{K,2} \\ 0 & 0 & N_{K,3} & tg_{13}^K N_{K,3} & tg_{23}^K N_{K,3} \\ N_{K,2} & N_{K,1} & 0 & t(g_{11}^K N_{K,2} + g_{12}^K N_{K,1}) & t(g_{21}^K N_{K,2} + g_{22}^K N_{K,1}) \\ 0 & N_{K,3} & N_{K,2} & t(g_{12}^K N_{K,3} + g_{13}^K N_{K,2}) & t(g_{22}^K N_{K,3} + g_{23}^K N_{K,2}) \\ N_{K,3} & 0 & N_{K,1} & t(g_{11}^K N_{K,3} + g_{13}^K N_{K,1}) & t(g_{21}^K N_{K,3} + g_{23}^K N_{K,1}) \end{bmatrix} \quad (4.5.10)$$

where: $N_{K,i} \equiv \partial N_K / \partial x_i$, $g_{1i}^K = -(1/2) h_K V_{2i}^K$, $g_{2i}^K = (1/2) h_K V_{1i}^K$, $i = 1, 2, 3$. The derivatives $\partial N_K / \partial x_i$ can be computed using (4.3.13). Note that the submatrix $\mathbf{B}^K$ is a 6×5 matrix. The Jacobian matrix terms J_{ij} follow from (4.5.6),

$$J_{1i} = \sum_{K=1}^{N} N_{K,r} \left(X_i^K + \frac{t}{2} h_K V_{ni}^K \right), \quad J_{2i} = \sum_{K=1}^{N} N_{K,s} \left(X_i^K + \frac{t}{2} h_K V_{ni}^K \right), \quad (4.5.11)$$

$$J_{3i} = \frac{1}{2} \sum_{K=1}^{N} h_K N_K V_{ni}^K$$

In order to determine the stiffness matrix we use (4.3.15). The matrix $\mathbf{C}$ corresponds to the global strain and stress components e_i and σ_i defined in (2.2.1). However, the matrix $\mathbf{C}$ must be such that it reduces to the shell matrix $\overline{\mathbf{C}}$ given in (2.2.8). Therefore, we obtain $\mathbf{C}$ by applying the transformation (2.2.17) to the shell matrix $\overline{\mathbf{C}}$,

$$\mathbf{C} = \overline{\mathbf{T}}^\sigma \overline{\mathbf{C}} \left(\overline{\mathbf{T}}^\sigma \right)^T \quad (4.5.12)$$

where the terms of the matrix $\overline{\mathbf{T}}^\sigma$ depend on the angles between axes of the local shell and global coordinate systems. Details about this transformation and other options for the calculations of strains, stresses, stiffness matrix and nodal forces, including a generalization to multilayered shells, are given on the web – Theory, Chapter 4.

4.5.3 Examples

Example 4.5-1. Deformation of a squared plate subjected to normal concentrated force
A squared plate, clamped or free supported, is loaded by a concentrated force at the plate center C (Fig. E4.5-1A). One-quarter of the plate, ABDC, is modeled due to symmetry in geometry, boundary conditions and loading.

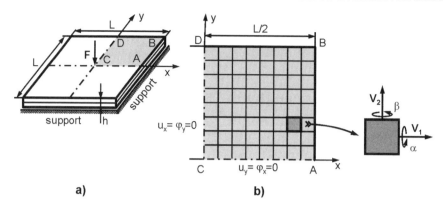

Fig. E4.5-1A Squared plate subjected to a concentrated force at the plate center. (a) Plate geometry; (b) Finite element model – one-quarter of the plate is modeled due to symmetry. The symmetry boundary conditions and the shell element local axes and rotations (see Fig. 4.5.2a) are shown in the figure

The deflection of the central point C can be obtained in the analytical form (Timoshenko & Vojinovski-Kriger 1962), and in the case of clamped plate edges it given as (for $v = 0.3$)

$$u_C = 0.0056 \frac{12FL^2(1-v^2)}{Eh^3} \quad \text{(E4.5-1.1)}$$

where E is Young's modulus and v is Poisson's ratio for elastic material of the plate. For the data: $L = 4\,cm$, $h = 0.01\,cm$, $E = 10^7 N/cm^2$, $v = 0.3$, $F = 4\,N$, the displacement field is shown in Fig. E4.5-1Ba. The normalized displacement (normalization with

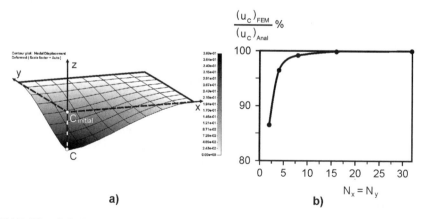

Fig. E4.5-1B Solutions for clamped plate under concentrated force. (a) Deformed shape and field of effective stress; (b) Deflection of central point C normalized with respect to the analytical solution, in terms of number of finite elements in one direction ($N_x = N_y$). Data given in the text

ISOPARAMETRIC FORMULATION OF FINITE ELEMENTS

respect to the analytical solution (E4.5-1.1)) in terms of the mesh density, expressed by the number of finite elements in one direction, is shown in Fig. E4.5-1Bb. It can be see that the FE solution approaches rapidly to the analytical solution with the increase of mesh density.

Solutions for other conditions: free support of the plate, loading by pressure, and change of geometrical and material parameters, as well as for various mesh densities, can be obtained by the Software on the web.

Example 4.5-2. Half of a sphere loaded by extension and compression forces
Due to symmetry conditions (symmetry planes are $x-y$ and $y-z$), one-quarter of the structure is modeled, with symmetry conditions shown in Fig. E4.5-2.

Deformed configuration with the field of effective stress is shown in the figure. Displacements at the points A and C are $u_{zA} = -u_{xC} = 0.08915$ cm, while the analytical solution is $u_{zA} = -u_{xC} = 0.0924$ cm (Simo et al. 1989). The data used for this solution are: sphere radius an thickness are $R = 10$ cm, $h = 0.04$ cm; Young's modulus and Poisson's ratio are $E = 6.825 \times 10^7$ N/cm^2, $v = 0.3$; and the force $F = 1$ N.

Detailed analysis of this example can be obtained using Software on the web.

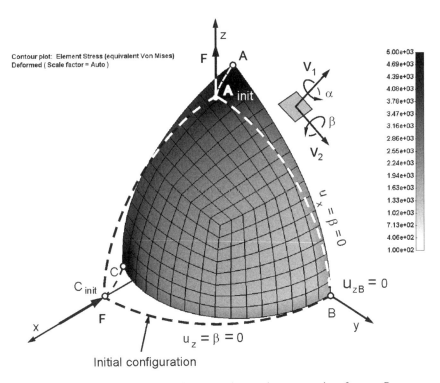

Fig. E4.5-2 Half of a sphere loaded by extension and compression forces. One-quarter of the structure is modeled, with symmetry boundary conditions; FE mesh, initial and deformed configurations with the field of effective stress. Note that point A moves in $y-z$ plane, while point C moves in the $x-y$ plane. Data given in the text

References

Bathe, K.J. (1996). *Finite Element Procedures*, Prentice-Hall, Englewood Cliffs, NJ.

Chapelle, D. & Bathe, K.J. (2003). *The Finite Element Analysis of Shells – Fundamentals*, Springer-Verlag, Berlin.

Crisfield, M.A. (1991). *Non-Linear Finite Element Analysis of Solids and Structures*. John Wiley & Sons, Ltd, Chichester, England.

Dvorkin, E.N. & Bathe, K.J. (1984). A continuum mechanics based four-node shell element for general nonlinear analysis, *Eng. Comp.*, **1**, 77–88.

Huebner, K.H. (1975). *The Finite Element Method for Engineers*, John Wiley & Sons, Inc., New York.

Hughes, T.J.R. (1987). *The Finite Element Method. Linear Static and Dynamic Finite Element Analysis*, Prentice-Hall, Englewood Cliffs, NJ.

Kojić, M. & Bathe, K.J. (2005). *Inelastic Analysis of Solids and Structures*, Springer-Verlag, Berlin.

Kojić, M., Slavković, R., Živković, M. & Grujovic, N. (1998). *The Finite Element Method – Linear Analysis* (in Serbian). Faculty of Mechanical Engineering, Univesity of Kragujevac, Serbia.

Sekulovic, M. (1984). *The Finite Element Method* (in Serbian), Gradjevinska knjiga, Belgrade.

Simo, J.C., Fox, D.D. & Rifai, M.S. (1989). On a stress resultant of geometrically exact shell model. Part II: The linear theory; computational aspects, *Comp. Meth. Appl. Mech. Eng.*, **73**, 53–92.

Timoshenko, S. & Goodier, J.N. (1951). *Theory of Elasticity*, McGraw-Hill, New York.

Timoshenko, S. & Vojinovski-Kriger, S. (1962). *Theory of Plates and Shells* (in Serbian, translation from English), Gradjevinska knjiga, Belgrade.

5

Dynamic Finite Element Analysis

Fundamentals of the finite element dynamic analysis of solids are presented in this chapter. We first formulate the differential equations of motion for isoparametric finite elements assuming dynamic loading and the presence of viscous damping within the material. Then the Newmark method is presented for integration of differential equations of motion. We also give the basic relations for determination of structural frequencies and vibration modes. Finally through examples we present the dynamic response of simple structures.

5.1 Introduction to dynamics of structures

In the previous chapter we considered static deformation of solids and structures, where the material deforms under loading which changes over time quasi-statically, producing slow (quasi-static) displacements. The dynamic effects, which are basically due to inertial forces, can be neglected under these conditions. However, in engineering practice as well as in biological solids, there are situations where dynamic effects must be included, such as in the case of structures under earthquake loading, impact loading, or fast muscle contraction.

As in static analysis, the finite element method can be used to find the dynamic responses of complex solids, structures and biosolids. We present the basis of the FE dynamics that will be used subsequently in bioengineering applications. The FE discrete differential equations of motion and a procedure for integration of these differential equations are presented to the extent needed for this book.

Very important information about the dynamic behavior of a solid can be obtained by finding the system frequencies and modal shapes. This is achieved by the so-called eigen-analysis. It can be shown that when the frequency of loading is close to the system (natural) frequency, the displacements may become very large even under small load intensity. For example, if an earthquake wave is close to the natural frequency of a structure, the amplitudes of displacements of the structure may become very large and cause structural collapse. Here, we only give the fundamental relations of the eigen-analysis.

Computer Modeling in Bioengineering   Edited by M. Kojić, N. Filipović, B. Stojanović, N. Kojić
© 2008 John Wiley & Sons, Ltd

5.2 Differential equations of motion

Consider a material body subjected to external time-dependent forces $\mathbf{F}_1(t), \mathbf{F}_2(t), \ldots$ producing motion and deformation, as schematically shown in Fig. 5.2.1. The motion is such that we must take into account inertial forces. The inertial force $d\mathbf{F}^{in}$ of a mass dm is

$$d\mathbf{F}^{in} = -\ddot{\mathbf{u}} dm = -\rho \ddot{\mathbf{u}} dV \qquad (5.2.1)$$

where $\ddot{\mathbf{u}} \equiv d^2\mathbf{u}/dt^2$ is the acceleration, ρ is material density, and dV is the elementary volume. The inertial force is a volumetric force and the equivalent nodal inertial force vector $\mathbf{F}^{in}$ follows from (4.3.16),

$$\mathbf{F}^{in} = -\int_V \rho \mathbf{N}^T \ddot{\mathbf{u}} dV = -\int_V \rho \mathbf{N}^T \mathbf{N} dV \ddot{\mathbf{U}} = -\mathbf{M}\ddot{\mathbf{U}} \qquad (5.2.2)$$

where $\mathbf{M}$ is the element *mass matrix*,

$$\mathbf{M} = \int_V \rho \mathbf{N}^T \mathbf{N} dV \qquad (5.2.3)$$

and $\ddot{\mathbf{U}}$ is the *nodal acceleration* vector. In derivation of (5.2.3) the relationship $\ddot{\mathbf{u}} = \mathbf{N}\ddot{\mathbf{U}}$ is used, which follows from the interpolation of displacements within the finite element, see (4.3.2) for the 3D solid element. Note that the mass matrix is symmetric, with dimension $3N \times 3N$ for a 3D finite element.

The derived mass matrix is called the *consistent mass matrix*. In practical applications of dynamic FE analysis, a simplified, so-called *lumped mass* matrix is used. The lumped mass matrix is a diagonal matrix with the nonzero terms equal to the element mass divided by the number of element nodes.

When damping (viscous) effects are present within the material, the elementary damping force can be expressed as

$$d\mathbf{F}^w = -b\dot{\mathbf{u}} dV \qquad (5.2.4)$$

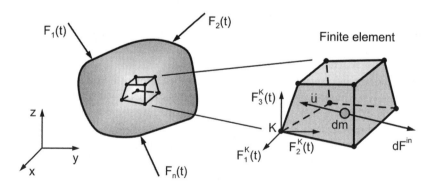

Fig. 5.2.1 A schematic representation of the dynamics of a deformable body. Elementary mass dm within the finite element; elementary inertial force $d\mathbf{F}^{in}$ and the time variable nodal force components at a node K

DYNAMIC FINITE ELEMENT ANALYSIS

where b is the damping (viscous) coefficient. Then, following the above derivation for the element inertial nodal force vector, we obtain the element nodal damping vector $\mathbf{F}^w$ as

$$\mathbf{F}^w = -\mathbf{B}^w \dot{\mathbf{U}} \tag{5.2.5}$$

where $\mathbf{B}^w$ is the element damping matrix,

$$\mathbf{B}^w = \int_V b \mathbf{N}^T \mathbf{N} dV \tag{5.2.6}$$

We now substitute the inertial and damping nodal force vectors (5.2.2) and (5.2.5) into the element equilibrium equation (4.2.21), and further assemble the equilibrium equations to obtain

$$\mathbf{M}_{sys} \ddot{\mathbf{U}}_{sys} + \mathbf{B}^w_{sys} \dot{\mathbf{U}}_{sys} + \mathbf{K}_{sys} \mathbf{U}_{sys} = \mathbf{F}^{ext}_{sys} \tag{5.2.7}$$

where $\mathbf{M}_{sys}$, $\mathbf{B}^w_{sys}$ and $\mathbf{K}_{sys}$ are the mass, damping and stiffness matrices of the system, respectively; and $\mathbf{F}^{ext}_{sys}$ is the system external force vector that includes the external concentrated, surface and body forces. Equation (5.2.7) represents the differential equation of motion of a material system discretized into finite elements.

5.3 Integration of differential equations of motion

Differential equations of motion (5.2.7) represent a system of linear differential equations of the second order. They can be integrated numerically to give the solution for the displacements, velocities and accelerations of nodal points and then of material points during a selected time period. Numerical methods of integration assume the incremental solutions with a time step Δt. Integration is performed within a time step, using the solution at the start of the time step and by employing certain approximations for changes of displacements, velocities and accelerations within the time-step period Δt. A variety of incremental integration schemes have been introduced (see, e.g. Bathe 1996). We present one of these methods, the widely used Newmark method, which is also employed in later applications.

Newmark Method
In order to derive the incremental relations for a finite element we will use the following notation: the current step is denoted by 'n'; all quantities which change with time will have the left superscript 'n' or '$n+1$', for values at the start and end of the time step, respectively.

The basic approximation used in the Newmark method is that the acceleration within the time step is considered constant, and is given as

$$\ddot{\mathbf{U}}(\tau) = (1-\delta)\,{}^n\ddot{\mathbf{U}} + \delta\,{}^{n+1}\ddot{\mathbf{U}}, \quad 0 \leq \tau \leq \Delta t \tag{5.3.1}$$

where $0 \leq \delta \leq 1$ is a parameter; $\delta = 0$, $\delta = 1$ and $\delta = 0.5$ correspond, respectively, to the Euler forward, Euler backward, and trapezoidal integration scheme. From the expression (5.3.1) it follows that the velocity changes linearly within the time step (see Fig. 5.3.1).

Integrating this equation with respect to time τ, the velocity ${}^{n+1}\dot{\mathbf{U}}$ and displacement ${}^{n+1}\mathbf{U}$ at end of time step are obtained as

$${}^{n+1}\dot{\mathbf{U}} = {}^n\dot{\mathbf{U}} + \left[(1-\delta)\,{}^n\ddot{\mathbf{U}} + \delta\,{}^{n+1}\ddot{\mathbf{U}}\right] \Delta t \tag{5.3.2}$$

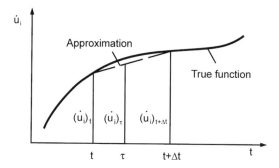

Fig. 5.3.1 Approximation for velocity of a material point within a time step

and

$$^{n+1}\mathbf{U} = {}^{n}\mathbf{U} + {}^{n}\dot{\mathbf{U}}\Delta t + \frac{1}{2}\left[(1-\delta)\,{}^{n}\ddot{\mathbf{U}} + \delta\,{}^{n+1}\ddot{\mathbf{U}}\right](\Delta t)^{2} \qquad (5.3.3)$$

In order to improve the solution accuracy and stability (see Bathe 1996), instead of (5.3.3) the following expression for the displacement $^{n+1}\mathbf{U}$ is used:

$$^{n+1}\mathbf{U} = {}^{n}\mathbf{U} + {}^{n}\dot{\mathbf{U}}\Delta t + \left[\left(\frac{1}{2}-\alpha\right){}^{n}\ddot{\mathbf{U}} + \alpha\,{}^{n+1}\ddot{\mathbf{U}}\right](\Delta t)^{2} \qquad (5.3.4)$$

where α is another integration parameter. It can be shown that the best solution accuracy is obtained for $\delta = 0.5$ and $\alpha = 0.25$. Now we substitute $^{n+1}\ddot{\mathbf{U}}$ from (5.3.4) into (5.3.3) and express $^{n+1}\ddot{\mathbf{U}}$ in terms of the displacement $^{n+1}\mathbf{U}$ as

$$^{n+1}\ddot{\mathbf{U}} = \frac{1}{\alpha(\Delta t)^{2}}\left[{}^{n+1}\mathbf{U} - {}^{n}\mathbf{U} - {}^{n}\dot{\mathbf{U}}\Delta t - \left(\frac{1}{2}-\alpha\right)(\Delta t)^{2}\,{}^{n}\ddot{\mathbf{U}}\right] \qquad (5.3.5)$$

Then, by substituting this expression for $^{n+1}\ddot{\mathbf{U}}$ into (5.3.3) it follows that

$$^{n+1}\dot{\mathbf{U}} = \frac{\delta}{\alpha\Delta t}\left({}^{n+1}\mathbf{U} - {}^{n}\mathbf{U}\right) - \left(\frac{\delta}{\alpha}-1\right){}^{n}\dot{\mathbf{U}} - \left(\frac{\delta}{2\alpha}-1\right)\Delta t\,{}^{n}\ddot{\mathbf{U}} \qquad (5.3.6)$$

Finally, the differential equation of motion of the finite element (see (5.2.7)) for the end of the time step can be written as

$$\mathbf{M}\,{}^{n+1}\ddot{\mathbf{U}} + \mathbf{B}^{w}\,{}^{n+1}\dot{\mathbf{U}} + \mathbf{K}\,{}^{n+1}\mathbf{U} = {}^{n+1}\mathbf{F}^{ext} \qquad (5.3.7)$$

Then we substitute the expressions for $^{n+1}\ddot{\mathbf{U}}$ from (5.3.5) and $^{n+1}\dot{\mathbf{U}}$ from (5.3.6) into (5.3.7) and obtain the system of algebraic equations

$$\hat{\mathbf{K}}\,{}^{n+1}\mathbf{U} = {}^{n+1}\hat{\mathbf{F}} \qquad (5.3.8)$$

where

$$\hat{\mathbf{K}} = \mathbf{K} + a_{0}\mathbf{M} + a_{1}\mathbf{B}^{w} \qquad (5.3.9)$$

$$^{n+1}\hat{\mathbf{F}} = {}^{n+1}\mathbf{F} + \mathbf{M}\left(a_{0}\,{}^{n}\mathbf{U} + a_{2}\,{}^{n}\dot{\mathbf{U}} + a_{3}\,{}^{n}\ddot{\mathbf{U}}\right) + \mathbf{B}^{w}\left(a_{1}\,{}^{n}\mathbf{U} + a_{4}\,{}^{n}\dot{\mathbf{U}} + a_{5}\,{}^{n}\ddot{\mathbf{U}}\right) \qquad (5.3.10)$$

DYNAMIC FINITE ELEMENT ANALYSIS

Here, **K** is the element stiffness matrix. The coefficients in (5.3.9) and (5.3.10) are

$$a_0 = \frac{1}{\alpha(\Delta t)^2}, \quad a_1 = \frac{\delta}{\alpha \Delta t}, \quad a_2 = \frac{1}{\alpha \Delta t}$$

$$a_3 = \frac{1}{2\alpha} - 1, \quad a_4 = \frac{\delta}{\alpha} - 1, \quad a_5 = \left(\frac{\delta}{2\alpha} - 1\right)\Delta t \tag{5.3.11}$$

Since we imposed the condition of satisfying the differential equations at the end of the time step (see equations (5.3.7) and (5.3.8)), the *Newmark method is thus implicit*.

The solution procedure for the current time step consists of the following: we form the matrix $\hat{\mathbf{K}}$ and the nodal vector $^{n+1}\hat{\mathbf{F}}$ according to (5.3.9) and (5.3.10) for each finite element, assemble them into the system matrix and vector in a usual manner (see Section 4.2.2) and solve the system (5.3.8) for the nodal displacements $^{n+1}\mathbf{U}$. Then we calculate $^{n+1}\ddot{\mathbf{U}}$ and $^{n+1}\dot{\mathbf{U}}$ from (5.3.5) and (5.3.6) to be used as $^n\ddot{\mathbf{U}}$ and $^n\dot{\mathbf{U}}$ for the next time step. Of course, the displacements $^{n+1}\mathbf{u}$, velocities $^{n+1}\dot{\mathbf{u}}$ and accelerations $^{n+1}\ddot{\mathbf{u}}$ at any material point of a finite element 'e' can be determined by using, respectively, the interpolations from the nodal values: $^{n+1}\mathbf{u} = \mathbf{N}\,^{n+1}\mathbf{U}^e$, $\dot{\mathbf{u}} = \mathbf{N}\dot{\mathbf{U}}^e$ and $\ddot{\mathbf{u}} = \mathbf{N}\ddot{\mathbf{U}}^e$. Also, the strains $^{n+1}\mathbf{e}$ and stresses $^{n+1}\boldsymbol{\sigma}$ within the finite elements can be calculated by employing (4.3.7) and (2.2.2), $^{n+1}\mathbf{e} = \mathbf{B}\,^{n+1}\mathbf{U}^e$ and $^{n+1}\boldsymbol{\sigma} = \mathbf{C}\,^{n+1}\mathbf{e}$.

5.4 System frequencies and modal shapes

Important information about the dynamic structural response can be obtained by calculating the system frequencies and modal shapes. We here outline the basic idea about the determination of these dynamic parameters.

Calculation of the system frequencies and modal shapes is called the eigen-analysis. It consists of the following. Consider a system without damping and under no external loadings. Then the differential equations of motion (5.2.7) reduce to (matrices and vectors correspond to the assemblage of finite elements)

$$\mathbf{M}\ddot{\mathbf{U}} + \mathbf{K}\mathbf{U} = \mathbf{0} \tag{5.4.1}$$

Assume the solution of this system in the form

$$\mathbf{U} = \mathbf{A}\sin(\omega t + \phi) \tag{5.4.2}$$

where **A** is the vector of amplitudes, ω is the angular frequency, and ϕ is a phase. Substituting (5.4.2) into (5.4.1), the system of equations reduces to

$$\left(\mathbf{K} - \omega^2 \mathbf{M}\right)\mathbf{A} = \mathbf{0} \tag{5.4.3}$$

since the common term $\sin(\omega t + \phi)$ is different from zero. In order to have the nonzero vector **A** (nontrivial solution), the following scalar equation must be satisfied:

$$\det\left(\mathbf{K} - \omega^2 \mathbf{M}\right) = 0 \tag{5.4.4}$$

104   COMPUTER MODELING IN BIOENGINEERING

This equation is called the *characteristic equation* of the system. It is of order n with respect to ω^2, and the solutions $\omega_1^2, \omega_2^2, \ldots, \omega_n^2$ are the squares of the system frequencies. Here 'n' is the number of degrees of freedom of the system. For the dynamic response of a mechanical system the most dominant frequencies are the lower ones, starting with the smallest value ω_1.

Substituting a value for ω_k into (5.4.3) it is possible to determine the ratios $\overline{A}_i^{(k)}$ of the amplitude vector components with respect to one of the components (say A_1), so that the solution (5.4.2) can be written as

$$\mathbf{U}^{(k)} = \overline{\mathbf{A}}^{(k)} \sin(\omega_k t + \phi_k) \qquad (5.4.5)$$

The vector $\overline{\mathbf{A}}^{(k)}$ represents the *modal shape vector* corresponding to the system frequency ω_k. If an excitation force acting on the system has the frequency ω_k, it will produce the motion of the form (5.4.5). The system then enters into the *resonant regime* in which a small excitation force can produce large displacements. In practice, due to damping effects, the resonant amplitudes are lowered and the resonant motion decays. We show modal shapes in one example (Example 5.5-1).

5.5 Examples

Example 5.5-1. Eigenvalue analysis of a quadratic plate
A quadratic plate is modeled using four-node shell elements. In order to determine symmetric and nonsymmetric eigenvalues and eigenvectors, the entire plate is modeled, without symmetry conditions (Fig. E5.5-1A).

Analytical solution for natural frequencies of simply supported rectangular plate (Bolotin et al. 1968) is given as

$$f = \frac{\pi}{2}\left(\frac{m_1^2}{a_1^2} + \frac{m_2^2}{a_2^2}\right)\sqrt{\frac{D}{\rho h}} \qquad (E5.5\text{-}1.1)$$

where D is the plate stiffness,

$$D = \frac{Eh^3}{12(1-\nu^2)} \qquad (E5.5\text{-}1.2)$$

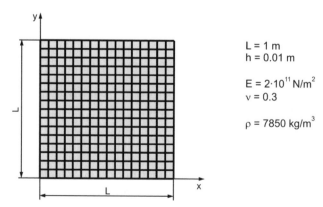

L = 1 m
h = 0.01 m

E = 2·10^{11} N/m^2
ν = 0.3

ρ = 7850 kg/m^3

Fig. E5.5-1A Simply supported quadratic plate modeled by four-node shell finite elements in eigenvalue analysis (the mesh is 16 × 16 elements)

DYNAMIC FINITE ELEMENT ANALYSIS

and a_1, a_2 are the plate dimensions, h is the plate thickness, and m_1 and m_2 are natural numbers. For quadratic plate $a_1 = a_2 = a$, so that (E5.5-1.1) is reduced to

$$f = \frac{\pi}{2a^2}\sqrt{\frac{D}{\rho h}}\left(m_1^2 + m_2^2\right) = f^*\left(m_1^2 + m_2^2\right) \quad \text{(E5.5-1.3)}$$

It can be seen from this equation that f^* is a constant dependent on material properties and plate dimensions. Natural frequencies can be obtained by choosing various values for m_1 and m_2, see Table E5.5-1.

Several modal shapes, corresponding to the smallest frequencies are shown in Fig. E5.5-1B.

Table E5.5-1 Natural frequencies of a simply supported quadratic plate, for data given in Fig. E5.5-1A. Numbers in parentheses are the FE solutions

Analytical solutions $f = f^*(m_1^2 + m_2^2)\ [s^{-1}]$

m_2 \ m_1	1	2	3
1	47.9865	119.966	239.932
	(48.1389)	(121.686)	(248.967)
2	119.966	191.946	311.912
	(121.686)	(194.849)	(321.533)
3	239.932	311.912	431.878
	(248.967)	(321.533)	(437.516)

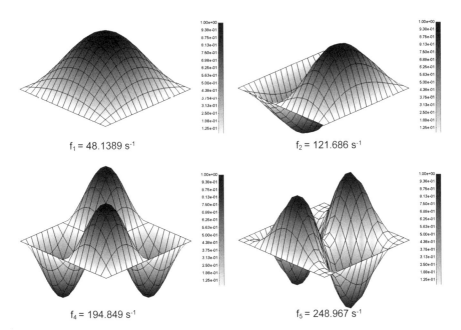

$f_1 = 48.1389\ s^{-1}$ $f_2 = 121.686\ s^{-1}$

$f_4 = 194.849\ s^{-1}$ $f_5 = 248.967\ s^{-1}$

Fig. E5.5-1B Modal shapes for four smallest natural frequencies (f_1 to f_5) of simply supported quadratic plate (note that $f_2 = f_3$ and the mode for f_2 is shown)

Finite element solutions for various plate dimensions and material data, as well as for a clamped plate, can be obtained by using software (see web – Software).

Example 5.5-2. Dynamic response of the cylindrical pipe subjected to the step pressure
A cylindrical pipe is subjected to the step pressure as show in the Fig. E5.5-2a. Due to symmetry, only one-half of the pipe is modeled using 2D axisymmetric elements. The

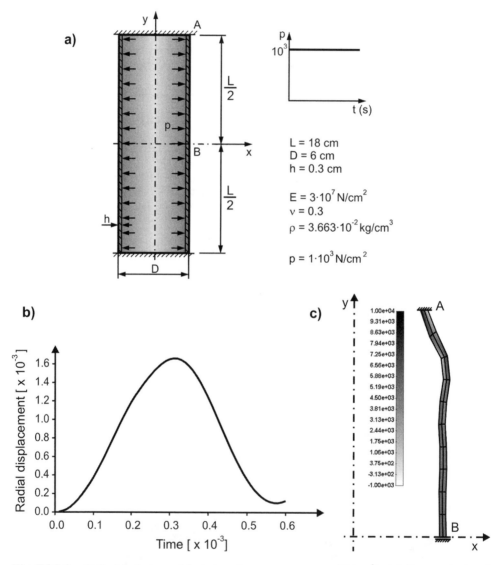

Fig. E5.5-2 Cylindrical pipe subjected to the step pressure p (N/cm^2). (a) Geometrical and material data; (b) Radial displacement of the point B at middle of the pipe; (c) Deformed configuration and distribution of axial stress (N/cm^2) at time $t = 3 \times 10^{-4}$ s

Newmark integration method is used to determine the dynamic response of the structure (60 time steps, $\Delta t = 10^{-5} s$, FE code PAK, Kojić *et al.* 1998a,b).

Detailed analysis of this example can be performed using the software (see web – Examples and Software).

References

Bathe, K.J. (1996). *Finite Element Procedures*, Prentice-Hall, Englewood Cliffs, NJ.

Bolotin, V.V. *et al.* (1968). *Stiffness – Durability – Vibrations* (in Russian), Izdateljstvo Masinostojenije, Moscow.

Kojić, M., Slavković, R., Živković, M. & Grujovic, N. (1998a). *PAK-Finite Element Program for Structural Analysis and Field Problems*, Faculty of Mechanical Engineering, University of Kragujevac, Serbia.

Kojić, M., Slavković, R., Živković, M. & Grujovic, N. (1998b). *Finite Element Method* (in Serbian), Faculty of Mechanical Engineering, University of Kragujevac, Serbia.

6

Introduction to Nonlinear Finite Element Analysis

In this chapter we present the basic relations of nonlinear finite element analysis of solids. The concept of linearization and forming an incremental-iterative scheme for nonlinear problems is introduced in a simple example. Then a linearized form of the principle of virtual work is derived as the basis for the finite element nonlinear analysis. The nonlinear finite element problems can in principle be divided into geometrically and materially nonlinear problems; we describe them with derivations of the corresponding matrices of 2D and 3D finite elements. Finally, we give specifics when large strain finite element formulation is used. A typical structural nonlinear example is presented, while other examples can be solved using the Software on the web.

6.1 Introduction

In an analysis of the deformation of solids or structures, a problem is considered nonlinear when the displacements due to mechanical action are not linearly proportional to loads. Therefore, for a discretized system a relation of the form (4.1.2) is not applicable. Instead of this linear force–displacement relationship, an incremental solution approach is introduced in which the total loads are divided into a number of increments and solutions for the displacement increments are obtained successively. Hence, a system of equations

$$^{n}\mathbf{K}\Delta\mathbf{U} = \Delta\mathbf{F}^{ext}, \qquad {}^{n}K_{JK}\Delta U_{K} = \Delta F_{J}^{ext}, \qquad J, K = 1, 2, \ldots, N_{total} \qquad (6.1.1)$$

is formed, corresponding to the load step n (denoted by left upper index) in the incremental procedure. Here, $\Delta\mathbf{U}$ is the vector of nodal displacement increments, $^{n}\mathbf{K}$ is the system stiffness matrix corresponding to the current displacements $^n\mathbf{U}$; $\Delta\mathbf{F}^{ext}$ is the vector of external force increments, and N_{total} is the number of system degrees of freedom. Note that here the stiffness matrix $\mathbf{K}$ changes during deformation, therefore we have the dependence $\mathbf{K}(\mathbf{U})$, i.e. $K_{JK}(U_1\ U_2.\ldots.U_{N_{total}})$.

In order to illustrate the basic strategy in solving nonlinear problems, consider a simple structure shown in Fig. 6.1.1a. A bar is fixed at one end, while the other end can slide along the x-axis under the action of a force F. We assume that the material is linear elastic, so that the axial force F_a generated within the bar is $F_a = AE(1 - L/L_0)$, see (4.2.1), where A and E are the cross-sectional area and Young's modulus of material, $L_0\ (= R\sqrt{1+r^2},\ r = H/R)$ is the initial length and L is the current length of the bar. From the balance between the force F and the component of F_a along the x-axis (denoted here as F^{int}), with use of the geometry shown in the figure, the following relation between the force and the displacement u is obtained:

$$\overline{F} = \frac{F}{AE} = \overline{F}^{int} = \frac{F^{int}}{AE} = \left(1 - \frac{\overline{L}}{\overline{L}_0}\right)\frac{1 - u/R}{\overline{L}} =$$

$$= \left(1 - \frac{\sqrt{r^2 + (1-u/R)^2}}{\sqrt{r^2+1}}\right)\frac{1-u/R}{\sqrt{r^2+(1-u/R)^2}}, \quad \overline{L}_0 = L_0/R, \overline{L} = L/R \quad (6.1.2)$$

This dependence of the force on the displacement is shown in Fig. 6.1.1b, for $r = 0.5$ and for the displacement range $0 \leq u/R \leq 2$.

Assume now that the whole loading interval is divided into a number of steps. The external load (here it is the force F) changes from $^nF^{ext}$ to $^{n+1}F^{ext}$ within step 'n', producing a change of the internal force from $^nF^{int}$ to $^{n+1}F^{int}$. Assume that the equilibrium state at the beginning of a load step is determined, hence the equation $^nF^{int} = {^nF^{ext}}$ is satisfied. Further, we seek the increment of displacement Δu, corresponding to the force increment $\Delta F = {^{n+1}F} - {^nF}$, such that the equation of balance is satisfied,

$$^{n+1}F^{int} = {^{n+1}F^{ext}} \quad (6.1.3)$$

The internal force due to deformation of the bar is the nonlinear function of displacement u according to (6.1.2) so that a Taylor series of the first order can be written as

$$^{n+1}F^{int} \approx {^nF^{int}} + {^n\left(\frac{\partial F^{int}}{\partial u}\right)}\Delta u = {^nF^{int}} + {^nK\Delta u} \quad (6.1.4)$$

where $^nK = {^n(\partial F^{int}/\partial u)}$ is the structural stiffness evaluated for displacement ^nu. We now substitute this approximation for the internal force into (6.1.3) and obtain

$$^nK\Delta u^{(1)} = {^{n+1}F^{ext}} - {^nF^{int}} \quad (6.1.5)$$

where $\Delta u^{(1)}$ is the first approximation for the displacement increment Δu. The first approximation for the displacement $^{n+1}u^{(1)}$, the internal force $^{n+1}F^{int(1)}$ and the stiffness $^{n+1}K^{(1)}$, are

$$^{n+1}u^{(1)} = {^nu} + \Delta u^{(1)}, \quad {^{n+1}F^{int(1)}} = F^{int}\big|_{u={^{n+1}u^{(1)}}} \quad \text{and} \quad {^{n+1}K^{(1)}} = K\big|_{u={^{n+1}u^{(1)}}} \quad (6.1.6)$$

NONLINEAR FINITE ELEMENT ANALYSIS

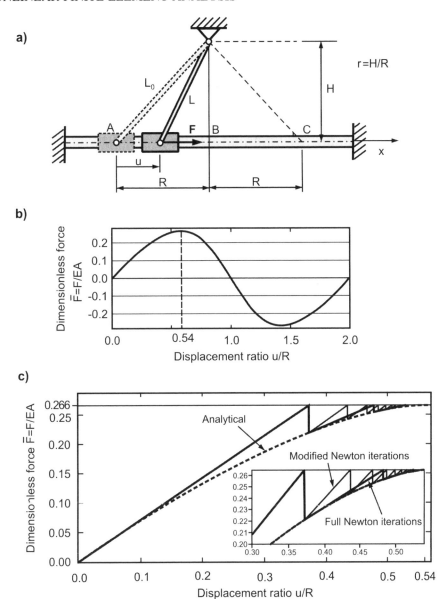

Fig. 6.1.1 A bar loaded by force F. (a) The bar end moves along x between points A and C, the initial length is L_0, the current length is L, and the displacement is u; (b) The force–displacement curve for $r = H/R = 0.5$; (c) Analytical relationship $\overline{F}(u/R)$ and iterative solutions to obtain the true displacement $u/R = 0.54$ corresponding to force $\overline{F} = 0.266$ (maximum force is $\overline{F} = 0.2675$ for displacement $u/R = 0.58$)

Substituting now the stiffness $^{n+1}K^{(1)}$ and internal force $^{n+1}F^{\text{int}(1)}$ into (6.1.3), the following equation is obtained:

$$^{n+1}K^{(1)} \Delta u^{(2)} = {}^{n+1}F^{ext} - {}^{n+1}F^{\text{int}(1)} \tag{6.1.7}$$

from which the increment $\Delta u^{(2)}$ can be calculated. Continuing this iterative procedure, we find the following equilibrium equation for the iteration 'i',

$$^{n+1}K^{(i-1)}\Delta u^{(i)} = {}^{n+1}F^{ext} - {}^{n+1}F^{int(i-1)} \tag{6.1.8}$$

and then the displacement $^{n+1}u^{(i)}$ is

$$^{n+1}u^{(i)} = {}^{n}u + \Delta u^{(1)} + \Delta u^{(2)} + \ldots + \Delta u^{(i)} \tag{6.1.9}$$

The iteration scheme (6.1.8) represents the so-called *Newton iteration* or *full Newton iteration* scheme. In the case of convergence, the 'unbalanced force' $\left({}^{n+1}F^{ext} - {}^{n+1}F^{int(i-1)}\right)$, as well as the increments $\Delta u^{(i)}$ decrease during the iterations.

Another approach in iterations is the so-called *modified Newton iteration* scheme. Then, instead of calculating the stiffness K at each iteration, K is evaluated at selected iterations, or only at the beginning of load steps (then $K = {}^{n}K$). In this approach, the computational time is saved since K is not calculated, but the convergence rate might be slow.

The stiffness in our example is

$$K = \frac{AE}{R} \frac{1}{\sqrt{1+r^2}} \left[1 + \frac{\overline{L}_0}{\overline{L}^2}\left(L'(u/R-1) - \overline{L}\right)\right] \tag{6.1.10}$$

where $L' = dL/du$. To demonstrate the convergence rate for the two iterative schemes, we used one load step – from the loading equal to zero to $\overline{F}^{ext} = 0.266$ (which corresponds to displacement $u/R = 0.54$). If the full Newton iterative scheme is used, the number of iterations was seven, while 48 iterations were necessary when the initial stiffness was used to achieve $\Delta u^{(i)} \leq 10^{-5}$ (see Fig. 6.1.1c). The example shows that the Newton iteration algorithm provides the *quadratic convergence rate*; this can be seen from the values for the displacement increments, unbalanced force and stiffness ($\overline{K} = KR/(AE)$, $R = 1$) given in Table 6.1.1. The displacement increments are also included when the modified Newton iterations are used. A significant difference in the convergence rate can be noticed for the two types of iteration schemes. Additional aspects of analysis of this example are given on the web (see Software – Chapter 6).

Table 6.1.1 Increments of displacement $\Delta u^{(i)}$, unbalanced force $\overline{F}^{ext} - \overline{F}^{(i-1)}$ and stiffness $\overline{K}^{(i-1)}$ during iterations

Full Newton				Modified Newton
Iteration i	$\Delta u^{(i)}/R$	$\overline{F}^{ext} - \overline{F}^{(i-1)}$	$\overline{K}^{(i-1)}$	$\Delta u^{(i)}/R$
1	3.71×10^{-1}	2.66×10^{-1}	7.15×10^{-1}	3.71×10^{-1}
2	1.10×10^{-1}	4.53×10^{-2}	4.12×10^{-1}	6.33×10^{-2}
3	4.26×10^{-2}	9.63×10^{-3}	2.26×10^{-1}	3.11×10^{-2}
4	1.40×10^{-2}	1.89×10^{-3}	1.35×10^{-1}	1.87×10^{-2}
5	2.21×10^{-3}	2.27×10^{-4}	1.03×10^{-1}	1.25×10^{-2}
6	5.93×10^{-5}	5.78×10^{-6}	9.74×10^{-2}	8.86×10^{-3}
7	4.31×10^{-8}	4.19×10^{-9}	9.72×10^{-2}	6.57×10^{-3}

6.2 Principle of virtual work and equilibrium equations in nonlinear incremental analysis

In Section 2.3 we presented the principle of virtual work as one of the most fundamental principles in mechanics. The basis of this principle relies on the equilibrium condition for stresses within a deformable body and at the boundary. Also, it was assumed that the displacements were small. Furthermore, the *principle is also applicable to any mechanical system*, composed of deformable and rigid bodies, with small and large displacements and deformation, under static or dynamic conditions. We here write this principle assuming discrete or continuum mechanical system in a form suitable for the incremental analysis.

According to the principle of virtual work we have that for any system configuration $^n\mathcal{B}$, the internal and external virtual works are equal,

$$\delta\,^n W^{int} = \delta\,^n W^{ext} \tag{6.2.1}$$

This fundamental relation will be further implemented.

6.2.1 Discrete system

Consider a discrete nonlinear system with N_{total} degrees of freedom. The system can be composed of rigid and deformable bodies, where deformable bodies can be discretized by a method such as the finite element method. Assume that we are solving the problem incrementally, as in the case of the simple example in Section 6.1. Hence, for an incremental step n we start with the known system configuration $^n\mathcal{B}$ and seek the unknown configuration $^{n+1}\mathcal{B}$. The principle of virtual work (6.2.1) for the configuration $^{n+1}\mathcal{B}$ can be written as

$$\delta\,^{n+1}\mathbf{U}^T\,^{n+1}\mathbf{F}^{int} = \delta\,^{n+1}\mathbf{U}^T\,^{n+1}\mathbf{F}^{ext} \tag{6.2.2}$$

where $^{n+1}\mathbf{U}$, $^{n+1}\mathbf{F}^{int}$ and $^{n+1}\mathbf{F}^{ext}$ are the displacement vector (translations and rotations), internal and external forces (forces and moments), respectively. The dimension of these vectors is N_{total}. Since the virtual displacement is an arbitrary vector (see Section 2.3), from (6.2.2) follows the equilibrium equation of the form (6.1.3),

$$^{n+1}\mathbf{F}^{int} = \,^{n+1}\mathbf{F}^{ext} \tag{6.2.3}$$

Note that in the case of dynamic analysis (see Chapter 5), the inertial forces must be included into the vector $^{n+1}\mathbf{F}^{int}$. Here, for simplicity of the presentation, we assume a quasi-static motion of the system and neglect the inertial effects.

Next, an incremental-iterative solution scheme can be formulated by linearization of equation (6.2.3), as in Section 6.1 for the one-degree-of-freedom example. Following the procedure for the derivation of (6.1.8), we obtain

$$^{n+1}\mathbf{K}^{(i-1)}\Delta\mathbf{U}^{(i)} = \,^{n+1}\mathbf{F}^{ext} - \,^{n+1}\mathbf{F}^{int(i-1)}, \quad \text{or} \quad ^{n+1}K_{jk}^{(i-1)}\Delta U_k^{(i)} = \,^{n+1}F_j^{ext} - \,^{n+1}F_j^{int(i-1)} \tag{6.2.4}$$

$$^{n+1}\mathbf{U}^{(i)} = \,^n\mathbf{U} + \Delta\mathbf{U}^{(1)} + \Delta\mathbf{U}^{(2)} + \ldots + \Delta\mathbf{U}^{(i)} \tag{6.2.5}$$

The stiffness matrix $^{n+1}\mathbf{K}^{(i-1)}$ is

$$^{n+1}\mathbf{K}^{(i-1)} = \,^{(n+1)}\left(\frac{\partial\mathbf{F}^{int}}{\partial\mathbf{U}}\right)^{(i-1)}, \quad \text{or} \quad ^{n+1}K_{jk}^{(i-1)} = \,^{(n+1)}\left(\frac{\partial F_j^{int}}{\partial U_k}\right)^{(i-1)} \tag{6.2.6}$$

6.2.2 Principle of virtual work for a continuum

We here derive the equation which represents the principle of virtual work for a deformable solid, assuming an incremental solution scheme. Consider first mechanical work of stresses due to virtual strains (Bathe 1996). If the configuration ${}^n\mathcal{B}$ is employed as the reference configuration, then the Green–Lagrange strain ${}^{n+1}_n\mathbf{E}^{GL}$ and its work-conjugate second Piola–Kirchhoff stress ${}^{n+1}_n\mathbf{S}$ for the end of the step (see (2.4.15) and (2.4.19)) are the strain and stress measures for expressing the internal virtual work. The internal virtual work per unit volume $\delta{}^{n+1}\overline{W}^{\text{int}}$ is

$$\delta{}^{n+1}\overline{W}^{\text{int}} = \delta{}^{n+1}_n\mathbf{E}^{GL} \cdot {}^{n+1}_n\mathbf{S} = \delta{}^{n+1}_nE^{GL}_{jk}\,{}^{n+1}_nS_{jk} = \left(\delta{}^{n+1}e_{jk} + \delta{}^{n+1}\eta_{jk}\right)\left({}^n\sigma_{jk} + \Delta\sigma_{jk}\right)$$
$$\approx \delta{}^{n+1}e_{jk}\,{}^n\sigma_{jk} + \delta{}^{n+1}e_{jk}\Delta\sigma_{jk} + \delta{}^{n+1}\eta_{jk}\,{}^n\sigma_{jk} \quad (6.2.7)$$

Here ${}^{n+1}e_{jk} \equiv {}^{n+1}_ne_{jk}$ are the small strains defined in (2.1.25) with displacements Δu_i and with respect to the reference configuration ${}^n\mathcal{B}$, ${}^{n+1}e_{ij} = 0.5\left(\partial(\Delta u_i)/\partial{}^n x_j + \partial(\Delta u_j)/\partial{}^n x_i\right)$; while ${}^{n+1}\eta_j \equiv {}^{n+1}_n\eta_j$ are the nonlinear (small) strains expressed in terms of displacement increments as

$$^{n+1}\eta_{ij} = \frac{1}{2}\frac{\partial(\Delta u_k)}{\partial{}^n x_i}\frac{\partial(\Delta u_k)}{\partial{}^n x_j}, \quad \text{sum on } k: \ k = 1, 2, 3 \quad (6.2.8)$$

Also, in (6.2.7) we used the fact that the Piola–Kirchhoff stress reduces to the Cauchy stress when two configurations ${}^n\mathcal{B}$ and ${}^{n+1}\mathcal{B}$ are close; and, further, the product $\delta{}^{n+1}\eta_{jk}\Delta\sigma_{jk}$ as the second-order small term is neglected.

Next, we use the constitutive relations for the stress increments $\Delta\sigma_{jk}$ as

$$\Delta\sigma_{jk} = {}^nC_{jkpq}\,{}^{n+1}e_{pq}, \quad \text{or} \quad \Delta\sigma_j = {}^nC_{jk}\,{}^{n+1}e_k \quad (6.2.9)$$

where ${}^nC_{jkpq}$ is the constitutive tensor evaluated at the start of the step. When the one-index notation for stresses and strains is employed (see (2.1.13) and (2.1.23)), the constitutive matrix ${}^nC_{jk}$ stands for the constitutive tensor ${}^nC_{jkpq}$. Substituting (6.2.9) into (6.2.7), a linearized form of the internal virtual work for the configuration ${}^{n+1}\mathcal{B}$ is obtained,

$$\delta{}^{n+1}\overline{W}^{\text{int}} \approx \delta{}^{n+1}e_j\,{}^n\sigma_j + \delta{}^{n+1}e_j\,{}^nC_{jk}\,{}^{n+1}e_k + \delta{}^{n+1}\eta_j\,{}^n\sigma_j \quad (6.2.10)$$

Here and in further presentation, the one-index notation is used as a more convenient form (the one-index form ${}^{n+1}\eta_j$ follows the notation in (2.1.23)).

Finally, substituting (6.2.10) into (6.2.1) a *linearized form of the principle of virtual work* for a deformable solid is obtained,

$$\int_{{}^nV}\left[\delta{}^{n+1}e_j\,{}^n\sigma_j + \delta{}^{n+1}e_j\,{}^nC_{jk}\,{}^{n+1}e_k + \delta{}^{n+1}\eta_j\,{}^n\sigma_j\right]dV = \delta{}^{n+1}W^{\text{ext}} \quad (6.2.11)$$

with the integration over the volume ${}^nV$. When an iterative solution scheme is employed, as in (6.1.8) or (6.2.4), the reference configuration for the iteration 'i' becomes the last known configuration ${}^{n+1}\mathcal{B}^{(i-1)}$ instead of ${}^n\mathcal{B}$, and the integration is performed over the volume ${}^{n+1}V^{(i-1)}$.

6.2.3 Finite element model

Equilibrium equations for a finite element can be obtained by application of (6.2.11) and with the use of the interpolation for the displacement field within the element. We assume an isoparametric interpolation (4.1.1) and then express the strains in terms of the nodal displacement increments $\Delta \mathbf{U}$. For linear strains and variations of linear strains we obtain (see, for example, expressions (4.2.8), (4.3.7), (4.4.3) and (4.5.9))

$${}^{n+1}_{n}\mathbf{e} = {}^{n}_{n}\mathbf{B}_{L}\Delta \mathbf{U} \quad \text{and} \quad \delta{}^{n+1}_{n}\mathbf{e} = {}^{n}_{n}\mathbf{B}_{L}\delta{}^{n+1}\mathbf{U} \qquad (6.2.12)$$

where the ${}^{n}_{n}\mathbf{B}_{L}$ is the *linear strain–displacement matrix* and the left indices 'n' mean that the matrix corresponds to the configuration ${}^{n}\mathcal{B}$ with derivatives of the interpolation functions N_{K} with respect to the coordinates ${}^{n}x_{i}$ (i.e. $\partial N_{K}/\partial {}^{n}x_{i}$); and vector $\delta{}^{n+1}\mathbf{U} = \delta(\Delta \mathbf{U})$ is the vector of variation of nodal displacements. It can be shown that, with use of the definition (6.2.8) of ${}^{n+1}_{n}\eta_{ij}$, the product $\delta{}^{n+1}_{n}\eta_{j}{}^{n}\sigma_{j}$ can be written in a matrix form as

$$\delta{}^{n+1}_{n}\eta_{k}{}^{n}\sigma_{k} = \delta{}^{n+1}\mathbf{U}^{T}{}^{n}_{n}\mathbf{B}_{NL}^{T}{}^{n}\tilde{\boldsymbol{\sigma}}{}^{n}_{n}\mathbf{B}_{NL}\Delta \mathbf{U} \qquad (6.2.13)$$

where ${}^{n}_{n}\mathbf{B}_{NL}$ is *the nonlinear strain–displacement matrix*, containing derivatives of the interpolation functions with respect to the coordinates ${}^{n}x_{i}$, and ${}^{n}\tilde{\boldsymbol{\sigma}}$ is the matrix with stresses components ${}^{n}\sigma_{j}$ (details of derivation of (6.2.13) are given on the web – Theory, Chapter 6).

Substituting (6.2.12) and (6.2.13) into (6.2.11) and, taking that variations $\delta{}^{n+1}\mathbf{U}$ of the nodal displacement vector are arbitrary, the equilibrium equation of a finite element is obtained as

$$({}^{n}\mathbf{K}_{L} + {}^{n}\mathbf{K}_{NL})\Delta \mathbf{U} = {}^{n+1}\mathbf{F}^{ext} - {}^{n}\mathbf{F}^{int} \qquad (6.2.14)$$

Here ${}^{n}\mathbf{K}_{L}$ and ${}^{n}\mathbf{K}_{NL}$ are linear and nonlinear (geometrical nonlinearity) element stiffness matrices,

$${}^{n}\mathbf{K}_{L} = \int_{{}^{n}V} {}^{n}_{n}\mathbf{B}_{L}^{T}{}^{n}\mathbf{C}{}^{n}_{n}\mathbf{B}_{L}dV \quad \text{and} \quad {}^{n}\mathbf{K}_{NL} = \int_{{}^{n}V} {}^{n}_{n}\mathbf{B}_{NL}^{T}{}^{n}\tilde{\boldsymbol{\sigma}}{}^{n}_{n}\mathbf{B}_{NL}dV \qquad (6.2.15)$$

and ${}^{n}\mathbf{F}^{int}$ is the vector of element internal forces (due to stresses within the material),

$${}^{n}\mathbf{F}^{int} = \int_{{}^{n}V} {}^{n}_{n}\mathbf{B}_{L}^{T}{}^{n}\boldsymbol{\sigma}dV \qquad (6.2.16)$$

The external forces are the nodal forces from the action of surrounding finite elements or the external structural loadings (see Sections 4.2.and 4.3). The constitutive matrix ${}^{n}\mathbf{C}$ is

$${}^{n}\mathbf{C} = \frac{\partial {}^{n}\boldsymbol{\sigma}}{\partial {}^{n}\mathbf{e}}, \quad \text{or} \quad {}^{n}C_{ij} = \frac{\partial {}^{n}\sigma_{i}}{\partial {}^{n}e_{j}} \qquad (6.2.17)$$

A usual assemblage procedure is performed (see Section 4.2.2) to obtain the incremental system of algebraic equations, which is then solved for the nodal displacement increments $(\Delta \mathbf{U})_{sys}$ for the entire material system. An iterative scheme of the form (6.2.4) can be formed to solve for the increments $\Delta \mathbf{U}_{sys}^{(i)}$. Therefore, the incremental-iterative equilibrium equations for a finite element are now

$$\left(^{n+1}\mathbf{K}_L + {^{n+1}\mathbf{K}_{NL}}\right)^{(i-1)} \Delta \mathbf{U}^{(i)} = {^{n+1}\mathbf{F}^{ext}} - {^{n+1}\mathbf{F}^{int(i-1)}} \quad (6.2.18)$$

where the last known configuration $^{n+1}\mathcal{B}^{(i-1)}$ is used as the reference configuration; also the constitutive matrix corresponds to the last known values for stresses and strains, i.e. $^{n+1}\mathbf{C}^{(i-1)}$ is used,

$$^{n+1}\mathbf{C}^{(i-1)} = \frac{\partial\, ^{n+1}\boldsymbol{\sigma}^{(i-1)}}{\partial\, ^{n+1}\mathbf{e}^{(i-1)}}, \quad \text{or} \quad ^{n+1}C_{ij}^{(i-1)} = \frac{\partial\, ^{n+1}\sigma_i^{(i-1)}}{\partial\, ^{n+1}e_j^{(i-1)}} \quad (6.2.19)$$

Considering the nonlinearities of the matrices in (6.2.18), two types of nonlinearities can be distinguished: geometric and material. The *geometric nonlinearity* comes from large displacements during deformation. This nonlinearity enters into the evaluation of the derivatives of the interpolation functions $\left(\partial N_K / \left(\partial\, ^{n+1}x_j\right)\right)^{(i-1)}$ for the iteration 'i'), and also into the volume of the finite elements. We note that the external loading may also depend on displacements, when the external force vector $^{n+1}\mathbf{F}^{ext}$ changes within load steps and within equilibrium iterations. The *material nonlinearity* is due to nonlinearity of the constitutive law which leads to changes of the constitutive matrix $^{n+1}\mathbf{C}^{(i-1)}$ during deformation history of the material. A problem is considered *geometrically nonlinear only* if the constitutive relations are linear and displacements are large. A problem is *materially nonlinear only* (MNO) when the constitutive law is nonlinear and the displacements are small. In the case of an MNO problem, the equilibrium equation (6.2.18) reduces to

$$\left(^{n+1}\mathbf{K}_L\right)^{(i-1)} \Delta \mathbf{U}^{(i)} = {^{n+1}\mathbf{F}^{ext}} - {^{n+1}\mathbf{F}^{int(i-1)}} \quad (6.2.20)$$

where

$$\left(^{n+1}\mathbf{K}_L\right)^{(i-1)} = \int_V \mathbf{B}_L^T\, {^{n+1}\mathbf{C}^{(i-1)}}\, \mathbf{B}_L\, dV, \quad \left(^{n+1}\mathbf{F}^{int}\right)^{(i-1)} = \int_V \mathbf{B}_L^T\, {^{n+1}\boldsymbol{\sigma}^{(i-1)}}\, dV \quad (6.2.21)$$

In summarizing the above presentation of the FE nonlinear analysis we emphasize the following important facts. First, in order to obtain the internal nodal force vector at the end of step, $\left(^{n+1}\mathbf{F}^{int}\right)^{(i-1)}$, the stresses $^{n+1}\boldsymbol{\sigma}^{(i-1)}$ must be evaluated. In a displacement-based FE formulation, used in this book, it is assumed that the strains $^{n+1}\mathbf{e}^{(i-1)}$ are known from the last known displacements $^{n+1}\mathbf{u}^{(i-1)}$. Then, for given strains, the stresses depend on the material properties expressed by the constitutive relationships $\boldsymbol{\sigma}(\mathbf{e})$. Therefore, the evaluation of the $^{n+1}\boldsymbol{\sigma}^{(i-1)}$ in an incremental analysis requires integration of the constitutive law $\boldsymbol{\sigma}(\mathbf{e})$ within the step, according to

$$^{n+1}\boldsymbol{\sigma}^{(i-1)} = {^n\boldsymbol{\sigma}} + \int_{^n\mathbf{e}}^{^{n+1}\mathbf{e}^{(i-1)}} \boldsymbol{\sigma}(\mathbf{e})\, d\mathbf{e} \quad (6.2.22)$$

where the integral represents the *stress integration* in the incremental step.

The second fact is that for a nonlinear constitutive law, it is necessary to calculate the tangent constitutive matrix $^{n+1}\mathbf{C}^{(i-1)}$. If this matrix is obtained by the differentiation of the governing relations used within the stress integration algorithm, the matrix is called the *consistent tangent constitutive matrix*.

Finally, we emphasize that *the accuracy of the solution depends on the algorithm for stress integration, and not on the matrix* $^{n+1}\mathbf{C}^{(i-1)}$. This follows from the equation (6.2.3) which has to be satisfied through the equilibrium iteration scheme, and from the expression (6.2.21) (see also (6.2.16)) for the internal nodal forces. On the other hand, *the matrix* $^{n+1}\mathbf{C}^{(i-1)}$ *affects the convergence rate only*, and not the solution accuracy. The best convergence rate is obtained when a consistent tangent matrix is computed (Kojić & Bathe 2005).

At the end of this section we give the expressions for the nonlinear strain–displacement matrix $\mathbf{B}_{NL}$ and the stress matrix $\tilde{\sigma}$ in (6.2.15) for a 3D isoparametric element (the linear strain–displacement matrix is given in (4.3.7)). For simplicity we omit indices for the reference configuration and incremental step. The matrix $\mathbf{B}_{NL}$ is (Bathe 1996, Kojić & Bathe 2005)

$$\mathbf{B}_{NL} = \begin{bmatrix} \hat{\mathbf{B}}_{NL} & \hat{\mathbf{0}} & \hat{\mathbf{0}} \\ \hat{\mathbf{0}} & \hat{\mathbf{B}}_{NL} & \hat{\mathbf{0}} \\ \hat{\mathbf{0}} & \hat{\mathbf{0}} & \hat{\mathbf{B}}_{NL} \end{bmatrix} \qquad (6.2.23)$$

where

$$\hat{\mathbf{B}}_{NL} = \begin{bmatrix} N_{1,1} & 0 & 0 & N_{2,1} & . & . & N_{N,1} \\ N_{1,2} & 0 & 0 & N_{2,2} & . & . & N_{N,2} \\ N_{1,3} & 0 & 0 & N_{2,3} & . & . & N_{N,3} \end{bmatrix}, \quad \hat{\mathbf{0}} = \begin{bmatrix} 0 \\ 0 \\ 0 \end{bmatrix} \qquad (6.2.24)$$

The stress matrix is

$$\tilde{\sigma} = \begin{bmatrix} \sigma & 0 & 0 \\ 0 & \sigma & 0 \\ 0 & 0 & \sigma \end{bmatrix} \qquad (6.2.25)$$

where σ is the 3 × 3 matrix given in (2.1.3), and $\mathbf{0}$ is the 3 × 3 zero matrix. The matrix $\mathbf{B}_{NL}$ and the stress matrix $\tilde{\sigma}$ for other elements can be found elsewhere (e.g. Bathe 1996).

Note that the above equilibrium equations (6.2.18) can be extended to include inertial and resistance effects (see Section 5.2). Then, the inertial and resistance forces $^{n+1}\mathbf{F}^{in} = -\mathbf{M}\,^{n+1}\ddot{\mathbf{U}}$ and $^{n+1}\mathbf{F}^{w} = -\left(^{n+1}\mathbf{B}^{w(i-1)}\right)\,^{n+1}\dot{\mathbf{U}}$ must be added on the right-hand side; $\mathbf{M}$ is the constant mass matrix, $^{n+1}\mathbf{B}^{w(i-1)}$ is the resistance matrix, $^{n+1}\ddot{\mathbf{U}}$ and $^{n+1}\dot{\mathbf{U}}$ are the nodal acceleration and velocity vectors. The integration of differential equations can be performed as for linear problems (Section 5.3); details are given on web – Theory, Chapter 6.

6.2.4 Finite element model with logarithmic strains

The logarithmic strains are often used in large strain analysis in engineering because the experiments are usually performed employing this strain measure. The logarithmic strain is the 'true strain' because it can be considered, in uniaxial extension, as a sum of the ratios of the increments of length with respect to the current length. Namely, if we extend a material sample from the initial length ^0L to the final length ^NL by N increments ΔL_n, $n = 1, 2, \ldots, N$, the strain increments can be calculated as $\Delta^n e = \Delta L_n/^n L$, where $^n L$ is the length at the n-th step; and the strain $^N e$ is the sum of the strain increments. Hence, we have

$$^N e = \sum_{n=1}^{N} \frac{\Delta L_n}{^n L}, \quad \lim_{N \to \infty} \left(^N e\right) = \ln \frac{^N L}{^0 L} = \ln \left(_0^N \lambda\right) = {}^N E^H \quad (6.2.26)$$

where $_0^N \lambda$ is the stretch and $^N E^H$ is the Hencky (logarithmic) strain (see Section 2.4.1). Note that $dE^H = de$, i.e. increment of the logarithmic strain is equal to the increment of small strain in the current configuration.

If logarithmic strains are used in general deformation conditions, it is necessary for the current incremental step 'n' to perform the following computational steps (see Section 2.4.1):

1. Determine deformation gradient $_0^n\mathbf{F}$ and the left Cauchy–Green deformation tensor $_0^n\mathbf{B} = {}_0^n\mathbf{F}\,_0^n\mathbf{F}^T$.

2. Calculate the principal stretches $_0^n\lambda_1, {}_0^n\lambda_2, {}_0^n\lambda_3$ and the principal unit vectors of the left basis $^n\bar{\mathbf{p}}_1, {}^n\bar{\mathbf{p}}_2, {}^n\bar{\mathbf{p}}_3$; calculate logarithmic strains $_0^n E_j^H = \ln \left(_0^n \lambda_j\right)$, $j = 1, 2, 3$.

3. Calculate principal stresses $\bar{\sigma}_j$, $j = 1, 2, 3$, using the corresponding constitutive law. Transform the principal stresses to the common coordinate system x_i using the transformation (2.1.14).

It is assumed here that the material is isotropic elastic.

The logarithmic strains are used in modeling biological tissue (Chapter 11). Also we employ the logarithmic strains for calculation of strains of material fibers, such as in the modeling of muscle (see Chapter 12).

Logarithmic strains are particularly used for modeling inelastic large strain deformations, as in the case of plastic deformation of metals in technological operations. Then, a fictitious elastic configuration is determined first and principal logarithmic elastic strains are calculated prior to the stress integration within the load step (see Kojić & Bathe 2005 for details).

6.3 Examples

Example 6.3-1. Large displacement of a channel section cantilever with local buckling
A cantilever channel-section beam is loaded by a tip force F at the free end, as shown in Fig. E6.3-1a. Due to the slenderness of the structure, the force causes regions of local buckling. The four-node shell elements (see Section 4.5.2) are used to model the cantilever.

The solution is determined using prescribed vertical displacement w of the point where the load is acting. Then, the force F necessary to cause this displacement is calculated. Deformed configuration of the cantilever, with the effective stress (von Mises stress, see web – Theory, Chapter 2), is shown in Fig. E6.3-1b, while the dependence of the force on the deflection w is given in Fig. E6.3-1c. Detailed analysis of this example can be performed using the Software (see web, Software).

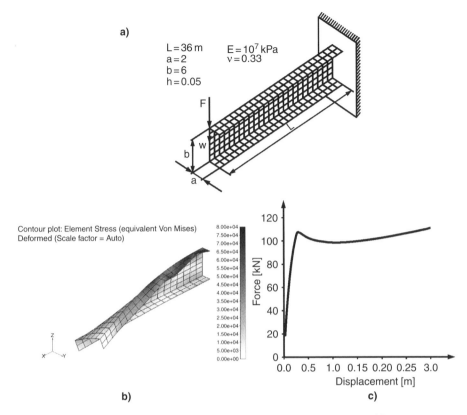

Fig. E6.3-1 Bending of a channel-section beam with local buckling. (a) Geometrical and material data; (b) Deformed configuration with local buckling regions and field of von Mises stress; (c) Dependence of vertical force F on the vertical displacement w of the point under the force

References

Bathe, K.J. (1996). *Finite Element Procedures*, Prentice-Hall, Englewood Cliffs, NJ.
Kojić, M. & Bathe, K.J. (2005). *Inelastic Analysis of Solids and Structures*, Springer-Verlag, Berlin.

7

Finite Element Modeling of Field Problems

In this chapter we present a finite element formulation for modeling field problems. General three-dimensional conditions are assumed and the isoparametric interpolation of physical fields is used. The FE formulation relies on the Galerkin method used to transform the governing partial differential equations of Chapter 3 into the FE algebraic equations. This general method is first implemented to linear problems of heat conduction and diffusion, and then to laminar incompressible viscous flow with and without heat and mass transfer. Next, the FE equations are derived in the case of fluid flow with large changes of the fluid domain; the arbitrary Lagrangian–Eulerian (ALE) formulation is used. Also, the FE modeling of the solid–fluid interaction is presented, according to a weak coupling approach. Finally, the FE balance equations are derived for fluid flow through porous deformable media. Each section is followed by typical example solutions.

7.1 Introduction

In many engineering and bioengineering problems the task is to determine the field of physical quantities, such as temperature, concentration, fluid pressure and velocities. Here we give some general considerations about solving field problems and then present the Galerkin method.

7.1.1 General considerations

In Chapter 3 we presented the governing equations of heat conduction, diffusion, laminar fluid flow, and flow through porous deformable media. They all rely on the physical law of balance of energy, mass and linear momentum, and have the form of partial differential equations with time and spatial derivatives.

Analytical methods were developed in the past as the first methods to solve the governing equations, summarized in textbooks (e.g. Incropera & DeWitt 1996, Munson et al. 1998, Mills 1999, 2001). In these methods, an analytical form of solutions is obtained so that the governing equations are satisfied within the domain, as well as for initial and boundary conditions. Usually, the analytical functions of time and space are such that the governing equations and boundary conditions are satisfied exactly. There are also approximate analytical solutions in which the functions are represented by series in which the higher order terms can be neglected.

The main limitation of the analytical methods is that they are only applicable to special, noncomplex geometrical shapes, with material parameters independent of the solution variables and other simplified conditions. They are not applicable to general, complex nonlinear problems, such as, for example airflow in alveolated structures with large motions of the boundaries, or blood flow in microvessels.

When modeling field problems by the finite element method (FEM) we seek the solution which satisfies the governing equations within the finite elements (subdomains) in a weighted sense (details are given in the next section). Also, the physical field within each element is approximated and expressed in terms of the nodal values, as in the case of solids where the displacements of points within a finite element are approximated from the nodal point displacements. The FE balance equations are commonly derived using the Galerkin method which is presented next.

7.1.2 The Galerkin method

To introduce the Galerkin method, consider the differential equation corresponding to heat transfer or diffusion (without convection) derived in Chapter 3:

$$-c\frac{\partial \phi}{\partial t} + \frac{\partial}{\partial x_i}\left(D\frac{\partial \phi}{\partial x_i}\right) + f^V = 0 \qquad (7.1.1)$$

where $\phi(x_i, t)$ is the field variable; c and D are the material coefficients, dependent in general on the coordinates x_i and on ϕ; and $f^V(x_i, t)$ is a source term.

The basic idea of weighted methods is to multiply the differential equation by a function of coordinates and impose the condition that the equation is satisfied over a selected domain in a weighted sense,

$$\int_V \psi(x_1, x_2, x_3)\left[-c\frac{\partial \phi}{\partial t} + \frac{\partial}{\partial x_i}\left(D\frac{\partial \phi}{\partial x_i}\right) + f^V\right]dV = 0 \qquad (7.1.2)$$

where $\psi(x_i)$ is a weighted function, and V is a selected domain. The equation (7.1.2) represents the so-called *weak form* of the differential equation since the differential equation (7.1.1) is not necessarily satisfied at each point of the domain.

FINITE ELEMENT MODELING OF FIELD PROBLEMS

In the Galerkin method we use here the domain V of an isoparamertric (3D) finite element and interpolation functions $N_K(r,s,t)$ as the weighted functions; see Section 4.3 for the description of N_K (Huebner 1975). The capital letters are used for indices (as N_K) corresponding to node numbering in order to make a distinction from lower case indices which refer to a Cartesian system x_i. Therefore, N weak form equations for a finite element are obtained,

$$\int_V N_K \left[-c\frac{\partial \phi}{\partial t} + \frac{\partial}{\partial x_i}\left(D\frac{\partial \phi}{\partial x_i}\right) + f^V \right] dV = 0 \qquad K = 1, 2, \ldots, N \qquad (7.1.3)$$

where N is the number of nodes.

In forming the equations of balance for a finite element, the key step is to transform the volume integral into a form suitable for application. Namely, the Gauss theorem (e.g. Fung 1965; see also Section 1.4) can be applied to the second term under the volume integral to obtain

$$\int_V N_K \frac{\partial}{\partial x_i}\left(D\frac{\partial \phi}{\partial x_i}\right) dV = \int_S N_K D \frac{\partial \phi}{\partial x_i} n_i dS - \int_V D N_{K,i} \phi_{,i} dV \qquad (7.1.4)$$

where S is the element surface; notation $_{,i} \equiv \partial/\partial x_i$ is used for shorter writing. The term $D\phi_{,i} n_i$ represents the flux q_S of ϕ through the element surface, therefore the surface integral can be written as

$$F_K^S = \int_S N_K D \frac{\partial \phi}{\partial x_i} n_i dS = \int_S N_K q_S dS \qquad (7.1.5)$$

where $F_K^S(t)$ is the surface flux corresponding to the node 'K'.

Next, the interpolation of the variable ϕ within the element can be used:

$$\phi = N_K \Phi^K = N_1 \Phi^1 + N_2 \Phi^2 + \ldots + N_N \Phi^N \qquad (7.1.6)$$

where Φ^K are the nodal values of ϕ. Now, substitute (7.1.6) and (7.1.4) into (7.1.3) and also use (7.1.5), to obtain

$$\mathbf{M}\dot{\boldsymbol{\Phi}} + \mathbf{K}\boldsymbol{\Phi} = \mathbf{F}^S + \mathbf{F}^V, \quad \text{or}$$
$$M_{KJ}\dot{\Phi}^J + K_{KJ}\Phi^J = F_K^S + F_K^V, \quad K, J = 1, 2, \ldots, N \qquad (7.1.7)$$

Here $\boldsymbol{\Phi}$ is the vector of nodal values Φ^K, $\dot{\boldsymbol{\Phi}}$ is the vector of time derivatives at nodal points, and the matrices and vectors are:

$$M_{KJ} = \int_V cN_K N_J dV, \quad K_{KJ} = \int_V DN_{K,i} N_{J,i} dV,$$
$$F_K^S = \int_S N_K q_S dS, \quad F_K^V = \int_S N_K f^V dS \qquad (7.1.8)$$

The system of equations (7.1.7) represents the equations of balance for a finite element. Assemblage of the element equations is then performed as in the case of a solid (Section 4.2) and the system of equations having the form (7.1.7) is obtained. Appropriate boundary conditions must be implemented prior to solving the equations of the entire system. Note that the surface fluxes F_K^S cancel over the internal surfaces of the finite elements and the contribution to the system surface vector $\mathbf{F}^S$ comes from the flux on the surface of the entire domain.

The incremental form for time integration of (7.1.7) is

$$\left(\frac{1}{\Delta t}\mathbf{M}+\mathbf{K}\right){}^{n+1}\mathbf{\Phi} = {}^{n+1}\mathbf{F}^S + {}^{n+1}\mathbf{F}^V + \frac{1}{\Delta t}\mathbf{M}^n\mathbf{\Phi} \qquad (7.1.9)$$

where Δt is the time step, and the upper left indices n and $n+1$ denote values at the start and end of the n-th time step.

When the material parameters depend on the variable ϕ and/or the domain changes significantly, an iterative solution scheme must be developed. As in Section 6.2 (see equations (6.2.4) and (6.2.5)), we write the solution in a form

$$^{n+1}\mathbf{\Phi}^{(i)} = {}^n\mathbf{\Phi} + \Delta\mathbf{\Phi}^{(1)} + \Delta\mathbf{\Phi}^{(2)} + \ldots + \Delta\mathbf{\Phi}^{(i)} \qquad (7.1.10)$$

which corresponds to the iteration 'i'. The incremental-iterative equation for the step n and iteration i is

$$^{n+1}\left(\frac{1}{\Delta t}\mathbf{M}+\mathbf{K}\right)^{(i-1)}\Delta\mathbf{\Phi}^{(i)} = {}^{n+1}\mathbf{F}^{S(i-1)} + {}^{n+1}\mathbf{F}^{V(i-1)} - {}^{n+1}\mathbf{K}^{(i-1)}\,{}^{n+1}\mathbf{\Phi}^{(i-1)} - \frac{1}{\Delta t}\,{}^{n+1}\mathbf{M}^{(i-1)}\left({}^{n+1}\mathbf{\Phi}^{(i-1)} - {}^n\mathbf{\Phi}\right) \qquad (7.1.11)$$

where the matrices and vectors are evaluated using the last known values $^{n+1}\mathbf{\Phi}^{(i-1)}$. The element equations are assembled and solved for the increments $\Delta\mathbf{\Phi}^{(i)}$. The iterations continue until a selected accuracy criterion for $\|\Delta\mathbf{\Phi}^{(i)}\|$ is satisfied.

The Galerkin method is further implemented in heat transfer, diffusion, fluid flow and flow through porous deformable media.

7.2 Heat conduction

7.2.1 The finite element equations

The governing equation for heat conduction is given in (3.1.5),

$$-\rho c \frac{\partial T}{\partial t} + \frac{\partial}{\partial x_i}\left(k_i \frac{\partial T}{\partial x_i}\right) + q_V = 0 \qquad (7.2.1)$$

This equation has the same form as (7.1.1), hence the FE balance equation (7.1.7) becomes:

$$\mathbf{M}\dot{\mathbf{T}} + \mathbf{K}^k \mathbf{T} = \mathbf{Q}^S + \mathbf{Q}^V \qquad (7.2.2)$$

FINITE ELEMENT MODELING OF FIELD PROBLEMS

where $\mathbf{T}$ is the vector of nodal temperatures, while $\mathbf{Q}^S$ and $\mathbf{Q}^V$ are the surface and volume nodal fluxes, respectively. The matrix $\mathbf{M}$ is given in (7.1.8), and the conductivity matrix $\mathbf{K}^k$ is

$$K_{IJ}^k = \int_V \left(\sum_{j=1}^{3} k_j N_{I,j} N_{J,j} \right) dV = \int_V \left(k_1 N_{I,1} N_{J,1} + k_2 N_{I,2} N_{J,2} + k_2 N_{I,3} N_{J,3} \right) dV \quad (7.2.3)$$

The heat flux through the element surface consists, in general, of three parts:

(a) given, prescribed flux q_n;

(b) flux due to convection, $q_c = h_c(T_{surr} - T_S)$, where $h_c(W/m^2\ K)$ is the convective heat transfer coefficient, T_{surr} is the surrounding temperature, and T_S is the body temperature at the surface; and

(c) flux due to radiation $q_r = h_r(T_{rad} - T_S)$, where $h_r(W/m^2 K)$ is the heat radiation coefficient, and T_{rad} is the radiation temperature.

Then, the nodal heat flux for the node 'I' is obtained as

$$Q_I^n = \int_{S^e} N_I q_n dS, \quad Q_I^c = \int_S N_I h_c (T_{surr} - T_S)\, dS = -K_{IJ}^c T^J + Q_I^{surr},$$

$$Q_I^{rad} = \int_S N_I h_r (T_{rad} - T_S)\, dS = -K_{IJ}^{rad} T^J + Q_I^{rad} \quad (7.2.4)$$

where the expressions for K_{IJ}^c, Q_I^{surr}, K_{IJ}^{rad}, and Q_I^{rad} follow from the surface integrals.

Therefore, for the elements having the surface on the boundary, the equation of balance with the terms (7.2.4) and (7.2.2) becomes:

$$\mathbf{M}\dot{\mathbf{T}} + \mathbf{K}\mathbf{T} = \mathbf{Q},$$
$$\mathbf{K} = \mathbf{K}^k + \mathbf{K}^c + \mathbf{K}^{rad}, \quad \mathbf{Q} - \mathbf{Q}^n + \mathbf{Q}^{surr} + \mathbf{Q}^{rad} + \mathbf{Q}^V \quad (7.2.5)$$

Time integration of the system of differential equations is performed according to (7.1.9), and the iterative scheme (7.1.11) must be used when the material parameters depend on temperature and/or the heat conduction domain changes over time.

7.2.2 Examples

Finite element solutions for two examples are presented here, for which the analytical solutions are given in Section 3.1.2. Other examples are available on the web – Section 7.2.

Example 7.2-1. Steady heat conduction laterally through a long column
The definition of this example is given in Section 3.1.2. Here, we show the FE solution. Due to symmetry conditions with respect to the plane $y = a/2$, a half of the cross-section is discretized, Fig. E7.2-1b. The boundary condition that the flux is equal to zero through the symmetry plane is imposed.

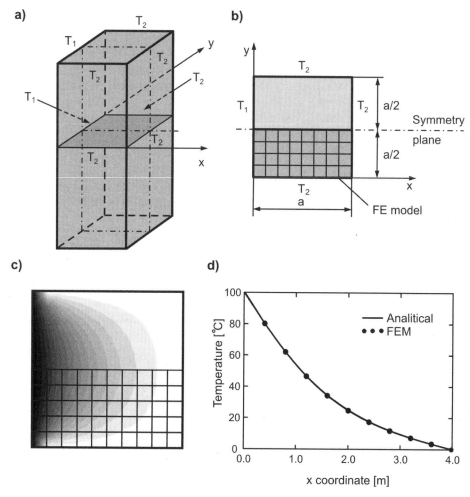

Fig. E7.2-1 Steady heat conduction laterally through a long column. (a) Geometry of the column and boundary conditions; (b) Column cross-section and a sketch of FE mesh; (c) Temperature field for data: $a = 4\,m$, $T_1 = 100\,°C$, $T_2 = 0\,°C$, $k = 1\,W/mK$; (d) Temperature distribution along the symmetry plane (analytical and FE solutions)

Temperature field and distribution of temperature along the symmetry plane are shown in Figs. E7.2-1c,d. A study of accuracy of the FE solution in terms of the mesh density can be performed using Software on the web.

Example 7.2-2. Unsteady heat conduction through a semi-infinite medium
This example is defined in Section 3.1.2 (Example 3.1-2). Here, we find the temperature field by using an FE model. We use 3D finite elements as shown in Fig. E7.2-2a, with the length model L.

FINITE ELEMENT MODELING OF FIELD PROBLEMS

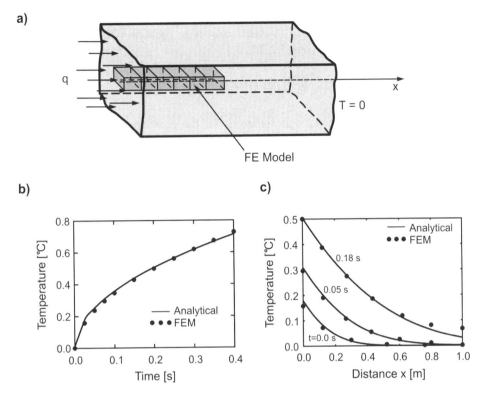

Fig. E7.2-2 Unsteady heat conduction through semi-infinite solid. (a) Schematic representation of the solid and FE mesh (3D finite elements) to model 1D heat conduction; (b) Increase of temperature with time at the surface where constant flux is given; (c) Temperature profiles along the length L for several times. Data: flux $q = 1\,W/m^2$, heat conduction coefficient $k = 1\,W/mK$, mass density $\rho = 1\,kg/m^3$, specific heat $c = 1\,J/kgK$, initial temperature $T_{initial} = 0\,°C$

Temperature at the boundary and distribution along the length L are shown in Figs. E7.2-2b,c. It can be seen that the FE solutions agree well with the analytical solutions. An insight into effects of the model length, heat conduction coefficient and lateral dimensions of the FE model, can be obtained using Software on the web.

7.3 Diffusion

7.3.1 The finite element equations

Consider the governing equation (3.2.8)

$$-\frac{\partial c_j}{\partial t} + \frac{\partial}{\partial x_i}\left(D_j \frac{\partial c_j}{\partial x_i}\right) + c_j^V = 0 \qquad (7.3.1)$$

for a constituent 'j', where c_j stands for the volumetric concentration and c_j^V is the source term. This equation is analogous to the heat conduction equation (7.2.1) and the balance equation for a finite element is (see (7.2.2)):

$$\mathbf{M}_j \dot{\mathbf{C}}_j + \mathbf{K}_j^d \mathbf{C}_j = \mathbf{Q}_j^S + \mathbf{Q}_j^V \quad \text{no sum on } j \quad (7.3.2)$$

where $\mathbf{C}_j$ is the vector of nodal concentrations for the constituent 'j'; the vector $\mathbf{Q}_j^V$ is defined as the vector $\mathbf{F}^V$ in (7.1.8) where c_j^V should be used for f^V. The matrix $\mathbf{M}_j$ is given in (7.1.8) with $c = 1$, while the matrix $\mathbf{K}^d$ is

$$\left(K_j^d\right)_{IJ} = \int_V D_j N_{I,k} N_{J,k} dV = \int_V D_j \left(N_{I,1} N_{J,1} + N_{I,2} N_{J,2} + N_{I,3} N_{J,3}\right) dV \quad (7.3.3)$$

If the element surface flux is $(c_n)_j$ kg/m²s, the nodal flux $\mathbf{Q}_j^S$ is

$$Q_{jI}^S = \int_S N_I (c_n)_j dS \quad (7.3.4)$$

As in the case of heat conduction, the FE equation of balance includes the nonzero terms (7.3.4) when the element has a surface on the boundary.

The equations of balance (7.3.2) are applicable to each of the constituents. Therefore, for each constituent we perform assemblage of the element equations of balance and then integrate these equations following (7.1.9). When the material parameters depend on the concentrations and/or the diffusion domain changes, the iterative procedure (7.1.11) should be followed. Note that this FE modeling is applicable to dilute solutions where concentrations of constituents do not affect the characteristics of the mixture.

7.3.2 Examples

Example 7.3-1. One-dimensional unsteady diffusion
This problem is described in Section 3.2.2, Example 3.2-1. Here, we present the FE solution using four-node 2D finite elements (see Section 4.4.1 for the description of these elements). One-half of the total domain is modeled ($0 \leq x \leq x_0$) due to symmetry with respect to $x = 0$, as schematically shown in Fig. E7.3-1a. Initial concentrations at the FE nodes are calculated from (E3.2-1.2). The imposed boundary conditions include: given concentration at the end nodes from (E3.2-1.3); and the symmetry (no flux) condition at the nodes with coordinate $x = 0$.

The solutions shown in Fig. E7.3-1b are obtained using 200 finite elements and time step $\Delta t = 1$ s. From the graphs in the figure it can be seen that the deviation of the FE from analytical solutions is maximum at $x = 0$ and is around 3% for $t = 2$ s. The FE solution accuracy increases with decreasing time step and with increase of number of finite elements. Investigation of the solution accuracy can be performed using Software (see web). Note that this problem can be modeled using 1D or 3D finite elements, as well.

FINITE ELEMENT MODELING OF FIELD PROBLEMS

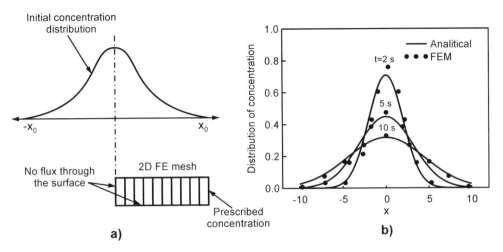

Fig. E7.3-1 Unsteady one-dimensional diffusion through infinite medium. (a) Schematic representation of the FE model; (b) FEM and analytical solutions for three times

7.4 Fluid flow with heat and mass transfer

In this section the finite element equations for incompressible viscous fluid flow are derived, considering that fluid domain does not change. It is first assumed that there is no heat and mass transfer, and subsequently heat and mass transfer are included.

7.4.1 The finite element equations

Velocity–Pressure Interpolation
Let us first consider the equation of balance of linear momentum (3.3.6) which we write as

$$\rho\left(\frac{\partial v_i}{\partial t} + v_{i,k}v_k\right) + \frac{\partial p}{\partial x_i} - \frac{\partial \tau_{ik}}{\partial x_k} - f_i^V = 0 \quad i = 1, 2, 3; \quad \text{sum on } k: \quad k = 1, 2, 3 \quad (7.4.1)$$

The fluid domain is discretized into finite elements with the *FE mesh fixed in space* in which the flow occurs. A 3D fluid flow is assumed within the domain which *does not change*. The following interpolation for velocity **v** is adopted:

$$\mathbf{v} = \mathbf{NV}, \quad \text{or} \quad v_i = N_K V_i^K; \quad \text{sum on } K, \quad K = 1, 2, \ldots, N; \quad i = 1, 2, 3 \quad (7.4.2)$$

where $N_K(r, s, t)$ are the interpolation functions, used also for interpolation of element geometry (see Section 4.3.1, equation (4.3.1)); and **V** is the vector of nodal velocities – its transpose is $\mathbf{V}^T \left[V_1^1 V_1^2 \ldots V_1^N; V_2^1 V_2^2 \ldots V_2^N; V_3^1 V_3^2 \ldots V_3^N \right]$, see web – Theory, Chapter 7. The pressure is also interpolated within finite elements. It is necessary to use another interpolation functions $\hat{N}_K(r, s, t)$ for the pressure (Bathe 1996) so that the pressure p is

$$p = \hat{\mathbf{N}}\mathbf{P}, \quad \text{or} \quad p = \hat{N}_K P^K, \quad \text{sum on } K, \quad K = 1, 2, \ldots, N \quad (7.4.3)$$

where **P** is the nodal pressure vector.

Now we apply the Galerkin method (Section 7.1) with multiplying (7.4.1) by the interpolation functions $N_K(r,s,t)$. Also, the Gauss theorem (7.1.4) is used for the terms $\partial p/\partial x_i$ and $\partial \tau_{ik}/\partial x_k$ and the following balance equation is obtained:

$$\mathbf{M}\dot{\mathbf{V}} + \hat{\mathbf{K}}_{vv}\mathbf{V} + \mathbf{K}_{vp}\mathbf{P} = \mathbf{F}_v - \mathbf{F}_\tau \tag{7.4.4}$$

where the matrices can be written in component form as (with no sum on i and sum on j; $i,j = 1,2,3$)

$$[M_{KJ}]_{ii} = \int_V \rho N_K N_J dV \tag{7.4.5}$$

$$\left[\left(\hat{\mathbf{K}}_{vv}\right)_{KJ}\right]_{ii} = \left[\rho \int_V N_K N_{J,j} v_j dV\right]_{ii} \tag{7.4.6}$$

$$\left[\left(\mathbf{K}_{vp}\right)_{KJ}\right]_{ii} = -\int_V N_{K,i} \hat{N}_J dV \tag{7.4.7}$$

Here, as in general within the text, the capital indices represent the node numbers, while the indices i and j denote the coordinate numbers ($= 1,2,3$). The terms of the nodal force vectors are:

$$(\mathbf{F}_v)_{Ki} = \int_V N_K f_i^V dV + \int_S N_K\left(-p\delta_{ij} + \tau_{ij}\right)n_j dS, \quad (\mathbf{F}_\tau)_{Ki} = \int_V N_{K,j}\tau_{ij}dV \tag{7.4.8}$$

Note that the integral over the element surface in the nodal vector $\mathbf{F}_v$ represents the integral over the normal stress on the surface. In the FE assembling process, only the surface nodal forces at the external boundaries remain, while those corresponding to the internal element surfaces cancel (as in the case of solids, see Section 4.2.2 where the internal FE nodal forces cancel).

A more common form of equation (7.4.4), which will be used in most applications, is obtained by substituting the constitutive law for the viscous stress (3.3.5), i.e. $\tau_{ij} = 2\mu\dot{e}_{ij} = \mu(v_{i,j} + v_{j,i})$. Then equation (7.4.4) becomes

$$\mathbf{M}\dot{\mathbf{V}} + \mathbf{K}_{vv}\mathbf{V} + \mathbf{K}_{vp}\mathbf{P} = \mathbf{F}_v \tag{7.4.9}$$

where now the matrix $\mathbf{K}_{vv}$ is (with no sum on i and sum on j; $i,j = 1,2,3$)

$$[(\mathbf{K}_{vv})_{KJ}]_{ii} = \left[\hat{K}_{KJ}\right]_{ii} + \left[K_{\mu KJ}\right]_{ii}, \quad \text{with}$$
$$\left[K_{\mu KJ}\right]_{ii} = \int_V \mu N_{K,j} N_{J,j} dV \tag{7.4.10}$$

Also, the nodal force vector can be expressed in terms of pressure and velocity on the surface,

$$(\mathbf{F}_v)_{Ki} = \int_V N_K f_i^V dV + \int_S N_K\left(-p\delta_{ij} + \mu v_{i,j}\right)n_j dS \tag{7.4.11}$$

FINITE ELEMENT MODELING OF FIELD PROBLEMS

Comparing (7.4.9) with (7.4.4) we note that in (7.4.9) there is no term $-\mathbf{F}_\tau$ on the right-hand side; it is taken by the additional term in the element matrix $\mathbf{K}_{vv}$ as given in (7.4.10). Equation (7.4.9) thus represents *the Navier–Stokes equation for a finite element*.

Consider now the continuity equation (3.3.3). Multiplying this equation by the interpolation functions $\hat{N}_K$, a *weak form of the continuity equation* is obtained as

$$\left(\int_V \hat{N}_K N_{J,j} dV\right) V_j^J = 0, \quad \text{or} \quad \mathbf{K}_{vp}^T \mathbf{V} = \mathbf{0} \tag{7.4.12}$$

where the matrix $\mathbf{K}_{vp}$ is given in (7.4.7).

The system of equations (7.4.9) (or (7.4.4)) and (7.4.12) represent the system of FE equations which are assembled in a usual manner (see Section 4.2). The system is nonlinear since the matrix $\mathbf{K}_{vv}$ is nonlinear: it contains the velocity as the coefficient in $[K_{KJ}]_{ii}$. Therefore, an iterative scheme (7.1.11) must be employed. For a time step 'n' we have the following iterative form (no sum on i)

$$\begin{bmatrix} \dfrac{1}{\Delta t}\mathbf{M} + {}^{n+1}\tilde{\mathbf{K}}_{vv}^{(i-1)} & \mathbf{K}_{vp} \\ \mathbf{K}_{vp}^T & \mathbf{0} \end{bmatrix} \begin{Bmatrix} \Delta\mathbf{V}^{(i)} \\ \Delta\mathbf{P}^{(i)} \end{Bmatrix} =$$

$$\begin{Bmatrix} {}^{n+1}\mathbf{F}_{ext}^{(i-1)} \\ \mathbf{0} \end{Bmatrix} - \begin{bmatrix} \dfrac{1}{\Delta t}\mathbf{M} + {}^{n+1}\mathbf{K}_{vv}^{(i-1)} & \mathbf{K}_{vp} \\ \mathbf{K}_{vp}^T & \mathbf{0} \end{bmatrix} \begin{Bmatrix} {}^{n+1}\mathbf{V}^{(i-1)} \\ {}^{n+1}\mathbf{P}^{(i-1)} \end{Bmatrix} + \begin{Bmatrix} \dfrac{1}{\Delta t}\mathbf{M}^n\mathbf{V} \\ \mathbf{0} \end{Bmatrix} \tag{7.4.13}$$

with $\left[{}^{n+1}\left(\tilde{\mathbf{K}}_{vv}^{(i-1)}\right)_{KJ}\right]_{ik} = \left[{}^{n+1}K_{KJ}^{(i-1)}\right]_{ii} + \left[{}^{n+1}J_{KJ}^{(i-1)}\right]_{ik}$

$$\left[{}^{n+1}J_{KJ}^{(i-1)}\right]_{ik} = \rho \int_V N_K {}^{n+1}v_{i,k}^{(i-1)} N_J dV$$

As in other iterative schemes, the iterations stop when the norm of the incremental vector of the left-hand side, or the norm of the right-hand side (the 'unbalanced force') is smaller than a selected error tolerance.

Penalty Method

The above formulation assumes interpolations of velocities and pressure, also called the mixed formulation, resulting in a solution for nodal velocities and pressures. However, it is possible to eliminate pressure calculation in the system (7.4.9) using a penalty method. The procedure is as follows. The continuity equation (3.3.3) is approximated as

$$v_{i,i} + \frac{p}{\lambda} = 0 \tag{7.4.14}$$

where λ is a selected large number, the penalty parameter. Substituting the pressure p from (7.4.14) into the Navier–Stokes equations (3.3.7) we obtain

$$\rho\left(\frac{\partial v_i}{\partial t} + v_{i,k} v_k\right) - \lambda v_{j,ij} - \mu v_{i,kk} - f_i^V = 0 \tag{7.4.15}$$

Then the FE equation of balance (7.4.9) becomes

$$\mathbf{M}\dot{\mathbf{V}} + \left(\mathbf{K}_{vv} + \mathbf{K}_{vv}^{\lambda}\right)\mathbf{V} = \mathbf{F}_v + \mathbf{F}_{\lambda} \qquad (7.4.16)$$

where

$$\left[K_{KJ}^{\lambda}\right]_{ik} = \lambda \int_V N_{K,i} N_{J,k} dV, \quad (\mathbf{F}_{\lambda})_{Ki} = \lambda \int_S N_K v_{j,j} n_i dS \qquad (7.4.17)$$

In examples we show a selection of the range of the penalty parameter λ and its effect on the solution (see web – Software).

Fluid Flow with Heat Transfer
The additional term in the governing equation for heat transfer in fluids (3.3.8), with respect to the heat conduction equation (3.1.5), is $-\rho c T_{,i} v_i$. Then, in the FE equation for heat conduction (7.2.2) we have the additional term, and this equation becomes

$$\mathbf{M}_T \dot{\mathbf{T}} + \mathbf{K}^k \mathbf{T} + \mathbf{K}_{Tv} \mathbf{V} = \mathbf{Q} \qquad (7.4.18)$$

where

$$\mathbf{K}_{Tv} = [\mathbf{K}_{Tv1} \mathbf{K}_{Tv2} \mathbf{K}_{Tv3}]; \quad (\mathbf{K}_{Tv\ell})_{KJ} = \rho \int_V c T_{,\ell} N_K N_J dV, \quad \ell = 1, 2, 3 \qquad (7.4.19)$$

The matrix $\mathbf{M}_T$ is the matrix $\mathbf{M}$ in (7.2.2), but this notation is used here to emphasize that this matrix is not the matrix $\mathbf{M}$ for fluid used above. Therefore, in solving fluid flow with heat transfer, we have the system of equations: (7.4.9) or (7.4.16) – and (7.4.18) coupled with respect to velocities, pressures and temperatures. Note that in using an iterative scheme of solution (7.4.13), we have the temperature-dependent matrices $\mathbf{M}_T$ and $\mathbf{K}^k$ as in the heat conduction case, but also the temperature-dependent matrix $\mathbf{K}_{Tv}$ in which the terms $c^{(i-1)}$ and $T_{,\ell}^{(i-1)}$ are used for iteration 'i'.

Fluid Flow with Mass Transfer
Completely analogous to the FE equation (7.4.18) for heat transfer, the FE equation for mass transfer is obtained: equation (7.3.2) is expanded to include an additional convection term $(\partial c_j / \partial x_i) v_i$ (see (3.3.9), and (3.2.8) or (7.3.1)).
Hence,

$$\mathbf{M}_j \dot{\mathbf{C}}_j + \mathbf{K}_j \mathbf{C}_j + \mathbf{K}_{Cjv} \mathbf{V} = \mathbf{Q}_j^S + \mathbf{Q}_j^V \quad \text{no sum on } j \qquad (7.4.20)$$

where

$$\mathbf{K}_{Cjv} = \left[(\mathbf{K}_{Cjv})_1 (\mathbf{K}_{Cjv})_2 (\mathbf{K}_{Cjv})_3\right]$$
$$(\mathbf{K}_{Cvj})_{\ell KJ} = \int_V C_{j,\ell} N_K N_J dV, \quad \ell = 1, 2, 3 \qquad (7.4.21)$$

FINITE ELEMENT MODELING OF FIELD PROBLEMS

Of course, equation (7.4.21) is solved together with (7.4.9) or (7.4.16), with the concentration boundary conditions (see web – Theory, Chapter 7).

7.4.2 Examples

Example 7.4-1. Two-dimensional steady flow between two parallel walls

A description of this example is given in Example 3.3-1. We here solve this example using nine-node 2D finite elements (Fig. E7.4-1a) and compare the numerical and analytical solutions. In this finite element, velocities are calculated at all nodes, while the pressure is only associated with the corner nodes (pressure is linear over the element). The boundary conditions imposed here are the same as in the analytical model: zero velocity at the walls and given pressures at the inlet and the outlet of the channel.

Solutions for other geometrical and material data and pressures, as well as with a change of mesh density, can be obtained using Software on the web.

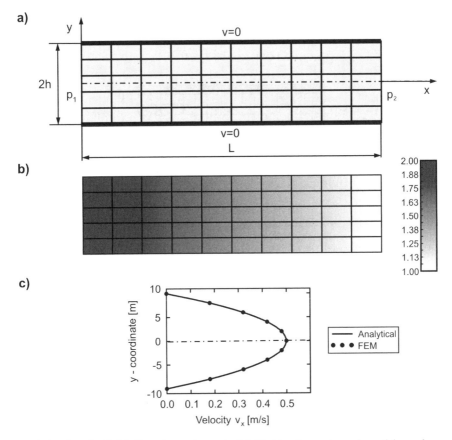

Fig. E7.4-1 Steady fluid flow in a channel. (a) Finite element mesh and boundary conditions; (b) Pressure field (linear along the x-axis); (c) Velocity profile (analytical and FE solutions). Data: $h = 10\,m$, $L = 100\,m$, $p_1 = 2\,Pa$, $p_2 = 1\,Pa$, $\mu = 0.1\,kg/ms$, $\rho = 1\,kg/m^3$; number of elements in x and y directions are 10 and 5, respectively

Example 7.4-2. Unsteady fluid flow on a plate

The example is described as Example 3.3-2 in Section 3.3.2. Here we present the solution obtained by using the FE modeling. The boundary conditions used here are that the FE nodes on the top surface have constant velocity v_0, while the nodes at the bottom surface do not have any constraint.

The velocity distribution in the domain L near the moving plate, for several times, is shown in Fig. E7.4-2b, obtained analytically and by FE method. It can be seen that the velocity tends to the plate velocity v_0 in the entire domain as time increases. Finite element solutions agree well with the analytical solutions.

Solutions with other data for the model can be obtained using Software on the web.

Example 7.4-3. Linear convection–diffusion equation (Burger's viscous equation)

This example is described in Section 3.3.2 (Example 3.3-3). Here we solve the 1D equation (E3.3-3.1) as well as the FE equation (7.4.18) or (7.4.20).

The finite element model used here consists of 100 four-node 2D finite elements, in a row along the x-direction, and the solution is obtained using time step $\Delta t = 0.001\,s$.

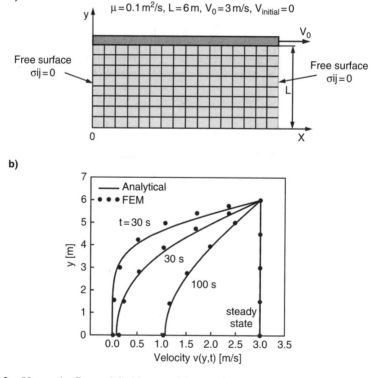

Fig. E7.4-2 Unsteady flow of fluid caused by motion of a plate on the top surface. Plate moves by constant velocity. (a) Finite element model, data for the model, and boundary conditions; (b) Velocity distribution along the fluid depth, analytical and FEM solutions, are shown for several times. Time step is $\Delta t = 10\,s$ and the division in the x and y directions is 4×8

FINITE ELEMENT MODELING OF FIELD PROBLEMS 135

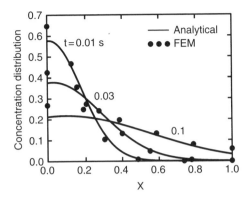

Fig. E7.4-3 Concentration (temperature) distribution along x-axis for three times. Numerical (FEM) and analytical solutions of linear convection–diffusion equation (E3-3.1)

The solutions compare well with the analytical solutions, as can be seen from Fig. E7.4-3. Solutions for other times and time steps can be obtained using Software (see web – Software, Chapter 7).

7.5 FE equations for modeling large change of fluid domain – Arbitrary Lagrangian–Eulerian (ALE) formulation

In this section we present a method of the FE modeling fluid flow when the fluid domain boundaries change in time. The basic equations of the so-called arbitrary Lagrangian–Eulerian formulation (ALE formulation) are derived and transformed to the FE equations of balance of linear momentum (Donea *et al.* 1982, Donea 1983, Nitikitpaiboon & Bathe 1993, Filipović 1999, Filipović *et al.* 2006).

7.5.1 The ALE formulation

The fluid flow is modeled here using a moving mesh. The reference domain – the FE mesh, in which usual FE calculations are performed – is moving in space. The control volume, which in this case is the volume of a finite element, changes with time. At a point G of the FE mesh in Fig. 7.5.1, the fluid velocity is **v** and the mesh velocity is $\mathbf{v}_m$. Consequently, in deriving the FE equations for mass balance and balance of linear momentum, the fact that the reference domain is not stationary must be taken into account. One of the approaches to appropriately account for motion of the reference domain is the ALE formulation which is briefly presented here.

The Navier–Stokes equations of balance of linear momentum (3.3.7) can be written in the ALE formulation as

$$\rho \left[v_i^* + \left(v_j - v_j^m \right) v_{i,j} \right] = -p_{,i} + \mu v_{i,jj} + f_i^V \qquad (7.5.1)$$

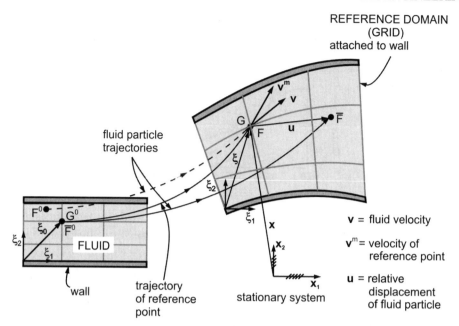

Fig. 7.5.1 Schematics of the FE modeling according to the ALE formulation (2D representation). An FE mesh attached to the solid wall is moving in space, changing also its size and shape. The current position of the fluid particle is F and the point of the mesh (grid) is G, with velocities $\mathbf{v}$ and $\mathbf{v}_m$; their initial positions are F^0 and G^0. Note that the fluid point $\bar{F}^0$ initially at G^0 is at space position $\bar{F}$ different from F, displaced by a vector $\mathbf{u}$. The coordinate system in the reference domain ξ_1, ξ_2 moves with the reference domain, without rotation

while the continuity equation remains as in (3.3.3). Here v_i are the velocity components of a generic fluid particle and v_i^m are the velocity components of the point on the moving mesh occupied by the fluid particle; other quantities are as in (3.3.7). The symbol '*' denotes the mesh-referential time derivative, i.e. the time derivative at a considered point on the mesh,

$$(\bullet)^* = \frac{\partial(\bullet)}{\partial t}\Big|_{\xi_i = const} \tag{7.5.2}$$

The Cartesian spatial coordinates of a generic fluid particle are x_i and of the corresponding point on the mesh are ξ_i. In deriving (7.5.1) we used the following expression for the material derivative $D(\rho v_i)/Dt$ (for a fixed material point, see (3.3.1)),

$$\frac{D(\rho v_i)}{Dt} = \frac{\partial(\rho v_i)}{\partial t}\Big|_\xi + (v_j - v_j^m)\frac{\partial(\rho v_j)}{\partial x_i} \tag{7.5.3}$$

The first term on the right-hand side is the so-called 'mesh-referential time derivative', while the second is the convective term.

The Galerkin method for the space discretization of the fluid domain can now be applied, as in Section 7.4.1. The finite element equations for a 3D domain that follow from (7.5.1) and (3.3.3) are (see (7.4.4 and (7.4.12)):

FINITE ELEMENT MODELING OF FIELD PROBLEMS

$$\rho \int_V N_K v_i^* dV + \rho \int_V N_K (v_j - v_j^m) v_{i,j} dV =$$
$$= -\int_V N_K p_{,i} dV + \int_V \mu N_K v_{i,jj} dV + \int_V N_K f_i^V dV \tag{7.5.4}$$

$$\int_V \hat{N}_K v_{i,i} dV = 0 \tag{7.5.5}$$

The integration is performed over the volume V of a finite element, which now is time dependent, using the Gauss theorem as in Section 7.1.2.

Consider first the system of equations (7.5.4) which is nonlinear with respect to the velocities, but also with the element volume change. In an incremental analysis a linearization with respect to time must be performed using the known values at the start of time step n. The approximation for a quantity F can be written as

$$^{n+1}F\big|_{n\xi} = {}^nF\big|_{n\xi} + F^* \Delta t \tag{7.5.6}$$

This relation is further applied to the left- and right-hand sides, (LHS) and (RHS), of (7.5.4) to obtain

$$^n(LHS) + (LHS)^* \Delta t = {}^{n+1}(RHS) \tag{7.5.7}$$

In calculating the mesh-referential time derivatives we use the relations:

$$\left(\frac{\partial F}{\partial x_i}\right)^* = \frac{\partial F^*}{\partial x_i} - \left(\frac{\partial v_k^m}{\partial x_i}\right)\frac{\partial F}{\partial x_k} \tag{7.5.8}$$

and

$$(dV)^* = \frac{\partial v_k^m}{\partial x_k} dV \tag{7.5.9}$$

With these linearizations the equations (7.5.4) and (7.5.5) can be written as

$$^n\mathbf{M}_{(1)} \mathbf{V}^* + {}^n\mathbf{K}_{(1)vv} \Delta \mathbf{V} + {}^n\mathbf{K}_{vp} \Delta \mathbf{P} = {}^{n+1}\mathbf{F}_{(1)} - {}^n\mathbf{F}_{(1)} \tag{7.5.10}$$

and

$$^n\mathbf{M}_{(2)} \mathbf{V}^* + {}^n\mathbf{K}_{(2)vv} \Delta \mathbf{V} = {}^{n+1}\mathbf{F}_{(2)} - {}^n\mathbf{F}_{(2)} \tag{7.5.11}$$

Expressions for these matrices and vectors are given on the web, Theory – Chapter 7. The integrals are evaluated over the known FE volumes and surfaces at start of time step. Further, some of the terms are calculated using the values at the last iteration (see (7.1.11) and (7.4.13)). Of course, the mesh-referential time derivatives $\mathbf{V}^*$ and $\mathbf{P}^*$ are replaced by $\mathbf{V}^* = \Delta \mathbf{V}/\Delta t$ and $\mathbf{P}^* = \Delta \mathbf{P}/\Delta t$ to obtain the incremental algebraic equations of the form (7.1.11) or (7.4.13).

138 COMPUTER MODELING IN BIOENGINEERING

The presented formulation of the FE modeling is necessary when the fluid boundaries change significantly over the time period used in the analysis. It is particularly convenient when the boundary of the fluid represents a deformable solid, as Fig. 7.5.1 suggests, for appropriate modeling of the solid–fluid interactions. Finally, note that the mesh motion is arbitrary and for each problem can be specifically designed. Also, it is important to emphasize that the solution for the fluid flow does not depend on the FE mesh motion (Filipović et al. 2006).

7.5.2 Examples

Example 7.5-1. Unsteady flow in a contracting or expanding pipe
This example is presented to compare the FE solution (using the ALE formulation of this section, Filipović et al. 2006) and the analytical solution (Uchida & Aoki 1977).

The fluid flow within a very long (infinite) pipe, symmetric with respect to the plane $x = 0$, is induced by the wall motion. Only half of the flow domain is modeled due to symmetry of the fluid flow field, as shown in Fig. E7.5-1A.

The radial velocity of fluid at the wall, v_w, is equal to the wall velocity $da/dt = \dot{a}$. The boundary conditions are shown in the figure. The wall motion is defined as (Uchida & Aoki 1977)

$$\dot{a}a/\nu = \alpha = \dot{a}_0 a_0/\nu \qquad (E7.5\text{-}1.1)$$

where a_0 and $\dot{a}_0$ are the initial radius and wall velocity, respectively. The absolute value of the parameter α represents the Reynolds number of the problem. For $\alpha > 0$ the pipe expands and for $\alpha < 0$ the pipe contracts over time. Integrating (E7.5-1.1), the dependence $a(t)$ can be obtained (and the wall velocity):

$$a/a_0 = \left[1+2\alpha\left(\nu t/a_0^2\right)\right]^{1/2}, \quad v_w/(v_w)_{t=0} = \dot{a}/\dot{a}_0 = \left[1+2\alpha\left(\nu t/a_0^2\right)\right]^{-1/2} \qquad (E7.5\text{-}1.2)$$

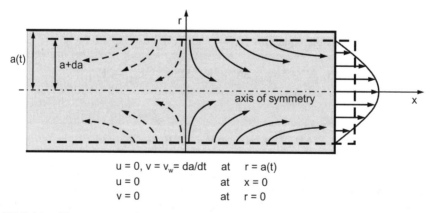

Fig. E7.5-1A Flow patterns in a contracting pipe and boundary conditions. The x and y velocity components are u and v; the current pipe radius is $a(t)$ and is equal along the pipe and the flow field is symmetric with respect to the axial coordinate x

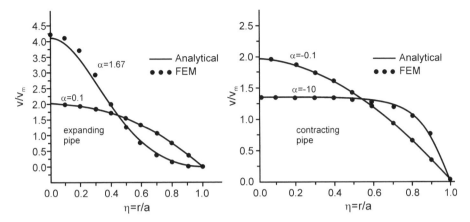

Fig. E7.5-1B Axial velocity distribution (v_m = mean axial velocity) for two values of the Reynolds number α. These solutions correspond to doubled radius (expanding pipe), or half of the initial radius (contracting pipe). The FEM solutions for $|\alpha| = 0.1$ are close to the analytical ones (not shown in the figure) (according to Filipović et al. 2006)

The FE model consists of 810 two-dimensional axisymmetric elements (see Section 4.4). The initial radius of the pipe is $a = 0.025\ m$ and the infinite pipe length is replaced by axial length of the model equal to $1.0\ m$. The outflow boundary conditions at the pipe end are used (zero stresses). The calculations are performed until the radius becomes two times larger with respect to the initial radius (expanding pipe), or it becomes two times smaller than the initial radius (contracting pipe). The initial conditions for the velocities are taken to be the steady solutions of this problem, obtained numerically for the nonmoving walls. The computed distributions of axial velocity for expanding and contracting pipes for several values of α are shown in Fig. E7.5-1B. Solutions for radial velocities and other geometrical and material data, as well as with a change of mesh density, can be obtained using Software on the web.

It can be seen that the numerical solutions are close to the analytical results, even in cases of very large wall motions.

7.6 Solid–fluid interaction

There are many conditions in science, engineering and bioengineering where fluid is acting on a solid producing surface loads and deformation of the solid material. The opposite also occurs, i.e. deformation of a solid affects the fluid flow. In these cases, the solid and fluid form a coupled mechanical system, so that in modeling the solid and fluid domains must be considered together. Therefore, we need to solve the solid–fluid interaction problem. An example of this coupled solid–fluid system is when blood flows through a deformable blood vessel (see a schematics of a blood vessel in Fig. 7.6.1).

There are, in principle, two approaches for the FE modeling of solid–fluid interaction problems: (a) strong coupling method, and (b) loose coupling method. In the first method, the solid and fluid domains are modeled as one mechanical system. In the second approach, the

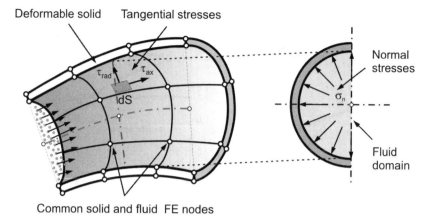

Fig. 7.6.1 Illustration of the solid–fluid interaction. A fluid is flowing through the deformable vessel producing tangential and normal stresses on the vessel surface. The common solid and fluid FE nodes have the same displacements and velocities

solid and fluid are modeled separately, the solutions are obtained with different FE solvers, but the parameters from one solution which affect the solution for the other medium are transferred successively.

Consider the example shown in Fig. 7.6.1. Assuming that there is no slip between the fluid and solid at the common boundary, we have that the nodes at the solid–fluid boundary have the same displacements and velocities for the solid and fluid domains. If the strong coupling approach is used, the terms of the finite element matrices and forces corresponding to these common nodes are summed, as is usual in the finite element assembling procedure (see Section 4.2.2). However, the values of matrix terms for the fluid and solid domains are usually several orders of magnitude different, so that computational difficulties may arise in the solution process. On the other hand, if the loose coupling method is employed, the systems of balance equations for the two domains are formed separately and there are no such computational difficulties. Hence, the loose coupling method is advantageous from the practical point of view and we further describe this method.

7.6.1 Loose coupling method

As stated above, the loose coupling approach consists in the successive solutions for the solid and fluid domains. Consider fluid flow within a deformable vessel shown schematically Fig. 7.6.1. The vessel deforms due to loading from the fluid which generates surface forces that are transferred to the solid. The stresses acting on the vessel surface are the tangential stresses τ_{rad} and τ_{ax} and the normal stress σ_n. The flow domain changes due to the vessel deformation, while the common nodes have the same displacements and velocities for the fluid and solid.

The computational steps within a time step 'n' employed when solving the solid–fluid interaction are given in Table 7.6.1. The solution is obtained iteratively, and the iteration counter in the solid–fluid interaction loop is 'I'. We have denoted by $^{n+1}\boldsymbol{\sigma}_{Sf}^{(I)}$ the stress within the fluid, at the common boundary S. Also, ε_{disp} and $\varepsilon_{velocity}$ are the error tolerances, respectively, for the norms of displacement increments of the solid and for the velocity

FINITE ELEMENT MODELING OF FIELD PROBLEMS

Table 7.6.1 Iteration scheme for the solid–fluid interaction, loose coupling approach

---

1. **Initial conditions for the time step 'n'**
   Iteration counter $I = 0$:
   configuration of solid $^{n+1}\mathcal{B}^{(0)} = {}^{n}\mathcal{B}$; common velocities $^{n+1}\mathbf{V}^{(0)} = {}^{n}\mathbf{V}$.

2. **Iterations for both domains**: $I = I + 1$

   a) Calculate fluid flow velocities and pressures $^{n+1}\mathbf{V}_f^{(I)}$ and pressures $^{n+1}\mathbf{P}^{(I)}$ by an iterative scheme, see (7.4.13).

   b) Calculate interaction nodal forces from the fluid acting on the solid as

   $$^{n+1}\mathbf{F}_S^{(I)} = -\int_S \mathbf{N}^\mathrm{T}\, ^{n+1}\boldsymbol{\sigma}_{Sf}^{(I)}\, dS \tag{7.6.1}$$

   c) Transfer the load from the fluid to solid. Find a new deformed configuration of the solid $^{n+1}\mathcal{B}^{(I)}$. Calculate velocities of the common nodes with the fluid $^{n+1}\mathbf{V}^{(I)}$ to be used for the fluid domain.

3. **Convergence check**. Check for the convergence on the solid displacement and fluid velocity increments for the loop on I:

   $$\left\|\Delta \mathbf{U}_{solid}^{(I)}\right\| \leq \varepsilon_{disp}, \quad \left\|\Delta \mathbf{V}_{fluid}^{(I)}\right\| \leq \varepsilon_{velocity} \tag{7.6.2}$$

If the convergence criteria are not satisfied, go to the next iteration, step 2. Otherwise, use the solutions from the last iteration as the initial solutions for the next time step and go to step 1.

---

increments. Other convergence criteria may also be used, such as the error in the 'unbalanced force' or energy (Bathe 1996, Kojić & Bathe 2005).

A graphical interpretation of the algorithm for the solid–fluid interaction problem is shown in Fig. 7.6.2 (Filipović 1999).

7.6.2 Examples

Example 7.6-1. Viscous flow in a collapsible tube

Fluid flow through collapsible tubes is a complex problem due to the interaction between the tube wall and the flowing fluid. It is usually used to simulate biological flows such as blood flow in arteries or veins and airflow in the bronchial airways.

It is assumed that the collapse is symmetric with respect to both x–y and x–z planes. One-quarter of the fluid flow field is modeled due to symmetry with respect to the two coordinate planes. Boundary conditions consist of prescribed velocity at the inlet nodes (zero velocity at the interface surface with the shell elements) and the symmetry conditions at the symmetry planes. Fluid flow is calculated by using 1250 eight-node 3D elements, and 500 four-node shell elements for the model of the tube wall, with the wall thickness ratio $h/R_0 = 1/20$ (Filipović 1999).

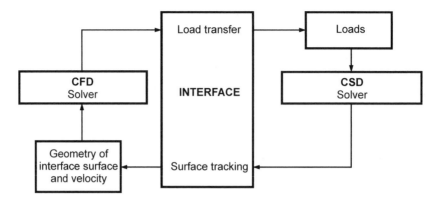

Fig. 7.6.2 Block diagram of the solid–fluid interaction algorithm. Information and transfer of parameters between the CSD (computational solid dynamics) and CFD (computational fluid dynamics) solvers through the interface block (Filipović 1999)

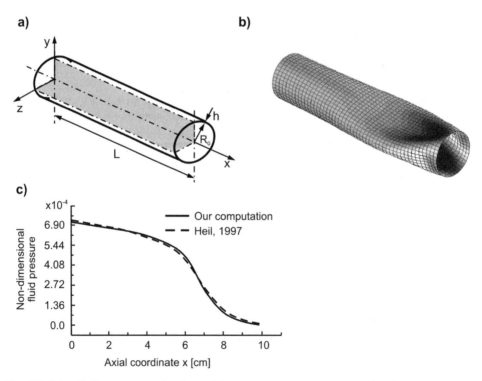

Fig. E7.6-1 Collapse of a tube loaded by external pressure p^*_{ext} and with flowing fluid through the tube. (a) Tube geometry before collapse; (b) Shape of the tube after collapse; (c) Pressure distribution along the collapsed tube (at the tube center line). Data: Lengths (cm) $R_0 = 1$, $L = 10$, $h = 0.05$; nondimensional volume flux $q = 8 \, \mu \dot{V} L / \pi R_o^4 E = 15 \times 10^{-5}$, dimensionless external pressure $P_{ext} = p^*_{ext}/E = 6.38 \times 10^{-4}$

We keep the fluid pressure equal to zero at the tube outer end, and induce the tube collapse by increasing the chamber pressure, p^*_{ext}. The tube is first inflated when $p^*_{\text{ext}} = 0$, deforming axisymmetrically. A small geometric irregularity is introduced to initiate the collapse. Then, the external pressure is increased and when it exceeds a critical value, the wall locally loses its stability and the tube buckles as shown in Fig.E7.6-1b. Figure E7.6-1c shows the pressure drop along on the tube center line (together with solution given in Heil 1997).

7.7 Fluid flow through porous deformable media

7.7.1 Finite element balance equations

In this section we transform the fundamental relations of Section 3.4 into the finite element equations. Large displacements and large strains of the solid, and elastic material with the constitutive relation (3.4.7) are assumed. First, by employing the principle of virtual work and assuming that the material is elastic, the following equation is obtained from (3.4.5):

$$\int_{^nV} \delta \mathbf{e}^T \mathbf{C}^E \mathbf{e} dV + \int_{^nV} \delta \mathbf{e}^T \left(\frac{\mathbf{C}^E \mathbf{m}}{3K_s} - \mathbf{m} \right) p dV + \int_{^nV} \delta \mathbf{u}^T \rho \ddot{\mathbf{u}} dV + \int_{^nV} \delta \mathbf{u}^T \rho_f \dot{\mathbf{q}} dV$$

$$= \int_{^nV} \delta \mathbf{u}^T \rho \mathbf{b} dV + \int_{^nA} \delta \mathbf{u}^T \mathbf{t} dA \tag{7.7.1}$$

where the left upper index 'n' denotes that the configuration $^n\mathcal{B}$ (at start of a time step in incremental analysis) is considered; on the right-hand side is the virtual work of the external body force $\mathbf{b}$ and the surface loading $\mathbf{t}$ on the area ^nA (see Sections 4.3 and 6.2.2). Next, following the Galerkin method we multiply (3.4.4) by the interpolation matrix $\mathbf{N}_q^T$ for the relative velocity of fluid $\mathbf{q}$, and integrate over the finite element volume ^nV. The resulting equation is

$$-\int_{^nV} \mathbf{N}_q^T \nabla p dV + \int_{^nV} \mathbf{N}_q^T \rho_f \mathbf{b} dV - \int_{^nV} \mathbf{N}_q^T \mathbf{k}^{-1} \mathbf{q} dV - \int_{^nV} \mathbf{N}_q^T \rho_f \ddot{\mathbf{u}} dV - \int_{^nV} \mathbf{N}_q^T \frac{\rho_f}{n} \dot{\mathbf{q}} dV = \mathbf{0} \tag{7.7.2}$$

Further, we multiply the continuity equation (3.4.10) by the interpolation matrix $\mathbf{N}_p^T$ for pressure (which is the vector-column) and obtain

$$\int_{^nV} \mathbf{N}_p^T \nabla^T \mathbf{q} dV + \int_{^nV} \mathbf{N}_p^T \left(\mathbf{m}^T - \frac{\mathbf{m}^T \mathbf{C}^E}{3K_s} \right) \dot{\mathbf{e}} dV + \int_{^nV} \mathbf{N}_p^T \left(\frac{1-n}{K_s} + \frac{n}{K_f} - \frac{\mathbf{m}^T \mathbf{C}^E \mathbf{m}}{9K_s^2} \right) \dot{p} dV = \mathbf{0}$$

$$\tag{7.7.3}$$

The standard procedure of integration over the element volume in equations (7.7.1)–(7.7.3) is employed, with use of the Gauss theorem (see Section 7.1). The resulting FE system of equations is solved incrementally, with time step Δt. The condition that the balance equations are to be satisfied at the end of each time step (denoted by the left upper index '$n+1$'; at time $t+\Delta t$) is imposed. This leads to the following system of equations

$$\begin{bmatrix} \mathbf{M}_{uu} & 0 & 0 \\ 0 & 0 & 0 \\ \mathbf{M}_{qu} & 0 & 0 \end{bmatrix} \begin{Bmatrix} {}^{n+1}\ddot{\mathbf{U}} \\ {}^{n+1}\ddot{\mathbf{P}} \\ {}^{n+1}\ddot{\mathbf{Q}} \end{Bmatrix} + \begin{bmatrix} 0 & 0 & {}^{n}\mathbf{C}_{uq} \\ {}^{n}\mathbf{C}_{pu} & \mathbf{C}_{pp} & 0 \\ 0 & 0 & \mathbf{C}_{qq} \end{bmatrix} \begin{Bmatrix} {}^{n+1}\dot{\mathbf{U}} \\ {}^{n+1}\dot{\mathbf{P}} \\ {}^{n+1}\dot{\mathbf{Q}} \end{Bmatrix}$$

$$+ \begin{bmatrix} {}^{n}\mathbf{K}_{uu} & {}^{n}\mathbf{K}_{up} & 0 \\ 0 & 0 & \mathbf{K}_{pq} \\ 0 & \mathbf{K}_{qp} & \mathbf{K}_{qq} \end{bmatrix} \begin{Bmatrix} \Delta\mathbf{U} \\ \Delta\mathbf{P} \\ \Delta\mathbf{Q} \end{Bmatrix} = \begin{Bmatrix} {}^{n+1}\mathbf{F}_u \\ {}^{n+1}\mathbf{F}_p \\ {}^{n+1}\mathbf{F}_q \end{Bmatrix} \qquad (7.7.4)$$

The matrices and vectors in this equation are:

$$\mathbf{M}_{uu} = \int_{{}^n V} \mathbf{N}_u^T \rho \mathbf{N}_u dV, \quad \mathbf{M}_{qu} = \int_{{}^n V} \mathbf{N}_q^T \rho_f \mathbf{N}_u dV,$$

$$\mathbf{C}_{uq} = \mathbf{M}_{qu}^T = \int_{{}^n V} \mathbf{N}_u^T \rho_f \mathbf{N}_q dV, \quad \mathbf{C}_{pu} = -\int_{{}^n V} \mathbf{N}_p^T \left(\mathbf{m}^T - \frac{\mathbf{m}^T \mathbf{C}^E}{3K_s} \right) {}^n\mathbf{B} dV,$$

$$\mathbf{C}_{pp} = -\int_{{}^n V} \mathbf{N}_p^T \left(\frac{1-n}{K_s} + \frac{n}{K_f} - \frac{\mathbf{m}^T \mathbf{C}^E \mathbf{m}}{9K_s^2} \right) \mathbf{N}_p dV,$$

$$\mathbf{C}_{qq} = \int_{{}^n V} \mathbf{N}_q^T \frac{\rho_f}{n} \mathbf{N}_q dV,$$

$${}^n\mathbf{K}_{uu} = \int_{{}^n V} {}^n\mathbf{B}^T \mathbf{C}^E {}^n\mathbf{B} dV, \quad {}^n\mathbf{K}_{up} = {}^n\mathbf{C}_{pu}^T = \int_{{}^n V} {}^n\mathbf{B}^T \left(\frac{\mathbf{C}^E \mathbf{m}}{3K_s} - \mathbf{m} \right) \mathbf{N}_p dV, \qquad (7.7.5)$$

$$\mathbf{K}_{pq} = \int_{{}^n V} \mathbf{N}_p^T \mathbf{N}_{q,\mathbf{x}} dV, \quad \mathbf{K}_{qp} = \mathbf{K}_{pq}^T = \int_{{}^n V} \mathbf{N}_{q,\mathbf{x}}^T \mathbf{N}_p dV, \quad \mathbf{K}_{qq} = \int_{{}^n V} \mathbf{N}_q^T \mathbf{k}^{-1} \mathbf{N}_q dV,$$

$${}^{n+1}\mathbf{F}_u = \int_{{}^n V} \mathbf{N}_u^T \rho \,{}^{n+1}\mathbf{b} dV + \int_{{}^n A} \mathbf{N}_u^T \,{}^{n+1}\mathbf{t} dA - \int_{{}^n V} {}^n\mathbf{B}^T \,{}^n\boldsymbol{\sigma} dV - {}^n\mathbf{K}_{up} \,{}^n\mathbf{P},$$

$${}^{n+1}\mathbf{F}_p = \int_{{}^n A} \mathbf{N}_p^T \,{}^n\mathbf{n}^T \,{}^n\mathbf{q} dA - \mathbf{K}_{pq} \,{}^n\mathbf{Q},$$

$${}^{n+1}\mathbf{F}_q = \int_{{}^n V} \mathbf{N}_q^T \rho_f \,{}^{n+1}\mathbf{b} dV - \mathbf{K}_{qp} \,{}^n\mathbf{P} - \mathbf{K}_{qq} \,{}^n\mathbf{Q}$$

In these expressions $\mathbf{N}_u \equiv \mathbf{N}$ is the interpolation matrix for displacements (see Section 4.3), ${}^n\mathbf{n}$ is the normal vector to the boundary and ${}^n\mathbf{B}$ is the strain–displacement transformation matrix (note that in the case of large displacements ${}^n\mathbf{B}$ can be decomposed into linear and nonlinear matrices ${}^n\mathbf{B}_L$ and ${}^n\mathbf{B}_{NL}$, see Section 6.2.3). With the left upper index 'n' for matrices it is emphasized that they change over time steps (and iterations, see (7.1.11)). As can be seen from (7.7.4), the nodal point variables are: displacements of solid $\mathbf{U}$, relative velocities $\mathbf{Q}$ and pressures $\mathbf{P}$. Boundary conditions include: general boundary conditions for the solid, relative velocities and surface pressures. Details of the above derivations are given on the web – Theory, Chapter 7.

In analyzing the system of differential equations we highlight the following important facts for its solution. The system is nonsymmetric and nonlinear. Only in the case of small displacements and constant porosity, the system becomes linear. The system is symmetric when inertial forces are neglected. For each of these cases a standard Newmark method can

FINITE ELEMENT MODELING OF FIELD PROBLEMS

be employed for time integration (Section 5.3). In general, the system is nonlinear and an iterative procedure must be employed (see Section 7.1 and web – Theory, Chapter 7). During iterations we use the density of the mixture, ρ, and the porosity, n, from the start of a time step, i.e., $^n\rho$ and $^n n$, respectively. After convergence is reached, the porosity is updated by using (3.4.12), from which it follows:

$$^{n+1}n = {}^n n - \Delta t \left[{}^n n \left(\frac{\partial p}{\partial t}\right) \frac{1}{K_f} + \nabla^{T\,n} \mathbf{q} \right] \quad (7.7.6)$$

In deriving this equation the spatial changes of the fluid density are neglected, i.e. $\partial \rho_f / \partial x_i = 0$ is used. This is a physically acceptable approximation since the fluid velocity is small, the fluid is nearly incompressible, and the fully saturated conditions are considered.

7.7.2 Examples

Example 7.7-1. Creep deformation of human spinal motion segment (SMS)
The one-dimensional model description of the SMS is given in Example 3.4-1. Here, the FE model is used to obtain creep deformation of the SMS over time. Since there is no lateral deformation and flow, the plane strain four-node finite elements are used, with the boundary conditions shown in Fig. E7.7-1a. The axial deformation and fluid flow are allowed in the model.

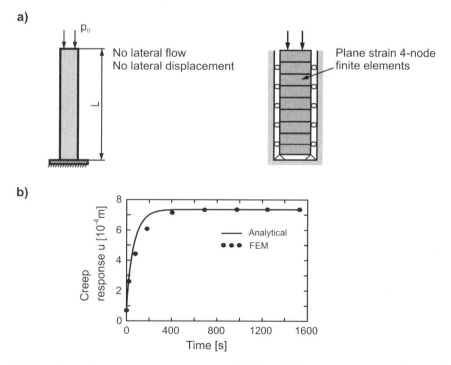

Fig. E7.7-1 One-dimensional creep response of SMS. (a) The column representing the SMS and FE model; (b) Column settlement over time due to creep response under step pressure at the top surface (data given in Example 3.4.1), analytical (Biot 1941) and FEM solutions

The solution shown in Fig. E7.7-1b is obtained using 10 finite elements. A small deviation between numerical and analytical solutions, which can be seen in Fig. E7.7-1b, might be due to simplification in the analytical solution, where it is assumed that the fluid pressure is constant over the whole column length.

Solutions for various model parameters can be obtained using the Software on the web.

References

Bathe, K.J. (1996). *Finite Element Procedures*, Prentice-Hall, Englewood Cliffs, NJ.
Biot, M.A. (1941). General theory of three-dimensional consolidation, *J. Appl. Physics*, **12**, 155–64.
Donea, J. (1983). Arbitrary Lagrangian–Eulerian finite elements methods. In T. Belytschko & T.J.R. Hughes (eds.), *Computational Methods for Transient Analysis* (pp. 473–516), Elsevier, Amsterdam.
Donea, J., Giuliani, S. & Halleux, J.P. (1982). An arbitrary Lagrangian–Eulerian finite element method for transient dynamic fluid–structure interactions, *Comp. Meth. Appl. Mech. Eng.*, **33**, 689–723.
Filipović, N. (1999). *Numerical Analysis of Coupled Problem: Deformable Body and Fluid Flow*, Ph.D. Thesis, Faculty of Mechanical Engineering, University of Kragujevac, Serbia.
Filipović, N., Mijailović, S., Tsuda, A. & Kojić, M. (2006). An implicit algorithm within the arbitrary Lagrangian–Eulerian formulation for solving incompressible fluid flow with large boundary motions, *Comp. Meth. Appl. Mech. Eng.*, **195**, 6347–61.
Fung, Y.C. (1965). *Foundations of Solid Mechanics*, Prentice-Hall, Englewood Cliffs, NJ.
Heil, M. (1997). Stokes flow in collapsible tubes – computational and experiment, *J. Fluid Mech.*, **353**, 285–312.
Huebner, K.H. (1975). *The Finite Element method for Engineers*, John Wiley & Sons, Inc., New York.
Incropera, F.P. & DeWitt, D.P. (1996). *Fundamentals of Heat and Mass Transfer*, John Wiley & Sons, Inc., New York.
Kojić, M. & Bathe, K.J. (2005). *Inelastic Analysis of Solids and Structures*, Springer-Verlag, Berlin.
Mills, A.F. (1999). *Heat Transfer*, Prentice-Hall, Englewood Cliffs, NJ.
Mills, A.F. (2001). *Mass Transfer*, Prentice-Hall, Englewood Cliffs, NJ.
Munson, B.R., Young, D.F. & Okiishi, T.H. (1998). *Fundamentals of Fluid Mechanics*, John Wiley & Sons, Inc., New York.
Nitikitpaiboon, C. & Bathe, K.J. (1993). An arbitrary Lagrangian–Eulerian velocity potential formulation for fluid–structure interaction, *Comp. Struc.*, **47**, 871–91.
Uchida, S. & Aoki, H. (1977). Unsteady flows in a semi-infinite contracting or expanding pipe, *J. Fluid Mech.*, **82**, 371–87.

8

Discrete Particle Methods for Modeling of Solids and Fluids

A selection of discrete particle methods is presented in this chapter. These methods are further implemented in the numerical simulations of bioengineering problems. We begin with molecular dynamics (MD) as the most fundamental method in the discrete particle approach, and then introduce the dissipative particle dynamics (DPD) mesoscale method, a recently developed method which is suitable for modeling flows of complex fluids. Coupling the DPD with the standard FE method in a multiscale modeling scheme is also presented. Further, the basic relations for the smoothed particle hydrodynamics (SPH) method – as the true discrete particle method – is described, followed by a meshless (continuum-based) element free-Galerkin (EFG) method which relies on 'free particles' representation. These methods have a potential for applications in modeling complex bioengineering problems (e.g. in Chapters 13 and 14).

Each section ends with typical examples and references to others included in Software on the web.

8.1 Molecular dynamics

8.1.1 Introduction

The molecular dynamics (MD) method is one of the major computational tools for industrial applications and scientific studies in statistical mechanics, condensed-matter physics, chemistry and materials science.

The simulation of fluids in MD dates from the mid-1950s (Fermi *et al.* 1955, Alder & Wainwright 1957) in which the phase diagram of a hard sphere system was investigated.

Today, MD simulations are increasingly popular in the field of solid and fluid mechanics and in the last decade several review articles have appeared, starting with the work of Koplik and Banavar (1995), who discussed the formulation of continuum flow deductions from atomistic simulations.

8.1.2 Differential equations of motion and boundary conditions

Interaction Forces

The most rudimentary microscopic model is based on spherical particles (atoms) that interact with one another. The two principal features of an interatomic force are resistance to compression, and the binding of atoms together in the solid and liquid states. Potential functions are usually used for describing these characteristics of the interaction forces.

One of the best-known potential functions is the Lennard-Jones (LJ) potential (Rapaport 2004). For a pair of atoms i and j located at $\mathbf{r}_i$ and $\mathbf{r}_j$ position vectors, the (LJ) potential energy is

$$V(r_{ij}) = \begin{cases} 4\varepsilon\left[\left(\dfrac{\sigma}{r_{ij}}\right)^{12} - \left(\dfrac{\sigma}{r_{ij}}\right)^{6}\right] & r_{ij} < r_c \\ 0 & r_{ij} \geq r_c \end{cases} \qquad (8.1.1)$$

where $\mathbf{r}_{ij} = \mathbf{r}_i - \mathbf{r}_j$ and $r_{ij} = |\mathbf{r}_{ij}|$. The parameter ε governs the strength of the interaction and $\sigma = (\sigma_i + \sigma_j)/2$ is the effective mean particle diameter for particles i and j, with diameters σ_i and σ_j, and $r_c = 2^{1/6}\sigma$ is the cutoff limiting separation.

The force corresponding to $V(r_{ij})$ is

$$\mathbf{f}_{ij} = -\nabla V(r_{ij}), \quad \text{with components} \quad (\mathbf{f}_{ij})_k = -\frac{\partial V(r_{ij})}{\partial x_k} \qquad (8.1.2)$$

where x_k, $k = 1, 2, 3$ are the coordinates of the particle 'i'.

Differential Equations of Motion

In order to obtain motions of MD particles (atoms), which include particle trajectories, velocities and accelerations, the Newton law is applied for each particle. Therefore, we have that the differential equations of motion are:

$$m\ddot{\mathbf{r}}_i = \mathbf{f}_i + \mathbf{f}_i^{ext} = \sum_{j(j \neq i)} \mathbf{f}_{ij} + \mathbf{f}_i^{ext} \qquad i = 1, 2, \ldots, N \qquad (8.1.3)$$

Here N is the number of particles, and m_i is mass of the particle i. Also $\mathbf{f}_i$ is the total interaction force exerted on particle i by all other particles j; practically, only particles within a domain of influence (such as r_c in (8.1.1)) are used. And $\mathbf{f}_i^{ext}$ is the external force acting on the particle i.

In these equations Newton's third law implies that $\mathbf{f}_{ji} = -\mathbf{f}_{ij}$, so each atom pair need only be examined once. Usually r_c is small compared with the size of the model domain, which significantly reduces the computational effort.

Boundary Conditions

The macroscopic system size (measured in mm) is of the order of 10^{24} particles, which is still much larger than 10^9 memory size order of a modern supercomputer, so that we

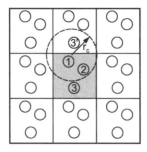

Fig. 8.1.1 A two-dimensional periodic system with the meaning of period boundary condition. The boxes surrounding the shaded box are considered to be periodic images of this box. A wraparound effect: it is taken that particles 1 and 2 interact with particle 3' which is an adjacent copy of particle 3

have a real computer simulation limitation. Hence, only a small domain of the space can be modeled by MD.

To overcome the MD main limitation, the so-called periodic boundary conditions are used, giving satisfactory solutions. Namely, it is possible to model one space 'box' by MD and assume that the solutions in the surrounding boxes are the same. Use of periodic boundary conditions in two dimensions is illustrated in Fig. 8.1.1. The central simulation box, which is shaded in the figure, is surrounded by its periodic images. When a moving particle leaves the simulation box at one boundary, one of its images simultaneously enters this box at the opposite boundary. Therefore the total number of particles in the system is conserved. There is also a wraparound effect when the particles lying within the distance r_c of a boundary, interact with particles in an adjacent copy of the box, or, equivalently, with particles near the opposite boundary. For example, we have in Fig. 8.1.1 that the particles 1 and 2 interact with the particle 3' which is an adjacent copy of particle 3.

MD Algorithm

We here summarize the main steps of an MD algorithm. They are given in Table 8.1.1.

Table 8.1.1 Main steps of an MD algorithm

Input phase
Define initial velocities and coordinates.

Data structure initialization
Perform various auxiliary calculations, such as memory array allocation, dividing the simulation box into small cells, with definition of neighboring cells.

Time stepping
For each discrete value of time $t = n \, \Delta t$, $n = 1, 2, \ldots, N_{steps}$
  Loop over all particles, $j = 1, 2, \ldots N$;
  Calculate forces;
  Apply boundary conditions where necessary;
  Calculate new coordinates $r_j \, (t + \Delta t)$ and new velocities $v_j \, (t + \Delta t)$

Output
Store information necessary for consequent analyses and for a possible continuation run.

A simple choice to start with an MD simulation is to use a regular lattice, such as the square in 2D or cubic lattice in 3D, with uniform atom positions and desired density. The initial velocities are assigned to have random directions, with a fixed magnitude based on temperature.

The equations of motion are usually integrated by the so-called leapfrog method (Rapaport 2004). This two-step integration method is applied to each component of particle coordinates and velocities, as

First step:

$$^{t+\Delta t/2}v = {}^{t}v + \Delta t/2 \, {}^{t}a, \quad {}^{t+\Delta t}r = {}^{t}r + \Delta t \, {}^{t+\Delta t/2}v \quad (8.1.4)$$

Second step:

$$^{t+\Delta t}v = {}^{t+\Delta t/2}v + \Delta t/2 \, {}^{t+\Delta t}a \quad (8.1.5)$$

where r stands for coordinate, v for velocity and a for acceleration components. According to this algorithm, we first (first step) calculate velocities at the half-time step ($\Delta t/2$) using the accelerations at start of the time step (first of equations (8.1.4)); and then determine coordinates at the end of the time step using the half-time velocity (second equation in (8.1.4)). In the second step, to obtain velocities at end of the time step, velocities at the half-time are corrected by adding the term with the use acceleration at the end of the time step.

8.1.3 Examples

Example 8.1-1. Planar Poiseuille flow – solution by MD method
It is assumed that a pressure gradient acts along the channel causing the fluid flow. Initially, the fluid is at rest. The following boundary conditions are used: when an atom reaches either of the rigid walls, it is reflected back into the interior with velocity having a fixed value (corresponding to the wall temperature), but with a random direction. Also, periodic boundary conditions are used: the number of atoms leaving the outlet is equal to the number of atoms entering the inlet.

Steady solution is reached after 32 000 time steps of size $\Delta t = 5.0 \times 10^{-3}$. The results for the flow velocity for two different times are presented in Fig. E8.1-1.

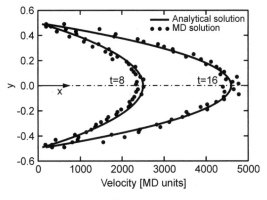

Fig. E8.1-1 Velocity profile for planar Poiseuille flow for two different times $t = 8$ and $t = 16$ (MD units)

8.2 Dissipative Particle Dynamics (DPD) method

8.2.1 Introduction to mesoscale DPD modeling

Although computer technology is continuously advancing, still there are limitations facing the ever increasing demands in computer modeling of scientific and engineering problems. As stated in Section 8.1, we have such limitations in MD modeling of domains of only several millimeters in size and over a time span of seconds. An illustrative example is related to successful modeling of protein conformation changes (Gerstein & Levitt 2005). However, the MD model is still very far from satisfying the needs in molecular biology of living cells: In an MD modeling of a small protein in water, half a million sets of Cartesian coordinates are generated in a nanosecond time period for the positions of 10 000 atoms. The handling of such large amounts of data is beyond the practical capabilities of computer hardware and software currently available.

One approach to overcome the limitations of the MD is so-called coarse graining, i.e. discretization of continuum (fluids and solids) into mesoscale particles of micron length scale and micro-seconds time scale, considering these particles as clusters of atoms. This change of scales can be seen in Fig. 8.2.1 where the length and time scale domains are shown, starting from the quantum mechanics scale to the macroscale. The mesoscale is typically in the range 10–1000 nm and 1 ns–10 ms.

The discretization into mesoscale particles relies on the Voronoi cell division (tessellation) of a continuum (e.g. Flekkoy *et al.* 2000, Serrano *et al.* 2002, Boryczko *et al.* 2003, Espanol 1998), see Fig. 8.2.2. As in the case of MD, the Lagrangian description of motion is employed, with appropriate quantification of interaction forces. One of the most developed mesoscale discrete particle methods is the dissipative particle dynamics (DPD) method, originating from work of Hoogerbrugge and Koelman (1992). The DPD method is particularly suitable for modeling polymeric and other complex fluid systems. We further present the basics of the DPD.

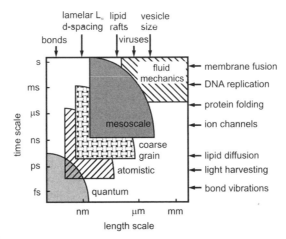

Fig. 8.2.1 Temporal and spatial scale domains in modeling of biofluids and biosolids, accessible by computational techniques (according to Nielsen *et al.* 2004)

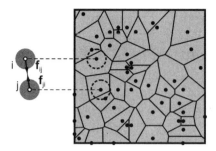

Fig. 8.2.2 Division of space into Voronoi cells and representation of cells by mesoscale discrete particles. A discrete particle 'i' is a cluster of MD particles (atoms) such that all MD particles within the cell are closer to point 'i' than to any other discrete particle. The interaction forces between particles 'i' and 'j' are $\mathbf{f}_{ij} = -\mathbf{f}_{ji}$

8.2.2 Basic DPD equations

Differential Equations of Motion and Interaction Forces

In DPD, as in molecular dynamics, the evolution of the particle position, $\mathbf{r}_i$, can be obtained by application of Newton's second law:

$$\dot{\mathbf{r}}_i = \mathbf{v}_i$$

$$\dot{\mathbf{v}}_i = \frac{1}{m_i} \sum_{j \neq i}^{N} \hat{\mathbf{f}}_{ij} = \sum_{j \neq i}^{N} \mathbf{f}_{ij} \qquad (8.2.1)$$

where $\mathbf{v}_i$ is the particle velocity, m_i is the particle mass; $\hat{\mathbf{f}}_{ij}$ is the force acting on particle i due to particle j, while $\mathbf{f}_{ij}$ is this force per unit mass (see Fig. 8.2.2); and the dot indicates a time derivative. Here, we have neglected the external forces for simplicity of further presentation.

The interaction forces can be represented as the sum of three forces (Espanol & Warren 1995): conservative (repulsion) $\mathbf{f}_{ij}^C$, dissipative $\mathbf{f}_{ij}^D$, and random force $\mathbf{f}_{ij}^R$,

$$\mathbf{f}_{ij} = \mathbf{f}_{ij}^C + \mathbf{f}_{ij}^D + \mathbf{f}_{ij}^R \qquad (8.2.2)$$

These forces can be expressed as (with no sum on i, j)

$$\mathbf{f}_{ij}^C = a_{ij}(1 - r_{ij}/r_c)\mathbf{r}_{ij}^0, \quad \mathbf{f}_{ij}^D = -\gamma w^D (\mathbf{v}_{ij} \cdot \mathbf{r}_{ij}^0)\mathbf{r}_{ij}^0, \quad \mathbf{f}_{ij}^R = a^R w^R (\Delta t)^{-1/2} \xi_{ij} \mathbf{r}_{ij}^0 \qquad (8.2.3)$$

Here, a_{ij} is the maximum repulsion force per unit mass, r_{ij} is the distance between particles i and j, $\mathbf{r}_{ij}^0 = \mathbf{r}_{ij}/r_{ij}$ is the unit vector pointing in direction from j to i, γ stands for the friction coefficient, and a^R is the amplitude of the random force. Also, w^D and w^R are weight functions for dissipative and random forces, dependent on the distance from the particle i; and ξ_{ij} is a random number with zero mean and unit variance. The domain of influence of the interaction forces is r_c, hence $\mathbf{f}_{ij} = \mathbf{0}$ for $r_{ij} > r_c$. The coefficient $(\Delta t)^{-1/2}$ multiplying the random force comes from the integration of the stochastic equations of motion (for physical interpretation of this coefficient see Español & Warren 1995, and Groot & Warren 1997).

Further, in order that a DPD fluid system possess a Gibbs–Boltzmann equilibrium state, the following relation between the weight functions of the dissipative and random forces, w^D and w^R, must hold (Español & Warren 1995):

$$w^D = \left(w^R\right)^2 \qquad (8.2.4)$$

Also the amplitude of the random force a^R is related to the absolute temperature T,

$$a^R = (2k_B T \gamma / m_i)^{1/2} \tag{8.2.5}$$

where k_B is the Boltzmann constant. The weight functions can be expressed in the form (Groot & Warren 1997)

$$w^D = (1 - r_{ij}/r_c)^2, \quad w^R = 1 - r_{ij}/r_c \tag{8.2.6}$$

DPD Boundary Conditions

Implementation of boundary conditions in DPD is not simple and straightforward. There are several approaches for imposing boundary conditions. For example, to impose planar shear conditions, according to the so-called Lees–Edwards method, it is assumed that the upper wall in a periodic box is moving with velocity $V_x/2$ and lower wall with $-V_x/2$. A particle crossing the upper boundary of the box at time t is reintroduced through the lower boundary with its x-coordinate shifted by $-V_x t$ and the x-velocity decreased by V_x. For a particle crossing the lower boundary of the box, the x-coordinate shift is $V_x t$ and the x-velocity is increased by V_x.

In the method of freezing particles, the layer of particles at the walls is attached to the walls, but continues to interact with other particles.

A no-slip condition at the rigid walls can be imposed by employing the specular and bounce-back reflections boundary conditions (Haber et al. 2006). The local specular reflection is expressed by the equations:

$$\begin{aligned} \mathbf{v}_{Wj}^+ \cdot \mathbf{n}_W &= -\mathbf{v}_{Wj}^- \cdot \mathbf{n}_W \\ \mathbf{v}_{Wj}^+ \cdot (\mathbf{I} - \mathbf{n}_W \mathbf{n}_W) &= \mathbf{v}_{Wj}^- \cdot (\mathbf{I} - \mathbf{n}_W \mathbf{n}_W) \end{aligned} \tag{8.2.7}$$

where $\mathbf{I}$ is the idem dyadic, the minus and plus superscripts denote velocities before and after collision with the wall, $\mathbf{v}_{Wj}$ stands for the velocity of particle j relative to the wall velocity, and $\mathbf{n}_W$ is a unit vector perpendicular to the wall. Thus, only the normal component is reflected while the tangential components remain unaltered. The bounce-back reflection is expressed by the relation

$$\mathbf{v}_{Wj}^+ = -\mathbf{v}_{Wj}^- \tag{8.2.8}$$

The Maxwellian reflection method has also been used (e.g. Revenga et al. 1998). According to this method, it is assumed that particles crossing or hitting a wall are randomly reflected back to the flow, with a Maxwellian velocity distribution centered around the wall velocity.

Finally, we cite a method for imposing no-slip boundary conditions that includes a frozen layer of particles near the wall, bounce-back reflection and a modification of the maximum conservative force $(a_{ij})_w$ for particles interacting with particles within the frozen layer (Pivkin & Karniadakis 2005).

In examples given on the web the above boundary conditions are implemented to examine their suitability for applications.

Table 8.2.1 The Verlet algorithm for integration of the DPD differential equations

$$^{n+1}\mathbf{r}_i = {}^n\mathbf{r}_i + \Delta t\, {}^n\mathbf{v}_i + \frac{1}{2}(\Delta t)^2\, {}^n\mathbf{f}_i \quad \Rightarrow \quad {}^{n+1}\tilde{\mathbf{v}}_i = {}^n\mathbf{v}_i + \frac{1}{2}\Delta t\, {}^n\mathbf{f}_i$$

$$^{n+1}\mathbf{f}_i = \mathbf{f}_i\left({}^{n+1}\mathbf{r},\, {}^{n+1}\tilde{\mathbf{v}}\right) \quad \Rightarrow \quad {}^{n+1}\mathbf{v}_i = {}^n\mathbf{v}_i + \frac{1}{2}\Delta t\left({}^n\mathbf{f}_i + {}^{n+1}\mathbf{f}_i\right)$$

Integration of DPD Equations

In integrating differential equations of motion (8.2.1) with a time step Δt, the resulting interaction force $\mathbf{f}_i$ is expressed as

$$\mathbf{f}_i = \sum_{j \neq i} \left(\mathbf{f}_{ij}^C + \mathbf{f}_{ij}^D + \mathbf{f}_{ij}^R\right) \tag{8.2.9}$$

Note that the time step size enters into the evaluation of the random force (see (8.2.3)).

A simple approach to perform the integration of differential equations (8.2.1) is to implement a simple Euler forward method. It was, however, shown that a so-called velocity-Verlet algorithm (Groot & Warren 1997) gives more accurate results. The integration scheme for a time step 'n' is as given in Table 8.2.1, where the left upper indices 'n' and '$n+1$' correspond, respectively, to the start and end of time step. Note that the particle position $^{n+1}\mathbf{r}_i$ at the end of time step is calculated by the Euler forward method, while the force $^{n+1}\mathbf{f}_i$ and velocity $^{n+1}\mathbf{v}_i$ are determined with use of a mid-step velocity $^{n+1}\tilde{\mathbf{v}}_i$.

Calculation of motion of particles according to Table 8.2.1 requires determination of particles which are in a neighborhood of the current particle 'i'. The overall efficiency of the solution depends, among other computational procedures, on the efficiency of the determination of neighboring particles. One of commonly used methods is the *neighbor-list* method (Rapaport 2004). There, the total domain is subdivided into the cells of the size r_c and searching is performed using these cells.

8.2.3 Examples

Example 8.2-1. Modeling of Poiseuille flow in a channel

In the simulation of fluid flow between two parallel walls (Poiseuille flow, see Example 3.3-1), a total of 600 simple DPD particles is used. The fluid domain in $x - y$ plane is defined by $-25 < x < 25$ and $-10 < y < 10$. The periodic boundary conditions are imposed along the x-direction. The gravity force in the flow direction, $g = 0.05$, is applied to each fluid particle and this drives the flow. Details about this example are given on the web (see Software).

The region is divided into 10 bins across the channel, while the simulations were run for 1 000 000 time steps (time step is $\Delta t = 10^{-4}$, DPD units) and the results were averaged over the last 100 000 time steps. Figure E8.2-1b shows agreement between the DPD and analytical solutions.

EXERCISE

Use the Software on the web to find solution by varying the model parameters.

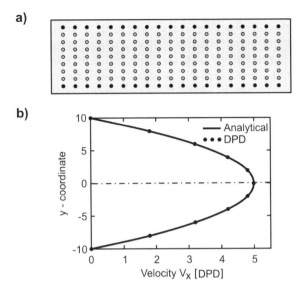

Fig. E8.2-1 DPD modeling of Poiseuille flow. (a) Schematics of discretization into DPD particles; (b) Velocity profile (flow is in x-direction)

8.3 Multiscale modeling, coupling DPD-FE for fluid flow

In this section we present a methodology of coupling the two scales for fluid flow, mesoscopic and macroscale, modeled by the DPD method (Section 8.2) and continuum-based FE method (Section 7.4). This approach is called the mesoscopic bridging scale method (MBS) (Kojić et al. 2008). The basic equations are given, with application on a simple example.

8.3.1 Introduction to multiscale modeling

As discussed in Section 8.2, a possibility to overcome limitations of the MD model is to use a multiscale approach which appropriately couples the MD and continuum methods. A review of the multiscale methods is given by Curtin and Miller (2003), Liu et al. (2004a) (see also Liu et al. 2004b and references therein) and Nielsen et al. (2004). An extension of this multiscale approach to further couple the mesoscale and macroscale modeling is presented in this section. It relies on the bridging scale (BS) method (W.K. Liu and co-workers, see e.g. Wagner & Liu 2003, Tang et al. 2006) of coupling MD and FE models.

The main idea of the MBS method is that the fluid velocity is decomposed into the coarse-scale mean velocity and fine-scale velocity fluctuation of a mesoscopic particle. The mean velocity can be calculated by a continuum-based method, such as the FE method, and the fine-scale correction velocity is determined by a mesoscopic discrete particle method (e.g. DPD). Use of the appropriate projection operator provides the orthogonality of the fine-scale velocities and coarse-scale (FE) interpolation functions. The most significant result is that this orthogonality allows the total kinetic energy of the material system to be represented as a sum of the coarse- and fine-scale kinetic energies, uncoupled with respect to the velocities in the two scales. Finally, this form of the kinetic energy leads to the two systems of differential equations of motion coupled in the force terms only.

156   COMPUTER MODELING IN BIOENGINEERING

The MBS approach is particularly attractive for modeling a dilute mixture flow with a detailed insight into flow in certain local regions, as in the case of, for example, blood flow in a large artery with growing thrombus at the wall caused by adhesion of platelets. Development of the thrombus is dependent on both the global hemodynamics within the artery, and local flow and interactions between blood constituents within a small region around the thrombus. Continuum methods are applicable for modeling global artery hemodynamics, but are inadequate for determination of local flows which involve platelet activation, aggregation and adhesion. Details of this modeling are given in Section 14.5. We here give one simple example in Section 8.3.3 (see also Example 14.4-1).

8.3.2 Basic equations and boundary conditions

Differential Equations of Motion

A fluid domain is discretized into the mesoscale discrete particles, further called 'particles', representing the fine-scale model; and into finite elements as the coarse-scale model. One finite element is shown in Fig. 8.3.1. The basic assumption is that the velocity of a particle 'i', $\mathbf{v}_i$, at any time, can be expressed as

$$\mathbf{v}_i = \bar{\mathbf{v}}_i + \mathbf{v}'_i \tag{8.3.1}$$

where $\bar{\mathbf{v}}_i$ is the coarse-scale velocity, representing the mean particle velocity, obtained by the FE method; and $\mathbf{v}'_i$ is the velocity correction, or fine-scale velocity fluctuation, obtained from the fine-scale solution. According to (7.4.2), the coarse-scale velocity $\bar{\mathbf{v}}_i$ can be expressed in terms of nodal velocities, $\mathbf{V}$, as

$$\bar{\mathbf{v}}_i = \mathbf{N}_i \mathbf{V} \tag{8.3.2}$$

where $\mathbf{N}_i(r^i, s^i, t^i)$ is the matrix of interpolation functions for velocities within the finite element, with the natural coordinates of particle 'i'; and $\mathbf{V}$ is the nodal velocity vector. The relations (8.3.1) and (8.3.2) can be written for all particles within the finite element,

$$\mathbf{v} = \bar{\mathbf{v}} + \mathbf{v}' \tag{8.3.3}$$

$$\bar{\mathbf{v}} = \mathbf{N}\mathbf{V} \tag{8.3.4}$$

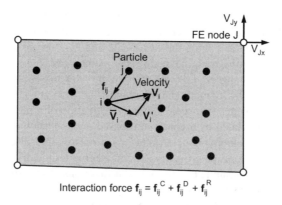

Fig. 8.3.1 Discretization of fluid within a finite element into mesoscopic discrete particles; velocities and interaction forces (2D representation) (according to Kojić et al. 2008)

DISCRETE PARTICLE METHODS FOR MODELING OF SOLIDS AND FLUIDS

The vector **v** and matrix **N** are defined by (we use $\mathbf{v}^T$ and $\mathbf{N}^T$ for more compact writing):

$$\mathbf{v}^T = \{\mathbf{v}_1^T, \mathbf{v}_2^T \ldots \ldots, \mathbf{v}_{n_a}^T\}^T \tag{8.3.5}$$

and

$$\mathbf{N}^T = [\mathbf{N}_1^T, \mathbf{N}_2^T, \ldots \ldots, \mathbf{N}_{n_a}^T] \tag{8.3.6}$$

Here n_a is the number of particles within the finite element. Note that dimensions of the vector **v** (and **v**′) and matrix **N** are $3n_a$ and $3n_a \times 3N$, respectively, where N is the number of element nodes. Obviously, general 3D flow conditions are considered.

We now use a projection operator (a matrix) **Q** to express the velocity vector **v**′ in terms of the nodal velocity **V**, as

$$\mathbf{v}' = \mathbf{Q}\mathbf{v} \tag{8.3.7}$$

The projection operator is obtained by the minimization of a properly defined residual (see Wagner & Liu 2003, Tang et al. 2006, Kojić et al. 2008), so that the kinetic energy of a finite element, E_k, can be expressed as the sum of two terms, kinetic energy of the coarse (macroscale), $\overline{E}_k$, and kinetic energy of the velocity corrections from the fine scale (mesoscale), E'_k,

$$E_k = \overline{E}_k + E'_k \tag{8.3.8}$$

where

$$\overline{E}_k = \frac{1}{2}\overline{\mathbf{v}}^T \mathbf{M}_A \overline{\mathbf{v}} = \frac{1}{2}\mathbf{V}^T \mathbf{M} \mathbf{V} = \frac{1}{2}\mathbf{V}^T \overline{\mathbf{M}} \mathbf{V} \tag{8.3.9}$$

and

$$E'_k = \frac{1}{2}\mathbf{v}'^T \mathbf{M}_A \mathbf{v}' \tag{8.3.10}$$

Here, $\mathbf{M}_A$ and $\mathbf{M} \equiv \overline{\mathbf{M}}$ are the mass matrices corresponding to all particles and FE model, respectively.

Note that the terms $\overline{E}_k$ and E'_k are decoupled with respect to the velocities of the two scales. Details of the derivations resulting in (8.3.8) are given on the web – Theory, Chapter 8. This decomposition of kinetic energy is the result of the fundamental importance for the MBS method for fluids.

From the principle of virtual power (With 2005) follow the differential equations of motion of fluid within one finite element. This mechanical system possesses $3n_a + 3N$ degrees of freedom, corresponding to particle fluctuation velocities **v**′ and FE nodal velocities **V**, respectively. The system is subjected to external and internal forces. The differential equations are

$$\mathbf{M}_A \dot{\mathbf{v}}' = \mathbf{f}'^{\text{ext}} + \mathbf{f}'^{\text{int}} \tag{8.3.11}$$

where $\mathbf{f}'^{\text{ext}}$ and $\mathbf{f}'^{\text{int}}$ are the external force (such as gravity, or inertial forces due to motion of the reference coordinate system) and internal force – from action of surrounding particles, respectively; and

$$\mathbf{M}\dot{\mathbf{V}} = \mathbf{F}^{\text{ext}} + \mathbf{F}^{\text{int}} \tag{8.3.12}$$

where the vectors $\mathbf{F}^{ext}$ and $\mathbf{F}^{int}$ are the external and internal forces corresponding to the FE nodal velocity vector $\mathbf{V}$ (see Section 7.4). The forces $\mathbf{f}'^{ext}$ and $\mathbf{f}'^{int}$ can be further expressed in terms of the forces $\mathbf{f}^{ext}$ and $\mathbf{f}^{int}$ acting on particles (see Section 8.2.2) as

$$\mathbf{f}'^{ext} + \mathbf{f}'^{int} = \mathbf{Q}^T \left(\mathbf{f}^{ext} + \mathbf{f}^{int} \right) \tag{8.3.13}$$

The system of equations (8.3.11) and (8.3.12) looks fully uncoupled and independent. However, the internal finite element forces $\mathbf{F}^{int}$ are evaluated using the stresses within the fluid, as described in Section 7.4. The stresses, on the other hand, can be calculated from the interaction forces among particles, using the Irving–Kirkwood model (Fan et al. 2003, Ren 2005, Kojić et al. 2006)

$$\boldsymbol{\sigma} = -n \left\langle \left[\sum_i m_i \hat{\mathbf{v}}_i \hat{\mathbf{v}}_i + \frac{1}{2} \sum_i \sum_{j \neq i} \mathbf{r}_{ij} \mathbf{f}_{ij} \right] \right\rangle \tag{8.3.14}$$

where n is the number density of particles; m_i is the particle mass; the vector $\hat{\mathbf{v}}_i$ is defined as $\hat{\mathbf{v}}_i = \mathbf{v}_i - \overline{\mathbf{v}}(\mathbf{x})$; $\overline{\mathbf{v}}(\mathbf{x})$ is the stream velocity at the position $\mathbf{x}$; $\mathbf{r}_{ij} = \mathbf{r}_i - \mathbf{r}_j$; and $\langle \ldots \rangle$ denotes the ensemble average. Therefore, the following functional relationship can be written:

$$\mathbf{F}^{int} = \mathbf{F}^{int}\left(\mathbf{f}^{int} \right) \tag{8.3.15}$$

This leads to the result that the differential equations of motion (8.3.11) and (8.3.12) are coupled through the force terms.

Instead to evaluate the forces $\mathbf{f}'^{ext}$ and $\mathbf{f}'^{int}$ and then calculate the fluctuation velocities $\mathbf{v}'$ from (8.3.11), it is more efficient to determine the forces $\mathbf{f}^{ext}$ and $\mathbf{f}^{int}$, use them to find the stresses within the fluid from (8.3.14), and then determine the nodal forces of the finite element needed in (8.3.12).

Coupling Navier–Stokes and DPD Equations

In order to couple the finite element balance equations and the mesoscale model, we write the incremental-iterative equation (7.4.13) as

$$\begin{bmatrix} \frac{1}{\Delta t}\mathbf{M} + {}^{n+1}\tilde{\mathbf{K}}_{vv}^{(i-1)} & \mathbf{K}_{vp} \\ \mathbf{K}_{vp}^T & 0 \end{bmatrix} \begin{Bmatrix} \Delta \mathbf{V}^{(i)} \\ \Delta \mathbf{P}^{(i)} \end{Bmatrix}$$
$$= \begin{Bmatrix} {}^{n+1}\mathbf{F}^{int(i-1)} \\ 0 \end{Bmatrix} + \begin{Bmatrix} {}^{n+1}\mathbf{F}_{ext}^{(i-1)} \\ 0 \end{Bmatrix} - \begin{bmatrix} \frac{1}{\Delta t}\mathbf{M} + {}^{n+1}\hat{\mathbf{K}}_{vv}^{(i-1)} & \mathbf{K}_{vp} \\ \mathbf{K}_{vp}^T & 0 \end{bmatrix} \begin{Bmatrix} {}^{n+1}\mathbf{V}^{(i-1)} \\ {}^{n+1}\mathbf{P}^{(i-1)} \end{Bmatrix} + \begin{Bmatrix} \frac{1}{\Delta t}\mathbf{M}^n\mathbf{V} \\ 0 \end{Bmatrix}$$
(8.3.16)

Here, the internal nodal force vector is

$${}^{n+1}F_{Ki}^{int(i-1)} = -\int_V N_{K,j} \, {}^{n+1}\tau_{ij}^{(i-1)} dV \tag{8.3.17}$$

where ${}^{n+1}\tau_{ij}^{(i-1)}$ are the viscous stresses at end of time step. As can be seen, the modification of (7.4.13) is that we here did not express the viscous stress by a continuum-based constitutive law, as given in equation (3.3.5). Instead, we leave this stress to be evaluated using a mesoscale discrete particle method, such as the DPD method, while the pressure

DISCRETE PARTICLE METHODS FOR MODELING OF SOLIDS AND FLUIDS   159

part $^{n+1}p^{(i-1)}$ of the total stress $^{n+1}\sigma_{ij}^{(i-1)}$ remains in the equation. Note that the internal force $^{t+\Delta t}\mathbf{F}^{\text{int}(i-1)}$ due to viscous stresses $^{t+\Delta t}\boldsymbol{\tau}^{(i-1)}$ replaces the viscous nodal force equal to $\mathbf{K}_\mu{}^{t+\Delta t}\mathbf{V}^{(i-1)}$ on the right-hand side in equation (7.4.13); and the matrix $^{n+1}\widehat{\mathbf{K}}_{vv}^{(i-1)}$ is used as the tangent matrix with respect to the changes of the velocities (see discussion in Section 6.2.3 on the convergence rate in solving nonlinear problems).

Therefore, the basis of coupling between the mesoscale model (DPD) and macroscale model (FE) is established through the equation (8.3.16). In an incremental-iterative solution scheme, we evaluate the nodal FE viscous forces using the mesoscale interaction forces and perform iterations on the coarse scale to satisfy equilibrium of the fluid continuum and given boundary conditions. A detailed computational scheme is given on the web – Theory, Chapter 8. Note that the Navier–Stokes equations express the balance of linear momentum of the fluid within a finite element in which the internal nodal viscous forces are evaluated using the interaction forces among the mesoscale particles; therefore the Lagrangian background, employed in the DPD method, is preserved in this formulation of coupling the two scales.

Local and Global Domains and Boundary Conditions

Finally, in order to achieve the main goals of this multiscale coupling specified in Section 8.3.1, we divide the entire fluid domain into the local domain (one or more), where the both mesoscopic (DPD) and continuum (FE) models are used, and the global domain discretized by the continuum (FE) method only. A schematic representation of this division is shown in Fig. 8.3.2.

In order to calculate the flow field, it is necessary to satisfy the boundary conditions. External boundary conditions are imposed on the coarse scale, i.e. the velocities or the stresses can be prescribed and used in the FE model. However, the boundary conditions at the common boundary between the local and global domains must also be specified. As shown in Fig. 8.3.2, for the DPD model we use the particle velocities at the common boundary to be equal to the coarse-scale FE velocities. Also, the periodic boundary conditions are imposed on the common boundary to keep the number of particles constant.

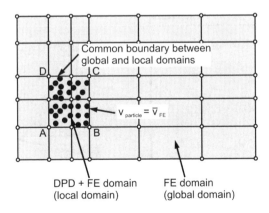

Fig. 8.3.2 Two domains within a flow field: local domain modeled by both DPD and FE methods; global domain modeled by FE method only. Boundary conditions at the common boundary between the local and global domains (velocities of particles are equal to those calculated using the FE model) (according to Kojić et al. 2008)

Various approaches in implementation of the periodic boundary conditions can be found elsewhere (e.g. Revenga *et al.* 1998, Pivkin & Karniadakis 2005, Haber *et al.* 2006, Filipović *et al.* 2008); see also web – Theory, Chapter 8.

8.3.3 Examples

Example 8.3-1. Cavity flow

The flow in the cavity is induced by a horizontal motion of the upper wall with a constant velocity of V (Fig. E8.3-1A). A local DPD domain (a small square region in the figure) is selected at the right upper corner of the cavity and the flow is solved both by the FE method alone and by the FE-DPD multiscale MBS method. In the DPD simulation, 900 mesoscale DPD particles are used. The boundary conditions for the DPD model are as follows: at the no-slip rigid wall, the bounce-back condition (see Section 8.2) was employed; at the moving wall it is assumed that particles have the prescribed velocity $V = 1\,\mu m/s$; at the FE-DPD common boundaries, the velocities of particles are taken to be equal to those of the FE model. Also, the periodic boundary conditions are used for the entering (lower horizontal) and the outlet (left vertical) boundaries to keep number of particles within the DPD box constant. The boundary conditions for the FE model consist of the prescribed velocity at the moving wall and zero velocity at the rigid walls. It is assumed that initial pressures and velocities in the whole domain are equal to zero.

The velocity components (v_x, v_y) along the vertical line, which goes through the middle of the local (FE+DPD) domain and extends through the global (FE) domain to the bottom of the cavity (see the right panel of Fig. E8.3-1B), are calculated both by the FE method alone and by the MBS multiscale method.

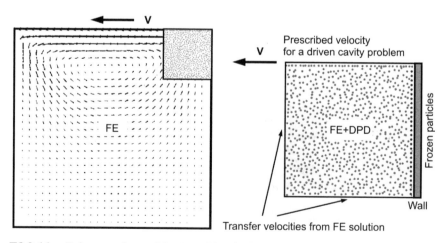

Fig. E8.3-1A Driven cavity problem: multiscale FE-DPD model. Global domain (FE) and local domain (FE+DPD). Dimensions of the cavity are $30 \times 30\,\mu m$, and dimensions of the DPD box are $7.5 \times 7.5\,\mu m$. Velocities are prescribed ($V = 1\,\mu m/s$) on the top boundary, from the right to the left. Other data: $a_{ij} = 25$, $a^R = 3$, $\gamma = 4.5$ (see Section 8.2). The time step used for the DPD simulation was $\Delta t_{DPD} = 0.001$ and $\Delta t_{FE} = 0.1\,s$ for the macro-scale FE model (according to Kojić *et al.* 2008)

DISCRETE PARTICLE METHODS FOR MODELING OF SOLIDS AND FLUIDS

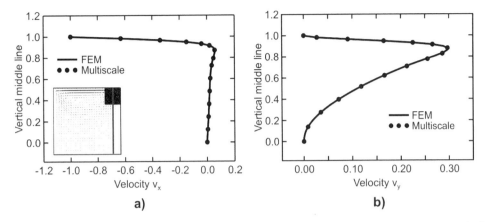

Fig. E8.3-1B Velocity profiles along the vertical middle-line (shown by vertical line) passing through the local (FE + DPD) and global (FE) domains. (a) v_x component; (b) v_y component. The final time, $t = 10\,s$, is reached after total 2 000 000 DPD time steps (and 100 FE time steps) (according to Kojić et al. 2008)

8.4 Smoothed Particle Hydrodynamics (SPH)

8.4.1 Introduction

The smoothed particle hydrodynamics (SPH) is also a truly discrete particle method as the MD and DPD described in the previous sections. The basic idea in the SPH is a representation of a physical field within a continuum by values at discrete points, considered as discrete material particles, using the so-called kernel approximation function. Then the continuum-based partial differential equations of balance are transformed into discrete particle equations. These discrete balance equations do not require space integration and use of a space mesh. The original version of SPH was developed for modeling compressible fluid flows in astrophysical problems (Ginhold & Monaghan 1977, Lucy 1977). Nowadays, applications of this Lagrangian method range from compressible/incompressible fluid flows to the structural mechanics.

The SPH has a potential of applications to complex bioengineering problems, such as blood flow in capillaries (see Section 14.3). We here present the fundamental equations of the SPH and show an example solution of a simple flow of incompressible fluid. The main advantage of this method is that it does not require any mesh, while the shortcoming is a complexity of implementation of boundary conditions.

8.4.2 The basic equations of the SPH method

The fundamental concept of the SPH is expressed by the relation

$$\langle f(\mathbf{r}) \rangle = \int_V f(\mathbf{r}')W(|\mathbf{r} - \mathbf{r}'|, h)\,dV \qquad (8.4.1)$$

Here $\langle f(\mathbf{r}) \rangle$ is the kernel approximation of a function at a space point defined by the position vector $\mathbf{r}$; $W(|\mathbf{r} - \mathbf{r}'|, h)$ is called the *smoothing kernel function* or just *kernel* in

SPH literature (Belytschko *et al.* 1996, Vignjevic 2004), while h defines the size of the kernel support domain with the spatial volume V; and $\mathbf{r}'$ is the position vector of a point within the spatial domain. The SPH kernels have the 'compact support', which means that their value is equal to zero outside the support domain around $\mathbf{r}$:

$$W(|\mathbf{r}-\mathbf{r}'|,h) = 0 \quad \text{for} \quad |\mathbf{r}-\mathbf{r}'| \geq 2h \quad (8.4.2)$$

The kernel has to be normalized, i.e. it has to satisfy the condition:

$$\int_V W(|\mathbf{r}-\mathbf{r}'|,h)\, dV = 1 \quad (8.4.3)$$

The above requirements ensure that the kernel is approaching Dirac's delta function when h tends to zero, and the approximation of the function tends to the exact value:

$$\lim_{h \to 0} W(|\mathbf{r}-\mathbf{r}'|,h) = \delta(|\mathbf{r}-\mathbf{r}'|), \quad \text{and} \quad \lim_{h \to 0} \langle f(\mathbf{r}) \rangle = f(\mathbf{r}) \quad (8.4.4)$$

When the function $f(\mathbf{r})$ is only known at a set of the discrete points, the integral equation (8.4.1) becomes the sum, and we have for $\mathbf{r} = \mathbf{r}^i$:

$$\langle f(\mathbf{r}^i) \rangle \equiv f(\mathbf{r}^i) \equiv f^i = \sum_{j=1}^{N} \frac{m^j}{\rho^j} f^j W^{ij} \quad (8.4.5)$$

where $f^j \equiv f(\mathbf{r}^j)$, $W^{ij} = W(|\mathbf{r}^i - \mathbf{r}^j|, h)$. Also, $m^j/\rho^j = \Delta V^j$ is the volume associated to particle j, and N is the number of particles within the support domain h. The term 'particle' in SPH has the same meaning as in the DPD method: the particle replaces its surrounding material and mass of the particle is constant during motion. A schematics of equation (8.4.5) is shown in Fig. 8.4.1.

In order to express the balance equations which have the form of partial differential equations (e.g. (2.1.12), or (7.1.1)) using the SPH method, it is necessary to derive the expression for partial derivatives or gradient of a function with respect to space coordinates x_α, $\alpha = 1, 2, 3$. Following the basic approximation (8.4.1) we have that

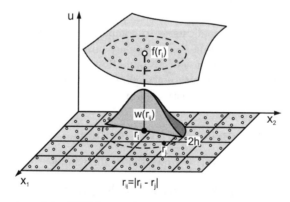

Fig. 8.4.1 Schematics of the SPH interpolation. Value of function f at a discrete point 'i' is interpolated from the values of surrounding points within the support domain of radius $2h$ by use of the kernel function $W(r_{ij})$

$$\left\langle \frac{\partial f(\mathbf{r})}{\partial x_\alpha} \right\rangle = \int_V W(|\mathbf{r}-\mathbf{r}'|, h) \frac{\partial f(\mathbf{r}')}{\partial x'_\alpha} dV \tag{8.4.6}$$

and integrating by parts finally follows

$$\frac{\partial f^i}{\partial x_\alpha} = \sum_{j=1}^{N} \frac{m^j}{\rho^j} f^j \frac{\partial W^{ij}}{\partial x^i_\alpha} \tag{8.4.7}$$

Details of this derivation are given on the web – Theory, Chapter 8. Although the derivation of this expression is done correctly, an empirical relation for the derivatives is recommended (Monaghan 1994):

$$\frac{\partial f}{\partial x_\alpha} = \frac{1}{\rho}\left[\frac{\partial}{\partial x_\alpha}(\rho f) - f\frac{\partial \rho}{\partial x_\alpha}\right] \tag{8.4.8}$$

Next, we list two kernel functions. The most common is the B-spline function (Monaghan & Ginhold 1983):

$$W(\nu, h) = \frac{C}{h^D}\begin{cases} \left(1-\frac{3}{2}\nu^2+\frac{3}{4}\nu^3\right) & \nu < 1 \\ \frac{1}{4}(2-\nu)^3 & 1 \le \nu \le 2, \quad \nu = |\mathbf{r}-\mathbf{r}'|/h \\ 0 & \nu > 2 \end{cases} \tag{8.4.9}$$

where D is the number of the dimensions of the problem (1, 2 or 3). The constant C is the scaling factor which has to provide that equations (8.4.2) and (8.4.3) are satisfied: $C = 2/3$, $C = 10/(7\pi)$, $C = 1/\pi$ for $D = 1, 2, 3$, respectively. The second is a quintic spline, which for 2D problems is:

$$W(\nu, h) = \frac{7}{478\pi}\begin{cases} (3-\nu)^5 - 6(2-\nu)^5 + 15(1-\nu)^5 & 0 \le \nu < 1 \\ (3-\nu)^5 - 6(2-\nu)^5 & 1 \le \nu < 2 \\ (3-\nu)^5 & 2 \le \nu < 3 \\ 0 & \nu \ge 3 \end{cases}, \quad \nu = |\mathbf{r}-\mathbf{r}'|/h \tag{8.4.10}$$

The usage of quintic kernel approximately doubles the computational time, but gives more stable results.

Finally, we give the SPH equations of balance of mass (continuity equation) and balance of linear momentum. They are:

$$\frac{d\rho^i}{dt} = \sum_{j=1}^{N} m^j (v^i_\beta - v^j_\beta) \frac{\partial W^{ij}}{\partial x^i_\beta} \tag{8.4.11}$$

$$\frac{dv^i_\alpha}{dt} = \sum_{j=1}^{N} m^j \left[\left(\frac{\sigma^j_{\alpha\beta}}{(\rho^j)^2} + \frac{\sigma^i_{\alpha\beta}}{(\rho^i)^2}\right)\frac{\partial W^{ij}}{\partial x^i_\beta} + \frac{f^{Vj}_\alpha}{\rho^j} W^{ij}\right] \tag{8.4.12}$$

sum on β : $\beta = 1, 2, 3$

no sum on i

Details of derivation of these equations are given on the web – Theory, Chapter 8.

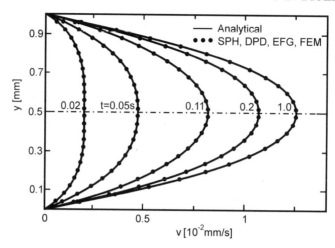

Fig. E8.4-1 Velocity profiles for several times. Analytical and numerical solutions (methods: SPH – smoothed particle hydrodynamics; DPD – dissipative particle dynamics; EFG – element free Galerkin; FEM – finite element)

8.4.3 Examples

Example 8.4-1. Unsteady flow between two plates
We consider flow of incompressible fluid between two stationary infinite plates located at $y = 0$ and $y = H$. The fluid is initially at rest and it is driven by body force (given here as acceleration a) parallel to the x-axis. The analytical solution ($v \equiv v_x$) is given as a series (Joseph *et al.* 1997):

$$v(y,t) = \frac{a}{2\nu}y(y-H) +$$

$$+ \sum_{n=0}^{\infty} \frac{4aH^2}{\nu\pi^3(2n+1)^3} \sin\left(\frac{\pi y}{H}(2n+1)\right) \exp\left(-\frac{(2n+1)^2\pi^2\nu}{H^2}t\right) \quad \text{(E8.4-1.1)}$$

Parameters used in the SPH model are: kinematic viscosity $\nu = 10^{-6} m^2 s^{-1}$, fluid density $\rho = 10^3 kgm^{-3}$, $H = 10^{-3}m$, $a = 10^{-4} ms^{-2}$, $\Delta t = 10^{-4}s$ and 30 particles spanning the space between plates.

Velocity profiles for several times are shown in Fig. E8.4-1. It can be seen that there is very good agreement between the SPH and analytical solutions. Practically the same results are obtained using other numerical methods: DPD, EFG and FE (Detailed analysis of applicability of these methods, including the multiscale method of Section 8.3 is given in Kojic 2008).

8.5 Element-Free Galerkin (EFG) method

8.5.1 Introduction

It can be seen from Chapters 4 to 7 that the finite element (FE) method is very well established and generally applicable. It is also used in bioengineering as the basic computational technique in modeling and is mainly used in this book. However, the method has some shortcomings, such as a need for remising (creation of a new finite element mesh) when the finite elements are too distorted due to large deformations. Or, the method is not suitable for modeling of colloidal fluid flow (such as blood flow in capillaries), where discrete particle methods of Sections 8.3 and 8.4 may be more appropriate. But these discrete methods suffer from, for example, complexity of handling boundary conditions.

The so-called element-free Galerkin (EFG) method can be considered as a computational procedure which overcomes the shortcomings of the abovementioned methods. The fundamental idea of the EFG method is to represent the field of a physical quantity by values at a set of discrete points which are not associated with a mesh as in the FE method, i.e. the points are *element-free*. These points are usually called the *free points*. The approximate value of a quantity at a material point within the domain is obtained by use of the weight-type interpolation functions within a domain of influence. The weight functions decay with the distance from the material point and are negligible outside the domain of influence, as schematically shown in Fig. 8.5.1a.

Finally, note that the EFG is in essence a continuum method. Discretization of the continuum leads to discrete (free) points, but evaluation of the matrices and vectors within the discrete balance equations is performed over the continuous subdomains.

8.5.2 Formulation of the EFG method

The interpolated value $u(\mathbf{r})$ of a physical quantity (spatial function) at a point, which we call the material point, with the position vector $\mathbf{r}$, is given as (Belytschko et al. 1994)

$$u(\mathbf{r}) = \sum_{j}^{m} p_j(\mathbf{r}) a_j(\mathbf{r}) \equiv \mathbf{p}^T(\mathbf{r}) \mathbf{a}(\mathbf{r}) \tag{8.5.1}$$

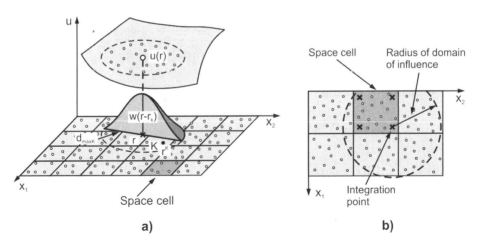

Fig. 8.5.1 Interpolation by EFG method (2D domain). (a) Function $u(\mathbf{r})$ at a material point with the position vector $\mathbf{r}$. The weight function $w(\mathbf{r} - \mathbf{r}^K)$ decreases with the distance between the material point and the free points K; $d_{max\ K}$ defines the size of the domain of influence for the weight function $w(\mathbf{r} - \mathbf{r}^K)$; (b) Space cell used for integration and domain of influence around the integration point

where $p_j(\mathbf{r})$ are the components of the base vector $\mathbf{p}(\mathbf{r})$, expressed as monomials in the coordinates of $\mathbf{r}[x, y, z]$ so that the basis is complete; and m is the basis size. The coefficients $a_j(\mathbf{r})$ are to be determined. The linear and quadratic bases for one-dimensional space are

$$\mathbf{p}^T(\mathbf{r}) = [1, x], \text{ linear, } m = 2; \quad \mathbf{p}^T(\mathbf{r}) = [1, x, x^2], \text{ quadratic, } m = 3 \quad (8.5.2)$$

These bases for the 2D space are

$$\mathbf{p}^T(\mathbf{r}) = [1, x, y], \qquad \text{linear,} \qquad m = 3;$$
$$\mathbf{p}^T(\mathbf{r}) = [1, x, y, x^2, xy, y^2] \qquad \text{quadratic} \qquad m = 6 \qquad (8.5.3)$$

The coefficients $a_j(\mathbf{r})$ are functions of the position vector and are determined by minimizing a weighted quadratic form

$$J = \sum_{K=1}^{n} w\left(\mathbf{r} - \mathbf{r}^K\right) \left[\mathbf{p}^T\left(\mathbf{r}^K\right) \mathbf{a}(\mathbf{r}) - U^K\right]^2 \quad (8.5.4)$$

where K denotes the free point number with the position vectors $\mathbf{r}^K$ and with the value of the function U^K; $w(\mathbf{r} - \mathbf{r}^K)$ is the weight function which depends on the distance between the material point and free point; and n is the number of free points in the domain of influence around the material point. Minimizing J with respect to the coefficients $a_j(\mathbf{r})$, the system of equations is obtained:

$$\frac{\partial J}{\partial a_i} = 2 \sum_{K=1}^{n} w^K \left(p_j^K a_j - U^K\right) p_i^K = 0, \quad \text{sum on } j: \quad j = 1, 2, \ldots, m \quad (8.5.5)$$

where $w^K \equiv w(\mathbf{r} - \mathbf{r}^K)$, $p_j^K \equiv p_j(\mathbf{r}^K)$. Note that w^K also depends on the position vector $\mathbf{r}$ of the material point. This system of equations can be written in the form

$$A_{ij} a_j - B_{iK} U^K = 0, \quad \text{sum on} \quad K \text{ and } j: \quad K = 1, 2, \ldots, n; j = 1, 2, \ldots, m \quad (8.5.6)$$

where the matrices $\mathbf{A}$ and $\mathbf{B}$ are:

$$A_{ij} = \sum_{I=1}^{n} w^I p_i^I p_j^I; \quad B_{iK} = p_i^K w^K, \quad \text{no sum on } K \quad (8.5.7)$$

The matrices $\mathbf{A}$ and $\mathbf{B}$ are of order $m \times m$ and $m \times n$, respectively. Equation (8.5.6) can be solved for the coefficients $a_j(\mathbf{r})$, hence

$$\mathbf{a}(\mathbf{r}) = \mathbf{A}^{-1} \mathbf{B} \mathbf{U}; \quad \text{or} \quad a_j(\mathbf{r}) = \sum_{K=1}^{n} \left(\mathbf{A}^{-1} \mathbf{B}\right)_{jK} U^K \quad (8.5.8)$$

Now, substitution of $\mathbf{a}(\mathbf{r})$ from (8.5.8) into (8.5.1) follows

$$u(\mathbf{r}) = \mathbf{p}^T \mathbf{A}^{-1} \mathbf{B} \mathbf{U} = \sum_{K=1}^{n} \Phi_K(\mathbf{r}) U^K \quad (8.5.9)$$

DISCRETE PARTICLE METHODS FOR MODELING OF SOLIDS AND FLUIDS

where the interpolation function $\Phi_K(\mathbf{r})$ corresponding to the free point K, is

$$\Phi_K(\mathbf{r}) = \sum_{j=1}^{m} (\mathbf{A}^{-1}\mathbf{B})_{jK} p_j \qquad (8.5.10)$$

The form (8.5.9) is the same as in the FE method, equation (4.2.5) or (4.3.2), where now we have the interpolation functions $\Phi_K(\mathbf{r})$ instead of the FE isoparametric interpolation functions N_K. Here, the interpolation functions are expressed in terms of the Cartesian coordinates x_i and x_i^K of the material and free point, while N_K are expressed in terms of the natural coordinates of a finite element.

It is necessary in applications to calculate the derivatives with respect to the coordinates x_i. From (8.5.10) follows

$$\Phi_{K,i} = \sum_{j=1}^{m} \left[(\mathbf{A}^{-1}\mathbf{B})_{jK} p_{j,i} + (\mathbf{A}_{,i}^{-1}\mathbf{B} + \mathbf{A}^{-1}\mathbf{B}_{,i})_{jK} p_j \right] \qquad (8.5.11)$$

where $_{,i} \equiv \partial/\partial x_i$. By differentiation of the equation $\mathbf{A}^{-1}\mathbf{A} = \mathbf{I}$ (see (1.2.9)) we obtain

$$\mathbf{A}_{,i}^{-1} = -\mathbf{A}^{-1}\mathbf{A}_{,i}\mathbf{A}^{-1} \qquad (8.5.12)$$

which can be used to evaluate $\mathbf{A}_{,i}^{-1}$.

We further give one of the weight functions, of the exponential form (Belytschko et al. 1994),

$$w^I\left(d_I^{2k}\right) = \begin{cases} \dfrac{\exp(-d_I/c)^{2k} - \exp(-d_{\max I}/c)^{2k}}{1 - \exp(-d_{\max I}/c)^{2k}}, & d_I \leq d_{\max I} \\ 0, & d_I \leq d_{\max I} \end{cases} \qquad (8.5.13)$$

where $d_I = \|\mathbf{r} - \mathbf{r}^I\|$ is the distance between the material point and the free point I; $d_{\max I}$ is the domain of influence for the weight function w^I; k is the parameter (in our applications we use $k = 1$); and

$$c = \alpha \max \|\mathbf{r}_K - \mathbf{r}_J\| \quad \text{for all free points} \qquad (8.5.14)$$

where $1 \leq \alpha \leq 2$ is recommended. Another form of the weight function and graphical representations are shown on web – Theory, Chapter 8.

Once the interpolation functions are formulated, the procedure is computationally similar to that of the FE method described in Chapter 4 for the solids, or in Chapter 7 for field problems. For an integration point, shown in Fig. E8.5.1b, the 'element' nodes include all free points within the domain of influence. Of course, the FE interpolation functions N_K are now replaced by the EFG interpolation functions Φ_K.

We perform integration over selected volumes as the subdomains (Figs. 8.5.1a,b) of the physical field to form the balance equations for the entire domain. In this case the subdomains are called volume cells and are defined independently of the free points. We simply integrate over the volume cells numerically by, say, Gauss quadrature, and add contributions to the matrices and vectors corresponding to variables at free points (degrees of freedom of free points) within the domain of influence, see Fig. E8.5.1b. The independence of volume cells on the free points is the main advantage of the EFG method. To increase solution accuracy, the number of integration points within a cell can be adjusted to the number of free points (Belytschko et al. 1994, Vlastelica 2003).

The EFG model can be coupled with an FE model. Details about this coupling are given on the web (see web – Theory, Chapter 8).

In one example we illustrate the robustness and accuracy of the EFG method (additional examples are given on the web, within Software).

8.5.3 Examples

Example 8.5-1. Plate with a hole subjected to uniaxial tension
This example is solved using the FE method as in Example 4.4.1, where a detailed description of the problem is given. Here, we model one-quarter of the plate (as in Example 4.4.2) by the EFG method. The mesh showing the EFG cells (132 cells) and free points (156 points) are shown in Fig. E8.5-1a. The geometrical and material data are taken from Belytschko et al. (1994), but replacing inches by centimeters, $R = 1$, $L = 5$; Young's modulus is $E = 3 \times 10^8$ kPa and tensional stress $\sigma = 1$ kPa. The linear basis (see (8.5.3)) is used for the displacement field and the exponential weight function (8.5.13) with $d_{max\,I} = 2$ cm.

The EFG solution for the stress σ_{xx} along the y-axis is shown in Fig. E8.5-1b, which agrees well with the analytical solution (in (E4.4-2.1) enters R/y instead of $d/2y$). Note that the analytical curves in E8.5-1b and Fig. E4.4-2c correspond to the same analytical expression (E4.4-2.1), but the ranges of the coordinate y relative to the hole radius are different ($b/d = 2$, while $L/R = 5$).

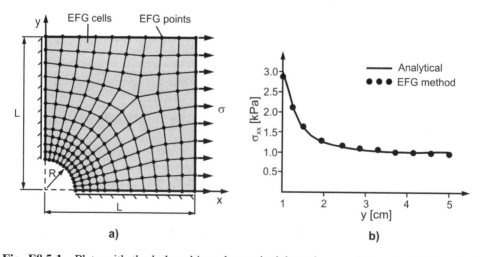

Fig. E8.5-1 Plate with the hole subjected to uniaxial tension – solution by EFG method. (a) Geometry of the EFG discretized domain, loading and boundary conditions (see also Fig. E4.4.2a); (b) Stress distribution along y-axis

References

Alder, B.J. & Wainwright, T.E. (1957). Phase transition for a hard sphere system, *J. Chem. Phys.*, **27**(5), 1208–9.
Belytschko, T., Krongauz, Y., Organ, D., Fleming, M. & Krysl, P. (1996). Meshless methods: an overview and recent developments, *Comp. Meth. Appl. Mech. Eng.*, **139**, 3–47.
Belytschko, T., Lu, Y.Y. & Gu, L. (1994). Element-free Galerkin methods, *Int. J. Num. Meth. Eng.*, **37**, 229–56.
Boryczko, K., Dzwinel, W. & Yuen, D.A. (2003). Dynamic clustering of red blood cells in capillary vessels, *J. Mol. Mod.*, **9**, 16–33.
Curtin, W.A. & Miller, R.E. (2003). Atomistic/continuum coupling in computational material science, *Mod. Simul. Mater. Sci. Eng.*, **11**, 33–68.
Espanol, P. (1998). Fluid particle model, *Phys. Rev. E*, **57**, 2930–48.
Espanol, P. & Warren, P. (1995). Statistical mechanics of dissipative particle dynamics, *Europhys. Lett.*, **30**(4), 191–6.
Fan, X., Phan-Thien, N., Yong, N.T., Wu, X. & Xu, D. (2003). Microchannel flow of a macromolecular suspension, *Phys. Fluids*, **15**, 11–21.
Fermi, E., Pasta, J. & Ulam, S. (1955). Studies in nonlinear problems, *Los Alamos Rep. LA-1940*.
Filipović, N., Ravnić, D., Kojić, M., Mentzer, S.J., Haber, S. & Tsuda, A. (2008). Interactions of blood cell constituents: Experimental investigation and computational modeling by discrete particle dynamics algorithm, *Microvascular Res.*, **75**, 279–284.
Flekkoy, E.G., Coveney, P.V. & De Fabritiis, G. (2000). Foundations of dissipative particle dynamics, *Phys. Rev. E*, **62**, 2140–57.
Gerstein, M. & Levitt, M. (2005). Simulating water and the molecules of life, *Scientific American* (The Water of Life, Special Issue), 24–9.
Ginhold, R.A. & Monaghan, J.J. (1977). Smoothed particle hydrodynamics: theory and application to non-spherical stars, *Mon. Not. R. Astron. Soc.*, **181**, 375–89.
Groot, R.D. & Warren, P.B. (1997). Dissipative particle dynamics: bridging the gap between atomistic and mesoscopic simulation, *J. Chem. Phys.*, **107**(11), 4423–35.
Haber, S., Filipović, N., Kojić, M. & Tsuda, A. (2006). Dissipative particle dynamics simulation of flow generated by two rotating concentric cylinders, *Phys. Rev. E*, **74**, 1–8.
Hoogerbrugge, P.J. & Koelman, J.M.V.A. (1992). Simulating microscopic hydrodynamic phenomena with dissipative particle dynamics, *Europhys. Lett.*, **19**, 155–160.
Joseph, P.M., Fox, P.J. & Zhu, Y. (1997). Modeling low Reynolds number incompressible flows using SPH, *J. Comp. Phys.*, **136**, 214–26.
Kojic, M. (2008). On the application of dicrete particle methods and their coupling to the continuum-based methods within a multiscale scheme, *Advances in Nonlinear Sciences*, Monograph 2, Yugoslavian Academy for Nonlinear Sciences (JANN), to appear.
Kojić, M., Filipović, N. & Tsuda, A. (2006). A multiscale method for bridging dissipative particle dynamics and Navier–Stokes finite element equations for incompressible fluid and its application in biomechanics. *Proc. First South-East European Conference on Comp. Mechanics* (eds. M. Kojić & M. Papadrakakis), Kragujevac, Serbia.
Kojić, M., Filipović, N. & Tsuda, A. (2008). A mesoscopic bridging scale method for fluids and coupling dissipative particle dynamics with continuum finite element method, *Comp. Meth. Appl. Mech. Eng.*, **197**, 821–33.
Koplik, J. & Banavar, J.R. (1995). Corner flow in the sliding plate problem, *Phys. Fluids*, **7**(12), 3118–125.
Liu, W.K., Karpov, E.G., Zhang, S. & Park, H.S. (2004a). An introduction to computational nanomechanics and materials, *Comput. Meth. Appl. Mech. Eng.*, **193**, 1529–78.
Liu, W.K., Qian, D. & Horstemeyer, M.F. (2004b). Preface, *Comput. Meth. Appl. Mech. Eng.*, Special Issue, **193**, iii–iv.
Lucy, L.B. (1977). A numerical approach to the testing of fusion process, *Astron. J.*, **88**, 1013–24.

Monaghan, J.J. (1994). Simulating free surface flows with SPH, *J. Comp. Phys.*, **110**, 399–406.

Monaghan, J.J. & Ginhold, R.A. (1983). Shock simulation by the particle method SPH, *J. Comp. Phys.*, **52**, 374–89.

Nielsen, S.O., Lopez, C.F., Srinivas, G. & Klein, M.L. (2004). Coarse grain models and computer simulation of soft materials, *J. Phys: Condens. Matter*, **16**, 481–512.

Pivkin, I. & Karniadakis, G.E. (2005). A new method to impose no-slip boundary conditions in dissipative particle dynamics, *J. Comp. Phys.*, **207**, 114–28.

Rapaport, D.C. (2004). *The Art of Molecular Dynamics Simulation*, Cambridge University Press, Cambridge, England.

Ren, W. (2005). Heterogeneous multiscale method for the modeling of complex fluids and microfluidics, *J. Comp. Phys.*, **204**, 1–26.

Revenga, M., Zuniga, I., Espanol, P. & Pagonabarraga, I. (1998). Boundary models in DPD, *Int. J. Mod. Phys. C*, **9**, 1319–31.

Serrano, M., De Fabritiis, G., Espanol, P., Flekoy, E.G. & Coveney, P.V. (2002). Mesoscopic dynamics of Voronoy fluid particles, *J. Phys. A*, **35**, 1605–25.

Tang, S., Hou, T.H. & Liu, W.K. (2006). A mathematical framework of the bridging scale method, *Int. J. Num. Meth. Eng.*, **65**, 1688–713.

Vignjevic, R. (2004). Review of development of the smooth particle hydrodynamics (SPH) method, Dynamics and Control of Systems and Structures in Space (DCSSS), 6th Conference, Riomaggiore, Italy.

Vlastelica, I. (2003). *Methods of solving elastoplastic deformation of non-porous and porous metals in fracture mechanics*. Ph.D. Thesis (in Serbian), Faculty of Mechanical Engineering., Univesity of Kragujevac, Serbia.

Wagner, G.J. & Liu, W.K. (2003). Coupling of atomistic and continuum simulations using a bridging scale decomposition, *J. Comp. Phys.*, **190**, 249–74.

With, G. de (2005). *Structure, Deformation, and Integrity of Materials I, II*. Wiley-VCH Verlag GmbH & Co. KGaA, Weinheim, Germany.

Part III

Computational Methods in Bioengineering

9

Introduction to Bioengineering

In this chapter we outline the subject and the scope of bioengineering. After a broad view we specify in general terms the role of computer modeling within bioengineering. The development of computer models is described and a few examples demonstrate some achievements in bioengineering and computer modeling.

9.1 The subject and scope of bioengineering

Bioengineering is a broad field of scientific, biological, medical and engineering disciplines in which living systems, processes and materials are investigated together with nonliving subjects, environment and materials, in order to advance fundamental knowledge and improve life. One of the goals of bioengineering is to couple technological advances with basic investigations in biology and medicine leading to novel discoveries. Due to its inherent interdisciplinary character, bioengineering is currently one of the most attractive areas in research, education and industry.

Bioengineering exploits new developments in the life sciences (biology, physiology, biochemistry, biophysics) and engineering and couples them in order to better understand living systems. Bioengineers are similar to biologists in that they study living organisms, but they also have a practical design aim in mind – they use research to create usable, tangible products.

Biomedical engineering within the broad field of bioengineering represents a subset of disciplines whose main purpose is to develop, design and manufacture products that will improve human health. The end user of this process is the *patient*.

As an illustration of the advances in bioengineering, the American Institute for Medical and Biological Engineering (AIMBE) published in 2005 a 'Hall of Fame' list of major technological achievements in the twentieth century that revolutionized health care and improved quality of human life (see http://www.aimbe.org/content/index.php?pid=127). These accomplishments include the artificial kidney, heart valve replacements, computer

---

Computer Modeling in Bioengineering   Edited by M. Kojić, N. Filipović, B. Stojanović, N. Kojić
© 2008 John Wiley & Sons, Ltd

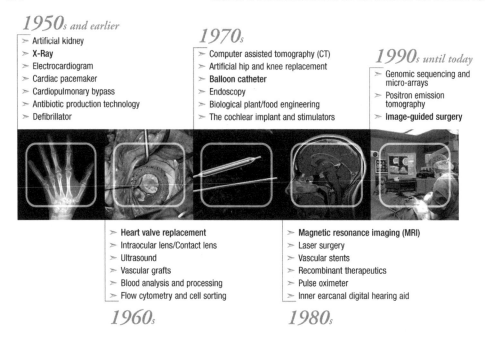

Fig. 9.1.1 The American Institute for Medical and Biological Engineering 'Hall of Fame' gives a perspective on the most significant technological advancements in bioengineering in the twentieth century (http://www.aimbe.org/content/index.php?pid=127) (see Plate 2)

aided tomography (CT), magnetic resonance imaging (MRI), genomic sequencing and microarrays (Fig. 9.1.1 – see color plate).

For all these accomplishments a broad base of knowledge, expertise and experience from different scientific disciplines had to be integrated, as illustrated in Fig. 9.1.2. Many biomedical problems are best addressed using a multidisciplinary approach that extends beyond the traditional biological and clinical sciences. Bioengineering integrates physical,

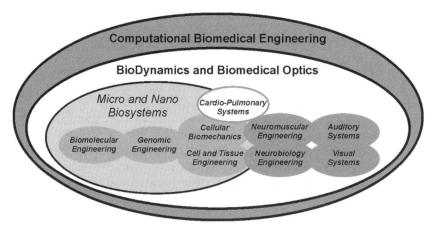

Fig. 9.1.2 A schematic view of an integrated bioengineering science. Adapted from the website of the Department of Biomedical Engineering at Boston University

engineering and computational science principles for the study of biology, medicine and health. It advances fundamental concepts and gives new insight into processes occurring from the molecular to the organ systems level. Furthermore, bioengineers develop innovative materials, implants, devices and informatics approaches for the prevention, diagnosis and treatment of disease.

The National Science Foundation and National Institutes of Health (USA) have identified bioengineering as 'an essential underpinning field for the 21st century'. Combining the traditional strengths of engineers (analytical and experimental methods) with those of biologists working on the molecular and cellular levels, the scope of bioengineering is as vast, and as intricate, as life itself.

There are already well-established specialty areas, such as biomechanics (the study of motion and forces affecting biological systems); biomaterials (both synthetic materials and living tissue); and bioinstrumentation (the development of measurement devices for diagnostic and treatment applications). Other fields within bioengineering include bioimaging and bioinformatics, which employ computer algorithms and mathematical methods to analyze large quantities of genetic data. Imaging technologies that deal with vision identification or face recognition belong to the field of biometrics.

9.2 The role of computer modeling in bioengineering

Since the topic of this book is computer modeling in bioengineering, we first present a summary of the theoretical background, then outline the role of computational methods and computer technology, and finally give two illustrative examples of potential future achievements in medical science and medial practice assisted by computer modeling.

9.2.1 Computational models

From a biomechanics standpoint, a model represents a mathematical interpretation of the mechanical behavior of a material body or system. This type of model is commonly called a *mechanical model* because it relies on physical laws or empirical relations which are relevant for the considered problem. The role of mechanical models is to elucidate the important factors and then simulate and predict the response of a mechanical system. Also, mechanical models can aid in the design of a system or process, and can serve as a tool accompanying experimental or clinical investigations. In both engineering and medicine the development of adequate models is becoming recognized as the cornerstone for future advancements in these fields.

Bioengineering models play the same role as do mechanical models in general fundamental and applied sciences. They are used in basic biomechanical research and laboratory investigations, as well as for medical and industrial applications.

For relatively simple conditions, models can be formulated using analytical approaches. However, in the analysis of more complex problems, numerical or computational methods must be employed. These computational models usually require significant software development and extensive use of computers. Today, complex computational models for a biological system or process assume the integration of fundamental disciplines (physics, biology, chemistry) with numerical methods, computer science and medicine. Altogether, creating a complex computational model is a challenging, multidimensional task spanning several, seemingly disparate, fields.

We now summarize the main components of a computational model.

Theoretical Foundation

For our purposes, it can be considered that in general terms computer modeling relies on biomechanics as the fundamental scientific discipline. Biomechanics (Humphrey & Delange 2004) is the development, extension and application of mechanics for the purpose of understanding the influence of mechanical loads on the structure, properties and function of biological systems. Biomechanics can be divided into two main branches. One is *bionics*, which deals with designing engineering systems that seek to mimic biological ones. An example would be Leonardo da Vinci's attempts to design a flying machine based on how birds fly. The other one is *mechanobiology* whose goal is to establish how the structure and function of living tissues, cells, organs and systems are influenced, controlled and regulated by mechanicals forces and motion. Galileo's study to relate the structural design of bones to the load that they carry would be one example.

In the past, several well-known scientists such as Hooke, Euler, Young, Poiseuille, and von Helmholz studied problems in biomechanics. However, biomechanics as a field began to rapidly expand in the second half of the twentieth century. There have been three major events that spurred this development of biomechanics. First, the beginning of outer space exploration where humans had to face the problem of a prolonged period in reduced gravity, which causes loss of bone and muscle mass. Second, the formulation of a nonlinear theory of continuum mechanics was influenced by new materials such as polymers. These theoretical developments were paralleled by advancements in powerful computational tools that could be used to numerically (e.g. by use of the finite element method) solve complex systems of differential equations. Third, the birth of modern biology, which involved the elucidation of the basic structure of proteins (Pauling) and DNA (Watson and Crick), and thus enabled the application of quantitative engineering methods to biology.

Biomechanics also plays an important role in healthcare delivery. For a long time, the most successful products of the biomechanics industry were for example artificial hips, surgical instruments (clamps, scissors), testing/diagnostic/aid equipment (respirators) and wheel chairs. Today the role of biomechanics in health delivery and prevention is rapidly expanding (e.g. artificial heart valves, artificial hearts, understanding and protection of injuries of athletes, angioplasty, robotic assisted surgery, mechanical exercises that support tissue healing). Certainly, beyond these 'classical' developments are modern trends in the application of nanotechnology in medicine (Ferrari 2005), with the potential of opening new frontiers in cancer drug delivery.

We have presented in Chapters 2 and 3 the key relations of general mechanics, including the specifics of biomechanics, which form the theoretical foundation for the computational methods used in this book.

Computational Methods

Development of these methods followed advances in computer technology and computer science. In the 1960s the finite element (FE) method emerged as a new approach, first in structural analysis, and then in general field problems, marking a revolution in solving various scientific and practical problems. Enormous advances in this field resulted in the formation of a new branch of the software industry devoted to computer modeling.

Today, generalizations of the FE method continue in addressing novel, complex problems. Also, new computational methods are emerging. Chapters 4–8 are devoted to the basics of the FE and several other methods used later in the book.

INTRODUCTION TO BIOENGINEERING 177

As previously emphasized, it is necessary to develop appropriate software for the models to be applicable. Presently, there is a great demand for more complex, sophisticated and efficient models. Therefore, modern achievements in computer science and computer technology must be utilized, such as parallel and distributed computing within a grid system of computers (summarized in Tirado-Ramos *et al.* 2004).

9.2.2 Future advances in computer modeling

In general, future advancements in computer modeling in bioengineering will likely include:

1. development of high-performance computational tools to make advances toward precise simulations of biomechanical problems, with realistic loading conditions, geometries and constitutive material properties;

2. multiscale biomechanics – application of physical principles and computational methods to couple various length and time scales, e.g. understanding relationships between cell-level organization and macroscopic organ functions.

Finally, we give two examples that demonstrate potential, future achievements in bioengineering and computer modeling.

Example 9.2-1. Virtual vascular surgery on the grid
In Fig. E9.2-1 (see color plate) a concept of virtual surgery is illustrated, with the use of a computer network (grid). The concept consists of several steps, starting with the

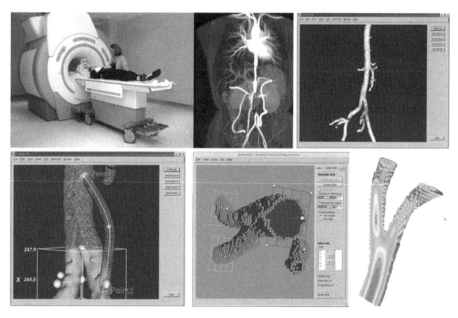

Fig. E9.2-1 Virtual vascular surgery on the grid: from the MRI or CT scan recording of the patient vascular surgery region, to automatic generation of the computational model and analysis of results, yielding options for the surgical procedure (Reproduced with permission from P.M.A. Sloot and A.G. Hoekstra: Virtual Vascular Surgery on the Grid, ERCIM news, October 2004) (see Plate 3)

178   COMPUTER MODELING IN BIOENGINEERING

MRI or CT scan recording the patient's vasculature in the region where the intervention is planned. These imaging scans further serve as input for automatic generation of the corresponding computer model. Then, by inputting appropriate parameter values into the model the surgeon can perform a 'virtual surgery' using the computer software. The model and the software allow the surgeon to explore various scenarios as part of the decision-making process.

In the future, it is conceivable that the surgeon will be able to interact in real-time with the software, thus having an 'online' computational model to assist the surgical procedure. Of course, in order to achieve this goal considerable advances are necessary in both computational methods (including parallel computing and efficient numerical methods) and simulation techniques.

Example 9.2-2. Nanoparticle delivery of therapeutic and imaging agents
Nanoparticles can be used to efficiently deliver drugs or imaging agents to the desired place, for example to cancer cells. Development of computer models to simulate nanoparticle motion, including binding processes between the ligand and receptor complexes, will help in the design of nanoparticles and their applications. Schematics of a nanoparticle releasing agents that pass through biological barriers are shown in Fig. E9.2-2 (see color plate).

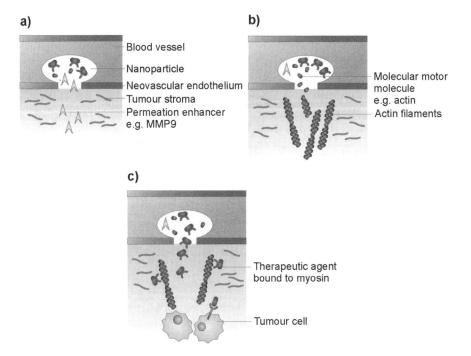

Fig. E9.2-2 A vision of a future multistage nanodevice. A nanoparticle selectively binds to the cancer neovascular endothelium releasing multiple agents that enable the drug to pass through biological barriers and reach the targeted tumor cell (Reprinted by permission from Macmillan Publishers Ltd: Cancer nanotechnology: opportunities and challenges, Mauro Ferrari, 2005) (see Plate 4)

References

Boston University (http://www.bu.edu/dbin/bme/), and the University of Notre Dame (http://www. nd.edu/~engineer/publications/signatures/2002/quality.html), web presentations of the bioengineering programs.

Ferrari, M. (2005). Cancer nanotechnology: opportunities and challenges, *Nature Rev. Cancer*, **5**, 161–71.

Humphrey, J.D. & Delange, S.L. (2004). *An Introduction to Biomechanics*, Springer-Verlag, New York.

Sloot, P.M.A. & Hoekstra, A.G. (2004). Virtual vascular surgery on the grid, *ERCIM News*, **59**, 50–1.

Tirado-Ramos, A., Sloot, P.M.A., Hoekstra, A.G. & Bubak, M. (2004). An integrative approach to high-performance biomedical problem solving environments on the grid, *Parallel Comp.*, **30**, 1037–55.

10

Bone Modeling

This chapter is devoted to finite element modeling of bones, and bone fractures in particular. The fundamentals of the structure and forms of bones are presented first. Then the mechanical properties of bone tissue used in the FE models are described. Also, the general FE dynamics equations for bone analysis are given. Typical bone fractures are described next, including medical aspects and several practical solutions. The FE models and solutions are presented for: (a) a femur comminuted fracture, with fixation by the neutralization plate, and by the intramedullary nail; and (b) a hip fracture with internal fixation, using the solutions by parallel screws and by dynamic hip implant. The solutions for fracture fixation are compared with respect to the advantages in medical practice applications.

10.1 The structure and forms of bones

In this section we present the basics of the bone structure and form of the bones which make the skeleton. The skeleton acts as a rigid framework for the protection of soft organs and allows locomotion.

10.1.1 The structure of bone tissue

The bone tissue represents the basic constituent of the skeleton. It belongs to the group of supportive connective tissues. Like others, this connective tissue has two components: cells and intercellular substance. The specific chemical compound of the intercellular substance and structural organization of the fibrous components determine the rigidity as the main characteristic of bone tissue.

Bone is functional tissue with permanent metabolic changes and continuous process of remodeling, resorption and tissue production. Homeostasis of bone tissue is regulated by systemic hormones and parathyroid glandula hormones. Any disbalance between resorption and production of bone tissue (homeostasis disorder) leads to major complications and loss of basic bone functions.

The Intercellular Substance of Bones
The intercellular substance of bones is made out of organic and nonorganic matter. The organic compound consists of collagen (90–95%) and interfibrillar substance (ground substance). Collagen, which we can find in bones, represents around 40% of all proteins in organisms and it appears in the form of fibrils. It comes in the group of collagen type I and is different from other types by its mechanical and physiochemical characteristics.

The interfibrillar basic substance, consisting of mucopolysaccharide and a small amount of intercellular liquid, appears in between collagen fibers as the osteoblast.

The nonorganic component of the intercellular substance of bones is made up of mineral salts. These salts in bones have a crystalline form – hydroxyapatite. The crystals are orientated towards a long shaft of the collagen fibers (long axis). They are mainly made out of calcium and phosphate, which is around 85% of the minerals in bones, with significantly less calcium carbonate (10%) and a very small amount of calcium chloride and magnesium sulphate (5%). All these salts surround the collagen fibers of bones and imbed inside the fibers. It is thought that 80% of the minerals of bones are deposited inside the collagen fibers (stable minerals), while there is around 20% of mineral salts in the ground substance.

The percentage and amount of mineral matter in bones depends on the age, functional ability of kidneys, hormonal and enzyme status, etc. The mineralization process of bones is very complex and is dependent on many factors and mechanisms involved in stimulation or inhibition.

The Histological Structure of Bones
Histologically, there are two types of bones:

1. Fibrous, nonlamellar, woven bone

2. Lamellar, mature bone.

Tough fibrous bone is the first, non-mature bone, which is present in the fetus and slowly grows into mature lamellar bone before the fetus is born. In infants, most of the bone tissue domains are transformed into lamellar bone.

Lamellar bone is a type of mature bone which contains collagen fibers spread into parallel strips and layers, lamellae. Lamellar bone represents the second phase in the maturing of bone tissue and is present in the area of compact, cortical bone and also in the spongy, trabecular area of the bone.

Histologically, cortical bone is made out of the lamellar bone which has the basic morphological and functional unit called osteon or the Haversian system. Osteon represents a cylinder made of 5–20 concentrated line up bone lamellae layers with the thickness of $3–7\,\mu m$. In the central area there is a vascular canal known as the Haversian canal (Fig 10.1.1 – see color plate).

Lamellar bone is made out of parallel collagen fibers that are buried in a mineralized basic bone substance, appearing altogether as a spiral flow. Parallel with lamellae, in osteons we have osteocytes which spread circumferentially around the Haversian canal and connect cytoplasmatic extensions.

The external surface of a cortical bone is covered by a compound vascular layer like a membrane. This layer is called the periosteum and contains collagen fibers parallel to

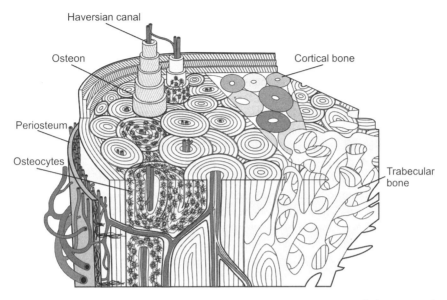

Fig. 10.1.1 Structure of a long bone (according to Remagen 1989) (see Plate 5)

the bone surface, blood vessels, nerve fibers, ligament fibers and muscle fiber insertion. Adjacent to the bone is a layer of cells of osteogenic potential, which are able to transform into osteoblasts during the phase of skeleton development or regeneration. There is a lack of periosteum in the domains of joints.

Spongy or trabecular bone is also made out of lamellar bone, but without the Haversian system. Trabecular spongy bone is composed of a row of thin joined bone layers, lamella, which point in the direction where most of the mechanical forces are applied. The bone layers cross at different angles, so they create gaps for the bone core. Lamellar bones do not contain blood vessels. Adjacent to the trabecular spongy bone there is a row of osteoblasts, which creates an inner layer of bone – the endost.

10.1.2 The form of bones

All bones of skeletons can be divided into:

(a) tubular bones

(b) short bones

(c) flat bones

(d) other bones

depending on their look and the interrelationship between spongy and compact bones.

(a) Tubular bones have three sections: diaphysis, metaphysis and epiphysis. In the middle section we have a region with a tubular appearance, which is called the diaphysis; at the bone ends there are epiphyses, and the regions between the diaphysis and epiphyses are called metaphysis. Tubular bones represent long bones of extremities (humerus, radius, ulna, femur, tibia and fibula), and short bones of the hand and foot (metacarpus, metatarsal and phalanges). These bones are composed of compact bone (substantia compacta) and spongy bone (substantia spongiosa). The relationship between compact and spongy bones is different in different bone domains. For example, in the shaft of the tubular bone (diaphysis) compact bone makes up the cortex, which surrounds a medullary cavity or bone marrow and it exists in this particular area in a significant amount. But, as we move toward the bone ends, the cortex decreases, hence the amount of compact bone decreases and the spongy bone increases.

Spongy bone is dominant in the area of epiphyses; going towards the middle trunk of the bone its amount decreases, and then in the diaphysis some of the tubular bone is lacking (fibula and short tubular bone). Spongy bone has a honeycomb, sponge-like appearance and is composed of a framework or trabeculae which are connected together. The trabeculae orientation and thickness depend on the direction of mechanical forces. In adults, there is a fatty marrow core in the spaces inside the spongy bone. In the proximal area of the femur and humerus, there remains a small red bone core during a major part of one's life. In adults, the red bone core, appearing in flat bones and at the ends of some other bones, contains cells which create the blood constituents. The fatty marrow core is made out of fat tissue and blood vessels.

(b) Short bones make up the parts of the base of the hands and foots and some superfluous bones. They are composed of spongy bone, which is surrounded by a thin layer of cortical bone in places where these bones are not covered by the articular cartilage.

(c) In the group of flat bones come ribs, sternum, scapula and most of the bones of the skull. They are different from other bones because they are thin and have a relatively small amount of spongy bone or they do not have it at all (lacrimal bone, for example). Here, the spongy bone is surrounded from the inside and outside by cortical bone, and contains a red bone core.

(d) Other bones are no name bones, some of which are the bones of the skull and spinal cord. Some domains of these bones can have the appearance of short or flat bones, but they do not fall within the groups mentioned above. For example, a vertebral body with its structure can belong to the group of tubular bones, however, some of the areas of the vertebral arches have the structure of flat bones. Due to the unequal structure of the spinal cord, it is usually considered that these bones fall into the group of other bones.

10.1.3 Osteoporosis and bone density

During the first three decade of human life bone mass continuously increases achieving maximum value at the age of 30 years. The bone mass starts to decrease after that age. This phenomenon is known as the osteoporosis. In latest stages of life the osteoporosis leads to low bone mass and microarchitectural changes causing enhanced bone fragility and increased risk of fracture (Fig 10.1.2).

In order to provide an efficient and unique method for osteoporosis assessment the World Health Organization established criteria based on measuring the bone density. According

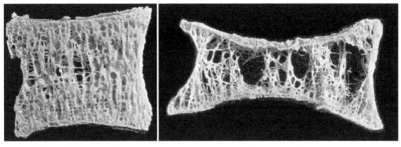

Normal bone marix · Osteoporosis

Fig. 10.1.2 Structure of normal and osteoporotic the trabecular bone (adopted from Remagen 1989)

to these criteria (Huiskes & Van Rietbergen 2005), if the bone density is 2.5 times the standard deviations below the average bone density for a 25-year-old female, it is diagnosed as 'osteoporotic'. Osteoporosis affects cortical bone as well as trabecular bone, but trabecular bone changes are predominant.

Bone density can be defined as bone mass per total volume of bone including any holes. The bone density calculated in this way represents the mean density of the apparent material specimen and is also known as the 'apparent density'. Note that the bone apparent density is not equal to the bone tissue density. Let us introduce the bone volume fraction V_V representing ratio of bone volume over total volume,

$$V_V = V_B/V_T \qquad (10.1.1)$$

where V_B and V_T are the bone volume and total volume, respectively. Assuming that the bone tissue is homogenous with density ρ_{tissue}, the relationship between apparent density and bone tissue density can be written as

$$\rho_{apparent} = V_V \rho_{tissue} \qquad (10.1.2)$$

10.2 The mechanical properties of bone and FE modeling

In this section basic mechanical characteristics of bone are described and the general concept of the finite element modelling is outlined.

Bone Mechanical Properties

For general physiological loading conditions the bone material can be considered linear elastic. The constitutive stress–strain relationship shows that bone material behaves in a manner similar to that of other engineering materials (Huiskes & Van Rietbergen 2005). Stress–strain curves in tension and compression consist of an initial elastic region, which is nearly linear. This region is followed by yielding and considerable, nonelastic, 'plastic' deformation before a failure. The nonelastic region of the stress–strain curve for the longitudinally oriented specimen reflects diffuse, irreversible microdamage created throughout the bone structure. Bone tissue that is loaded into this nonelastic region will not return to its original configuration after the load is removed.

As we have discussed in Section 10.1, the bone tissue is a two-phase material consisting of collagen and bone mineral, organized in matrix form. Bone mineral (hydroxyapatite) is very rigid and has bigger compression than tensile stiffness. On the other hand, bone collagen has only tensile stiffness. Generally, the bone mineral influence on the mechanical properties of bone tissue is predominant. Hence, bone tissue has bigger compression than tensile stiffness and strength. Also, bone tissue has anisotropic behavior caused by specific microstructural organization. Investigations show that bone tissue is stronger and stiffer in the direction of the osteon orientation than in the perpendicular direction. Because of that, bone stability depends not only on load value, but also on load direction.

Elastic modulus and strength of bone tissue are not constant. They are dependent on rate of deformation. Rate of deformation is usually quantified with material strain rate. Because of that, bone tissue is viscoelastic material. Studies in which bone specimens were exposed to loads of different strain rates showed that increasing of strain rate caused increases of the elastic modulus. The elastic modulus of bone tissue is approximately proportional to the strain rate raised to the 0.06 power. Using this relation it can be shown that over a very wide range of strain rates the elastic modulus increases by about a factor of two. Experimental analysis revealed that the apparent density is important for elastic modulus estimation and the following relationship was proposed (Carter & Hayes 1977)

$$E_{axial} = E_c \dot{e}^{0.06} \left(\frac{\rho}{\rho_c}\right)^3 \tag{10.2.1}$$

where E_{axial} is the elastic modulus of bone of apparent density ρ, tested at strain rate of $\dot{e}\,(s^{-1})$; and E_c elastic modulus of bone with an apparent density of ρ_c tested at strain rate of $1.0\,s^{-1}$.

With respect to composition and true tissue density, cortical and trabecular bone are very similar. However, their structural organization is very different. The basic difference of cortical and trabecular bone tissue is porosity. Cortical bone porosity ranges from 5 to 30%, while trabecular bone has porosity from 30 to 90%. The apparent density of bone tissue for both cortical and trabecular bones shows approximately linear dependence on bone porosity.

Finite Element Modeling

General principles of FE modeling of structures are applicable here. Therefore, the static or dynamic equations of balance presented in Chapters 4 and 5 can be used. The bone structure is usually modeled by 3D finite elements in order to capture the bone geometry, and we here write the dynamic equation of motion for a 3D finite element (see Sections 4.3 and 5.2)

$$\mathbf{M}^{n+1}\ddot{\mathbf{U}} + {}^n\mathbf{K}\mathbf{U} = {}^{n+1}\mathbf{F}^{ext} \tag{10.2.2}$$

where $\mathbf{M}$ and ${}^n\mathbf{K}$ are the element mass and stiffness matrices, ${}^{n+1}\ddot{\mathbf{U}}$ and $\mathbf{U}$ are nodal acceleration and displacement vectors, and ${}^{n+1}\mathbf{F}^{ext}$ is the external nodal force which includes structural external forces and action from the surrounding elements. The equation of motion corresponds to time step 'n', where the left upper indices 'n' and '$n+1$' indicate, respectively, start and end of time step. The stiffness matrix ${}^n\mathbf{K}$ can be written as (see (4.3.15))

$$^n\mathbf{K} = \int_V \mathbf{B}^T \, {}^n\mathbf{C}\mathbf{B}\,dV \tag{10.2.3}$$

where $\mathbf{B}$ is the strain–displacement matrix (4.3.7) and ${}^n\mathbf{C}$ is the constitutive matrix. In the case of isotropic material, the constitutive matrix is given in (2.2.5); the left upper index 'n' is used to show that the axial modulus (10.2.1) corresponding to the strain rate $^n\dot{e}$ may be used when the rate effects are important. Of course, if the bone tissue is considered anisotropic, the orthotropic constitutive matrix should be employed (see web, Theory – Chapter 2).

10.3 Bone fracture – medical treatment and computer modeling

Here several procedures of medical treatment of bone fractures and accompanied computer modeling are presented.

10.3.1 General considerations

A fracture means that the continuity of bone is disrupted. Force transmission through the bone is no longer possible in any direction. A fracture results in loss of the structural integrity of bone, and a loss of its weight carrying capacity. A fractured bone becomes mechanically functionless.

Whether or not a bone fractures due to stresses within the bone tissue depends on both extrinsic and intrinsic factors. The extrinsic factors important in the production of fractures are the magnitude, duration and direction of one bone acting on the other. The most important intrinsic factors in determining the bone fracture resistance are the size and geometry of the bone and material properties of the bone tissue (energy-absorbing capacity, Young's modulus of elasticity and density).

Fracture occurs when the stresses in one region of the bone exceed the ultimate strength of the bone material. Bone fracture can therefore be thought of as an event which is initiated at the level of the material. The size and shape of the bone under loading determine the distribution of stresses throughout the bone. A large bone is more resistant to fracture simply because it distributes the internal forces over the bone tissue, which are lower than those in a smaller bone loaded under similar conditions.

In laboratory conditions, specific fractures are produced under various modes of load. On the other hand, the fractures seen clinically are usually caused by complex loading conditions, and the resulting fracture patterns are therefore numerous.

Bone fracture occurs both from direct and indirect forces. Fractures produced by direct application of force to the fracture site can be divided into typing fractures, crush fractures and penetrating fractures. These are caused, respectively, by a small force acting on a small area, a large force acting on a large area, and a large force acting on a small area (Harkess & Ramsey 1991). Fractures produced by a force action at the distance from the fracture site are said to be caused by an indirect trauma.

When a bone is subjected to tensile forces, a fracture perpendicular to the direction of the applied loading occurs. This can happen to the patella or olecranon when the knee or elbow is forcibly flexed while the extensor muscles are contracting. The fracture line in tension fracture is transverse.

Compressive forces will generally fracture along a plane, which is at an oblique angle to the direction of the applied loading. Long bones only rarely are fractured by a pure compression force. A common site of compression fracture in trabecular bone is a vertebral body.

Torsional forces cause a spiral bone fracture. Then the fracture is usually initiated at a small crack on the surface of the bone that runs parallel to the bone axis, on a plane of high shear stress. After initiation of the fracture, the crack runs in a spiral manner through the bone, following the planes of high tensile stresses.

A bone specimen subjected to bending forces will be exposed to high tensile stress on one side of the long bone and high compressive stress on the other side. Because bone is stronger in compression than in tension, the fibers over convexity fail first. A transverse fracture will be present on the tensile side, while an oblique fracture surface may be created on the compressive side. The compressive side of the specimen may contain two oblique fracture patterns, creating a loose wedge of bone as the bone is fractured. This fracture pattern is sometimes referred to as a 'butterfly' fracture.

10.3.2 Fracture treatment

The main aim of fracture treatment is to obtain final function as close to pre-fracture conditions as possible and as soon as possible. In order to achieve this aim, good reposition of the bone and stable fixation (immobilization) of the fractured fragments have to be obtained. Different means of treatment exist which result in different degrees of immobilization. The choice of method depends on the type of fracture, local conditions of soft tissue, patient expectations and requirements, prognostic criteria etc.

All methods of fracture fixation must provide adequate stability in order to maintain length and correct joint alignment. Whatever the type of fixation is implemented, the implants must be sufficiently strong to withstand the early functional forces, and not fail due to mechanical overload. The repair must also be resilient enough to last until osseous union is achieved.

External Skeletal Fixation
The first option here is the *plaster-of-Paris cast* solution. The objective of applying a plaster cast is to keep the bone ends in apposition and the fracture aligned until the fracture heals. This type of fracture fixation allows a relatively large interfragmentary movement. It has been said that immobilization by plaster cast will work only where there is inherent stability of reduced fracture, and where the cast is properly applied using three-point fixation. Instability due to bending movements and torque must be limited by a good fit between the plaster cast and the outer shape of the extremity. The stability of the fixation is influenced by the circular compression of the plaster cast against the soft tissue under the envelope around the bone. Soft tissue can act as a hydrodynamic medium, and the compression helps to stabilize the fragments (Sarmiento & Latta 1981). The displacement under load is decreased steadily, because of the increasing stiffness of the callus (Sarmiento *et al.* 1996).

Another solution is by the *external fixator*. External stabilization of the fracture can be performed by using an external fixator. This is a system that allows the stabilization of fragments away from the fracture site with the aid of percutaneous screws or wires that are connected to one or more bars on the outside of the skin. There are several types of external fixators: simple pin fixator, clamp fixators and ring fixators. The stability of this fixation is mainly determined by the stiffness of the fixator design and the quality of the connection between the screws and bone. The stiffness is described by the interfragmentary movement occurring under external loads.

This technique is the only one that allows the surgeon to adapt the stability as needed not only intraoperatively, but also postoperatively.

Internal Fixation of Fractures

Since the time when Lord Lister used silver wire to repair broken patella, many devices have been developed for internal fixation of fractures. Here we describe two of the solutions of internal fixation of fractures: plates and screws, and intramedullary nail.

Plates and screws. In transverse or short oblique fractures of the diaphysis, stabilization can be performed with a compression plate. The plate has a special hole design with a slope on which the screw head slides (Müller *et al.* 1979). When the screw is inserted into the bone, it moves towards the bone cortex, which is only possible if the slope of the screw hole is pushed axially (Fig. 10.3.1a). This axial movement of the plate creates a compression between two fragments fixed by a compression plate (Fig. 10.3.1b).

The plate which conducts all the forces from one fragment to the other and protects the fracture from the all kind of forces is called the neutralization plate. It is often used in the cases of comminuted fractures where it is very difficult to achieve a stable fixation of the bone fragments (Fig. 10.3.2a).

Intramedullary nail. In the long tubular bone very effective fracture stabilization can be achieved by some intramedullary devices, such as the intramedullary nail (Fig. 10.3.2b).

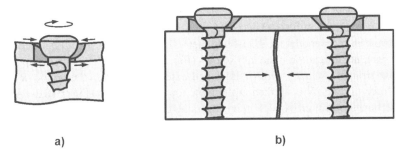

Fig. 10.3.1 Compression plate (according to Schauwecker 1981)

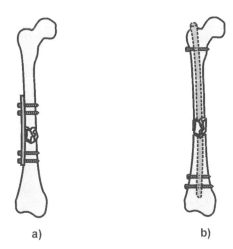

Fig. 10.3.2 Solutions for fractures of long bones. (a) Neutralization plate; (b) Intramedullary nail

The stability of fracture fixation by nailing mainly depends on the mechanical properties of the nail, the nail's fit in the medullary space, and the mechanical properties of the locking screws or bolts. The bending and torsional stiffness of the nailed bone mainly depends on the diameter of the nail. The bending stiffness can be obtained as $K_b = E \times I$, where E is Young's modulus of elasticity and I is the second moment of inertia for bending of the nail cross-section; while the torsional stiffness is $K_t = G \times I_t$, where G is the shear modulus and I_t is the second moment of inertia for torsion.

10.3.3 FE modeling of femur comminuted fracture

Two example solutions of femur fracture are presented: (a) when the neutralization plate is used; and (b) when the intramedullary nail is implemented.

Example 10.3-1. Fixation by the neutralization plate
During walking humans alternately stand on each leg with the whole weight. Hence, at the moment of standing on one leg, the axial force in the leg is equal to the human weight. In this example we consider a patient having diaphysial comminuted fracture, with the mass of $70\,kg$ which produces the axial force of $70\,daN$ in the femur. It is assumed that the fracture is fixed by a neutralization plate (Fig. E10.3-1A).

In the finite element model the 3D isoparametric finite elements are used (Section 4.3) for the bone tissue, neutralization plate and screws (Fig. E10.3-1Aa). The bone is approximated by a hollow straight cylinder with internal and external radii $R_i = 9\,mm$ and $R_e = 14\,mm$. The axial force of $70\,daN$ is applied at the bone top cross-section (Fig. E10.3-1Ab). Nodes lying at the bone bottom cross-section are restrained in all directions. Also, it is considered that there is no slip between the screws and plate or bone tissue.

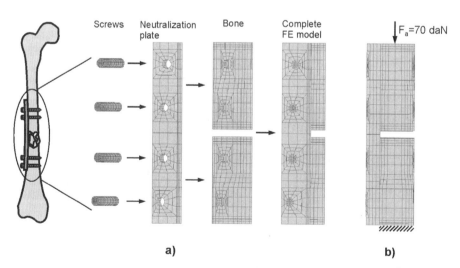

Fig. E10.3-1A (a) Finite element model of femur comminuted fracture fixed by neutralization plate. (b) Boundary condition and load. The load is transferred from the upper bone fragment to the upper set of screws, then by the plate to the lower set of screws and the lower bone fragment

BONE MODELING

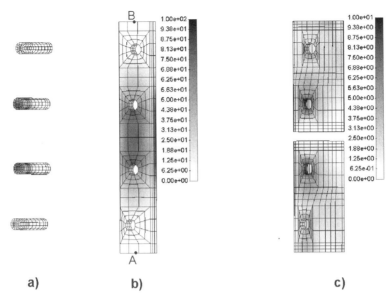

Fig. E10.3-1B FE solution for the femur comminuted fracture. Effective stress field within: (a) Screws; (b) Neutralization plate; (c) Bone tissue

The bone tissue is modeled using the material model defined by the relation (10.2.1). The solution is obtained using the following data for the bone tissue: $E_c = 22.1 \times 10^3 \, MPa$; $\rho = 2.1 \times 10^{-3} \, g/mm^3$; $\rho_c = 1.8 \times 10^{-3} \, g/mm^3$; $\dot{e} = 0.1 \, s^{-1}$, where it is assumed that the

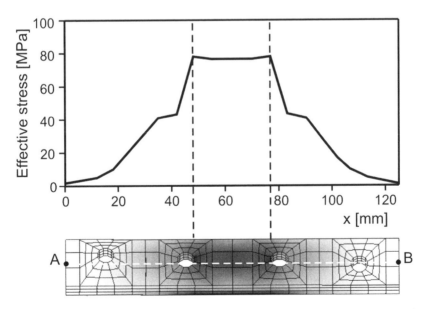

Fig. E10.3-1C Distribution of the effective stress within the neutralization plate along the line AB

strain rate is the same within the tissue. For the neutralization plate and screw materials stainless steel is used, with Young's modulus $E = 2.1 \times 10^5 \, MPa$ and Poisson's ratio $\nu = 0.3$.

The field of effective stress in the screws, neutralization plate and bone is shown in Fig. E10.3-1B. Also, distribution of the effective stress within the neutralization plate along the line AB is shown in Fig. E10.3-1C.

It can be noted from Figs. E10.3-1B,C that a significant stress elevation (stress concentration) appears within the neutralization plate in the region between the second and third screw. The extreme values of effective stress are below the critical values for stainless steel. However, the neutralization plate is subjected to the cycling loading during the walk which leads to the material fatigue. The region between the second and third screw is critical (locus minoris resistentiae) for the fatigue failure, which is frequently observed in clinical practice (Bucholz & Brumback 1991).

Example 10.3-2. Fixation by the intramedullary nail

The femoral shaft fracture fixed by an interlocking nail is modeled (Ranković et al. 2007). In this particular case we assume a comminuted type of fracture. Also, it is taken that the intramedullary nail is locked by two screws proximally and two screws distally.

In order to compare the values and stress distribution in a femur diaphysial comminuted fracture fixed by intramedullary nail and neutralization plate, the finite element model is generated using the same geometric and material parameters, FE type (3D isoparametric elements) and loading as is in the previous example. For the intramedullary nail material the stainless steel is taken, with the characteristics as for neutralization plate and screws. The finite element model is shown in Fig. E10.3-2A.

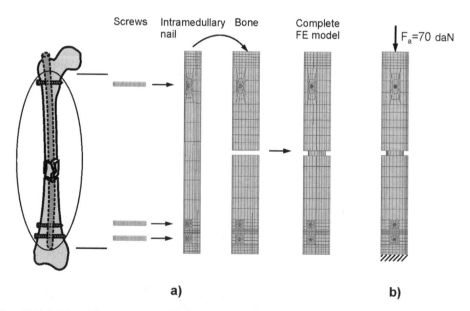

Fig. E10.3-2A (a) Finite element model of the femur comminuted fracture fixed by intramedullary nail. (b) Boundary condition and load. The load is transferred from the upper bone fragment to the upper screw, then by the intramedullary nail to the lower set of screws and the lower bone fragment

BONE MODELING 193

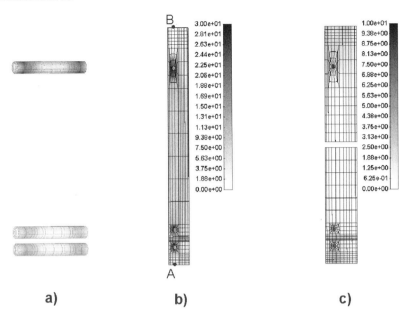

Fig. E10.3-2B Effective stress field within: (a) Screws; (b) Intramedullary nail; (c) Bone tissue

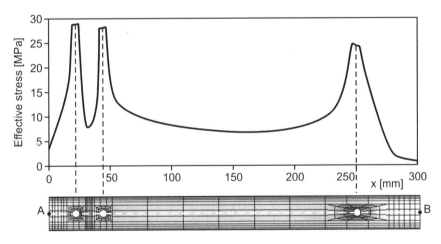

Fig. E10.3-2C Distribution of the effective stress within the intramedullary nail along the line AB

The field of effective stress in the screws, intramedullary nail and bone is shown in Fig. E10.3-2B. Also, distribution of the effective stress within the intramedullary nail along the line AB is shown in Fig. E10.3-2C.

Figures E10.3-2B, C show that a notable stress concentration occurs within the intramedullary nail in the region of the screw holes. But these values are appreciably lower than the effective stress generated in the neutralization plate for the same loading conditions.

Hence, the risk of intramedullary nail failure is significantly lower than it is in when using the neutralization plate, which also is clinically approved (Bucholz & Brumback 1991).

10.4 Internal fixation of hip fracture – two solutions and computer models

Hip fractures represent certainly the most important orthopaedic-traumatologic problem. This fracture has specific biomechanical and biological characteristics. In this section we present two solutions and compare them through the results obtained by FE modeling.

10.4.1 Solutions by parallel screws and by dynamic hip implant

The most important factor in healing of the femoral neck fracture is directly affected by the surgeon's achievement of stable fixation. The significance of the fixation stability for the healing of the fracture can be seen through the fact that the fractures without dislocation, when there is a good contact between fracture surfaces, result almost always in the fracture healing (Frandsen *et al.* 1984, Skinner & Powles 1986, Madsen *et al.* 1987, Stromqvist *et al.* 1987). It was proved that, by using instrumentalized intraoperative measuring (Rehnberg & Olerud 1989), the femoral neck fractures that have healed, had much bigger intraoperative stability then the fractures that did not heal.

Stability of the fracture depends on the quality of the bone tissue (osteoporosis and comminution), quality of the reposition of the fracture, the implants' design and the position of the fixation device on the femoral neck and the head (Stromqvist *et al.* 1987, Frandsen & Andersen 1981, Olerud *et al.* 1991).

Apart from different opinions, it is obvious that besides the knowledge and skills of the surgeon, the choice of implants plays the crucial role regarding the fracture stability. Some implants provide good fracture stability even in inconvenient situations (inappropriate reposition, bad quality of the bone tissue, stressed comminution of the fracture; see Swiontkowski *et al.* 1987).

The basic function of the implant is to keep the bone fragments in a reduced relation until the fracture heals, and on the other hand not to endanger the healing processes and the vascularization of the femur head (which is most often already damaged). Osteosynthetic devices have to sustain the stability of the fracture by neutralizing the effect of forces affecting the hip (load bearing) or redirect the force effect so that a positive impact of the fracture surfaces and the bone to bone (load sharing) is gained. The implant has to endure forces that are much larger than the body weight and by which the hip joint is affected while walking. By using rigid fixation of the fracture surfaces the implant fulfills its main role. It has to tightly keep reposition on the fracture site until fracture healing, because any relative movement of the bone surfaces withholds ingrowing of the vascular buds and the revascularization of the head. A relative movement on the fracture site also causes fibrose tissue creation in the fracture crack.

The implant should be designed so that an early loosening on the bone–implant connection does not occur, causing undesired head movement to varus and lateral angulation and rotation. A certain flexibility level of implant is needed to slow down the process of loosening which is constantly happening from the moment of the implant application. The difference between

BONE MODELING

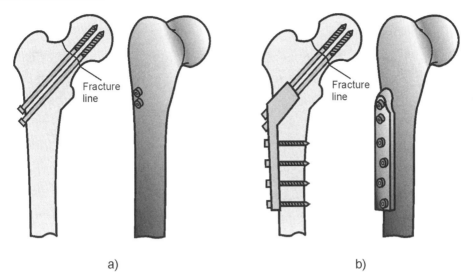

Fig. 10.4.1 Solutions for the hip fracture. (a) Parallel screws; (b) Dynamic hip implant

the module of elasticity of the bone tissue and the material from which the implant is made is an important factor in fixation loosening. There has recently been increasing research concerning the materials that could fulfill the requirements with respect to prevention of loosening of fixation.

Fixation by parallel screws (Fig. 10.4.1a) and the by the dynamic hip implants (Fig. 10.4.1b) (Ristic & Bogosavljevic 2004) are the two methods for internal fixation of intracapsular fractures of the femoral neck. The stability of the osteosynthetic structure is different for these two solutions due to different biomechanical relations between the bone and fixator devices. We consider that these solutions provide an insight into the problem of hip fracture and have generality in medical applications.

10.4.2 Finite element models of intracapsular fractures of the femoral neck

The finite element solutions for two types of internal fixation of the intracapsular fractures of femoral neck are presented.

Example 10.4-1. Fracture fixation by parallel screws

The bone and screws are modeled by 3D finite elements as in Section 10.3.3. The bone shape, FE model and boundary conditions are shown in Fig. E10.4-1A.

The structure is loaded by two forces F_A and F_R. The force F_A is generated by the gluteal muscles which connect the greater trochanter with the pelvis. Gluteal muscles have the important role in achieving the moment balance to gravity body forces at the instant when the human is standing on one leg. The second force (F_R) represents the force transferred from the pelvis to the femur head. The magnitude of these forces depends on the body weight and geometrical relations between the body mass center and proximal femur. In the FE model, the gluteal muscles force F_A is distributed over the top nodes of the greater trochanter,

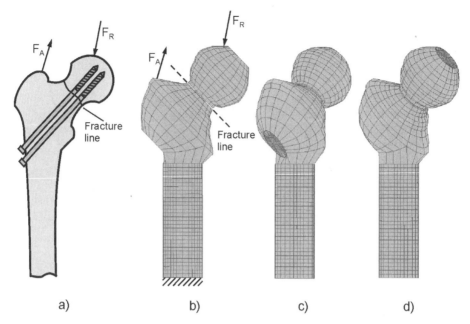

Fig. E10.4-1A Finite element model of the femoral neck fixation by parallel screws. (a) Schematic view; (b) Anterior aspect; (c) Anterolateral aspect; (d) Anteromedial aspect

while the force F_R is distributed over the top nodes of the femur head. The FE nodes lying at the bottom cross-section of the bone are assumed to be restrained in all directions.

It is considered that the loading is quasi-static and that there is no relative motion between the screws and the bone. The compact and trabecular bones are modeled using the constitutive model given in (10.2.1), with the same data as in Example 10.3-1, including the strain rate $\dot{e} = 0.1\,s^{-1}$. The screws are taken to be of stainless steel with the material characteristics as in Example 10.3-1. The body weight is taken to be $70\,daN$, and $F_A = 137\,daN$, $F_R = 199\,daN$.

The field of effective stress within the cortex bone tissue is shown in Fig. E10.4-1B. Distribution of the effective stress within the lateral cortex of the proximal femur along the line AF is given in Fig. E10.4-1C.

Figures E10.4-1B,C show that a significant stress concentration appears within the lateral cortex of the proximal femur in the region of the screw holes. And this region is critical (it is said that it is the locus minoris resistentiae) for fixation failure, which is confirmed in clinical practice. Fixation stability depends on the stiffness of the lateral cortex of the proximal femur which usually is thin and weak in patients with femur neck fracture.

Example 10.4-2. Fracture fixation by dynamic hip device

The idea of this fixation solution is shown in Fig. 10.4.1b. Here we analyze this solution by the FE modeling. Figure E10.4-2A shows the finite element model with loading and boundary conditions.

Using the same material properties and loads as in the previous example, we obtain the finite element solution for stresses and strains. The effective stress distribution in cortex

BONE MODELING

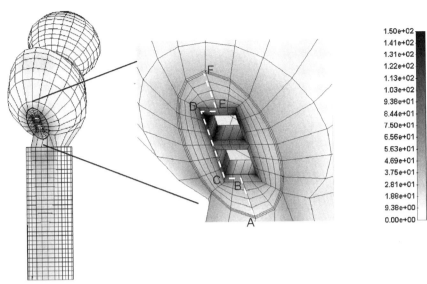

Fig. E10.4-1B Effective stress distribution in lateral cortex of the proximal femur for the intracapsular fracture of the femoral neck (solution by parallel screws)

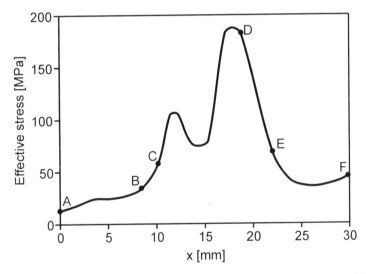

Fig. E10.4-1C Distribution of the effective stress within the lateral cortex of the proximal femur along the line AF (see Fig. E10.4-1B)

bone tissue is shown in Fig. E10.4-2B, while the distribution of the effective stress within the lateral cortex of the proximal femur along the line AF is given in Fig. E10.4-2C.

As can be seen from Figs. E10.4-2B,C the extreme effective stresses within the bone tissue appear in the lateral cortex of the proximal femur, in the region of the dynamic hip device insertion. However, these stresses are appreciably lower than stresses generated when using the solution by parallel screws (compare maximum stresses in Figs. E10.4-1C and

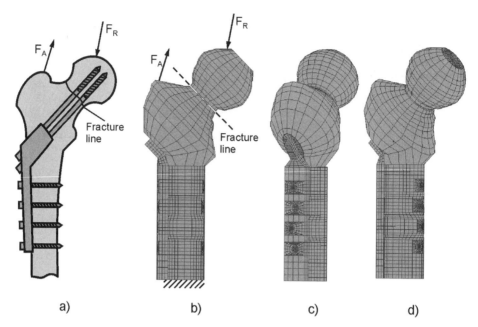

Fig. E10.4-2A Finite element model of the femoral neck fixation by dynamic hip device. (a) Schematic view; (b) Anterior aspect; (c) Anterolateral aspect; (d) Anteromedial aspect

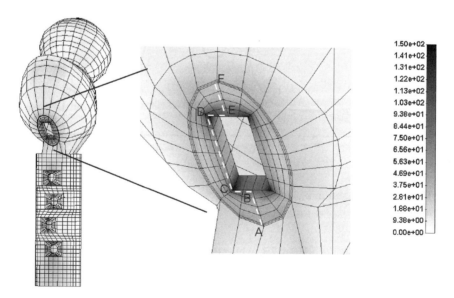

Fig. E10.4-2B Effective stress distribution in the lateral cortex of the proximal femur for the intracapsular fracture of the femoral neck (solution by dynamic hip device)

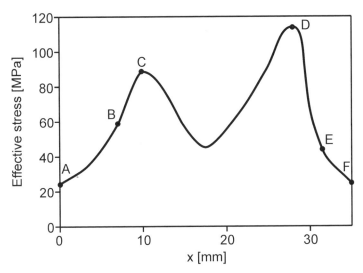

Fig. E10.4-2C Distribution of the effective stress within the lateral cortex of the proximal femur along the line AF (see Fig. E10.4-2B)

E10.4-2C). This is expected because the larger part of the hip load is carried by the dynamic hip device and not by the bone only. Finally, the fixation stability in general depends on the dynamic hip device stiffness. The effective stresses within the screws and dynamic hip device are low (Fig. E10.4-2D) and are significantly lower than within the parallel screws. Hence, the risk of the dynamic hip device fixation failure is lower than it is when using parallel screws; this is confirmed in clinical practice (Ristic & Bogosavljevic 2004).

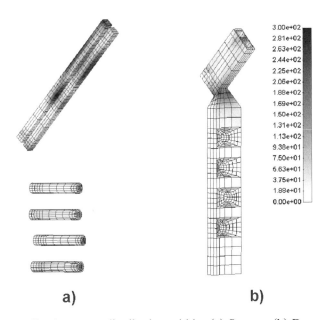

Fig. E10.4-2D Effective stress distribution within: (a) Screws; (b) Dynamic hip device

References

Bucholz, R.W. & Brumback, R.J. (1991). Fractures of the shaft of the femur. In C.A. Rockwood, D.P. Green & R.W. Bucholz (eds.), *Fractures in Adults*, 3rd edition (pp. 1653–724), J.B. Lippincott, New York.

Carter, D.R. & Hayes, W.C. (1977). The compressive behavior of bone as a two-phase porous structure, *J. Bone Joint Surg.*, **59**, 954–62.

Frandsen, P.A. & Andersen, P.E. Jr. (1981). Treatment of displaced fractures of the femoral neck. Smith–Petersen osteosynthesis versus sliding-nail-plate osteosynthesis, *Acta Orthop. Scand.*, **52**, 547–52.

Frandsen, P.A., Andersen, P.E. Jr., Christoffersen, H. & Thomsen, P.B. (1984). Osteosynthesis of femoral neck fracture: the sliding-screw-plate with or without compression, *Acta Orthop. Scand.*, **55**, 620–3.

Harkess, J.W. & Ramsey, W.C. (1991). Principles of fractures and dislocations. In C.A. Rockwood, D.P. Green & R.W. Bucholz (eds.), *Fractures in Adults*, 3rd edition (pp. 1–180), J.B. Lippincott, New York.

Huiskes, R. & Van Rietbergen, B. (2005). Biomechanics of bone. In V.C. Mow & R. Huiskes (eds.), *Basic Orthopaedic Biomechanics and Mechano-Biology*, 3rd edition (pp. 123–79), Lippincott Williams & Wilkins, Philadelphia.

Madsen, F., Linde, F., Andersen, E., Birke, H., Hvass, I. & Poulsen, T.D. (1987). Fixation of displaced femoral neck fractures. A comparison between sliding screw plate and four cancellous bone screws, *Acta Orthop. Scand.*, **58**, 212–16.

Müller, M.E., Allgöwer, M. & Schneider, R. (1979). *Manual of Internal Fixation*, 2nd edition, Spinger Verlag, Berlin.

Olerud, C., Rehnberg, L. & Hellquist, E. (1991). Internal fixation of femoral neck fractures. Two methods compared, *J. Bone Joint Surg.*, **73-B**(1), 16–19.

Ranković, V., Ristić, B. & Kojić, M. (2007). Internal fixation of femoral bone comminuted fracture – FE analysis, *J. Serbian Soc. Comp. Mech.*, **1**(1), 120–8.

Rehnberg, L. & Olerud, C. (1989). Subchondral screw fixation for femoral neck fractures. *J. Bone Joint Surg.*, **71-B**(2), 178–80.

Remagen, W. (1989). *Osteoporosis*, Sandoz, Basle, Switzerland.

Ristic, B. & Bogosavljevic, M. (2004). The femoral neck fracture: a biomechanical study of two internal fixation techniques, *Medicus*, **5**(2), 17–21.

Sarmiento, A. & Latta, L.L. (1981). *Closed Functional Treatment of Fractures*, Springer Verlag, Berlin.

Sarmiento, A., McKellop, H.A. & Llinas, A. (1996). Effect of loading and fracture motions on diaphyseal tibial fractures, *J. Orthop. Res.*, **14**, 80–4.

Schauwecker, F. (1981). *Osteosynthese Praxis*, Thieme, Verlag, Stuttgart.

Skinner, P.W. & Powles, D. (1986). Compression screw fixation for displaced subcapital fracture of the femur: success or failure?, *J. Bone Joint Surg.*, **68-B**, 78–82.

Stromqvist, B., Hansson, L.I., Nilsson, L.T. & Thorngren, K.-G. (1987). Hook-pin fixation in femoral neck fractures: a two-year follow-up study of 300 cases, *Clin. Orthop.*, **218**, 58–62.

Swiontkowski, M.F., Harrington, R.M., Keller, T.S. & Van Patten, P.K. (1987). Torsion and bending analysis of internal fixation techniques for femoral neck fractures: the role of implant design and bone density, *J. Orthop. Res.*, **5**(3), 433–44.

11

Biological Soft Tissue

This chapter is devoted to modeling of soft biological tissues. We first introduce the basics about the structure and function of biological tissue and then describe typical mechanical tests and tissue models. The mechanical models include: uniaxial, biaxial, hysteretic, viscoelastic, models based on the strain energy function, and models of surfactant. Further, specifics of the tissue computer modeling are presented within the general finite element analysis. These specifics are related to the stress integration and calculation of the consistent tangent constitutive matrix for large strain deformation of biological membranes and 3D tissue bodies, with use of the tissue models.

The solved examples include spherical and cylindrical membranes and a membrane with a hole subjected to cyclic loading, blood vessel, and urinary bladder. These examples are available on the web where the solutions for various model parameters can be obtained using the Software.

11.1 Introduction to mechanics of biological tissue

In this section we first present the structure of biological soft tissue and then describe basic experimental testing procedures for determining constitutive relations. Mostly common expressions for the constitutive laws are given, including both tissue and surfactant which usually covers the tissue.

11.1.1 Structure and function of biological tissue

The emphasis of this chapter is on connective biological tissue behavior and its modeling using appropriate constitutive laws. We focus on soft tissues of the planar (membrane) type which are met in hollow organs such as the lung parenchyma and pleura, stomach, mucous membranes, bladder, uterus, skin, eye, endocardium and pericardium. These tissues usually experience very large strains and stretches (stretches can be of order 2, meaning that a tissue fiber doubles its length) under normal physiological conditions.

Soft tissues display nonlinear mechanical response, i.e. nonlinear interdependence between the loadings and deformation within the range of the physiological working conditions. These nonlinearities come from shape change and also from the nonlinear constitutive relationships. Also, the time-dependent constitutive phenomena, such as creep of material under prolonged loadings, stress relaxation when the strains are held constant over time, or viscous effects under dynamic deformation. These mechanical characteristics can be attributed to the tissue constituents which are described next.

The main constituents of these tissues are the extracellular fibrous proteins collagen and elastin. These two constituents usually go together, as in lung parenchyma where their mutual ratio varies with location within the tissue (Lee & Hoppin 1972, Oldmixon & Hoppin 1989). Mechanical characteristics of the collagen and elastin are very different. Collagen is a relatively inextensible protein. It dominates in tendons and ligaments, as well as in bone and skin. Individual collagen fibers break at around 2% strain. However, within tissue these fibers have as significant initial slack with no stiffness. In practical applications the collagen fibers can be modeled by a nonlinear constitutive law shown in Fig. 11.1.1 expressed as (Kowe et al. 1986, Denny & Schroter 2000)

$$\sigma = c_1 \ln\left[1 - \frac{\exp(e) - 1}{c_2}\right] + c_3 e \qquad (11.1.1)$$

where $e = \lambda - 1$ is the strain and λ is the fiber stretch. The graph in the figure is drawn for: $c_1 = -22.5 \times 10^5\, Pa$; $c_2 = -1.26$; $c_3 = -17.8 \times 10^5\, Pa$.

Elastin is an extensible protein in connective tissues giving them the elastic mechanical behavior. It consists of polypeptide chains which are elongated and sparsely cross-linked and can experience large strains. It can be considered that the constitutive law of elastin bundles is linear even in the domain of large strains. Young's modulus of an elastin fiber is of order $10^5\, Pa$. A graphical representation of the elastin constitutive law is given in Fig. 11.1.1 for alveolar tissue, together with collagen.

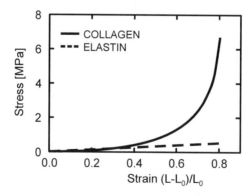

Fig. 11.1.1 The stress–strain relationships for a network of elastin and collagen fiber bundles (according to Denny & Schroter 2000; see also references therein). The curve for collagen is obtained using (11.1.1), while the liner stress–strain relationship for elastin is drawn for $E = 0.71\, MPa$

BIOLOGICAL SOFT TISSUE 203

11.1.2 Basic experiments and mechanical models

The basic experiments for the determination of tissue constitutive laws, or material models, are uniaxial and biaxial tests. Here are presented some typical experimental results for isotropic and orthotropic tissue characteristics and the corresponding material models. Also, the tissue models which describe the hysteretic, as well as viscoleastic behavior of tissue are included.

Uniaxial Test and Uniaxial Model
The uniaxial test is the basic mechanical test for biological tissue, as it is in general for other engineering materials. If a strip of a tissue, dissected from a membrane, is stretched quasi-statically, a typical stress–strain relationship shown in Fig. 11.1.2b is obtained. The main characteristic of tissue is that it has hardening behavior, namely the stress nonlinearly increases with strain. It can be seen that the tissue becomes very stiff at large strains. Also, tissue displays certain hysteretic behavior. The hysteresis is pronounced when muscle cells are dominant and we will define a material model when the hysteresis is large (see text below).

Therefore, the tissue material model for uniaxial loading is mathematically defined as dependence of the stress on strain, or, more conveniently, as dependence of the stress σ on stretch λ, i. e.

$$\sigma = \sigma(\lambda) \tag{11.1.2}$$

Biaxial Test and Biaxial Model
The uniaxial constitutive law is not sufficient to describe mechanical behavior of biological membranes under general loading conditions. If a membrane is loaded in two orthogonal directions, the stress–strain relationships for these directions are not the same as in the case of

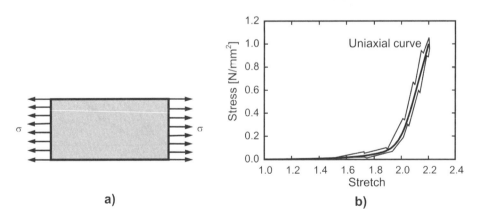

Fig. 11.1.2 Uniaxial constitutive law for tissue. (a) Schematic of uniaxial loading of material element; (b) Stress–strain relationship of a strip cut from alveolar tissue (according to Fukaya et al., 1968). The tissue has a small hysteresis when unloaded. The loading and unloading curves are not smooth because stress relaxations and stress recovery were allowed during the experiment. Idealized constitutive relationship $\sigma = \sigma(\lambda)$ is shown by the solid line as the idealization of the experiment

uniaxial loading. Characterization of the membrane mechanical response under stretching in both directions is obtained by performing biaxial tests: a membrane squared strip is stretched in two orthogonal directions, as schematically shown in Fig. 11.1.3a.

When the tissue can be considered isotropic, the loadings in two directions are to be the same (here called biaxial conditions). This loading can be achieved if a circular sample is fixed along the rim and loading by pressure. Then, in the central region the stress–strain state is the same in all directions (Hildebrandt et al. 1969). Uniaxial and biaxial experimental data for cat mesentery are shown in Fig. 11.1.3c. If the material is loaded by keeping the ratio of the smaller stress σ_2 to the larger stress σ_1, $r = \sigma_2/\sigma_1 = const$, a curve lying between the uniaxial and biaxial curves is obtained. Hence, the stress–stretch relationship at a material point is defined by the ratio of the stresses r. The model can be described by a family of curves

$$\sigma = \sigma(\lambda)|_{r=const}, \quad r = \sigma_2/\sigma_1, \quad \sigma_2 \leq \sigma_1 \qquad (11.1.3)$$

In practical application of this model, called further biaxial model, the two curves corresponding to experimental findings are the uniaxial curve ($\sigma_2 = 0$, $r = 0$) and biaxial

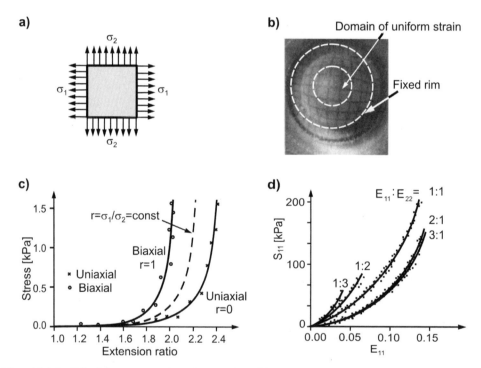

Fig. 11.1.3 Biaxial stress–strain relationships. (a) Schematics of biaxial test; (b) Experimental procedure for biaxial testing of isotropic membrane according to Hildebrandt et al. (1969): circular sample is fixed around the rim and subjected to pressure which produces biaxial stress–strain state (stresses and strains in all directions are the same in the central region); (c) Data for uniaxial and biaxial loading and fitted curves for cat mesentery according to Hildebrandt et al. (1969); (d) Test results and fitted curves (see (11.1.3)) for in-plane loading of bovine pericardium under several constant ratios of Green–Lagrange strain, E_{11} (preferred fiber direction), E_{22} (orthogonal to direction of E_{11}) – stress is the second Piola–Kirchhoff stress (according to Sacks 2000)

BIOLOGICAL SOFT TISSUE

curve ($\sigma_2 = \sigma_1$, $r = 1$) are used, while the other curves are obtained by linear interpolation between these two experimental curves.

Another approach to represent the membrane tissue mechanical behavior is to use models based on the strain energy function, see Section 2.4.2. The most common functions have exponential form (Fung's type) (2.4.21). We cite here a function used for the modeling of canine pericardium (Choi & Vito 1990)

$$W = b_0 \left[\exp\left(b_1 E_{11}^2\right) + \exp\left(b_2 E_{22}^2\right) + \exp\left(2 b_3 E_{11} E_{22}\right) - 3 \right] \qquad (11.1.4)$$

where $b_0 - b_3$ are constants, and E_{11} and E_{22} are the Green–Lagrange strains. This function is also used to fit the experimental data in Fig. 11.1.3d (Sacks 2000). Other forms of strain energy functions will be given in Section 11.2 within computational procedures and applications.

Hysteretic Model

It was found experimentally that the connective tissue has hysteretic behavior when subjected to cyclic loading, which is particularly significant when muscle cells are present (Sasaki & Hoppin 1979). The experimentally recorded dependence between the tensional force and the material strip length is shown in Fig. 11.1.4a. The constitutive law for the hysteretic tissue model is shown in Fig. 11.1.4b. It is represented by two relationships:

$$\sigma_\ell = \sigma_\ell(\lambda) \quad \text{and} \quad \sigma_u = \sigma_u(\lambda) \qquad (11.1.5)$$

where the first one corresponds to the loading part within the cycle, with increase of stretch, while the second is the unloading. It is assumed here and in Section 11.2 that the hysteretic curve does not change over time, hence the viscous effects are neglected in this model.

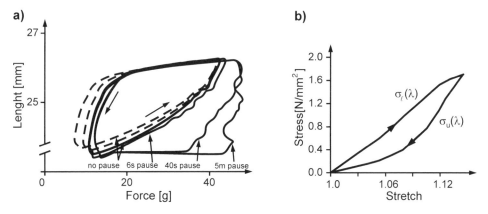

Fig. 11.1.4 Hysteretic response of smooth muscle tissue. (a) Experimental results on airway dog trachealis muscle strips (according to Sasaki & Hoppin 1979) (1 g = 0.00981 N); (b) Constitutive model for tissue histeresis expressed as the stress–stretch relationship: $\sigma_l(\lambda)$ is the loading curve and $\sigma_u(\lambda)$ is the unloading curve for one cycle. Note that the axes in the graph (b) are used as it is usual in engineering practice, opposite from figure (a)

Viscoelastic Models

In Section 2.2.2 a linear viscoelastic model was introduced based on the relaxation function (see (2.2.14) and (2.2.15)). Here, we specify another model, the fiber–fiber kinetics model (Mijailović 1991, Mijailović et al. 1993, 1994). The connective tissue is modeled by a system of fibers within an elastic medium (Fig. 11.1.5a, top panel). When the tissue deforms the force transfers among fibers and relative sliding among fibers occur generating internal frictional force

$$T(x,t) = [sign\ \nu(x,t)]\mu p(x,t) + b_w \nu(x,t) \quad (11.1.6)$$

where $T(x,t)$ is the force per unit length, $\nu(x,t)$ is the relative velocity at a point x along the fiber, $p(x,t)$ is the compressive stress between the fibers; and μ and b_w are the Coulomb and viscous friction coefficients. Due to assumption of symmetry of geometry and loading, one-half of the force is transferred to the other fiber through the traction which is generated in the region of sliding (Fig. 11.1.5a). In the case of Coulomb friction only, the traction is constant within the sliding region (Fig. 11.1.5a, bottom panel, $p = p_0$). The Coulomb friction contributes to the history-dependent deformation of tissue, while the viscous friction generates the response corresponding to a relaxation function (see

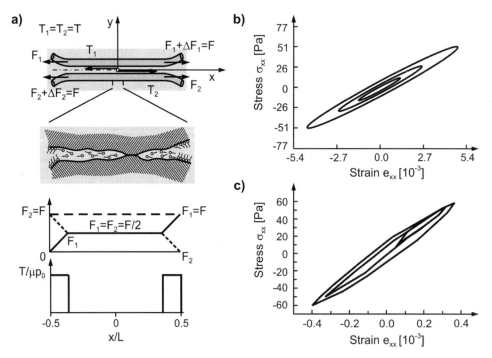

Fig. 11.1.5 Fiber–fiber kinetics model of connective tissue. (a) Schematics of the model, from top down: extensional force F transferred among fibers as forces F_1 and F_2, enlarged contact between fibers, distribution of force within fibers, distribution of contact traction T; (b) Experimental stress–strain hysteretic loops when the tissue is loaded cyclically (according to Mijailović 1991); (c) Computed hysteresis (according to Mijailović et al. 1993, 1994; Kojić et al. 1998, 2003)

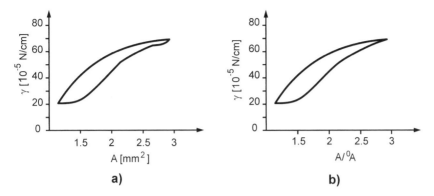

Fig. 11.1.6 Hysteretic characteristic of surface tension. (a) Experimentally determined dependence of surface tension on the surfactant area A (according to Ingenito et al. 1999); (b) Idealized dependence of surface tension on area ratio (^0A is the initial area)

(2.2.14)). The differential equations of force balance can be integrated and the hysteretic tissue response can be obtained, which is in agreement with experimental observations (Fig. 11.1.5b,c; see references given in the figure caption).

Models of Surfactant Covering Tissue
Biological membranes within organs are covered by surfactant. The surfactant plays an important role not only in biophysical processes, but also in biomechanical response of membranes due to the action of the surface tension of surfactant. This is true in particular for lung microstructure (e.g. Wilson 1982, Ingenito et al. 1999, Bachofen & Schurch 2001). Our description of the mechanical behavior of surfactant refers first of all to lung surfactant.

As can be seen from Fig. 11.1.6, surface tension depends on the surfactant area and has a hysteretic characteristic. This hysteresis plays an important role in lung functioning and gas exchange deep in the lung. In the mathematical description of surface tension γ we will use the relation

$$\gamma = \gamma\left(A/^0A\right) \qquad (11.1.7)$$

where $A/^0A$ is the ratio of the current area and the initial area of the surfactant (at a given point of the surfactant surface).

11.2 Modeling methods for isotropic tissue

In this section we first present a basic concept of numerical procedures used in the modeling soft tissue. Then we give specifics related to implementation of the mechanical models described in Section 11.1 into the finite element method and element-free Galerkin method.

11.2.1 General concept of computational procedures

Mechanical models for soft tissue described in Section 11.1.2 represent the nonlinear constitutive laws. Also, displacements of soft tissue can in general be large, therefore the problems of modeling soft tissue deformation are geometrically nonlinear (see Section 6.2.3). We here give a review of the basic finite element (FE) relations in the form used in tissue modeling. These relations are also applicable to the element-free Galerkin (EFG) method.

In many physiological conditions, inertial forces can be neglected, therefore the tissue deformation can be considered as a quasi-static problem. Solution is obtained by discretization of the deformable body using a discrete method, such as FE or EFG methods (Fig. 11.2.1). The basic equation of balance of forces for a finite element has the form (6.2.18),

$$\left(^{n+1}\mathbf{K}_L + {}^{n+1}\mathbf{K}_{NL}\right)^{(i-1)}_{tissue} \Delta \mathbf{U}^{(i)} = {}^{n+1}\mathbf{F}^{ext} - {}^{n+1}\mathbf{F}^{int(i-1)}_{tissue} \quad (11.2.1)$$

which corresponds to the step 'n' and iteration 'i' in an incremental-iterative solution procedure described in Section 6.2. Here, $\left(^{n+1}\mathbf{K}_L\right)_{tissue}$ and $\left(^{n+1}\mathbf{K}_{NL}\right)_{tissue}$ are the linear and geometrically nonlinear stiffness matrices; $^{n+1}\mathbf{F}^{ext}$ and $^{n+1}\mathbf{F}^{int(i-1)}_{tissue}$ are the external forces acting to the element, which include the action of the surrounding finite elements, and internal forces due to stresses within the tissue; $\Delta \mathbf{U}^{(i)}$ is the vector of increments of nodal displacements; and the left upper index '$n+1$' denotes the end of the incremental step. As is already indicated, the matrix $^{n+1}\mathbf{K}_{NL}$ is the same as for any solid, while the matrix $^{n+1}\mathbf{K}_L$ and the internal force vector contain the material tissue characteristics, and we further concentrate on them. They are

$$\left(^{n+1}\mathbf{K}_L\right)^{(i-1)}_{tissue} = \int_V {}^{n+1}\left(\mathbf{B}_L^T \mathbf{C}_{tissue} \mathbf{B}_L\right)^{(i-1)} dV,$$

$$^{n+1}\mathbf{F}^{int(i-1)}_{tissue} = \int_V {}^{n+1}\mathbf{B}_L^T \, {}^{n+1}\boldsymbol{\sigma}^{(i-1)}_{tissue} dV \quad (11.2.2)$$

where $^{n+1}\mathbf{B}_L^{(i-1)}$ is the linear strain–displacement matrix, see (6.2.12); $^{n+1}\mathbf{C}^{(i-1)}_{tissue}$ is the tangent constitutive matrix for tissue, defined in (6.2.19); and $^{n+1}\boldsymbol{\sigma}^{(i-1)}_{tissue}$ is the stress tensor within

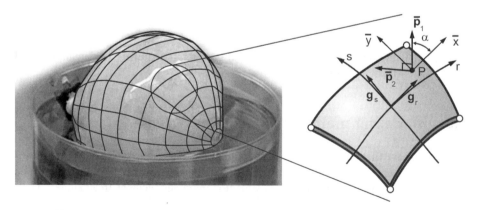

Fig. 11.2.1 Urinary bladder discretized into shell finite elements. Base vectors $\mathbf{g}_r$ and $\mathbf{g}_s$ are tangent to the isoparametric r and s lines. Unit vectors $\bar{\mathbf{p}}_1$ and $\bar{\mathbf{p}}_2$ (left basis, see (2.4.13)) of the principal stretches in the shell tangential plane, at a material point P; they are rotated for an angle α with respect to the local shell coordinate system $\bar{x}, \bar{y}$.

BIOLOGICAL SOFT TISSUE

tissue (written here in a matrix form, see (2.1.13)). Integration is performed over the last known element volume $V^{(i-1)}$. In Sections 11.2.2 and 11.2.3 we will show how the stress $^{n+1}\boldsymbol{\sigma}_{tissue}^{(i-1)}$ and the matrix $^{n+1}\mathbf{C}_{tissue}^{(i-1)}$ can be evaluated for several tissue models.

The above expressions in (11.2.2) assume a general 3D body. However, if a biological membrane, as shown in Fig.11.2.1, is considered, the matrix $^{n+1}\mathbf{C}_{tissue}^{(i-1)}$ and the stresses $^{n+1}\boldsymbol{\sigma}_{tissue}^{(i-1)}$ are calculated in the membrane tangential plane (shell finite elements must be used, see Section 4.5), and then they must be transformed to the global coordinate system using the transformation relationships (2.2.17) and (2.1.14). Details about these transformations are given on the web – Theory, Chapter 4, Section 4.5. Furthermore, if the membrane deformation is large and logarithmic strains are employed, the principal values and principal directions of the left Cauchy–Green deformation tensor must be calculated (Section 2.4.1). Namely, we first calculate the left Cauchy–Green deformation tensor from the displacement field, as in the case of a general 3D deformation: $^{n+1}_{0}\mathbf{B}^{(i-1)} = {}^{n+1}_{0}(\mathbf{FF}^T)^{(i-1)}$, where $^{n+1}_{0}\mathbf{F}^{(i-1)}$ is the deformation gradient; details are given in Section 2.4.1. Then, the tensor $^{n+1}_{0}\mathbf{B}^{(i-1)}$ is transformed to the local shell system $\bar{x}_i$ according to the tensorial transformation (1.3.4),

$$^{n+1}_{0}\bar{\mathbf{B}}^{(i-1)} = \left(^{n+1}\mathbf{T}\, ^{n+1}_{0}\mathbf{B}\, ^{n+1}\mathbf{T}^T \right)^{(i-1)} \quad (11.2.3)$$

where $^{n+1}\mathbf{T}^{(i-1)}$ is the transformation matrix containing the cosines between local and global coordinate systems $\bar{x}_i$ and x_j. Then, using $^{n+1}_{0}\bar{\mathbf{B}}^{(i-1)}$, we calculate the angle $^{n+1}\alpha^{(i-1)}$ for the principal directions $^{n+1}\bar{\mathbf{p}}_1^{(i-1)}$ and $^{n+1}\bar{\mathbf{p}}_2^{(i-1)}$ in the element tangential plane (see Fig. 11.2.1 and Example 2.1-1; also see web – Theory, Chapter 2) as

$$^{n+1}\alpha^{(i-1)} = \frac{1}{2}\tan^{-1}\left(\frac{2\, ^{n+1}\bar{B}_{12}}{^{n+1}\bar{B}_{11} - {}^{n+1}\bar{B}_{22}} \right)^{(i-1)} \quad (11.2.4)$$

The principal stretches are determined using these principal vectors and the relation (2.4.6) from which follows (see also Kojić & Bathe 2005):

$$^{n+1}_{0}\lambda_k^{(i-1)} = 1 / \left\| ^{n+1}_{0}\mathbf{F}^{-1}\, ^{n+1}\bar{\mathbf{p}}_k \right\|^{(i-1)}, \quad k = 1, 2 \quad (11.2.5)$$

From these stretches we calculate the principal stresses in the membrane tangential plane, neglecting transversal shear stresses (Kojić 2002).

Details about the above transformations and calculation of the principal stretches and directions are given on the web – Theory, Chapter 11, Section 11.2. We further present the stress integration within an incremental step and calculation of the tangent constitutive matrix, following Section 6.2, for most of the tissue material models described in Section 11.1.

The above computational scheme and the form of governing equations is the same when the element-free Galerkin (EFG) method is employed. This can be seen by inspecting the governing relations presented in Section 8.5.

11.2.2 Biaxial models of membranes, hardening and hysteretic behavior, action of surfactant

We here present stress integration, i.e. stress calculation at the end of a load step $^{n+1}\boldsymbol{\sigma} = {}^{n+1}\boldsymbol{\sigma}_{tissue}$ (see (6.2.22)), and calculation of the tangent constitutive matrix $^{n+1}\mathbf{C}_{tissue}$

(further denoted by $^{n+1}\mathbf{C}$) for the membrane biaxial model, without and with surfactant, and for hysteretic tissue model. In the notation below the iteration counter '$i-1$' is omitted for simpler writing, but it is implied.

Biaxial Model

The uniaxial and biaxial constitutive relations for membrane are shown in Fig. 11.2.2 (see also Fig. 11.1.3c) by the stress–stretch relations in the principal strain directions 1 and 2. In the case of a shell shape, we consider the stretches $^{n}\lambda_1$ and $^{n}\lambda_2$ and the principal directions $^{n}\bar{\mathbf{p}}_1$ and $^{n}\bar{\mathbf{p}}_2$ in the tangential plane of the shell element, as described above (Section 11.2.1). If the principal stresses at the start of step 'n' are $^{n}\sigma_1$ and $^{n}\sigma_2$, corresponding to stretches $^{n}\lambda_1 > {}^{n}\lambda_2$, then we introduce the stress ratio ^{n}r as (Kojić et al. 2006)

$$^{n}r = \frac{^{n}\sigma_2}{^{n}\sigma_1} \qquad (11.2.6)$$

The stress ratio is $r=1$ for biaxial loading, while for uniaxial case $r=0$. All curves $r=const.$ for $0 < r < 1$ lie between the biaxial and uniaxial curves. We assume that stresses are positive, i.e. $^{n}\lambda_1 \geq 1$ and $^{n}\lambda_2 \geq 1$, occurring under usual physiological conditions. In the case of a stretch less than 1 we use $1/\lambda$ and the corresponding constitutive curve in extension, in order to have the same stress in magnitude corresponding to the same ratio of material length change; this is an approximation which does not usually affect the solutions, since biological tissues in physiological conditions are subjected to extensional rather than compressive loadings.

In the stress integration procedure it is assumed that the known quantities at the current state of deformation are:

$$^{n}\sigma_1, {}^{n}\sigma_2, {}^{n}\lambda_1, {}^{n}\lambda_2, {}^{n+1}\lambda_1, {}^{n+1}\lambda_2 \qquad (11.2.7)$$

with $^{n}\sigma_1 > {}^{n}\sigma_2$, hence the stress ratio ^{n}r is given by the relation (11.2.6). The stress $^{n+1}\sigma_1$, corresponding to stretch $^{n+1}\lambda_1$, is obtained as a linear interpolation between stresses at uniaxial and biaxial curves $^{n+1}\sigma_u$ and $^{n+1}\sigma_b$,

$$^{n+1}\sigma_1\left(^{n+1}\lambda_1\right) = \left(1 - {}^{n+1}r\right) {}^{n+1}\sigma_u + {}^{n+1}r \, {}^{n+1}\sigma_b \qquad (11.2.8)$$

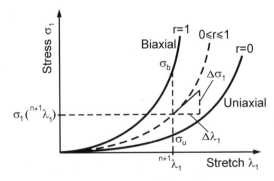

Fig. 11.2.2 Stress–stretch relations for uniaxial ($r=0$), biaxial ($r=1$) and $r = \sigma_2/\sigma_1 = const$ conditions. The stress $\sigma_1(^{n+1}\lambda_1)$ for stretch $^{n+1}\lambda_1$ is linearly interpolated from the stresses σ_b and σ_u on the biaxial and uniaxial curves. The tangent modulus $E_\lambda = \Delta\sigma_1/\Delta\lambda_1$

BIOLOGICAL SOFT TISSUE

The relation (11.2.8) also follows from a linear interpolation of tangent modulus between the values $E_{\lambda u}$ and $E_{\lambda b}$ on uniaxial and biaxial curves to obtain $^{n+1}E_\lambda$ on the curve with ^{n+1}r,

$$^{n+1}E_\lambda = \left(1 - {}^{n+1}r\right){}^{n+1}E_{\lambda u} + {}^{n+1}r\,{}^{n+1}E_{\lambda b} \quad (11.2.9)$$

and the stress increment

$$\Delta\sigma_1 = {}^{n+1}E_\lambda \Delta\lambda_1 \quad (11.2.10)$$

Using the secant modulus corresponding to the calculated stress $^{n+1}\sigma_1$ and the plane stress conditions, the stress $^{n+1}\sigma_2$ is obtained as

$$^{n+1}\sigma_2 = {}^{n+1}\sigma_1 \frac{\nu\,{}^{n+1}e_1 + {}^{n+1}e_2}{{}^{n+1}e_1 + \nu\,{}^{n+1}e_2} \quad (11.2.11)$$

where $^{n+1}e_1$ and $^{n+1}e_2$ are the total strains measured from the undeformed configuration. We here use the logarithmic strains, hence $^{n+1}e_1 = \ln\left({}^{n+1}\lambda_1\right)$ and $^{n+1}e_2 = \ln\left({}^{n+1}\lambda_2\right)$.

The computational steps are summarized in Table 11.2.1. Note that when the values of $^{t+\Delta t}r$ are (within a numerical tolerance) equal to 0 or 1, a solution check is performed to keep the solution for stress within the domain bounded by the biaxial and uniaxial curves.

We calculate the tangent constitutive matrix $^{n+1}\bar{C}$ in the principal directions 1 and 2. In order to derive the expressions for $^{n+1}\bar{C}_{ij}$, $i, j = 1, 2$, we use the elastic matrix for the plane stress conditions given in (2.2.7) with the tangent elastic modulus $^{n+1}E_T$. From the second of the constitutive stress–strain relations (2.2.2) it follows

$$de_2 = \frac{1 - \nu^2}{{}^{t+\Delta t}E_T} d\sigma_2 - \nu\,de_1 \quad (11.2.12)$$

Further, the relation $d\sigma_2/d\sigma_1 = {}^{n+1}r$ and the first constitutive relation can be used to obtain

$$^{t+\Delta t}E_T = \left(1 - {}^{n+1}r\nu\right){}^{n+1}\lambda_1\,{}^{n+1}E_\lambda \quad (11.2.13)$$

where $^{n+1}E_\lambda$ is the tangent modulus given in (11.2.9); also, the relation $de_1 = {}^{n+1}\lambda_1 d\lambda_1$ is used here which follows from the definition of the logarithmic strain. Then, the constitutive matrix for the shell in-plane terms is

$$\bar{C} = {}^{n+1}\lambda_1\,{}^{n+1}E_\lambda \frac{1 - {}^{n+1}r\nu}{1 - \nu^2} \begin{bmatrix} 1 & \nu & 0 \\ \nu & 1 & 0 \\ 0 & 0 & \frac{1-\nu}{2} \end{bmatrix} \quad (11.2.14)$$

Table 11.2.1 Computational steps for stress calculation for biaxial model

---

Calculate the principal stretches and principal directions in the tangential membrane plane.
Calculate stresses $^{n+1}\sigma_1$ and $^{n+1}\sigma_2$ in the principal directions.
Transform the stresses $^{n+1}\sigma_1$ and $^{n+1}\sigma_2$ first to the local membrane coordinate system $\bar{x}_1, \bar{x}_2$ (axes $\bar{x}, \bar{y}$ in Fig. 11.2.1) and then to the global coordinate system for use in equation (11.2.1).

The general form of the shell constitutive matrix is given in (2.2.8) which is the extension of this matrix. This matrix is isotropic with respect to the rotation around the shell normal (see web, Examples – Example 2.2-6), and the transformation to the local shell coordinate system is not needed. Note that the calculated tangent constitutive matrix affects the rate of convergence during equilibrium iterations, but not the solution (see Section 6.2).

Membranes Covered by Surfactant with Hysteretic Characteristic

We present a computational procedure to model the mechanical response of a biological membrane covered by surfactant. It is assumed that the surface tension depends on the change of the surfactant area, in particular with a hysteresis when the surfactant area changes cyclically. The presented method is applicable to any membrane geometry and will be illustrated through examples in Section 11.3.

A membrane covered by surfactant is schematically shown in Fig. 11.2.3. We assume that there is no slip between the surfactant layer and tissue, therefore change of the surfactant area during membrane deformation is the same as the change of the membrane area. Experiments show (Wilson 1982, Ingenito et al. 1999) that the surface tension γ (N/m) is a function of the ratio of the current surfactant area A and the initial area, $A/^0A$, see Fig. 11.2.4. At a membrane point on the surface P, the surface tension can be expressed in the form

$$\gamma = \gamma\left(\frac{dA}{d^0A}\right) = \gamma\left(\frac{\det \mathbf{J}}{\det {^0\mathbf{J}}}\right) \qquad (11.2.15)$$

where $\mathbf{J}$ and $^0\mathbf{J}$ are Jacobians of transformation between the Cartesian coordinate system and the natural coordinate system r, s of the membrane finite element, corresponding respectively, to the current configuration $\mathcal{B}$ and undeformed configuration $^0\mathcal{B}$ (see (4.3.9), as well as (4.5.6) in which $t = 0$ defines the mid-surface). The $\det \mathbf{J}$ is

$$\det \mathbf{J} = \left\| \frac{\partial \mathbf{x}}{\partial r} \times \frac{\partial \mathbf{x}}{\partial s} \right\| \qquad (11.2.16)$$

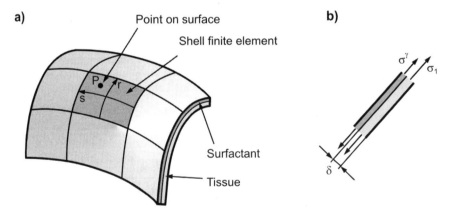

Fig. 11.2.3 Biological membrane covered with surfactant and modeled by shell finite elements. The surfactant deforms as the shell surface. (a) Geometry of the membrane and surfactant; (b) Stress in tissue and surfactant (in the first principal strain direction)

BIOLOGICAL SOFT TISSUE

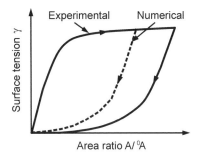

Fig. 11.2.4 Dependence of surface tension γ on the surfactant area ratio $A/^0A$ for the surfactant covering alveolar lung miscrostucture during inspiration–expiration cycles (Wilson 1982). Dashed curve (numerical) is obtained by a scaling procedure. Note that the surface tension is much higher during inspiration

where $\partial \mathbf{x}/\partial r$ and $\partial \mathbf{x}/\partial s$ are the base vectors evaluated at a shell point (see Fig 11.2.3), and $\|\bullet\|$ represents the vector modulus (see (1.3.10)).

Hence, for a current membrane deformation, the surface tension can be determined. It is of particular interest to consider a hysteretic characteristic of surface tension, such as shown in Fig.11.2.4 (Wilson 1982). We have also shown by dashed line a scaled curve $\gamma\left(A/^0A\right)$, corresponding to a smaller amplitude of deformation. Namely, membrane points have, in general, a different extent of deformation and therefore may reach the amplitudes of the ratio $\left(A/^0A\right)_{max}$ which defer from the experimental curve. Then, it is necessary to scale the experimental curve (Kojić et al. 2006). Details about the curve scaling are given on the web – Theory, Chapter 11.

Surface tension has the same action on the tissue in all directions within the surface (see Fig. 11.2.3). In order to simplify calculation of tissue external loading due to surfactant, we evaluate the stress $^{n+1}\sigma^\gamma$ equivalent to the surface tension $^{n+1}\gamma$, as

$$^{n+1}\sigma^\gamma = \frac{^{n+1}\gamma}{^{n+1}\delta} \qquad (11.2.17)$$

where $^{n+1}\delta$ is the membrane thickness at integration point. Then, according to (11.2.2), the nodal force due to surfactant (applicable to any equilibrium) is

$$^{n+1}\mathbf{F}^\gamma = \int_{^{n+1}V} {^{n+1}\mathbf{B}_L^T} \, {^{n+1}\boldsymbol{\sigma}^\gamma} dV \qquad (11.2.18)$$

Here the stress tensor $^{n+1}\boldsymbol{\sigma}^\gamma$ has the non-zero normal components only in the local shell system, i.e. $^{n+1}\bar{\sigma}_{km}^\gamma = \delta_{km} \, ^{n+1}\sigma^\gamma$ where δ_{km} is the Kronecker delta-symbol ($\delta_{km} = 1$ for $k=m$ and $\delta_{km} = 0$ for $k \neq m$).

Note that the force due to surfactant $^{n+1}\mathbf{F}^\gamma$ depends on the displacements since the stress $^{n+1}\boldsymbol{\sigma}^\gamma$ depends on $^{n+1}\gamma\left(^{n+1}A/^0A\right)$ according to (11.2.15). Hence, using (11.2.17) and (11.2.15) the nonzero constitutive terms in the local membrane coordinate system are

$$^{n+1}\bar{C}_{ij}^\gamma = \frac{^{n+1}E_\gamma}{^{n+1}\delta} \frac{\det {}^{n+1}\mathbf{J}}{\det {}^0\mathbf{J}}, \quad i,j = 1, 2 \qquad (11.2.19)$$

where $^{n+1}\bar{C}_{ij}^{\gamma} = \partial^{\,n+1}\bar{\sigma}_{ii}^{\gamma}/\partial^{\,n+1}\bar{e}_{jj}$ (no sum on i and j), and $^{n+1}E_{\gamma} = {}^{n+1}\left[d\gamma/\left(dA/d\,^{0}A\right)\right]$. These terms are added to the constitutive matrix for tissue. In deriving (11.2.19) the following relation is used

$$\frac{d(dA)}{d\,^{0}A} = (d\bar{e}_{11} + d\bar{e}_{22})\frac{\det \mathbf{J}}{\det {}^0\mathbf{J}} \quad (11.2.20)$$

In Section 11.3 it will be demonstrated how the surfactant affects the mechanical response of biological membranes.

Tissue with Hysteresis

Here, we present a computational procedure for modeling of biological tissue with hysteretic mechanical response when subjected to a cyclic loading. It is assumed that the hysteresis is displayed in the direction of tissue fibers, and described by a uniaxial stress–stretch relationship (see Section 11.1.2, hysteretic model). Consequently, we use simple 1D finite elements of Section 4.2.1 and 1D hysteretic constitutive models.

A uniaxial constitutive relationship is schematically shown in Fig. 11.2.5 (see also Fig. 11.1.4b) as a stress–stretch loop, with the same amplitude $(\lambda_{\max})_{\exp}$ over cycles. The main conditions which must be considered in the stress integration within an incremental-iterative FE procedure are: (a) to know whether the loading regime is 'loading' or 'unloading' within the cycle; and (b) to evaluate the stress in a way that the stress–stretch point at the end of an incremental step lies on the corresponding constitutive curve. The stress integration procedure to obtain the stress at the end of load step 'n', $^{n+1}\sigma_{\xi\xi} \equiv {}^{n+1}\sigma$, in the fiber direction of the fiber unit vector $^{n+1}\boldsymbol{\xi}_0$, is summarized in Table 11.2.2.

After the stress $^{n+1}\sigma$ is evaluated, this stress is transformed to the global coordinate system using the relation (2.1.14), see also web – Theory, Chapter 2, to obtain the components $^{n+1}\sigma_{(f)ij}$ of the fiber stress $^{n+1}\boldsymbol{\sigma}_f$,

$$^{n+1}\sigma_{(f)ij} = {}^{n+1}\sigma\,{}^{n+1}\xi_{0i}\,{}^{n+1}\xi_{0j} \quad (11.2.21)$$

where $^{n+1}\xi_{0i}$ and $^{n+1}\xi_{0j}$ are projections of the unit vector $^{n+1}\boldsymbol{\xi}_0$ to the global coordinate axes x_i. The stress $^{n+1}\boldsymbol{\sigma}_f$ is superimposed to the stress of the surrounding material (see also Section 12.2 where muscle fibers are considered). If the surrounding material is elastic we have (as in (12.2.20))

$$^{n+1}\boldsymbol{\sigma} = {}^{n+1}\boldsymbol{\sigma}^E(1-\phi) + \phi\,{}^{n+1}\boldsymbol{\sigma}_f \quad (11.2.22)$$

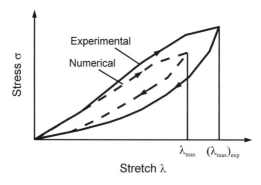

Fig. 11.2.5 Experimental and numerical stress–stretch loops

BIOLOGICAL SOFT TISSUE   215

Table 11.2.2 Iterative scheme for material with hysteretic characteristic

---

Initial solution (cycle counter $l = 0$)
We follow the experimental curve in the loading part and reach the amplitude $\lambda_{\max}^{(0)}$ for calculation of the scale factor $s_\lambda^{(0)} = \left(\lambda_{\max}^{(0)} - 1\right) / \left[(\lambda_{\max})_{\exp} - 1\right]$ for each material (integration) point. The unloading curve is determined using this initial scale factor and the scaling procedure given on the web – Theory, Chapter 11.

Iteration loop on cycles $l = l + 1$:
Compute the stresses for the entire cycle following the scaled constitutive loops for each integration point. The scaling has a 'symmetric' character (the loop 'thickness' is scaled). The scaled hysteretic characteristic for a material point is determined using the scale factors $s_\lambda^{(l-1)}$ and given functional $\sigma - \lambda$ relationship. Also, determine scale factors for the current loop iteration $s_\lambda^{(l)}$.

Convergence check for cycle l:
Difference between two successive amplitudes $\lambda_{\max}^{(l-1)}$ and $\lambda_{\max}^{(l)}$ for all material points must be within a selected numerical tolerance ε_λ, i. e. $\left|\lambda_{\max}^{(l)} - \lambda_{\max}^{(l-1)}\right| \leq \varepsilon_\lambda$. If this convergence criterion is not satisfied go to step 2 for the next cycle iteration.

---

where ϕ is the volumetric fraction of fibers, and $^{n+1}\boldsymbol{\sigma}^E$ is the stress of the elastic material ($= \mathbf{C}^E$ $^{n+1}\mathbf{e}$, where $\mathbf{C}^E$ is elastic matrix and $^{n+1}\mathbf{e}$ is the strain). In the case of a biological membrane, instead of elastic stress in (11.2.22), the membrane stress obtained by the stress integration shown above (biaxial model) should be used. The stresses $^{n+1}\boldsymbol{\sigma}$ enter the equation (11.2.2) for the current equilibrium iteration 'i'.

The constitutive coefficient E_T for the fiber direction follows from the relation

$$^{n+1}E_T = \frac{d\,^{n+1}\sigma}{d\,^{n+1}e} = {}^{n+1}\lambda\;^{n+1}E_\lambda \qquad (11.2.23)$$

where $^{n+1}E_\lambda$ is the slope on the current cyclic curve. The term $^{n+1}E_T$ is further used to perform superposition with the constitutive matrix of the surrounding material (see web – Theory, Chapter 11).

11.2.3 Use of strain energy functions

General expressions for stress calculation and calculations of the tangent constitutive matrix are given here for tissue material represented by the use of the strain energy functions. Also, some details are presented for two specific forms for the strain energy.

Consider the strain energy function for a nonlinear elastic material model (tissue), which has a general form (Holzapfel 2001):

$$W = \psi_{vol}(J) + \psi_{iso}(\bar{I}_1, \bar{I}_2) \qquad (11.2.24)$$

where the first and the second term on the right-hand side correspond, respectively, to the volumetric and isochoric deformation of the material; J is the determinant of the deformation gradient $\mathbf{F}$ (see (2.4.2)), i.e. $J = \det \mathbf{F} = \lambda_1 \lambda_2 \lambda_3$; and $\bar{I}_1$ and $\bar{I}_2$ are the first and the second invariant of the modified Cauchy–Green deformation tensors. The three invariants of these deformation tensors are (see (1.3.15))

$$\bar{I}_1 = J^{-2/3} I_1 = J^{-2/3} \left(\lambda_1^2 + \lambda_2^2 + \lambda_3^2 \right)$$
$$\bar{I}_2 = J^{-4/3} I_2 = J^{-4/3} \left(\lambda_1^2 \lambda_2^2 + \lambda_2^2 \lambda_3^2 + \lambda_3^2 \lambda_1^2 \right) \quad (11.2.25)$$
$$\bar{I}_3 = J^{-2} I_3 = J^{-2} \left(\lambda_1^2 \lambda_2^2 \lambda_3^2 \right) = J^{-2} J^2 = 1$$

The modified invariants (and modified Cauchy–Green deformation tensors) correspond to the modified deformation gradient,

$$\bar{\mathbf{F}} = J^{-1/3} \mathbf{F} \quad (11.2.26)$$

which gives incompressible deformation, since $\det \bar{\mathbf{F}} = J^{-1} \det \mathbf{F} = 1$. We consider the deformation in the principal stretch directions defined by the left basis $\bar{\mathbf{p}}_1, \bar{\mathbf{p}}_2, \bar{\mathbf{p}}_3$ (see (2.4.13)) with the principal stretches $\lambda_1, \lambda_2, \lambda_3$. The indices for the configurations used in Section 2.4.2 and above are omitted here for simpler writing.

The stresses follow from the relation (2.4.20). We use from (11.2.24), the definition of the Green–Lagrange strains (2.4.15) – which in the principal directions are: $E_i^{GL} = 0.5 \left(\lambda_i^2 - 1 \right), i = 1, 2, 3$; and the relationship (2.4.19) between the Piola–Kirchhoff and Cauchy stresses in the principal direction: $S_i = \sigma_i / \lambda_i^2$, where it is taken ${}^0\rho/{}^t\rho = 1$. Then, from (11.2.24) follows:

$$\sigma_i = \lambda_i \frac{\partial \psi}{\partial \lambda_i} = \lambda_i \frac{\partial \psi_{vol}(J)}{\partial \lambda_i} + \lambda_i \frac{\partial \psi_{iso}(\bar{I}_1, \bar{I}_2)}{\partial \lambda_i} = $$
$$= \sigma_i^{vol} + \sigma_i^{iso}, \quad \text{no sum on } i \quad (11.2.27)$$

where σ_i^{vol} and σ_i^{iso} are the volumetric and isochoric parts of the Cauchy stress.

The constitutive matrix in the principal stretch directions is obtained from the definition $C_{ij} = \partial \sigma_i / \partial e_j = \lambda_j \partial \sigma_i / \partial \lambda_j$. Then, from (11.2.27) we obtain (no sum on i, j)

$$C_{ij}^{vol} = \lambda_j \frac{\partial \sigma_i^{vol}}{\partial \lambda_j} = \lambda_i \lambda_j \frac{\partial^2 \psi_{vol}(J)}{\partial \lambda_i \partial \lambda_j}, \quad C_{ij}^{iso} = \lambda_j \frac{\partial \sigma_i^{iso}}{\partial \lambda_j} = \lambda_i \lambda_j \frac{\partial^2 \psi_{iso}(\bar{I}_1, \bar{I}_2)}{\partial \lambda_i \partial \lambda_j} \quad (11.2.28)$$

Note that the above expressions for the stresses and the constitutive matrix can be evaluated at the end of the incremental step and for the iteration '$i-1$' by using the stretches ${}^{n+1}\lambda_i^{(i-1)}$. These expressions (as well the matrix in (11.2.14)) do not take into account change of the principal stretch directions due to change of strains. The additional 'kinematic' terms of C_{ij} do not depend on the material moduli (derivatives of stresses with respect to strains) and can be found elsewhere (e.g. Simo & Taylor 1991, Holzapfel 2001).

Potential According to Delfino et al. (1997)

Here, the above general expressions are implemented to the model with the strain energy function (potential) according to Delfino et al. (1997) used for the tissue of carotid artery. The potentials are

$$\psi_{iso}(\bar{I}_1) = \frac{a}{b} \left\{ \exp\left[\frac{b}{2}(\bar{I}_1 - 3) \right] - 1 \right\}, \quad \psi_{vol}(J) = \frac{1}{2\alpha} \ln^2 J \quad (11.2.29)$$

BIOLOGICAL SOFT TISSUE

where a and b are material constants – the first has the dimension of stress, while the second is dimensionless; and α is a penalty parameter which can be determined.

The principal stresses σ_i^{iso} and σ_i^{vol} are obtained from (11.2.27) and (11.2.29), with use of (11.2.25),

$$\sigma_i^{iso} = aJ^{-2/3}(\lambda_i^2 - \frac{1}{3}I_1)\exp\left[\frac{b}{2}(\bar{I}_1 - 3)\right], \quad \sigma_i^{vol} = \frac{1}{\alpha}\ln J \quad (11.2.30)$$

From (11.2.28) and (11.2.29) follow the terms of the matrix C_{ij}^{iso} and C_{ij}^{vol} as

$$C_{ij}^{iso} = a\, J^{-2/3} \exp\left[\frac{b}{2}(\bar{I}_1 - 3)\right]$$
$$\times \left[2\left(\frac{1}{9}I_1 + \lambda_i\lambda_j\delta_{ij} - \frac{1}{3}(\lambda_i^2 + \lambda_j^2)\right) + bJ^{-2/3}(\lambda_i^2 - \frac{1}{3}I_1)(\lambda_j^2 - \frac{1}{3}I_1)\right], \quad (11.2.31)$$
$$C_{ij}^{vol} = 1/\alpha$$

where δ_{ij} are the Kronecker delta symbols.

Details about the derivation of the expressions (11.2.30) and (11.2.31) and determination of the penalty parameter α are given on the web – Theory, Chapter 11.

Two-Dimensional Fung's Potential

The strain energy potential function for 2D problems is given as (Fung *et al.* 1979, Fung 1990)

$$W = \frac{c}{2}\left[\exp\left(a_1 E_1^2 + a_2 E_2^2 + 2a_4 E_1 E_2\right) - 1\right] \quad (11.2.32)$$

where c (dimension of stress) and a_1, a_2, a_4 (dimensionless) are material constants; and E_1, E_2 are the Green–Lagrange strains (in the principal strain directions). The principal stresses follow from (2.4.20), (11.2.32) and the relation $S_i = \sigma_i/\lambda_i^2$, as

$$\sigma_1 = c\lambda_1^2 (a_1 E_1 + a_4 E_2)\exp\left(a_1 E_1^2 + a_2 E_2^2 + 2a_4 E_1 E_2\right)$$
$$\sigma_2 = c\lambda_2^2 (a_2 E_2 + a_4 E_1)\exp\left(a_1 E_1^2 + a_2 E_2^2 + 2a_4 E_1 E_2\right) \quad (11.2.33)$$

where λ_1, λ_2 are the stretches. The constitutive matrix $C_{ij} = \partial \sigma_i/\partial e_j$ follows from these expressions for stresses, with use of the relations: $dE_i/\partial \lambda_i = \lambda_i$ and $\partial e_i/\partial \lambda_i = \lambda_i$, no sum on i. The expressions for the matrix C_{ij} are given on the web – Theory, Chapter 11.

11.3 Examples

Several typical examples are solved where the models' computational procedures presented in Section 11.2 are used. For some of these examples the Software is provided on the web where solutions can be obtained for a range of the problem parameters.

Example 11.3-1. Deformation of spherical biological membrane under cyclic pressure loading

A spherical biological membrane (Fig. E11.3-1a) is subjected to internal pressure loading which has a cyclic character (Kojić *et al.* 2006). The cases when the tissue is not covered by surfactant and when it is covered on the internal side are considered. It is assumed that the membrane tissue behavior is defined by the biaxial model, with the two characteristic curves shown in Fig. E11.3-1b. Surfactant generates surface tension which depends on the area ratio and has a hysteretic character (Fig. E11.3-1c).

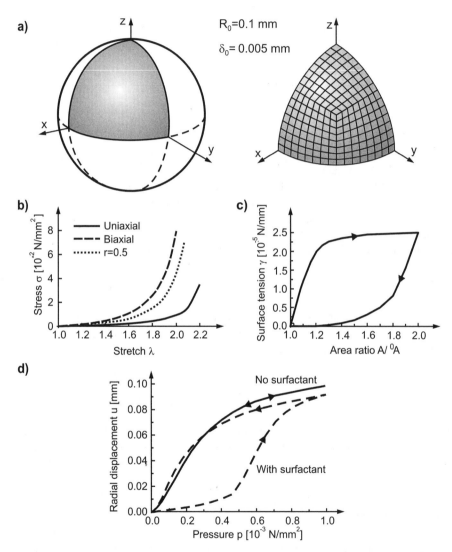

Fig. E11.3-1 Spherical biological membrane loaded by cyclic internal pressure. (a) Membrane geometry and FE mesh of part of the sphere which is modeled by shell finite elements (R_0 – initial radius, δ_0 – initial thickness); (b) Constitutive curves for tissue biaxial model (also shown the curve for the stress ratio $r = 0.5$); (c) Dependence of the surface tension of surfactant covering tissue on the area ratio (Wilson 1982); (d) Calculated dependence of radial displacement on pressure (it has a hysteretic character in case of action of surfactant)

BIOLOGICAL SOFT TISSUE   219

Part of the membrane in the first quadrant is modeled by shell finite elements (Section 4.5) due to symmetry in loading and geometry as shown in Fig. E11.3-1a. Computational procedures for stress integration of Section 11.2.2 is applied for the biaxial tissue model and for surfactant. The pressure is increased to a selected amplitude and then decreased to zero.

The calculated pressure–radial displacement relationship is shown in Fig. E11.3-1d. In the case with no surfactant action, the loading and unloading paths are along the same nonlinear curve. However, when the membrane is covered by surfactant, the loading curve is below the unloading curve and there is an energy loss within the cycle due to hysteresis in surface tension. This mechanical response can have important effects on the behavior of membrane-type organs, such as microstructure of the lung (Wilson 1982, Bachofen & Schurch 2001).

Solutions for various model parameters can be obtained using Software on the web.

Example 11.3-2. Cylindrical membrane without and with rings subjected to internal cyclic pressure

Here, deformation of a cylindrical biological membrane without and with rings is studied when subjected to cyclic internal pressure (Kojić et al. 2006), Fig. E11.3-2a. A cylinder restrained to deform axially (called here closed cylinder) is analyzed in the case when the tissue is covered from one side by surfactant or when there is no surfactant. Also, an open cylinder, free to deform axially (and without axial loading) is considered in the case when there are rings distributed at the same distances, or there are no rings. The tissue behavior is described by the biaxial model, and surfactant with hysteresis, with data as in Example E11.3-1. Hysteretic constitutive curve for the rings is given in Fig. E11.3-2c.

Four-node shell elements (Section 4.5) are used for tissue and line elements for rings. Symmetry boundary conditions are employed for the point lying in the symmetry planes (zero displacement through symmetry plane, and no rotations around normal to the symmetry plane). Radial displacements are allowed, while the axial displacements are restrained in the case of a closed cylinder.

The calculated dependences of radial displacement on the pressure are shown in Figs. E11.3-2d,e. In the case of a closed cylinder and no surfactant we have that the stress state is such that the stress ratio is close to 0.5 during the whole cycle (see details in Tutorial for this example on the web), following the dotted constitutive curve in Fig. E11.3-1b; while the radial displacement–pressure curve is nonlinear and loading–unloading paths follow the same curve (Fig. E11.3-2d). When surfactant is present, the displacement–pressure relationship has a pronounced hysteretic character.

When the cylinder is open we have the following solutions (Fig. E11.3-2e). If there is no surfactant and no rings, the stress state reduces to uniaxial and the loading–unloading paths follow the same nonlinear curve (top curve, the softest response). When surfactant is acting and no rings, the cylinder becomes stiffer and has hysteresis in the displacement–pressure relationship (dashed line). In the case with rings and no surfactant, the cylinder is stiffer than in the case with action of surfactant only (solid line) and the hysteresis in the displacement–pressure relationship is due to hysteretic characteristics of the rings. Finally, the largest stiffness of the cylinder is displayed when both surfactant and rings are present (the lowest dashed hysteretic curve). The partial hystereses due to actions of surfactant and rings (which act in the same sense) are not simply summed to obtain the resulting hysteresis in the last case, because the amplitude in displacement is decreased and the constitutive loops of surfactant and rings become smaller.

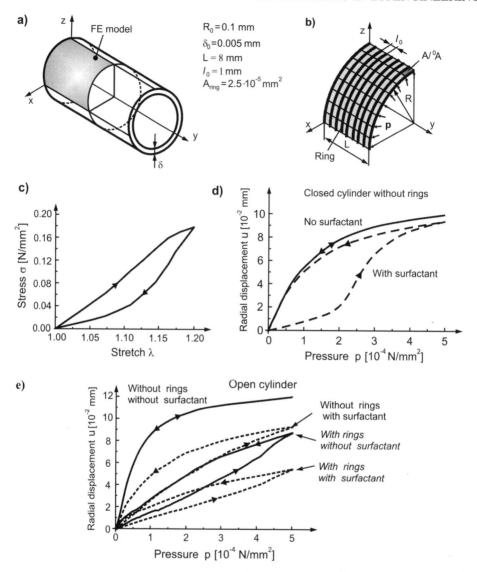

Fig. E11.3-2 Cylinder loaded by a cyclic internal pressure. (a) Cylinder geometry and geometrical data; (b) Open cylinder with rings; (c) Constitutive curve for ring (material model with hysteresis); (d) Dependence of radial displacement on internal pressure (closed cylinder); (e) Dependence of radial displacement on pressure when there are rings and surfactant

Example 11.3-3. Deformation of a biological membrane with a hole

In this example we calculate the deformation of squared plane membrane with a hole, stretched biaxially (Vlastelica *et al.* 2006). The membrane has a ring at the internal rim of the hole and is covered by surfactant on both sides and over the ring (Fig. E11.3-3a). It is assumed that the stretches are the same and uniform in both directions along the external

BIOLOGICAL SOFT TISSUE

boundary of the membrane, given by the cyclic displacement $\mathbf{u}_{ext}(t)$. Geometrical data are shown in the figure and material characteristics of tissue, surfactant and ring are the same as in the previous two examples.

The membrane is modeled by the four-node shell finite elements, as shown in the figure, the action of surfactant is calculated as presented in Section 11.2, and the ring is modeled by the line elements. One-quarter of the plate is modeled, with the symmetry boundary conditions at the coordinate axes (no displacements through the symmetry lines), with all rotations for the shell elements excluded. The external displacement increases, reaches a maximum and decreases. Since the material models used here are time independent and the inertial forces are neglected, the results presented in Figs. E11.3-3b,c do not depend on the shape of the loading–unloading functions of time.

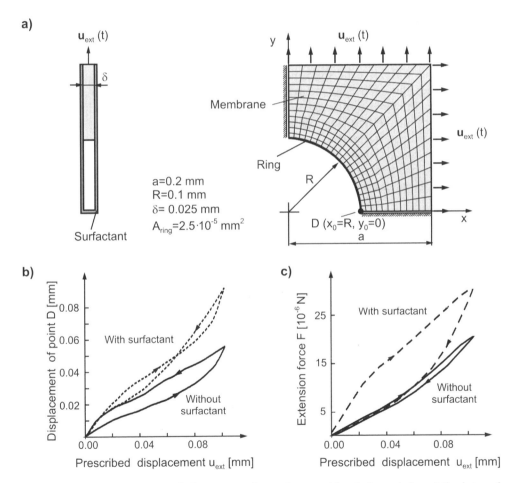

Fig. E11.3-3 Deformation of plane squared membrane with a hole and ring at the internal rim of the hole; membrane is covered by surfactant. (a) Geometry of the membrane (one-quarter) and the FE mesh; (b) Calculated displacement of the point D in terms of the external prescribed displacement. (c) Dependence of the external stretching force in one direction on the external displacement

In order to quantify the important characteristics of the displacement field of the membrane during cyclic stretching, consider displacement of a point D which moves along the x-axis (Fig. E11.3-3a). Characteristics of the displacement of this point are the same for the whole membrane and ring (except the external boundary where displacements are prescribed). When there is no surfactant, the displacement has a hysteretic character due to hysteresis in the ring material (Fig. E11.3-2c) with the counterclockwise direction (Fig. E11.3-3b). This loop direction is due to the fact that the resistance to the external stretching is larger in the period of loading than during unloading. When surfactant is present, it acts outward on the ring and tissue and increases the displacement. This action is larger in the loading period since the surface tension is larger during loading (Fig. E11.3-1c). With the data used in this example we have that the hysteretic action of surfactant is more dominant in the period of small stretches and the loop in displacement is clockwise, while in the domain of large stretches (near half of the cycle), hysteresis of ring dominates and the displacement loop is counterclockwise. Note that in the case of no hysteresis within tissue (no ring in this example), the hysteresis in displacement under action of surfactant will be clockwise (not shown in the figure). Therefore, in this example the hystereses of surfactant and tissue have opposite actions with respect displacements of the tissue.

Dependence of the stretching force (in one direction) on the displacement u_{ext} is shown in Fig. E11.3-3c. The force–displacement loop is clockwise since an energy dissipation occurs within the cycle due to hysteretic characteristics of tissue and surfactant. The hysteresis is smaller when there is no surfactant, and is significantly increased when surfactant is present. Note that here the hystereses due to tissue and due to surfactant have the same direction (clockwise) with respect to the force–displacement loop.

Solutions for other example parameters can be obtained using Software on the web.

Example 11.3-4. Blood vessel under pressure and extension
Deformation of a blood vessel under pressure and axial loading is modeled using the model of Delfino et al. (1997) described in Section 11.2.3. We use the strain energy functions defined

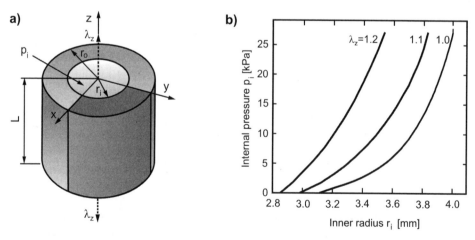

Fig. E11.3-4 Blood vessel under pressure and extension. (a) Geometry of blood vessel (one-quarter of the vessel, dark region, is modeled by 3D finite elements); (b) Dependence of the internal pressure on the inner radius which changes during pressure increase, for three values of initial axial stretch λ_z. Data (lengths in *mm*): $r_i = 3.1$; $r_0 = 4$; $L = 5$

BIOLOGICAL SOFT TISSUE

in (11.2.29), with the constants $a = 44.2\ kPa$, $b = 16.7$. The blood vessel wall is modeled by 3D finite elements (Section 4.3) and one quarter of the vessel is discretized (Fig. E11.3-4a) due to symmetry in geometry and loading, with symmetry boundary conditions at coordinate planes $x - z$ and $y - z$ (no displacements through these planes).

The vessel is first stretched axially until a desired axial stretch is reached and then loaded by internal pressure. The incremental-iterative procedure described in Section 6.2.3 is implemented, with use of the relations (11.2.30) and (11.2.31) for the stress and constitutive matrix calculations. Dependence of the internal radius on the pressure is given in Fig. E11.3-4b for three axial stretches. These results agree with the solutions in Holzapfel et al. (2000). Solutions for other model parameters can be obtained using the Software on the web.

Example 11.3-5. Modeling of the urinary bladder

Deformation of the urinary bladder under internal pressure during filling is considered. The geometry and material data of the rabbit urinary bladder are obtained from experiments (Zdravkovic 2000). A simplified shape with axial symmetry shown in Fig. E11.3-5a is

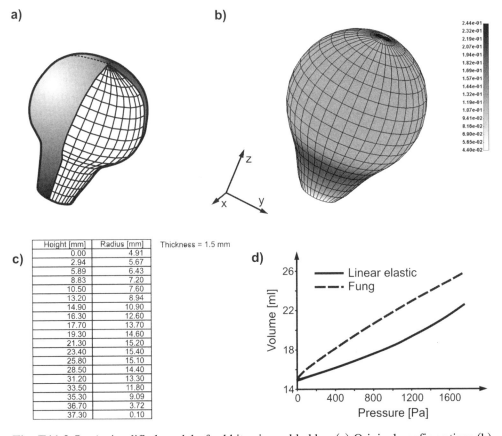

Fig. E11.3-5 A simplified model of rabbit urinary bladder. (a) Original configuration; (b) Deformed configuration with strain field (equivalent strain); (c) Geometrical data of the original configuration; (d) Pressure–volume relationship obtained using elastic and Fung's nonlinear elastic 2D model. Data: linear elastic material model ($E = 0.05\ MPa$, $\nu = 0.49$) and Fung's model (equation (11.2.32): $c = 3.372 \times 10^{-3} MPa$, $a_1 = 0.6$, $a_2 = 0.43$, $a_4 = 0.49$)

adopted (Vlastelica et al. 2007). Data about shape geometry and thickness are given in Fig. E11.3-5c.

Due to axial symmetry with respect to geometry and loading we model only one-quarter of the body. It is considered that the wall is thin, hence the membrane conditions (with zero stress in the direction to the membrane normal) and we use the four-node shell elements (see Section 4.5) of the FE model as shown in the figure. Nodes lying in $x-z$ and $y-z$ planes are constrained according to symmetry conditions: no displacement through the symmetry planes and no rotation around the axes normal to the symmetry planes. Also, it is taken that the nodes at the bottom rim have no displacements and rotations. We consider quasi-static deformation and increase the internal pressure assuming continuous filling.

Two material models are used: linear elastic and the Fung 2D model with the strain potential given in (11.2.32). The principal strains are in the axial and circumferential directions due to axial symmetry and Fung's model with only normal strains is applicable. Note that this model provides an orthotropic mechanical response which is observed experimentally. Material constants for the models are given in the figure caption. The elastic modulus is estimated from the stress–strain relationship obtained assuming biaxial stretching by the same strains in both membrane directions and Fung's model; also, it is taken that material is almost incompressible.

Deformed configuration is shown in Fig. 11.3-5b with the field of equivalent ($=(2/3 e_{ij} e_{ij})^{1/2}$) logarithmic strain. It can be seen that the strains close to 25% occur in the bottom region. The pressure–volume relationships for the two materials, Fig. 11.3-5d, show the nonlinear character. Note that, although the elastic modulus is fitted to a biaxial stretching with use of the Fung model, the elastic solution shows a larger stiffness of the bladder. Solutions for other parameters of the FE model can be obtained using the Software on the web.

The constitutive model used here corresponds to the loading (filling) regime. Experiments show that another constitutive law governs the unloading (voiding) process (Korkmaz & Rogg 2007).

References

Bachofen, H.P. & Schurch, H. (2001). Alveolar surface forces and lung architecture, *Comp. Biochem. Physiol. Part A*, **129**, 183–93.

Choi, H.S. & Vito, R.P. (1990). Two-dimensional stress-strain relationship for canine pericardium, *J. Biomech. Eng.*, **112**(2), 153–9.

Delfino, A., Stergiopulos, N., Moore, J.E. & Meister, J.J. (1997). Residual strain effects on the stress field in a thick wall finite element model of the human carotid bifurcation, *J. Biomech.*, **30**, 777–86.

Denny, E. & Schroter, R.C. (2000). Viscolelastic behavior of a lung alveolar duct model, *J. Biomech. Eng. Trans. ASME*, **122**, 143–51.

Fukaya, H., Martin, C.J., Young, A.C. & Katsura, S. (1968). Mechanical properties of alveolar walls, *J. Appl. Physiol.*, **25**(6), 689–95.

Fung, Y.C., Fronek, K. & Patitucci, P. (1979). Pseudoelasticity of arteries and the choice of its mathematical expression, *Am. J. Physiol.*, **237**, 620–31.

Fung, Y.C. (1990). *Biomechanics – Motion, Flow, Stress, and Growth*, Springer-Verlag, New York.

Hildebrandt, J., Fukaya, H. & Martin, C.J. (1969). Stress–strain relations of tissue sheets undergoing uniform two-dimensional stretch, *J. Appl. Physiol.*, **27**(5), 758–62.

Holzapfel, G.A. (2001). *Nonlinear Solid Mechanics, a Continuum Approach for Engineering*, John Wiley & Sons, Ltd., Chichester, England.

Holzapfel, G.A., Gasser, C.T. & Ogden, R.W. (2000). A new constitutive framework for arterial wall mechanics and comparative study of material models, *J. Elasticity*, **61**, 1–48.

Ingenito, E.P., Mark, L., Morris, J., Espinosa, F.F., Kamm, R.D. & Johnson, M. (1999). Biophysical chracterization and modeling of lung surfactant components, *J. Appl. Physiol.*, **86**(5), 1702–14.

Kojić, M. (2002). An extension of 3D procedure to large strain analysis of shells, *Comp. Meth. Appl. Mech. Eng.*, **191**, 2447–62.

Kojić, M. & Bathe, K.J. (2005). *Inelastic Analysis of Solids and Structures*, Springer-Verlag, Berlin.

Kojić, M., Mijailović, S. & Zdravkovic, N. (1998). A numerical algorithm for stress integration of a fiber–fiber kinetics model with Coulomb friction for connective tissue, *Comp. Mech.*, **21**(2), 189–98.

Kojić, M., Vlastelica, I., Stojanović, B., Ranković, V. & Tsuda, A. (2006). Stress integration procedures for a biaxial isotropic material model of biological membranes and for hysteretic models of muscle fibers and surfactant, *Int. J. Num. Meth. Eng.*, **68**, 893–909.

Kojić, M., Zdravkovic, N. & Mijailović, S. (2003). A numerical stress calculation procedure for a fiber–fiber kinetics model with Coulomb and viscous friction of connective tissue, *Comp. Mech.*, **30**, 185–95.

Korkmaz, I. & Rogg, B. (2007). A simple fluid-mechanical model for prediction of the stress-strain relation of the male urinary bladder, *J. Biomechanics*, **40**, 663–8.

Kowe, R., Schroter, R.C., Matthews, F.L. & Hitchings, D. (1986). Analysis of elastic and surface tension effects in the lung alveolus using finite element methods. *J. Biomech.*, **19**(7), 541–9.

Lee, C.G. & Hoppin, F.G. Jr. (1972). Lung elasticity. In Y.C. Fung, N, Perrone & M. Anliker (eds.), *Biomechanics – Its Foundations and Objectives* (pp. 317–35), Prentice-Hall, Englewood Cliffs, NJ.

Mijailović, S.M. (1991). *Elasticity and energy dissipation in lung connective tissue*. Ph.D. Thesis, MIT, Cambridge, MA.

Mijailović, S.M., Stamenović, D. & Fredberg, J.J. (1993). Toward a kinetic theory of connective tissue micromechanics, *J. Appl. Physiol.*, **74**(2), 665–81.

Mijailović, S.M., Stamenović, D., Brown, R., Leith, D.E. & Fredberg, J.J. (1994). Dynamic moduli of rabit lung tissue and pigeon ligamentum propatagiale undergoing uniaxial cyclic loading, *J. Appl. Physiol.*, **76**(2), 773–82.

Oldmixon, E.H. & Hoppin, F.G. Jr. (1989). Distribution of elastin and collagen in canine lung alveolar parenchyma, *J. Appl. Physiol.*, **67**(5), 1941–9.

Sacks, M.S. (2000). Biaxial mechanical evaluation of planar biological materials, *J. Elast.*, **61**, 194–246.

Sasaki, H. & Hoppin, F.G. Jr. (1979). Hysteresis of contracted airway smooth muscle, *J. Appl. Physiol: Respirat. Environ. Exercise Physiol.*, **47**(6), 1251–62.

Simo, J.C. & Taylor, R.L. (1991). Quasi-incompressible finite elasticity in principal stretches. Continuum basis and numerical algorithms, *Comp. Meth. Appl. Mech. Eng.*, **85**, 273–310.

Vlastelica, I., Kojić, M., Stojanović, B., Ranković, V. & Tsuda, A. (2006). On the superposition of hysteretic actions of tissue and surfactant. In M. Kojić & M. Papadrakakis (eds.), *Proc. First South-East European Conf. Comp. Mech* (pp. 462–8), Kragujevac, Serbia.

Vlastelica, I., Veljković, D., Ranković, V., Stojanović, B., Rosić, M. & Kojić, M. (2007). Modeling of urinary bladder deformation within passive and active regimes, *J. Serbian Soc. Comp. Mech.*, **1**(1), 129–34.

Wilson, T.A. (1982). Surface tension-surface area curves calculated from pressure-volume loops, *J. Appl. Physiol: Respirat. Environ. Exercise Physiol.*, **53**(6), 1512–20.

Zdravkovic, N. (2000). *Solution of mechanics of connective tissue and muscle problems by finite element method*, Ph.D. Thesis, Faculty of Mechanical Engineering, University of Kragujevac, Serbia.

12

Skeletal Muscles

In this chapter we present fundamentals of skeletal muscle (further called *muscle*) mechanics and a numerical algorithms, based on the finite element method, for the determination of the muscle response. Hill's three-component phenomenological model is used as a basis for our analysis. In order to include muscle fatigue, an extension of Hill's model is introduced. Considering muscle as a bundle of sarcomeres of various physiological properties, we present a modification of Hill's three-component model to take into account different fiber types. A cylindrical muscle is given as the example, with some representative solutions, for which the software on the web can be used to investigate muscle mechanical response for various geometrical and material parameters, as well as loading, activation, fatigue and relaxation conditions.

12.1 Introduction

12.1.1 Basic physiology of muscle mechanics

A single muscle fiber is a cylindrical, elongated cell. Each fiber is surrounded by a thin layer of connective tissue called endomysium (Fig. 12.1.1 – see color plate). Organizationally, thousands of muscle fibers are wrapped by a thin layer of connective tissue called the perimysium to form the muscle bundle. Groups of muscle bundles that join into a tendon at each end are called muscle groups, or simply muscles. The entire muscle is surrounded by a protective sheath called the epimysium.

The muscle cells exhibit a striking banding pattern responsible for their classification as *striated muscle*. The striations arise from a highly organized arrangement of subcellular structures. The muscle cell membrane is called the sarcolemma. Each *muscle fiber* (muscle cell) contains several hundred to several thousand *myofibrils* (Fig. 12.1.1). Each myofibril in turn has, lying side by side, about 1500 *thick filaments* and 3000 *thin filaments*, which are large polymerized protein molecules that are responsible for muscle contraction. The thick

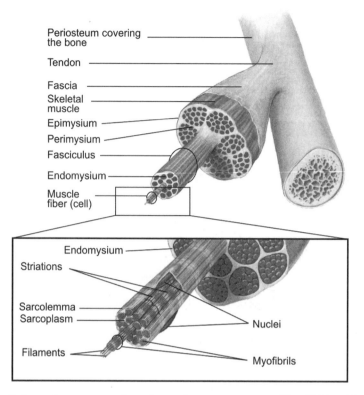

Fig. 12.1.1 Schematic representation of muscle macrostructure (Fox 2004, with permission from The McGraw-Hill Companies, January 14, 2008) From *Human physiology*, Fox, (2004), pp. 327 (see Plate 6)

and thin filaments partially interdigitate and thus cause the myofibrils to have alternate light and dark bands (Fig. 12.1.2 – see color plate). The light bands contain only thin filaments and are called 'I' bands because they are isotropic to polarized light. The dark bands contain the thick filaments as well as the ends of the thin filaments where they overlap the myosin, and are called 'A' bands because they are anisotropic to polarized light. The thin filaments are attached to a so-called 'Z' disc (or Z *line*) which itself is composed of filamentous proteins different from the thin and thick filaments. The myofibrils are suspended inside the muscle fibers in a matrix called the sarcoplasm, which is composed of usual intracellular constituents. In the sarcoplasm there are a tremendous number of organelles called mitochondria that lie between and parallel to the myofibrils. These organelles convert chemical energy contained in carbohydrate and fat to ATP, the only energy source that can be used directly by the cell to support contraction. Also in the sarcoplasm is an extensive endoplasmic reticulum, which in the muscle is called the sarcoplasmic reticulum. This reticulum is a reservoir of calcium ions and is extremely important in the control of muscle contraction.

The basic contractile unit of a striated muscle cell is the *sarcomere* (Fig. 12.1.2 – see color plate), which consists of a centrally located array of thick filaments that interdigitate with thin filaments attached to the cytoskeleton at each end of the sarcomere (the Z lines). Myofibrils contain many sarcomeres in series, and muscle cells contain large numbers of myofibrils in parallel. Thin filaments consist of polymers of actin, tropomyosin and nebulin, plus the Ca^{++}-binding regulatory protein troponin. Sometimes these filaments are simply

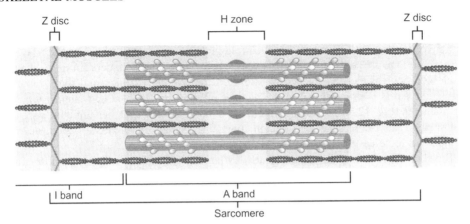

Fig. 12.1.2 Muscle microstructure (Fox 2004, with permission from The McGraw-Hill Companies, January 14, 2008) From *Human Physiology*, Fox, (2004), pp. 331 (see Plate 7)

called 'actin filaments'. Thick filaments consist mainly of myosin (often referred to as 'myosin filaments') and titin. The 'head' regions of individual myosin molecules project laterally from the filament. These projections, called *cross-bridges*, contain the actin- and ATP-binding sites. A high free energy state of the cross-bridges occurs after ATP binding and hydrolysis to form the *myosin–ADP–Pi complex*, which has a high affinity for actin. Rapid attachment to the thin filaments in a preferred 90-degree conformation follows. Subsequent release of bound Pi and ADP leads to a complex whose free energy is minimized after a conformational change to 45 degrees. This *conformational change* produces a force on the thin filament and movement toward the center of the sarcomere. The ATP binding reduces the affinity of myosin for actin, and the cross-bridges detach from the thin filaments. ATP hydrolysis regenerates the myosin–ADP–Pi complex in a 90-degree conformation to complete the cross-bridge cycle.

Muscle contraction is said to be *isometric* when the muscle does not shorten during contraction, while it is *isotonic* when muscle does shorten and the tension on the muscle remains constant.

The velocity–load relationship shows that muscles shorten more slowly as the load is increased in isotonic contraction (see Fig. 12.2.1). On the other hand, power output is maximized at moderate loads at which 40% to 45% of the free energy of ATP hydrolysis is converted into mechanical work.

Velocity depends on the number of sarcomeres in a muscle cell. Sarcomere–shortening velocities are a function of cross-bridge cycling rates and load. Maximal cycling rates at zero load (v_0) depend only on the molecular properties of the myosin isoform synthesized within a striated muscle cell. The direct proportionality between the ATPase activity of myosin isolated from a cell and v_0 for that cell illustrates this molecular diversity, which is responsible for physiological differences in the speed of contraction of muscle cells from different sources.

Force generation is a function of the number of cross-bridges that can interact with the thin filaments. In Fig. 12.2.2 maximum number of cross-bridges corresponds to the region B–C, where the maximum force is generated. Although the force per one cross-bridge is of the order of piconewtons, a muscle generates a large force per unit area (around $0.3\,N/mm^2$) due to the enormous number of cross-bridges.

Contracting muscles often lengthen when opposing forces are very high. Stresses in muscle, tendons or skeleton can be very high under such conditions, because cross-bridges can transiently bear loads that are about 1.6-fold higher than the stresses which they develop isometrically. Precise studies of the stress–length behavior of the sarcomeres in single skeletal muscle cells reveal the dependence of stress on the overlap of thick and thin filaments, because the stress is proportional to the number of cross-bridges that can interact with the thin filaments (Fig. 12.2.2). Changing the overlap between filaments has minimal effects in most skeletal muscle because the skeleton constrains changes in lengths to values near the optimum for force development.

The skeletal muscle fibers are innervated by large, *myelinated nerve fibers* that originate in the large alpha motoneurons of the anterior horns of the spinal cord. The nerve fiber branches at its end to form a complex of branching nerve terminals, which invaginate into the muscle fiber but lie outside the muscle membrane. The entire structure is called the *motor end plate*. A single alpha motoneuron, and all the muscle fibers it activates through its nerve terminals, is called a motor unit.

Excitation–contraction coupling involves binding of the neurotransmitter acetylcholine (which is released from the motor nerve endings) to its receptors in the sarcolemma. This receptor interaction increases the sarcolemma conductance and generates an action potential that propagates in both directions along the muscle cell followed by:

1. mobilization of Ca^{++} from sarcoplasmic reticulum;

2. a thin filament conformational change caused by the allosteric binding of Ca^{++}; and

3. cross-bridge attachment and cycling.

The sarcoplasmic reticulum that surrounds each myofibril contains a pool of Ca^{++} that is mobilized, when the action potential propagated along the sarcolemma depolarizes transverse T-tubules. This depolarization briefly opens Ca^{++} channels in the opposing sarcoplasmic reticulum membrane. A transient increase in the intracellular Ca^{++} concentration follows the action potential.

The binding of four Ca^{++} ions to troponin induces the conformational change in the thin filament that enables cross-bridges to bind and cycle. Cycling continues until intracellular Ca^{++} is returned to its low resting concentration by active transport into the sarcoplasmic reticulum. The Ca^{++} then dissociates from troponin, the thin filament is 'switched off', and relaxation ensues. The mechanical response to a single action potential is a *twitch*. The force of a twitch is considerably less than the maximum that can be developed, because the release of Ca^{++} is too brief to allow generation of maximal force.

The force of contraction is graded in a skeletal muscle cell by increasing the frequency of action potentials, and thereby maintaining the thin filaments in a state of prolonged cross-bridge cycling (tetanus). In a skeletal muscle, force is also increased by the recruitment of more motor units. The maximum strength of a *tetanic contraction* for a muscle operating at a normal muscle length, averages between 3 and 4 kilograms per square centimeter of muscle. At this *tetanized state* the muscle is maximally activated, which includes:

1. the thin filaments in a state of prolonged cross-bridge cycling at the level of sarcomere;

2. the highest frequency of action potentials that propagates along the sarcolemma, and thereby maximal intracellular concentration of Ca^{++} at the level of muscle fiber;

3. optimal recruitment of all motor units at the level of muscle.

SKELETAL MUSCLES

In the *passive (non-activated)* state, all events described above are reversed. It is important to underline that, even when muscles are at rest, a certain amount of tautness usually remains. This is called *muscle tone*. The skeletal muscle tone results entirely from a low rate of nerve impulses coming from the spinal cord.

Skeletal muscles have a high power output when they shorten, and they consume ATP at a rapid rate. The ATP cost is minimized by the efficient *conversion of chemical to mechanical energy*. ATP consumption is lower during isometric contractions, but considerable cycling still occurs. Consequently, the economy of force maintenance is poor in isometric contractions. The high free energy of the myosin–ADP–Pi complex is never released when a contracting muscle is stretched. The transformation of an attached cross-bridge from 90-degree to a 45-degree conformation cannot occur when sarcomeres are lengthening, therefore, no ATP is consumed by cross-bridge cycling.

All muscles have varying percentages of fast-twitch and slow-twitch muscle fibers. The basic differences between the *fast-twitch* and *slow-twitch fibers* are the following:

1. Fast-twitch fibers are about twice as large in diameter.

2. The enzymes that promote rapid release of energy are two to three times more active in fast-twitch fibers than in slow-twitch fibers, thus making the maximal power that can be achieved by fast-twitch fibers twice as large when compared to slow-twitch fibers.

3. Slow-twitch fibers are mainly organized for endurance, especially for generation of aerobic energy. They have far more mitochondria than the fast-twitch fibers. Also, they contain considerably more myoglobin, a hemoglobin-like protein that combines with oxygen within the muscle fiber. In addition, the enzymes of the aerobic metabolic system (which requires oxygen, but has slower release of energy) are considerably more active in slow-twitch fibers than in fast-twitch fibers.

4. The number of capillaries per mass of fibers is larger in the vicinity of slow-twitch fibers than in the vicinity of fast-twitch fibers.

In summary, fast-twitch fibers can deliver extreme amounts of power for a few seconds to about a minute. On the other hand, slow-twitch fibers provide endurance, delivering prolonged strength of contraction over many minutes to hours.

12.1.2 Basics of muscle finite element modeling

A muscle is a material body which moves and deforms under external and internal mechanical action. Thus, the basic mechanical principles of motion of deformable bodies are applicable. In order to determine the mechanical response of a muscle we can implement methods of solid mechanics analysis, such as the finite element method.

We here present the fundamental concept of finite element analysis of skeletal muscles, which in principle is the same as for any deformable solid or structure, and we also emphasize specifics of muscle modeling.

Fig. 12.1.3a shows a schematic representation of a muscle discretized into 3D finite elements. Muscle deforms under external loading and internal excitation, and in general has large displacements and strains. Muscle material has nonlinear constitutive relations

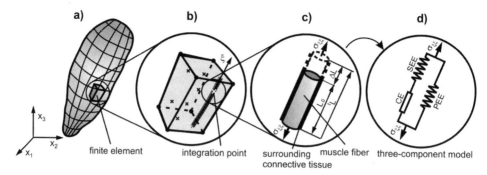

Fig. 12.1.3 Schematics of muscle FE modeling: from muscle as a deformable body to Hill's model. (a) Muscle discretization into finite elements; (b) A 3D finite element with integration points and muscle fiber; (c) Elongation of muscle fiber under the stress $\sigma_{\xi\xi}$; (d) Hill's three-component model

(see Section 6.2). Neglecting the inertial forces, an incremental-iterative scheme for determining muscle motion can be formed. Hence, we have the equilibrium equation of a finite element (6.2.18) for a load step 'n' and iteration 'i',

$$\left(^{n+1}\mathbf{K}_L + {}^{n+1}\mathbf{K}_{NL}\right)^{(i-1)} \Delta \mathbf{U}^{(i)} = {}^{n+1}\mathbf{F}^{ext} - {}^{n+1}\mathbf{F}^{int(i-1)} \tag{12.1.1}$$

where $^{n+1}\mathbf{K}_L^{(i-1)}$ and $^{n+1}\mathbf{K}_{NL}^{(i-1)}$ are the geometrically linear and geometrically nonlinear stiffness matrices for the end of step, $\Delta \mathbf{U}^{(i)}$ are the increments of nodal displacements; and $^{n+1}\mathbf{F}^{ext}$ and $^{n+1}\mathbf{F}^{int(i-1)}$ are the external and internal nodal forces. Description of these matrices and force vectors is given in Section 6.2. We here concentrate on the calculation of the stresses $^{n+1}\boldsymbol{\sigma}^{(i-1)}$ and the tangent constitutive matrix $^{n+1}\mathbf{C}^{(i-1)}$ within the finite element, since we have that (see (6.2.21))

$$^{n+1}\mathbf{F}^{int(i-1)} = \int_{^{n+1}V^{(i-1)}} \left(^{n+1}\mathbf{B}_L^T \, {}^{n+1}\boldsymbol{\sigma}\right)^{(i-1)} dV,$$

$$^{n+1}\mathbf{K}_L^{(i-1)} = \int_{^{n+1}V^{(i-1)}} \left(^{n+1}\mathbf{B}_L^T \, {}^{n+1}\mathbf{C} \, {}^{n+1}\mathbf{B}_L\right)^{(i-1)} dV \tag{12.1.2}$$

Here $^{n+1}\mathbf{B}_L^{(i-1)}$ is the linear strain–displacement matrix.

As described in Section 12.1.1, skeletal muscles have a fibrous structure and the muscle force is generated within the muscle fibers. Therefore, our task is to find the stresses in the directions of muscle fibers. In Fig. 12.1.3b we have schematically shown a finite element with a fiber, whose direction in space is defined by the unit vector $\boldsymbol{\xi}_0$ (axis ξ in the figure). This direction corresponds to an integration point used for the numerical evaluation of the finite element matrices and vectors in equations (12.1.1) and (12.1.2). The stress in the fiber direction, denoted as $\sigma_{\xi\xi}$ (consisting of active part σ_s and passive elastic part σ^E, see Section 12.2), depends on the elongation ΔL, or the stretch $\lambda = 1 + \Delta L/L_0$, of the muscle fiber. Here L_0 is the initial length of the fiber, when the stress is equal to zero. The

SKELETAL MUSCLES

dependence $\sigma_{\xi\xi}(\lambda)$ represents the constitutive law for muscle. This constitutive law can be defined by a phenomenological (experimentally established) material model, such as Hill's model schematically shown in Fig.12.1.3d.

We next present calculation of the stretch in the fiber direction for the current stage (configuration) of muscle within the incremental-iterative finite element scheme (equation (12.1.1)), which is needed for the stress determination. The last known configuration for the current step 'n' and iteration 'i' is specified by the position vectors $^{n+1}\mathbf{x}^{(i-1)}$ of material points. Then, the displacements are $^{n+1}\mathbf{u}^{(i-1)}$. The stretch $^{n+1}\lambda^{(i-1)}$ in the fiber direction can be calculated as (see Example 2.4-1 of Section 2.4.3)

$$^{n+1}\lambda^{(i-1)} = \left[\left({}_{n+1}^{0}B_{ij}\ {}^{n+1}\xi_{0i}\ {}^{n+1}\xi_{0j} \right)^{(i-1)} \right]^{-1/2} \tag{12.1.3}$$

where $^{n+1}\xi_{0i}^{(i-1)}$, $^{n+1}\xi_{0j}^{(i-1)}$ are components of the unit vector $^{n+1}\xi_{0}^{(i-1)}$, and $_{n+1}^{0}B_{ij}^{(i-1)}$ are the components of the inverse left Cauchy–Green deformation tensor $_{n+1}^{0}\mathbf{B}^{(i-1)}$. This tensor can be computed from the inverse deformation gradient $_{n+1}^{0}\mathbf{F}^{(i-1)}$ as

$$_{n+1}^{0}\mathbf{B}^{(i-1)} = \left({}_{n+1}^{0}\mathbf{F}^{T}\ {}_{n+1}^{0}\mathbf{F} \right)^{(i-1)} \tag{12.1.4}$$

while the tensor $_{n+1}^{0}\mathbf{F}^{(i-1)}$ can be evaluated from the displacements $^{n+1}\mathbf{u}^{(i-1)}$ (see (2.4.5)),

$$_{n+1}^{0}\mathbf{F}^{(i-1)} = \mathbf{I} - \frac{\partial\,^{n+1}\mathbf{u}^{(i-1)}}{\partial\,^{n+1}\mathbf{x}^{(i-1)}}, \quad \text{or} \quad _{n+1}^{0}F_{ij}^{(i-1)} = \delta_{ij} - \frac{\partial\,^{n+1}u_{i}^{(i-1)}}{\partial\,^{n+1}x_{j}^{(i-1)}} \tag{12.1.5}$$

The derivatives $\partial\,^{n+1}u_{i}^{(i-1)}/\partial\,^{n+1}x_{j}^{(i-1)}$ follow from the interpolation of displacements (4.3.2), hence

$$^{n+1}\left(\frac{\partial u_i}{\partial x_j}\right)^{(i-1)} = \sum_{K=1}^{N}\frac{\partial N_K}{\partial\,^{n+1}x^{(i-1)}}\,^{n+1}U_i^{K(i-1)}, \quad i=1,2,3 \tag{12.1.6}$$

where N_K are the interpolation functions and $^{n+1}U_i^{K(i-1)}$ are the components of the nodal displacements.

In summary, for the current configuration we calculate the unit vector $^{n+1}\xi_{0}^{(i-1)}$ and find the stretch in the fiber direction $^{n+1}\lambda^{(i-1)}$ from (12.1.3). Then we evaluate the stress $^{n+1}\sigma_s^{(i-1)}$ from the muscle material model, as well as the coefficient $^{n+1}C_s^{(i-1)} \equiv \partial\,^{n+1}\sigma_s^{(i-1)}/\partial\,^{n+1}e_{\xi\xi}$ necessary for the tangent constitutive matrix $^{n+1}\mathbf{C}^{(i-1)}$:

$$^{n+1}\sigma_s^{(i-1)} = \sigma_s\left(^{n+1}\lambda^{(i-1)}\right), \quad ^{n+1}C_s^{(i-1)} \equiv \frac{\partial\,^{n+1}\sigma_s^{(i-1)}}{\partial\,^{n+1}e_{\xi\xi}^{(i-1)}} = \left(^{n+1}\lambda\frac{\partial\,^{n+1}\sigma_s}{\partial\,^{n+1}\lambda}\right)^{(i-1)} \tag{12.1.7}$$

Here, we have used the relation $\partial(\bullet)/e_{\xi\xi} = \lambda\partial(\bullet)/\partial\lambda$, where $e_{\xi\xi} = \ln(L/L_0)$ is the fiber strain.

In the section below we present details for calculation of $^{n+1}\sigma_s^{(i-1)}$ and $^{n+1}C_s^{(i-1)}$ for Hill's model and its modifications.

12.2 Muscle modeling

12.2.1 Hill's phenomenological model

The mostly used relation in muscle mechanics is Hill's equation. It refers to mechanical behavior of skeletal muscle in the tetanized condition. This equation is given as

$$(v+b)(S+a) = b(S_0+a) \tag{12.2.1}$$

where S represents tension (tensional stress) in muscle; v is the velocity of the contraction; and a, b and S_0 are constants (Fung 1993). The constant S_0 is the maximum tension that can be produced under isometric tetanic contraction. The Hill equation can be rewritten in dimensionless form as

$$\frac{S}{S_0} = \frac{1-(v/v_0)}{1+c(v/v_0)} \tag{12.2.2}$$

in which the maximum velocity is $v_0 = \dfrac{bS_0}{a}$, and the constant $c = \dfrac{S_0}{a}$. A graphical representation of the equation is given in Fig. 12.2.1.

The maximum tension in the tetanized condition is strongly dependent on the muscle stretch ratio. The tension–sarcomere length (or stretch) relationship, shown in Fig. 12.2.2, is due to Gordon (Gordon et al. 1966) and corresponds to the intact skeletal muscle. It can be noticed that the maximum tension corresponds to the slack muscle length, i.e. to the extension ratio, or stretch, equal to 1.0. A decrease in stress under shortening and extension can be explained by a change of the number of cross-bridges between myosin and actin fibers under nonslack conditions (see also Section 12.1.1).

A simple model which reflects the mechanical behavior of muscle, described by Fig. 12.2.1 and Fig. 12.2.2, is Hill's three-component model. A graphical representation of this model is given in Fig. 12.2.3 (Fung 1993).

The contractile element (CE) has the characteristic that is described by Hill's equation and Gordon's curve, and will be adequately included in the model. The tension–stretch relationship for the nonlinear elastic element (SEE) is given by

$$S = (S^* + \beta) e^{\alpha(\lambda - \lambda^*)} - \beta \tag{12.2.3}$$

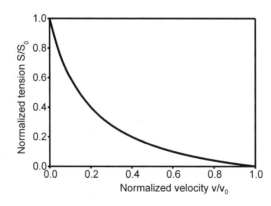

Fig. 12.2.1 Tension–velocity curve corresponding to a muscle in the tetanized condition

SKELETAL MUSCLES

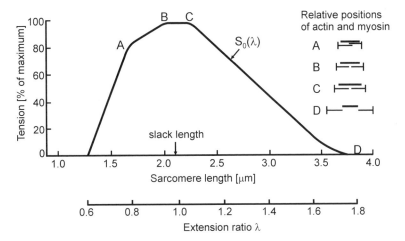

Fig. 12.2.2 Isometric tension–length curve (Gordon *et al.* 1966)

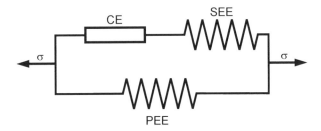

Fig. 12.2.3 Hill's functional model of a muscle. CE is the contractile element, SEE is the serial elastic element and PEE is the parallel elastic element; σ is the stress in the muscle fiber direction

in which S^* represents the tension corresponding to a stretch λ^*, while α and β are material constants. The surrounding connective tissue within a muscle is represented by a linear elastic element that is coupled in parallel to the series of contractile and nonlinear elastic elements.

12.2.2 Determination of stresses within muscle fiber

We here present a computational procedure for the stress calculation and evaluation of the tangent constitutive matrix, used in an incremental finite element analysis.

Stress Integration
The main task in stress calculation is to determine stresses in the direction of muscle fibers (Kojić *et al.* 1998). We start with the geometry shown in Fig. 12.2.4, corresponding to a fiber direction and arbitrary muscle state.

The next relation follows from the figure:

$$L_{p0} + {}^0U_p = L_{m0} + {}^0U_m + L_{s0} + {}^0U_s \qquad (12.2.4)$$

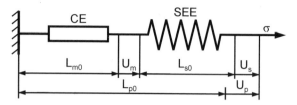

Fig. 12.2.4 Geometry of the contractile (CE) and nonlinear serial elastic elements (SEE)

where L_{p0} is the total initial length, while L_{m0} and L_{s0} are the initial lengths of the contractile and serial elastic elements; ^0U_p, ^0U_m and ^0U_s are the initial elongations of the corresponding elements (here taken to be zero).

We further suppose that the ratio of the initial lengths of serial to the contractile element is given as

$$k = \frac{^0L_s}{^0L_m} \qquad (12.2.5)$$

where k is a constant muscle parameter. Dividing equation (12.2.4) by L_{m0} and using $L_{p0} = L_{m0} + L_{s0}$, we obtain a relation for the initial stretches

$$^0\lambda_m = (1+k)\,^0\lambda_p - k \qquad (12.2.6)$$

It is usually considered that the initial state corresponds to the undeformed configuration, hence $^0\lambda_p = 1$, and then $^0\lambda_m = 1$.

At an arbitrary time t we have the following equation for the lengths:

$$^tL_p = L_{p0} + {}^tU_p = L_{m0} + {}^0U_m + \int_{t_a}^{t} v_m\,dt + L_{s0} + {}^tU_s \qquad (12.2.7)$$

where v_m is the rate of the muscle length change, and t_a is the activation time. Dividing this equation by L_{m0} we obtain

$$(1+k)\,{}^t\lambda_p = {}^0\lambda_m + \int_{t_a}^{t} \frac{v_m}{L_{m0}}\,dt + k\,{}^t\lambda_s \qquad (12.2.8)$$

Further we write this equation for the end of a time step 'n' as

$$(1+k)\,{}^{n+1}\lambda_p = {}^n\lambda_m + \Delta\lambda_m + k\,{}^n\lambda_s + k\Delta\lambda_s \qquad (12.2.9)$$

where $\Delta\lambda_m$ and $\Delta\lambda_s$ are the increments of stretches; and the indices 'n' and '$n+1$' denote the start and end of time step, respectively.

The stresses σ_m and σ_s in the contractile and the serial elastic element must be equal at any time, hence

$$^{n+1}\sigma_m = {}^{n+1}\sigma_s \qquad (12.2.10)$$

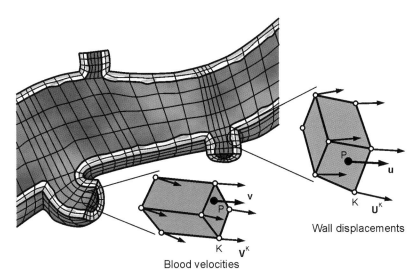

Plate 1 Discretization into finite elements of the blood velocity field and displacement field of blood vessel deformation. The velocity **v** and displacement **u** at a material point within a finite element are obtained, respectively, by interpolations from the nodal points vectors $\mathbf{V}^K$ and $\mathbf{U}^K$ (see Figure 4.1.1)

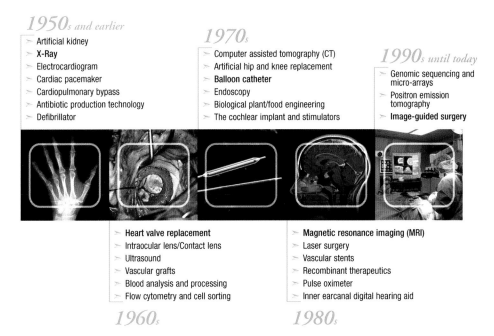

Plate 2 The American Institute for Medical and Biological Engineering 'Hall of Fame' gives a perspective on the most significant technological advancements in bioengineering in the twentieth century (http://www.aimbe.org/content/index.php?pid=127) (see Figure 9.1.1)

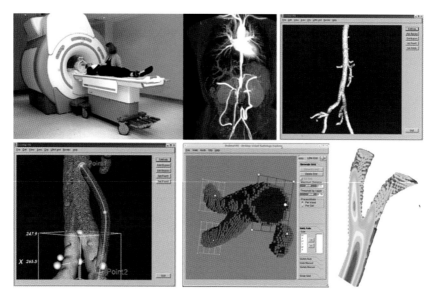

Plate 3 Virtual vascular surgery on the grid: from the MRI or CT scan recording of the patient vascular surgery region, to automatic generation of the computational model and analysis of results, yielding options for the surgical procedure (Reproduced with permission from P.M.A. Sloot and A.G. Hoekstra: Virtual Vascular Surgery on the Grid, ERCIM news, October 2004) (see Figure E9.2-1)

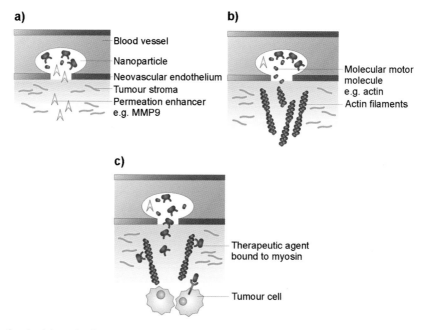

Plate 4 A vision of a future multistage nanodevice. A nanoparticle selectively binds to the cancer neovascular endothelium releasing multiple agents that enable the drug to pass through biological barriers and reach the targeted tumor cell (Reprinted by permission from Macmillan Publishers Ltd: Cancer nanotechnology: opportunities and challenges, Mauro Ferrari, 2005.) (see Figure E9.2-2)

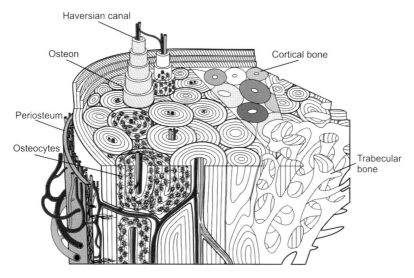

Plate 5 Structure of a long bone (according to Remagen 1989) (see Figure 10.1.1)

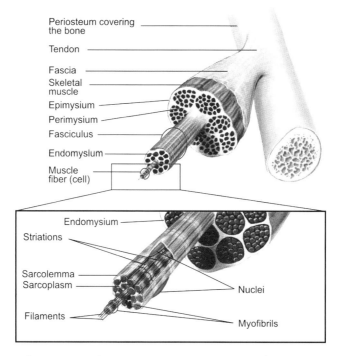

Plate 6 Schematic representation of muscle macrostructure (Fox 2004, with permission from The McGraw-Hill Companies, January 14, 2008) From *Human Physiology*, Fox, (2004), pp. 327 (see Figure 12.1.1)

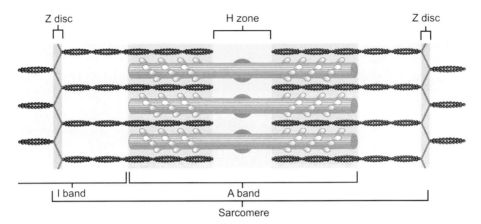

Plate 7 Muscle microstructure (Fox 2004, with permission from The McGraw-Hill Companies, January 14, 2008) From *Human Physiology*, Fox, (2004), pp. 331 (see Figure 12.1.2)

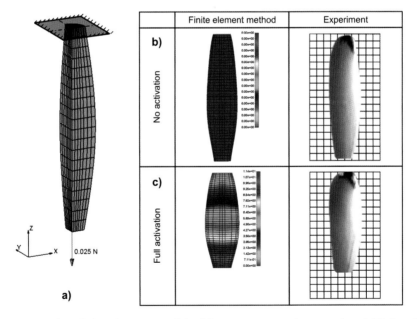

Plate 8 Simplified finite element model of frog *gastrocnemius* muscle. **a)** Finite element mesh, constraints and load; **b)** Left panel: undeformed configuration (zero displacement field) and real muscle without activation (right panel); **c)** Displacement field under full activation (left panel) and real deformed muscle (right panel) (see Figure E12.3-2)

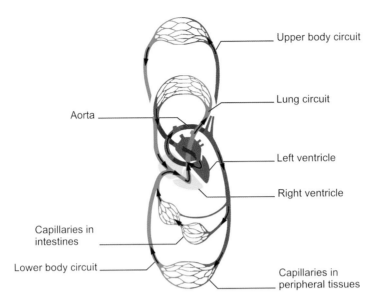

Plate 9 Schematic representation of the cardiovascular system (adapted from Mohr 2006) (see Figure 13.1.1)

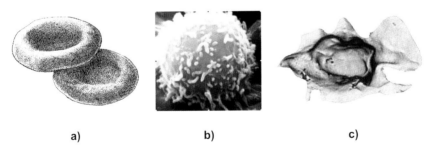

a)   b)   c)

Plate 10 Basic blood cells. (a) Erythocytes (http://www.mast.queensu.ca/~julia/sgc.html); (b) Leukocyte; (http://www.funsci.com/fun3_en/blood/blood.htm); (c) Activated platelet (according to Loscalzo & Schafer 2003) (see Figure 13.1.3)

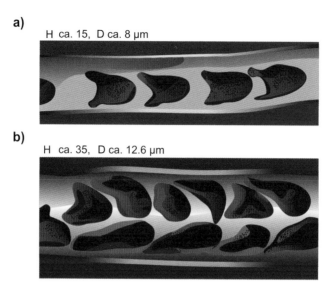

Plate 11 Motion of RBCs in microvessels. (a) RBCs flow in a single file form when the vessel diameter (D) is around the RBC size; (b) With increasing vessel diameter, RBCs tend to form multiple file flow (according to Pries & Secomb 2005) (see Figure 13.1.4)

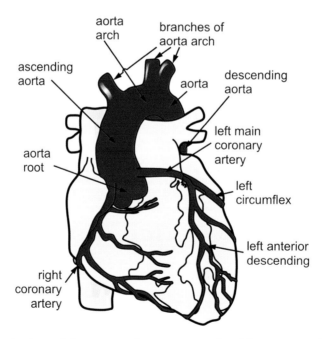

Plate 12 The structure of the aorta and coronary arteries. The aorta consists of the aorta root, the aorta arch, the ascending and descending parts, and aortic branches. The coronary arteries exit from the aorta root and lead to a branching network of small arteries, arterioles, capillaries, venules and veins (see Figure 13.3.1)

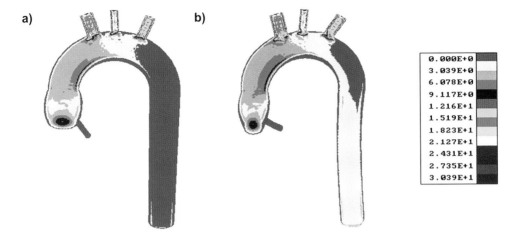

Plate 13 The velocity magnitude field in the human aorta for early systolic flow $t = 0.05$ s. (a) Rigid walls; (b) Deformable walls (note that diameters are smaller with respect to rigid aorta shown in (a)) (see Figure 13.3.3)

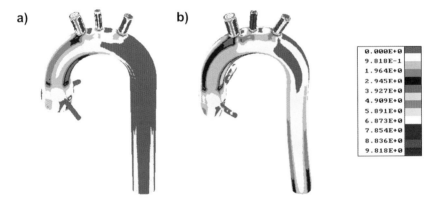

Plate 14 The wall shear stress in the human aorta for early systolic flow $t = 0.05$ s; (a) Rigid walls; (b) Deformable walls (see Figure 13.3.4)

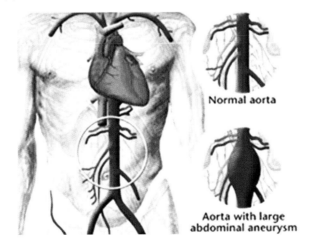

Plate 15 Normal physiological abdominal aorta and aorta with large abdominal aneurysm (http://www.lifelinescreening.com/Disease/AAA/Pages/Index.aspx) (see Figure 13.4.1)

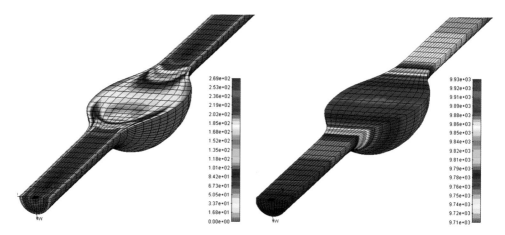

Plate 16 Velocity field (left panel) and pressure distribution (right panel) for peak systole $t/T = 0.16$ of AAA for the model with $D/d = 2/75$, $d = 12.7$ mm (Peattie et al. 2004) (see Figure 13.4.4)

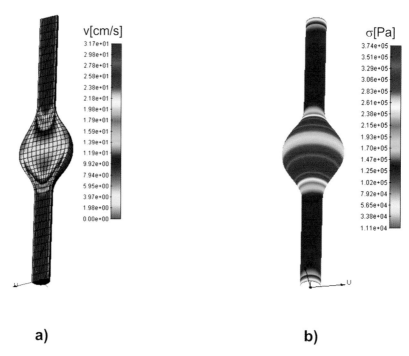

a) b)

Plate 17 Velocity magnitude field and von Mises wall stress distribution for symmetric AAA on a straight vessel. (a) Velocity field distribution for peak at $t = 0.305$ s; (b) von Mises wall stress distributions for blood pressure peak at $t = 0.4$ s (see Figure 13.4.6)

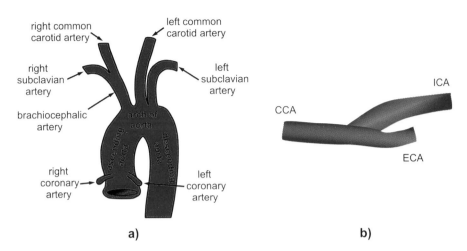

Plate 18 Carotid artery bifurcation. (a) Position of carotid arteries in the arterial system; (b) Typical carotid artery bifurcation. CCA – common carotid artery, ICA – internal carotid artery, ECA – external carotid artery (see Figure 13.5.1)

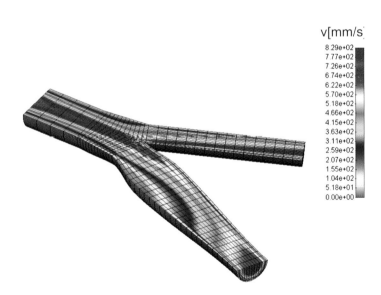

Plate 19 3D field of velocity magnitude at the maximum systolic flow (at relative time $t/T = 0.11$ within the period T); see also web – Software for solutions for the entire period of the cycle (see Figure 13.5.4)

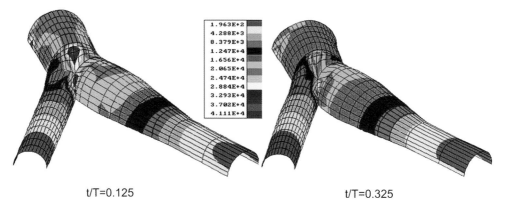

Plate 20 Distribution of von Mises stress (Pa) within the artery walls of the carotid artery bifurcation due to action of blood, for systolic deceleration flow ($t/T = 0.125$) and diastolic minimum flow ($t/T = 0.325$) (see Figure 13.5.5)

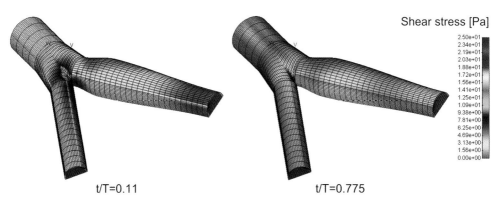

Plate 21 Wall shear stress field at two relative times (left panel – systole; right panel – diastole). Systolic shear stresses are much larger than diastolic (see Figure 13.5.6)

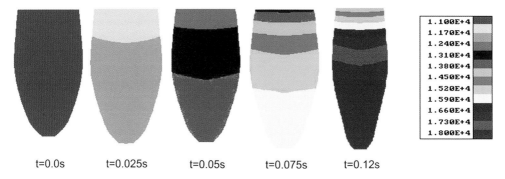

Plate 22 Intraventricular pressure distribution pattern in systolic phase for five times. Intravascular pressure peak of 18 kPa is reached at $t = 0.12\ s$ (see Figure 13.8.5)

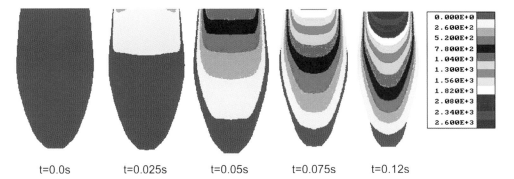

Plate 23 Velocity field at the five characteristic times. The blood velocity propagates from the aortic valve to the heart apex, from 0 mm/s to $V_{max} = 2600$ mm/s at time $t = 0.12$ s (see Figure 13.8.6)

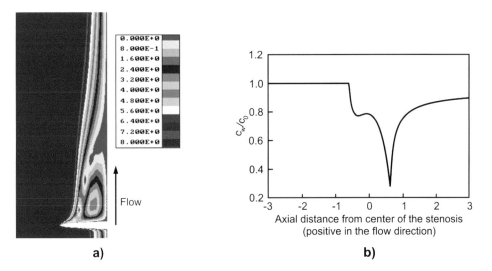

Plate 24 Albumin transport in stenosed artery, the stenotic artery part: (a) Velocity field; (b) Normalized concentration at the wall c_w/c_0 (see Figure E14.2-1B)

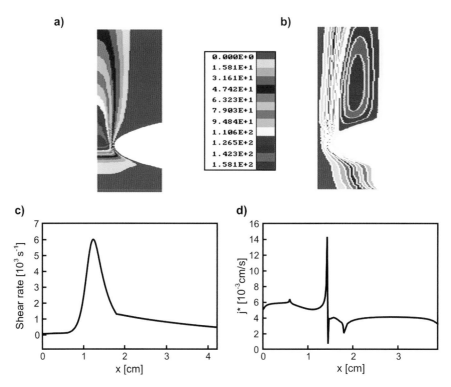

Plate 25 A straight artery with 75% stenosis. (a) Intensity of blood velocity field (in *cm/s*); (b) Streamline contours; (c) Wall shear strain rate along the wall; (d) Platelet accumulation rate along the wall ($j^* = j(x)/c_0$, $j(x) = k_t c_w(x)$, see (14.3.2); $k_t = 5 \times 10^{-3}$ *cm/s*; Wootton et al. 2001) (see Figure E14.3-1B)

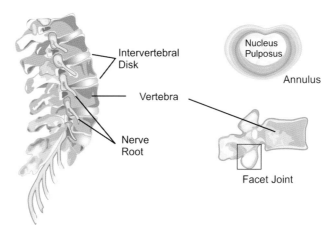

Plate 26 Spine anatomy. There are L1 to L5 spinal motion segments (SMS) which consist of vertebrae and intervertebral discs. The annulus and nucleus pulposus are two main materials of disc, while vertebrae have inside a soft spongy type of bone, called cancellous bone, and an outer shell called cortical bone which is much stronger to support the spinal cord (see Figure 15.1.2)

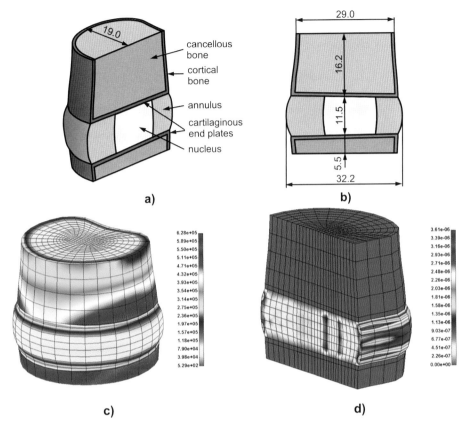

Plate 27 Dynamic response of human spinal motion segment (SMS). (a) Human SMS (intervertebral disk) and one-half of the model; (b) Geometrical data and material properties used for the FE model; (c) von Mises stress distribution (*MPa*); (d) Relative fluid velocity distribution (*m/s*) (see Figure E15.4-5)

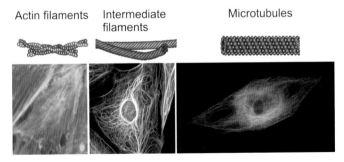

Plate 28 The immuno-fluorescent images of the principal stress-bearing components of the cytoskeleton: actin filaments, intermediate filaments and microtubules. The blue oval in the left panel is the nucleus. The artistic depiction of molecular structure of each filament is shown above the corresponding image (from Ingber 1998) (see Figure 16.1.1)

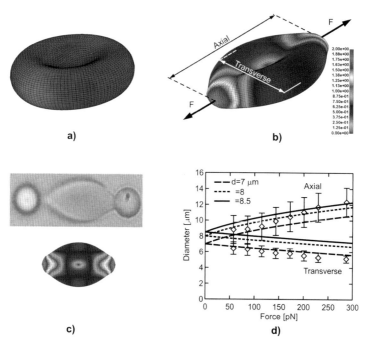

Plate 29 A biphasic FE model of RBC subjected to uniaxial extension forces. One-eighth of the cell is modeled due to symmetry (3D biphasic finite elements). (a) Initial biconcave shape; (b) Deformed shape at force of 300 pN; (c) Deformed shape (top view) experimentally recorded and computed; (d) Change of axial and transverse diameters in terms of extensional force for three initial diameters d (computed results are represented by lines, and experimental by bars) (see Figure E16.3-2A)

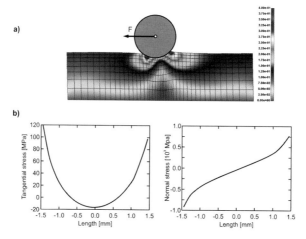

Plate 30 Cell deformation due to action of bead tightly bound to the cell surface. The bead is subjected to a force. (a) Field of effective stress within the cell represented by a 2D biphasic continuum; (b) Tangential and normal tractions along the contact surface between the cell and bead (see Figure E16.3-3)

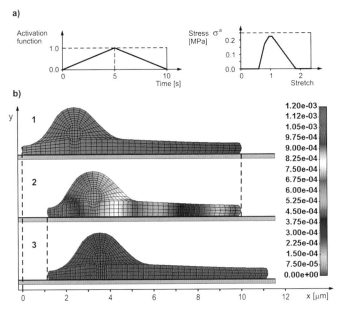

Plate 31 Crawling of cell over a flat surface (2D plane strain conditions in plane x–y). Biphasic model includes: cytoplasm with cytoskeleton and nucleus, and membrane. (a) Activation function of skeleton structure (left panel) and constitutive law for the active stress σ^a (right panel); (b) Three positions of the cell during crawling (1 – initial, after first step; 2 – middle, when detachment of the front and attachment of the rear part occur; 3-after relaxation) with the displacement field. Data: Young's moduli (MPa) for solid within solid-fluid mixture, and within nucleus $E = 0.1$ and $E = 0.3$, respectively; initial porosity $n = 0.7$; permeability $k = 10^2 \ \mu m^4/pN\,s$; solid and fluid density $\rho = 10^{-9} \ mg/\mu m^3$; bulk moduli of solid and fluid $K_s = 8.3333 \times 10^{-2} \ MPa$, $K_f = 1.0 \times 10^9 \ pN/\mu m^2$ (see Figure E16.3-4)

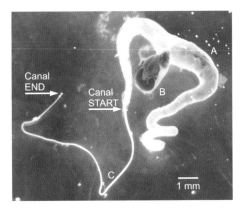

Plate 32 Major ampullate gland of a female *Nephila clavipes* spider. A: Ampulla, dope reservoir; B: A blob of dope used in the pan weighing experiment; C: Spinning canal, S-duct (see Figure 18.3.1)

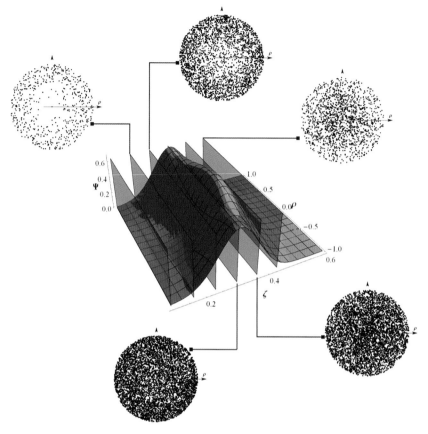

Plate 33 The local concentration Ψ ($\tau = 0.4$; $\Pi = 1$, $\Omega = -2$) (see Figure 19.4.6)

SKELETAL MUSCLES

According to Hill's equation (12.2.2) we have

$$^{n+1}\sigma_m = {}^n\sigma_0\, {}^{n+1}\alpha_a \frac{1+\Delta\lambda_m/\Delta\lambda_{m0}}{1-c\Delta\lambda_m/\Delta\lambda_{m0}} \quad (12.2.11)$$

Here, we have taken the contraction velocity to be negative, and $v/v_0 = \Delta\lambda_m/\Delta\lambda_{m0}$. Also, $^n\sigma_0$ is the tetanic stress corresponding to the stretch $^n\lambda_m$, according to the diagram in Fig. 12.2.2. Finally, $^{n+1}\alpha_a$ is the activation function of muscle.

We have introduced a time function $\alpha_a(t)$ to take into account muscle activation, as a scaling factor for stress with respect to the tetanized state, $0 \leq \alpha_a(t) \leq 1$. Hence, $\alpha_a = 0$ corresponds to a passive (non-activated) state, while $\alpha_a = 1$ for a tetanized state.

Further, we calculate $\Delta\lambda_{m0}$ as

$$\Delta\lambda_{m0} = \Delta t \dot{\lambda}_{m0} \quad (12.2.12)$$

where $\dot{\lambda}_{m0}$ is the stretch rate which corresponds to a maximum isometric tetanized force. The value $\dot{\lambda}_{m0}$ is taken as a characteristic of the contractile element, considered known in our muscle model.

The constitutive equation for the stress $^{t+1}\sigma_s$ follows from (12.2.3)

$$^{n+1}\sigma_s = \beta\left[e^{\alpha(^n\lambda_s - 1 + \Delta\lambda_s)} - 1\right] = e^{\alpha\Delta\lambda_s}(^n\sigma_s + \beta) - \beta \quad (12.2.13)$$

where

$$^n\sigma_s = \beta\left(e^{\alpha(^n\lambda_s - 1)} - 1\right) \quad (12.2.14)$$

is the stress corresponding start of time step.

From equation (12.2.9) we obtain

$$\Delta\lambda_m = a_1 - k\Delta\lambda_s \quad (12.2.15)$$

where

$$a_1 = (1+k)\,^{n+1}\lambda_p - {}^n\lambda_m - k\,^n\lambda_s \quad (12.2.16)$$

Finally, from (12.2.10), (12.2.11), (12.2.13), and (12.2.15) we obtain the following nonlinear equation

$$f(\Delta\lambda_s) = (a_2 + a_3\Delta\lambda_s)e^{\alpha\Delta\lambda_s} + a_4\Delta\lambda_s + a_5 = 0 \quad (12.2.17)$$

with one unknown quantity, $\Delta\lambda_s$. The coefficients in this equation are:

$$a_2 = (^n\sigma_s + \beta)\left(1 - \frac{a_1 c}{\Delta\lambda_{m0}}\right), \quad a_3 = (^n\sigma_s + \beta)\frac{kc}{\Delta\lambda_{m0}}$$

$$a_4 = k\frac{^n\sigma_0\,^{n+1}\alpha_a - \beta c}{\Delta\lambda_{m0}}, \quad a_5 = -{}^n\sigma_0\,^{n+1}\alpha_a - \beta - a_1\frac{^n\sigma_0\,^{n+1}\alpha_a - \beta c}{\Delta\lambda_{m0}} \quad (12.2.18)$$

We solve this equation numerically by a standard Newton's method.

Therefore, the problem of *stress calculation within a muscle fiber is reduced to solving the nonlinear equation* (12.2.17) with respect to $\Delta\lambda_s$. In an incremental finite element analysis, we have that the stresses and stretches at the start of the time step are known. Also, the total sarcomere stretch $^{n+1}\lambda_p$ at the end of the time step can be calculated from the displacements $^{n+1}\mathbf{u}$ (see Section 12.1.2). Hence, all coefficients $a_1, a_2, \ldots, a_5$ are known, and the only unknown quantity is the stretch increment $\Delta\lambda_s$ of the serial elastic element.

We assume that the surrounding connective tissue is a linear elastic isotropic medium. Then, the stress in the parallel elastic element is calculated as follows

$$^{n+1}\boldsymbol{\sigma}^E = \mathbf{C}^E \, {}^{n+1}\mathbf{e} \tag{12.2.19}$$

where $\mathbf{C}^E$ is the elastic constitutive matrix, and $^{n+1}\mathbf{e}$ is the strain at a material point of muscle. The strain $^{n+1}\mathbf{e}$ is calculated from displacements (see equation 6.2.12).

Finally, the total stress can be expressed as

$$^{n+1}\boldsymbol{\sigma} = {}^{n+1}\boldsymbol{\sigma}^E(1-\phi) + \phi \, {}^{n+1}\boldsymbol{\sigma}_s \tag{12.2.20}$$

where ϕ is the fraction of the muscle fibers (active part) in the total muscle volume. We note that the stress $^{n+1}\boldsymbol{\sigma}_s$ has only one nonzero component in the direction of the fiber.

Tangent Constitutive Matrix

The tangent constitutive matrix $^{n+1}\mathbf{C}$ (Kojić & Bathe 2005; see also (6.2.19)) of the muscle as a continuum can be calculated by using the expression (12.2.20),

$$^{n+1}\mathbf{C} = \frac{\partial \, {}^{n+1}\boldsymbol{\sigma}}{\partial \, {}^{n+1}\mathbf{e}} = (1-\phi)\mathbf{C}^E + \phi \frac{\partial \, {}^{n+1}\boldsymbol{\sigma}_s}{\partial \, {}^{n+1}\mathbf{e}} \tag{12.2.21}$$

This constitutive matrix corresponds to the global coordinate system. However, we can determine the derivatives $\partial \, {}^{n+1}\boldsymbol{\sigma}_s / \partial \, {}^{n+1}\mathbf{e}$ in the local coordinate system ξ, η, ζ where ξ is the fiber direction at the integration point, while η and ζ are the axes orthogonal to ξ. This is because we have the dependence $^{n+1}\boldsymbol{\sigma}_s\left(^{n+1}\lambda_p\right)$ defined by equations (12.2.13), (12.2.16) and (12.2.17) corresponding to the local coordinate system. Therefore, we form the local constitutive matrix $^{n+1}\overline{\mathbf{C}}$ in which the terms are:

$$^{n+1}\overline{C}_{11} = (1-\phi)C_{11}^E + \phi\frac{\partial \, {}^{n+1}\sigma_s}{\partial \, {}^{n+1}e_{\xi\xi}}; \quad {}^{n+1}\overline{C}_{ij} = (1-\phi)C_{ij}^E \quad \text{for other } i,j \tag{12.2.22}$$

where C_{ij}^E are the elastic matrix terms (see (2.2.5)) of the surrounding connective tissue.

Finally, we transform the matrix $^{n+1}\overline{\mathbf{C}}$ to the global coordinate system using the relations (2.2.17) (see also Example 2.2-4). We further present a procedure for determining the derivative $\partial \, {}^{n+1}\sigma_s / \partial \, {}^{n+1}e_{\xi\xi} = {}^{n+1}C_s$.

We express $^{n+1}C_s$ as

$$^{n+1}C_s = {}^{n+1}\lambda_p \frac{\partial \, {}^{n+1}\sigma_s}{\partial \, {}^{n+1}\lambda_p} \tag{12.2.23}$$

where the relationship

$$\frac{\partial \, {}^{n+1}\sigma_s}{\partial \, {}^{n+1}e_{\xi\xi}} = \lambda_p \frac{\partial \, {}^{n+1}\sigma_s}{\partial \, {}^{n+1}\lambda_p} \tag{12.2.24}$$

SKELETAL MUSCLES

is used. Next, from (12.2.13) it follows that

$$\frac{\partial^{n+1}\sigma_s}{\partial^{n+1}\lambda_p} = \alpha(^n\sigma_s + \beta) e^{\alpha\Delta\lambda_s} \frac{\partial\Delta\lambda_s}{\partial^{n+1}\lambda_p} \quad (12.2.25)$$

In order to calculate $\partial\Delta\lambda_s/\partial^{n+1}\lambda_p$ we differentiate the nonlinear equation (12.2.17) with respect to $^{n+1}\lambda_p$ and obtain

$$\frac{\partial f(\Delta\lambda_s)}{\partial^{n+1}\lambda_p} = (k_2 + a_3 x) e^{\alpha\Delta\lambda_s} + (a_2 + a_3\Delta\lambda_s) \alpha e^{\alpha\Delta\lambda_s} x + a_4 x + \frac{\partial a_5}{\partial^{n+1}\lambda_p} = 0 \quad (12.2.26)$$

with the unknown $x = \dfrac{\partial\Delta\lambda_s}{\partial^{n+1}\lambda_p}$. From this equation it follows that

$$\frac{\partial\Delta\lambda_s}{\partial^{n+1}\lambda_p} = -\frac{k_5 + k_2}{e^{\alpha\Delta\lambda_s}(a_3 + \alpha(a_2 + a_3\Delta\lambda_s)) + a_4} \quad (12.2.27)$$

where

$$\frac{\partial a_1}{\partial^{n+1}\lambda_p} = 1 + k \qquad \frac{\partial a_2}{\partial^{n+1}\lambda_p} = -(^n\sigma_s + \beta)\left(\frac{c}{\Delta\lambda_{m0}} \frac{\partial a_1}{\partial^{n+1}\lambda_p}\right) = k_2$$

$$\frac{\partial a_5}{\partial^{n+1}\lambda_p} = -\frac{a_4}{k}\frac{\partial a_1}{\partial^{n+1}\lambda_p} = k_5 \quad (12.2.28)$$

Hence, we find $\partial\Delta\lambda_s/\partial^{n+1}\lambda_p$ from (12.2.27), substitute into (12.2.25) to obtain $\partial^{n+1}\sigma_s/\partial^{n+1}\lambda_p$, and further substitute into (12.2.23) to determine $^{n+1}C_s$. Note that we could use the derivatives $\partial^{n+1}\sigma_m/\partial^{n+1}e_{\xi\xi}$ to obtain $^{n+1}C_m = {}^{n+1}C_s$, since $^{n+1}\sigma_m = {}^{n+1}\sigma_s$ (see (12.2.10)).

12.2.3 Hill's model which includes fatigue

There are many daily activities which occur over an extended period of time and during which the forces generated by muscles may be reduced due to fatigue. However, muscle fatigue is a complex physiological phenomenon, and the underlying mechanisms are not well understood. So far, different views on the most important mechanisms for fatigue, as well as other processes associated with fatigue, have been presented by a number of researchers (some references are cited below).

When muscle contraction is sustained for a certain period of time, the muscle becomes fatigued. The force production is affected by the underlying fatigue and recovery effects in the neuromuscular system. Many investigators have given various definitions of fatigue, and one of them is that muscle fatigue can be 'any exercise-induced reduction in the maximal capacity to generate force or power output' (Vøllestad 1997).

Several models have been developed to study muscle fatigue, including nonphysiological analytic models and models based on physiological information or metabolic measurements to establish fatigue parameters. Hawkins and Hull (1993) considered the fatigue effect and force output over a long period of time by incorporating several empirical fatigue parameters, such as fiber endurance times and fatigue rates, into their muscle fiber-based model. A model relying on accurate measurements of temporal changes on muscle metabolites was developed by Giat et al. (1996) and Levin and Mizrahi (1999). A motor unit recruitment function and a muscle fitness function were included into some models for describing the muscle fatigue and recovery effects. A dynamic model of muscle fatigue, which is governed by a fatigue factor, a recovery factor, number of motor units and one input parameter for brain effort, was developed by Liu et al. (2002). The difference between the course of fatigue in muscle activated by the CNS (central nervous system) and various types of FES (functional electrical stimulation) has also been studied. We here give a mathematical formulation of fatigue based on Hill's model.

A modification of Hill's model to include fatigue
Stress developed in the contractile element of intact muscle can be expressed in a form given in (12.2.11). However, muscle fatigue is not considered in this equation.

Fatigue can be induced in a muscle that is loaded for a long time by a constant or variable force. Due to a reduction in the capacity of muscle to produce force, tetanic stress of fatigued muscle $\sigma_{0f}(\lambda, t)$ is always less then tetanic stress of intact, nonfatigued muscle $\sigma_0(\lambda, t)$ (Tang et al. 2005). Assume that $F_0(\alpha_a, \lambda)$ is the force developed by an intact muscle under the activation α_a and the total stretch λ, and $F_f(\alpha_a, \lambda, t)$ is the force developed by the fatigued muscle under the same activation α_a. These forces correspond to the stresses in the tetanized conditions $\sigma_0(\lambda, t)$ and $\sigma_{0f}(\lambda, t)$, respectively. Now the *fitness level* can be defined as

$$\alpha_f(t) = \frac{F_f(\alpha_a, \lambda, t)}{F_0(\alpha_a, \lambda)} = \frac{\sigma_{0f}(\lambda, t)}{\sigma_0(\lambda)} \qquad (12.2.29)$$

Thus, the fitness level is the normalized maximal force that muscle can produce, ranging from 0 to 1. For an intact muscle the fitness level is equal to 1, while for a muscle under certain loading duration, the fitness level decreases with time.

Using the relation (12.2.29), the stress developed in the contractile element of a fatigued muscle, analogous to equation (12.2.11), can be written as

$$\begin{aligned} {}^{n+1}\sigma_m &= {}^{n+1}\alpha_a{}^n \sigma_{0f}(\lambda_p) \frac{1 + \Delta\lambda_m/\Delta\lambda_{m0}}{1 - c\Delta\lambda_m/\Delta\lambda_{m0}} \\ &= {}^{n+1}\alpha_a{}^n \alpha_f{}^n \sigma_0(\lambda_p) \frac{1 + \Delta\lambda_m/\Delta\lambda_{m0}}{1 - c\Delta\lambda_m/\Delta\lambda_{m0}} \end{aligned} \qquad (12.2.30)$$

Since the underlying mechanisms of fatigue are still not well understood, there is no reliable model that can predict fitness level of a muscle subjected to an arbitrary activation over a long time period. Several models given in the cited references can be used to determine fitness level $\alpha_f(t, \dots)$, which depends on time, activation and other physiological and nonphysiological parameters.

A time dependency of maximal output force under various sustained constant activations can be obtained experimentally. Normalizing the recorded force–time curves by the maximal output force of intact muscle, we find the fitness level under constant activations $\alpha_{fca}(\alpha_a, t)$. A typical set of functions $\alpha_{fca}(\alpha_a, t)$ is shown in Fig. 12.2.5. Note that the decrease of the functions $\alpha_{fca}(\alpha_a, t)$ is sharper for larger activations $\alpha_a = const$.

SKELETAL MUSCLES

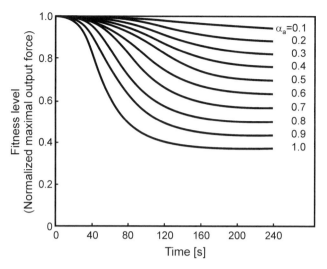

Fig. 12.2.5 Example of fatigue curves under a sustained constant activations $\alpha_a = \text{const}$

Let us assume that fatigue rate does not depend on the entire history of loading, but only on the current fitness level and activation. Consequently, muscle fatigue rate after arbitrary loading will be equal to the fatigue rate under sustained constant activation for the activation and fitness level equal to the current ones, which can be written as

$$\left.\frac{\partial \alpha_f(\alpha_a)}{\partial t}\right|_{\alpha_f} = \left.\frac{\partial \alpha_{fca}(\alpha_a)}{\partial t}\right|_{\alpha_f} \quad (12.2.31)$$

or in incremental form

$$^{n+1}\alpha_f(\alpha_a) = {}^n\alpha_f + \Delta t \frac{\partial \alpha_{fca}(\alpha_a, t_{fca} + \Delta t)}{\partial t} \quad (12.2.32)$$

where the time t_{fca} is obtained from the condition that $\alpha_{fca}(\alpha_a, t_{fca}) = {}^n\alpha_f$. For an activation which is between measured curves $\alpha_{fca}(\alpha_a, t)$, we use a linear interpolation.

Analogous to the fitness level curves of activated muscles, we can obtain time dependency of fitness level in the case when there is no activation, i.e. when a muscle is at rest. Such a curve represents a recovery function $\alpha_r(t)$ of the resting muscle. A typical recovery curve is shown in Fig. 12.2.6.

Thus, the incremental form (12.2.32) for the fitness level in a resting state is

$$^{n+1}\alpha_f(\alpha_a) = {}^n\alpha_f + \Delta t \frac{\partial \alpha_r(t_r + \Delta t)}{\partial t} \quad (12.2.33)$$

where the time t_r is obtained from the condition that $\alpha_r(t_r) = {}^n\alpha_f$.

In general, we have time periods of muscle loading with activation, followed by resting periods. Consequently, we use either equation (12.2.32) or (12.2.33) to calculate the fitness

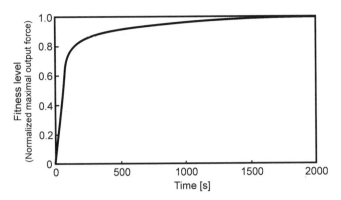

Fig. 12.2.6 Example of recovery curve

level $^{n+1}\alpha_f(\alpha_a)$, depending on the condition whether the muscle is activated or at rest. In summary, we have

$$^{n+1}\alpha_f(\alpha_a) = \begin{cases} {^n\alpha_f} + \Delta t \dfrac{\partial \alpha_{fca}(t_{fca} + \Delta t)}{\partial t}, & \alpha_a > 0 \\ {^n\alpha_f} + \Delta t \dfrac{\partial \alpha_r(t_r + \Delta t)}{\partial t}, & \alpha_a = 0 \end{cases} \qquad (12.2.34)$$

12.2.4 An extension of Hill's model to include different fiber types

Most of the previous muscle models based on Hill's model were established assuming a sliding-element theory and a single sarcomere. Muscle is regarded as an assemblage of identical sarcomeres. However, a number of authors have shown that muscle is a heterogenous material structure which induces nonuniform stress and strain distributions (Pappas *et al.* 2002), as well as different fatiguing of muscle regions.

An individual skeletal muscle, as a body organ, represents a collection of different fiber types. A large range in contractile properties among fiber types enables production of very diverse mechanical responses, from extremely rapid ballistic to slow sustained motions and postural support. A classification of skeletal muscle beyond simple fast-twitch and slow (tonic) fiber types was first proposed over 40 years ago by Lannergren and Smith (1966). According to these authors, the fast-twitch fibers could further be differentiated into three distinct types. For the sake of simplicity, but without significantly affecting the basic findings of muscle behavior, we will here refer to two basic muscle fiber types: fast- and slow-twitch fibers. Moreover, differences between muscles may be due to the spatial arrangement of different muscle fiber types, sample sites within a muscle, locations within the fascicle, and age of the subject.

The essential differentiation between fast- and slow-twitch muscle fibers is that the contraction time is shorter for fast-twitch muscle fibers, when compared to the slow-twitch muscle fibers. In determining which types of motor units participate in activation, one important aspect to be considered is that the two types of motor units have different activation thresholds. The slow-twitch type has a lower threshold than the fast-twitch one. Also, it

SKELETAL MUSCLES

is known that the fast-twitch fibers fatigue more easily than the slow-twitch ones, so that a selective recruitment of fast-twitch fibers during intermittent stimulation could explain a faster decline in the generated force.

Maximal force produced by a muscle depends, in general, on the percentage of slow and fast muscle fibers in the muscle volume. This seems to be determined almost entirely by genetic inheritance. Some people appear to be born for marathons, while others are born to be sprinters and jumpers. This could determine to some extent the athletic capabilities of different individuals. For example, marathoners have only 18% of the fast muscle fibers, while sprinters have up to 65% of fast fibers and therefore much larger output force then marathoners (Guyton 2005). Also, due to a large percentage of fast muscle fibers (also fast fatigued), sprinters can produce maximal force for several seconds to a minute only. According to these physiological facts, it follows that a modification of Hill's model is needed, in order to predict the total output force dependent on the participation of various fibers within the muscle (Stojanović 2007, Stojanović et al. 2007).

An extension of Hill's three-component model is here introduced to take into account different fiber types. This extended Hill's model consists of a number of sarcomeres of different types coupled in parallel with the connective tissue. Practically, we introduce a multi-fiber material model at the level of a material point of the continuum.

The Extended Hill's Model

The extended Hill's model consists of a number of series of contractile and serial elements, corresponding to various types of sarcomeres (active part of a muscle), coupled in parallel to the linear elastic element representing the connective tissue (passive part), as shown in Fig. 12.2.7. In this schematic, CE_i and SEE_i are, respectively, the contractile and serial elastic elements of the i-th type of muscle fiber, while PEE is the common parallel elastic element.

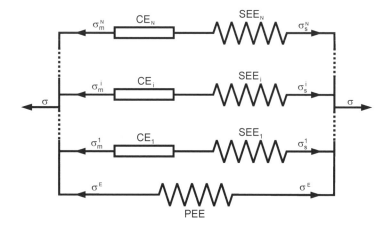

Fig. 12.2.7 Extended Hill's model of muscle

At a considered muscle point, all these fibers have the same spatial direction (which is the fiber direction) and the same fiber stretch λ_p.

Analogous to relations (12.2.11) and (12.2.13), we express the stresses in contractile and serial elements representing a sarcomere of the i-th fiber type as (with the upper index 'i' to all quantities)

$$^{n+1}\sigma_m^i = {^n\sigma_0^i} \, {^{n+1}\alpha_a^i} \frac{1 + \Delta\lambda_m^i/\Delta\lambda_{m0}^i}{1 - c^i \Delta\lambda_m^i/\Delta\lambda_{m0}^i} \tag{12.2.35}$$

$$^{n+1}\sigma_s^i = \left({^n\sigma_s^i} + \beta^i\right) e^{\alpha^i \Delta\lambda_s^i} - \beta^i \tag{12.2.36}$$

Following the computational procedure for stress integration of the basic Hill's model in Section 12.2.2, we find that the stress calculation for the i-th fiber type reduces to finding the stretch increment $\Delta\lambda_s^i$. The governing equation is now (see (12.2.17))

$$f(\Delta\lambda_s^i) = \left(a_2^i + a_3^i \Delta\lambda_s^i\right) e^{\alpha^i \Delta\lambda_s^i} + a_4^i \Delta\lambda_s^i + a_5^i = 0 \tag{12.2.37}$$

where

$$a_1^i = (1 + k^i)\, {^{n+1}\lambda_p} - {^{n+1}\lambda_m^i} - k\, {^{n+1}\lambda_s^i}, \qquad a_2^i = \left({^n\sigma_s^i} + \beta^i\right)\left(1 - \frac{a_1^i c^i}{\Delta\lambda_{m0}^i}\right)$$

$$a_3^i = \left({^n\sigma_s^i} + \beta^i\right)\frac{k^i c^i}{\Delta\lambda_{m0}^i}, \qquad a_4^i = k^i \frac{{^n\sigma_0^i}\, {^{n+1}\alpha_a^i} - \beta^i c^i}{\Delta\lambda_{m0}^i}, \tag{12.2.38}$$

$$a_5^i = -{^n\sigma_0^i}\, {^{n+1}\alpha_a^i} - \beta^i - a_1^i \frac{{^n\sigma_0^i}\, {^{n+1}\alpha_a^i} - \beta^i c^i}{\Delta\lambda_{m0}^i}$$

Therefore, for a given stretch in the muscle fiber direction, $^{n+1}\lambda_p$ (calculated from displacements $^{n+1}\mathbf{u}$ at a muscle material point), we solve equation (12.2.37) with respect to $\Delta\lambda_s^i$, and obtain the stress $^{n+1}\sigma_s^i$ from equation (12.2.36); therefore, we have the stress $^{n+1}\sigma_m^i$ at a material point.

Now, the component of stress in the fiber direction, $^{n+1}\boldsymbol{\sigma}_s$, defined in equation (12.2.20), can be written as

$$^{n+1}\sigma_s = \sum_{i=1}^{N} \varphi^i \, {^{n+1}\sigma_s^i} \tag{12.2.39}$$

where φ^i is the fraction of the i-th fiber type in the active part of muscle.

Since the fraction of the specific fiber type is nonuniform over the muscle volume, we have to take into account material heterogeneity. If the fraction of a specific fiber type in the active part is known at the finite element nodal points, we can interpolate these values over the entire finite element domain. The interpolation functions, used for displacements and geometry, can also be used for the material parameters. Hence, the interpolation of the fraction φ^i for the i-th fiber type is given by

$$\varphi^i = \sum_{J=1}^{m} N_J \varphi_J^i \tag{12.2.40}$$

SKELETAL MUSCLES

Here, N_J is the interpolation function and φ_J^i is the i-th fiber fraction, at J-th node of the element; also, m is the number of finite element nodes. It should be noted that the following equation is satisfied (by definition of the fractions φ^i)

$$\sum_{i=1}^{N} \varphi^i = 1 \tag{12.2.41}$$

Tangent Constitutive Matrix

The tangent constitutive matrix of the extended Hill's model can be obtained in the same manner as in the case of the basic Hill's model, with some minor modifications. Here, we can write the derivative of stretch increment $\Delta \lambda_s^i$ with respect to the total stretch $^{n+1}\lambda_p$, as (see (12.2.27))

$$\frac{\partial \Delta \lambda_s^i}{\partial^{n+1} \lambda_p} = -\frac{k_2^i + k_5^i}{e^{\alpha^i \Delta \lambda_s^i} \left(a_3^i + \alpha^i \left(a_2^i + a_3^i \Delta \lambda_s^i\right)\right) + a_4^i} \tag{12.2.42}$$

where

$$k_2^i = \frac{\partial a_2^i}{\partial^{n+1} \lambda_p}, \quad k_5^i = \frac{\partial a_5^i}{\partial^{n+1} \lambda_p} \tag{12.2.43}$$

Further, the derivative of $^{n+1}\sigma_s^i$ with respect to the sarcomere stretch $^{n+1}\lambda_p$ is (see (12.2.25))

$$\frac{\partial^{n+1} \sigma_s^i}{\partial^{n+1} \lambda_p} = \alpha^i \left(^n\sigma_s^i + \beta^i\right) e^{\alpha^i \Delta \lambda_s^i} \frac{\partial \Delta \lambda_s^i}{\partial^{n+1} \lambda_p} \tag{12.2.44}$$

and then

$$^{n+1}C_s = {}^{n+1}\lambda_p \sum_{i=1}^{N} \varphi^i \frac{\partial^{n+1} \sigma_s^i}{\partial^{n+1} \lambda_p} \tag{12.2.45}$$

12.3 Examples

Example 12.3-1. Modeling of a cylindrical muscle

Figure E12.3-1A shows a simplification of a typical cylindrical muscle. The muscle is represented by a cylinder with variable cross section, defined by three diameters (D_{orig}, D_{ins} and D_{max}) and a cubic Bezier spline between these diameters. The muscle is fixed at the top (origin), while the other end (insertion) can be free to move or it can be fixed. At the insertion we can have a force (Fig. E12.3-1Aa), a spring (Fig. E12.3-1Ab), or isometric conditions (Fig. E12.3-1Ac).

The above description of the cylindrical muscle model is used in the Software on the web (Example 12.1), where the solutions can be obtained for various geometric parameters, distribution of fast and slow fibers over the cross-section, and activation and time functions.

We here give some typical results as an illustration of the solutions. Fiber parameters are given in Fig. E12.3-1Ba, while the activation functions of slow and fast fibers are shown in Fig. E12.3-1Bb. Figure E12.3-1Bc shows muscle shortening over time depending of percentage of slow and fast fibers. It can be seen from the figure that muscles with higher percentage of the fast fibers have fast and strong response to the activation, while muscles consisting predominantly of slow fibers exhibit slower and minor response.

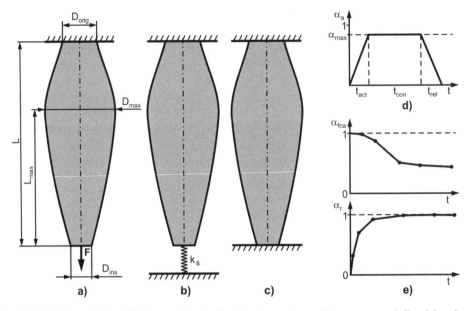

Fig. E12.3-1A A simplified model of cylindrical muscle, with geometry defined by three diameters (D_{orig}, D_{ins}, D_{max}) and lengths L and L_{max}. (a) Muscle loaded by a force F; (b) Muscle attached to a spring with stiffness k_s; (c) Isometric conditions; (d) Activation function $\alpha_a(t)$ increasing linearly to a value α_{max} in activation time period t_{act}, then remains constant within the period t_{con}, and linearly decreases to zero within the relaxation period t_{rel}; (e) Fatigue and recovery functions α_{fca} and α_r defined as multilinear curves

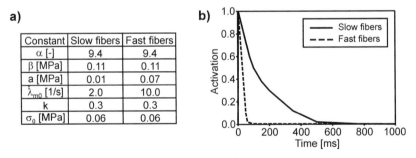

Fig. E12.3-1B Muscle response to single twitch activation depending on percentage of slow and fast fibers. (a) Constants for Hill's model (a is the constant in (12.2.1) and σ_0 is maximal tetanic stress); (b) Activation function for slow and fast fibers; (c) Shortening of the muscle depending of percentage of slow and fast fibers. (*Continued on page 247*)

Example 12.3-2. Modeling of the frog *gastrocnemius* muscle

Here, we present numerical and experimental results for frog *gastrocnemius* muscle. A simplified geometry of the muscle is defined by setting the geometric parameters as follows:

$$L = 45\,mm, \quad L_{max} = 23\,mm, \quad D_{orig} = 6.4\,mm, \quad D_{ins} = 6\,mm, \quad D_{max} = 11.6\,mm$$

SKELETAL MUSCLES

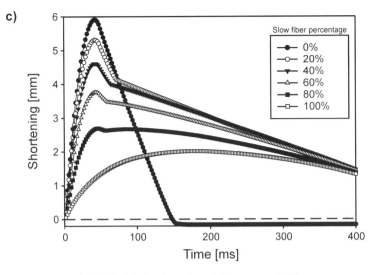

Fig. E12.3-1B (*continued from page 246*)

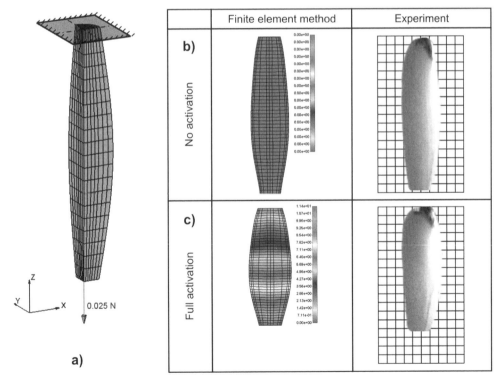

Fig. E12.3-2 Simplified finite element model of frog *gastrocnemius* muscle. **a)** Finite element mesh, constraints and load; **b)** Left panel: undeformed configuration (zero displacement field) and real muscle without activation (right panel); **c)** Displacement field under full activation (left panel) and real deformed muscle (right panel) (see Plate 8)

Due to symmetry, only one quarter of the muscle is considered as shown in Fig. E12.3-2a. A small concentrated load of $0.025\,N$ (i.e. $1/4 \times 0.1N$) was added to the bottom end of the quarter model to simulate the load weight of $0.1\,N$ (10 grams) attached to the muscle in the experiment.

Figure E12.3-2b shows original configuration of the finite element model (left panel) and real frog gastrocnemius muscle with its slack length (right panel). These are undeformed configurations (zero displacement field) corresponding to non-activated muscle without load. Calculated deformation of fully activated muscle is shown in Fig. E12.3-2c (left panel), with the muscle shortening of 11.4 *mm*. As can be seen from the figure, numerical results give satisfactory prediction of the experimental observation (Fig. E12.3-3c – right panel).

References

Fox, S.I. (2004). *Human Physiology*, 8th edition. McGraw-Hill, New York.
Fung, Y.C. (1993). *Biomechanics: Mechanical Properties of Living Tissue*, 2nd edition, Springer-Verlag, New York.
Giat, Y., Mizrahi, J. & Levy, M. (1996). A model of fatigue and recovery in paraplegic's quadriceps muscle subjected to intermittent FES, *J. Biomech. Eng.*, **118**, 357–66.
Gordon, A.M., Huxley, A.F. & Julian, F.J. (1966). The variation in isometric tension with sarcomere length in vertebrate muscle fibers, *J. Physiol.*, **269**, 441–515.
Guyton, A.C. (2005). *Medical Physiology*, Elsevier, Amsterdam.
Hawkins, D.A. & Hull, M.L. (1993). Muscle force as affected by fatigue: mathematical model and experimental verification, *J. Biomech.*, **26**, 1117–28.
Kojić, M. & Bathe, K.J. (2005). *Inelastic Analysis of Solids and Structures*, Springer-Verlag, Berlin.
Kojić, M., Mijailović, S. & Zdravkovic, N. (1998). Modelling of muscle behaviour by the finite element method using Hill's three-element model, *Int. J. Num. Meth. Eng.*, **43**, 941–53.
Lannergren, J. & Smith, R.S. (1966). Types of muscle fibres in toad skeletal muscle, *Acta Physiol. Scand.*, **68**, 263–74.
Levin, O. & Mizrahi, J. (1999). EMG and metabolite-based prediction of force in paralyzed quadriceps muscle under interrupted stimulation, *IEEE Trans. Rehab. Eng.*, **7**, 301–14.
Liu, J.Z., Brown, R.W. & Yue, G.H. (2002). A dynamical model of muscle activation, fatigue, and recovery, *Biophys. J.*, **82**, 2344–59.
Pappas, G.P., Asakawa, D.S., Delp, S.L., Zajac, F.E. & Drace, J.E. (2002). Nonuniform shortening in the biceps brachii during elbow flexion, *J. Appl. Physiol.*, **92**, 2381–9.
Stojanović, B. (2007). *Generalization of Hill's phenomenological model in order to investigate muscle fatigue*, Ph.D. Thesis, CIMSI, University of Kragujevac, Serbia.
Stojanović, B., Kojić, M., Rosić, M., Tsui, C.P. & Tang, C.Y. (2007). An extension of Hill's three-component model to include different fiber types in finite element modeling of muscle, *Int. J. Num. Meth. Eng.*, **71**, 801–17.
Tang, C.Y., Stojanović, B., Tsui, C.P. & Kojić, M. (2005). Modeling of muscle fatigue using Hill's model, *Bio-Med. Mater. Eng.*, **15**, 341–8.
Vøllestad, N.K. (1997). Measurement of human muscle fatigue, *J. Neurosci. Methods*, **74**, 219–27.

13

Blood Flow and Blood Vessels

Cardiovascular diseases are considered a leading cause of death in the developed world and are now becoming more prevalent in developing countries. The main objective of this chapter is to present finite element (FE) computer modeling for an insight into blood flow and deformation of blood vessels. An introduction to cardiovascular system with basic description of blood and blood vessels is given first. Then, the FE method described in Chapter 7 is further extended in order to simulate blood flow in large blood vessels including specific calculation of shear stress and interaction of blood flow with deformable blood vessels. Finally, we present some of the results for blood flow through rigid and deformable blood vessels in human aorta, abdominal aortic aneurysm, carotid artery bifurcation, femoral artery with stent, veins including compression therapy, and human heart.

13.1 Introduction to the cardiovascular system

In this section we introduce basic notions used to define the cardiovascular system, blood and blood vessels. The emphasis is on the mechanical characteristics which are further used in computer modeling.

13.1.1 The circulatory system

The circulatory system consists of *blood vessels* whose main role is transport by blood of nutrients and oxygen to periphery tissues, and elimination of harmful metabolites. Anatomically it is composed of two elements:

1. systemic circulation, transporting nutrients; and
2. pulmonary circulation, with main role in oxygenation of blood hemoglobin.

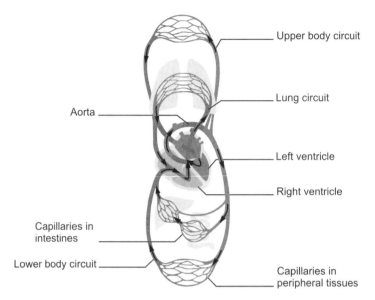

Fig. 13.1.1 Schematic representation of the cardiovascular system (adapted from Mohr 2006) (see Plate 9)

These two systems are connected via the heart acting as a pump of the circulatory system (Fig. 13.1.1 – see color plate).

Systemic circulation starts with the major blood vessel, the *aorta*, followed by large blood vessels with similar histological and biomechanical characteristics rich in elastic fibers. After division into smaller vessels, there is a gradual change in structure forming medium size vessels with different biomechanical characteristics, with rise of thick nonelastic fibers and more developed vascular muscle tissue. Finally, there are *small arteries* and a capillary system, with predominant vascular muscle tissue acting with *post-capillary vessels* as a valve controlling the volume of blood entering the peripheral tissue. Post-capillary vessels are called the venula (the smallest veins), and have their own structural characteristics. After the venula, blood is collected via veins with predominant fibrous tissue and valve systems, returning the blood into heart through two major venous vessels, the superior and inferior vena cava.

Pulmonary circulation begins with the pulmonary trunk, divided into two *pulmonary arteries* which enter the lung and after branching into multiple smaller vessels become in close contact with the respiratory system, resulting in oxygenation of deoxygenated blood from peripheral tissues. Oxygenized blood is collected via a system of *pulmonary veins* ending as four large pulmonary veins with their ostium in the left atrium. Both systemic and pulmonary circulation systems have their arterial and venous parts, determined not by the type of blood they carry, but by their structural characteristics. Arterial blood in the systemic circulation is oxygenized, while venous blood is deoxygenized and vice versa in the pulmonary circulation.

The structure of vessels is dependent on the role they have. Large vessels are conductive, with the role of distributing blood, and their elastic structure allows stretching in systole and contracting in diastole, thus making continuous flow of the blood possible. While systole blood flow is the result of the heart contraction, perfusion in diastole is retained by the elastic pressure of the vessels. The result is continuous and variable blood flow throughout the heart cycle. *Resistance* to the blood flow is also variable and in proportion to the diameter of the vessel and to the percentage of muscular tissue in vessel, with the peak at the level of small arteries (arteriola) that causes significant drop in the velocity of the blood flow. Approximately 50% of all resistance is due to the resistance in arteriola, 20% in large arteries and capillaries and only 10% in veins. Flow velocity depends on the phase of heart cycle and of the diameter of the vessel and varies from values $< 100\,cm$ per second in ascending aorta to $< 1\,cm$ per second at capillary level. The change in the tonus of the muscular layer of vessels results in the elevated resistance and it is related to the decrease in the perfusion. This mechanism is controlled by the sympathetic nervous system.

The *critical pressure* is the perfusion pressure, essential to allow continuous perfusion, and is different from the mechanical model of nonelastic pipes and from the model of ideal fluid, where the minimal pressure elicits fluid movement. Blood pressure is the measure of the blood energy as the fluid flows in various segments of the circulatory system. Arterial pressure has its maximum at systole and its value is known as systolic, and its minimum at diastole, named diastolic pressure. Average pressure is the mid-value of pressure throughout the heart cycle, and its value is smaller than the arithmetic average because of the longer duration of the diastole. In large vessels average value of pressure is $120/80\,mmHg$ systolic/diastolic, respectively. With the decrease of the vessel size and at the end of the vascular bed in capillary system mid-pressure is $17\,mmHg$. In pulmonary circulation values of the pressure are significantly smaller due to the less potent right ventricle contraction and somewhat different structure of vessels which have a higher amount of nonelastic fibers.

Distributions of cross-sectional area, velocity, pressure and blood volume in the circulatory system (aretial system, veins and capillaries) are shown in Fig. 13.1.2.

13.1.2 Blood

Blood represents a two-component system which consists of cells (formed cellular elements) and plasma. There are several blood cell types which are present in blood in a form of functional mature cells such as *erythrocytes*, *leucocytes* and *platelets*. In adult humans all of the cell types are formed in the bone marrow (except B-lymphocytes which are formed in lymphatic nodes) from pluripotent cells named stem cells or CFU-S (colony forming unit spleen).

Red blood cells (RBCs) or erythrocytes (Fig. 13.1.3a – see color plate) are mature, highly differentiated cells which consist of plasma membrane and citosol, an inner fluid with dissociated protein hemoglobin. Hemoglobin colors erythrocytes red and since 99% of total cell residue (hematocrit) are red blood cells, it gives the origin to the blood color. Red blood cells are of biconcave shape, diameter 6.7–7.7 μm, average thickness 1.7–2.5 μm and volume 76–96 fL. The shape of these cells is adjusted to pass through the smallest blood vessel (capillary vessel) without changing their original volume or surface area. Hemoglobin is the most important red cell protein and is placed adjacent to the cell membrane, with a major role in oxygen binding, transport and exchange with peripheral tissues. Normal hemoglobin concentration ranges 120–180 g/L and concentration of red blood cell varies from 3.7 to $5.8 \times 10^{12}/L$. Percentage of volume occupied by red blood cells (predominantly) is named *hematocrit* (H) and represents the ratio between the volume of RBC cellular part of the blood and total blood volume. It ranges between 0.4 ± 0.04 in females and 0.43 ± 0.04 in males. The major influence on hematocrit value comes (besides sex) from the vessel size

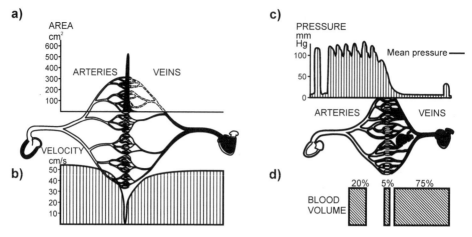

Fig. 13.1.2 Velocities, pressure and blood volume within blood circulatory system. (a) Cross-sectional areas of various segments of the systemic circulation computed for a 13 kg dog. The enormous area belongs to small blood vessels: arterioles, capillaries and venules; (b) The velocity of blood flow is inversely proportional to the cross-sectional area; it decreases from valuse of around 50 cm/s in large vessels to about 0.07 cm/s in the capillaries; (c) The pressure in the arterial system is high and pulsatile. The mean arterial pressure decreases gradually from the main branches toward the arterial tree. The pressure is small in small vessels and oscillations are damped out due to high resistance to flow. The pressure and pressure gradient are small in the major veins; (d) The blood volume distribution: around 20% in arteries, 75% in veins and only 5% in the capillaries (adapted from Rushmer 1976)

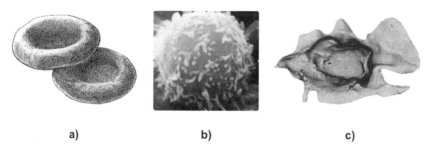

Fig. 13.1.3 Basic blood cells. (a) Erythrocytes (http://www.mast.queensu.ca/~julia/sgc.html); (b) Leukocyte; (http://www.funsci.com/fun3_en/blood/blood.htm); (c) Activated platelet (according to Loscalzo & Schafer 2003) (see Plate 10)

and the type of motion within the vessel. The World Health Organization defines anemia as the value of hemoglobin under 120 g/L in females and under 130 g/L in males.

Since the the RBCs are the dominant cells within plasma, total blood volume can be expressed as

$$V_{blood} = V_{plasma}/(1-H) \tag{13.1.1}$$

where H is the number of hematocrit. Total blood volume is 4–6 L or 6–8% of total body weight. The increase of blood volume is called 'hypervolemia', while the decrease is called 'hypovolemia' and is mainly due to bleeding.

Leukocytes are the only fully morphologically complete cells, because they contain all cell elements (nucleus, cytosol and cell organelles) (Fig. 13.1.3b – see color plate). They represent the base of the immune system. The basic division is into polymorph nuclears (containing cell granules) and mono nuclears (without granules and also divided into two large divisions: monocytes and lymphocytes). The number of leukocytes varies between $4.1–10 \times 10^9/L$. Their number in a particular part of the body depends not only on the cell turnover (number of produced and depleted cells), but also on the presence of local or systemic infection. The condition of total number of reduced white blood cells is leucopenia, and the increased total number of these is leucocytosis.

Platelets (Fig. 13.1.3c – see color plate) are cell fragments produced from giant precursor cells called megakariocytes during differentiation. Physiological concentration varies from 150 to $350 \times 10^9/L$. They are disc-shaped, enabling adherence to vessel walls as well to each other when activated. There is an abundance of various receptors imbedded in the surface of the membrane, participating in the interactions between vessel wall (endothelia), platelets and different coagulation factors. The integrity of the vessels is preserved and dependent on those interactions. Impaired function causes excessive bleeding or pathologic thrombosis within vessels. More details about platelets and platelet-mediated thrombosis are given in Chapter 14.

Plasma is liquid constituent of blood. It contains water, various electrolytes, small organic molecules (glucoses and amino acids) and also large proteins and lipids. Roughly, in blood per liter there is 70 g of proteins, 9 g of nonorganic electrolytes and 250 mg of lipids. Average density of plasma is 1.025–1.034 kg/L (1.050–1.060 kg/L in blood). Viscosity of plasma is two times the viscosity of plain water (in blood it is up to four times).

Blood Flow in Large Blood Vessels

In most arteries, blood behaves in a Newtonian fashion, and the viscosity can be taken as a constant of about 4 centipoise (cP) for a normal hematocrit. However, non-Newtonian mechanical behavior of blood is pronounced in smaller blood vessels, such as capillary systems, where the viscosity depends on the flow conditions. Various phenomenological mechanical models for blood, based on experimental biorheological investigations, have been introduced. A review of these models is given in, for example, Chien (1970) and Rodkiewicz et al. (1990). We here describe one of these models further used in modeling of blood flow.

In the case when the shear strain rates (D_{ij}, $i \neq j$, see eq. (2.1.27) in Chapter 2) of blood flow are not too low, as in medium size arteries and veins, the blood viscosity μ can be expressed as a function of the hematocrit H and shear strain rate. This functional relationship is called the *Cason relation*:

$$\mu = \frac{1}{2\sqrt{D_{II}}} \left(k_0(H) + k_1(H) \sqrt{2\sqrt{D_{II}}} \right)^2 \tag{13.1.2}$$

$k_0(H)$ and $k_1(H)$ are the functions determined experimentally (Perktold et al. 1998); and D_{II} is the second invariant of the strain rate (see web – Theory, Chapter 2) which can be written as

$$D_{II} = \frac{1}{2} D_{ij} D_{ij} \tag{13.1.3}$$

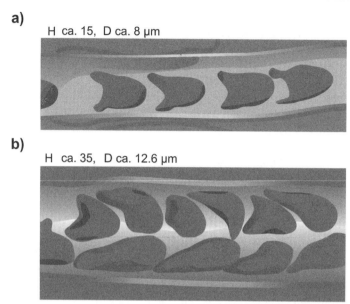

Fig. 13.1.4 Motion of RBCs in microvessels. (a) RBCs flow in a single file form when the vessel diameter (*D*) is around the RBC size; (b) With increasing vessel diameter, RBCs tend to form multiple file flow (according to Pries & Secomb 2005) (see Plate 11)

Blood Flow in Capillaries

In capillaries of a small diameter, of around RBC size, the red cells form the so-called *single file flow* in their motion, as shown in Fig. 13.1.4a (see color plate). With increasing diameter and hematocrit, RBCs form multi-file flow, Fig. 13.1.4b – see color plate (Pries & Secomb 2005).

In order to quantify the mechanical behavior of blood in flow through microvessels, an equivalent or *apparent viscosity* is introduced. Namely, for a steady laminar flow of a non-Newtonian fluid through a circular tube, the viscosity of a Newtonian fluid with the viscosity μ_{app} (apparent or equivalent viscosity) can be expressed as

$$\mu_{app} = \frac{\pi}{128} \frac{\Delta p}{Q} \frac{D^4}{L} \qquad (13.1.4)$$

where D and L are the tube diameter and length, respectively; Q is the volumetric flow rate induced by the pressure drop Δp between the tube ends. This relation follows from calculation of the flow resistance due to viscous stress along the tube surface caused by the fluid. Further, the *relative blood viscosity* μ_{rel} with respect to the plasma viscosity μ_p is defined as

$$\mu_{rel} = \frac{\mu_{app}}{\mu_p} \qquad (13.1.5)$$

Experimental investigations reveal the phenomenon known as Fåhraeus–Lindquist effect (Fåhraeus & Lindquist 1931) about the dependence of the blood relative viscosity on the

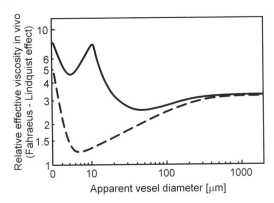

Fig. 13.1.5 Dependence of the relative apparent viscosity of blood on the microvessel diameter (according to Pries & Secomb 2005): solid line is for blood vessels *in vivo* measurement on rat mesentery; dashed line is for glass tube. The hematocrit is 45. Large differences between the results for blood vessels and tube are due to effects of the endothelial surface layer present in blood vessels. The existence of a pick value for the vessel diameter around 10 μm can be explained by a change of the endothelial surface layer thickness (detailed discussion is given in Pries & Secomb 2005)

vessel diameter (Fig. 13.1.5). It can be seen from the figure that the relative viscosity decreases and then increases with a decrease of the vessel diameter reaching two minima. In the case of a glass tube, the viscosity change is monotonic with the minimum at diameter around 7 μm. The viscosity changes can be explained by different RBC flow patterns, from multiple to single file flows. The notable discrepancy between viscosities for flow within blood vessels and glass tube is due to the existence of the endothelial surface layer whose thickness also depends on the microvessel diameter (Pries & Secomb 2005).

13.1.3 Blood vessels

As already described in Section 13.1.1, the circulatory system of blood vessels may be divided into those vessels that deliver oxygenated blood to tissues: the arteries, arterioles, and capillaries; and those vessels that return blood with carbon dioxide for gas exchange: the veins and venules. It can be considered that the basic structure of all these vessels consists of three layers: (a) the intima; (b) the media; and (c) the adventia (Fig. 13.1.6).

Comparing wall structures of arteries and veins, the following differences are notable. Arteries have a larger media layer than veins. Since smooth muscles are generally found in the media layer, this means that arteries have more smooth muscle to produce contraction than do the veins. Arteries have a higher amount of elastin than veins. Thus, the ratio of collagen to elastin is larger in veins than in arteries. In addition, veins have a thicker adventia layer (in proportion to the media layer) when compared with this layer in arteries.

The *media* is the middle layer of the artery, composed of smooth muscle cells, a network of elastic and collagen fibrils and elastic laminae which separate the media into a number of fiber-reinforced layers. The media consists of a highly organized three-dimensional network of elastin, vascular smooth muscle cells and collagen with extracelluar matrix protoglycans. As was found recently (Dobrin 1999), the media behaves mechanically as if its material properties were homogenous.

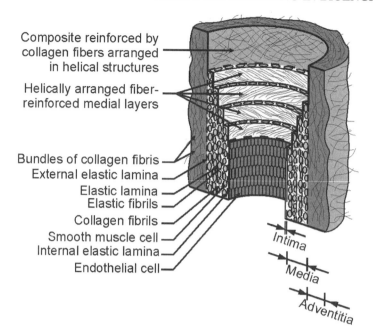

Fig. 13.1.6 The structure of the arterial wall. The major components of a healthy elastic artery consists of three layers: intima, media and adventitia (according to Holzapfel *et al.* 2000)

The *adventitia* is the outer layer composed primarily of thick bundles of collagen fibrils arranged in helical structures and fibroblast cells.

A significant amount of studies has been performed to elucidate the mechanical properties of the arterial wall. The general consensus of these studies is that the artery exhibits the following behavior: nonlinearly inelastic, anisotropic, radially heterogenous, load-rate insensitive, expansion upon cooling and shrinkage upon heating. A transient response of blood vessels was measured and dynamic mechanical properties of the vessel tissue were proposed in Rosic *et al.* (2007).

In modeling of the most arterial walls it is assumed that the wall material is homogenous and that the artery is a cylindrical membrane or a thick-walled tube. Associated material parameters are given for each layer based on a Fung-type strain–energy function (Fung, 1993), see Sections 2.4 and 11.1. Briefly, these results show that the media and adventitia are anisotropic; that the media is stiffer, and more nonlinear and both layers are stiffer in the axial than in the circumferential direction. Modeling of the arterial wall is very important also for mass transfer analysis and thrombosis processes (see Chapter 14).

13.2 Methods of modeling blood flow and blood vessels

13.2.1 Introduction

In recent years, computational techniques have been increasingly used as a support in understanding hemodynamics and mechanical behavior of blood vessels. Computer models are employed in research, but have also been used in medical practice. In research, modeling has become a necessary tool in supporting *in vitro* and *in vivo* experimental investigations, from the macro to micro and nano levels. Regarding the medical practice, use of computer models is continually being enhanced due to tremendous advances in computer technology, with the goal of establishing a direct coupling of imaging between a patient's organs and computer models, followed by the 'online' feedback of modeling results to a medical doctor (see Section 9.2). Computer simulations can be used to obtain detailed flow information including wall shear stress, pressure drops, stagnation and recirculations regions, particle residence times, turbulence; as well as the stress–strain states within the blood vessel walls, epithelial layers and RBCs.

We here summarize methods which are used in modeling blood flow and blood vessels. Basically, two types of physical conditions of blood flow can be considered as distinct: *macrocirculation* (blood flow in large vessels), and *microcirculation* (flow within capillaries). Accordingly, we distinguish two groups of computational methods for these two types of physical conditions: continuum and discrete particle methods.

Blood flow in the heart and large arteries has been investigated for decades. Dean (1928) made the first prediction of helical blood flow in a curved tube. Rotational blood flow *in vivo* was detected early by Doby and Lowman (1961), who used a radiopaque streamer technique. They presented detailed results on the velocity, pressure and wall shear stress. Lei *et al.* (1995) described the hemodynamic conditions in a model of a rabbit aorto–celiac junction and postulated a role of the wall shear stress gradient in atherogenesis. A large body of references is available in this field, for example Perktold *et al.* (1991a, 1998), Perktold & Rappitsch (1995). Recently, computational models have been generated directly from medical imaging data, most notably magnetic resonance imaging (Milner *et al.* 1998, Moore *et al.* 1998).

In many cases blood flow is affected by vessel wall and epithelial layer deformability, which depend on the tissue characteristics. A summary of material models for arterial walls is presented in Holzapfel *et al.* (2000). Some of these models, which will be used in the simulation of blood flow coupled with the biosolid boundary deformation, are already described in Chapter 11 and will be further discussed in Section 13.2.3. Under physiological conditions of the human arterial system, the equations which describe blood flow through elastic arteries are only weakly nonlinear. Two-dimensional equations for flow in straight circular elastic tubes can be linearized and the wave solution by Fourier techniques can be obtained (Womersley 1957). This linear model has become a standard model of waves in the arteries. However, this approach does not match many aspects of physiological waveforms. Also, the wall does have linear elastic properties. An alternative approach is to use the full fluid–structure interaction analysis for a particular zone of interest because there is a limit in computer power technology to simulate the entire human arterial system (Collins *et al.* 2004).

In microcirculation, the dominant effects are due to motion and deformation of red blood cells (RBCs). We described the main characteristics of microcirculation in Section 13.1.2, while computational procedures for modeling deformation of RBCs as living cells are given in Section 16.3 (Example 16.3-2).

13.2.2 Methods of blood flow modeling in large blood vessels

The basic description of blood as a complex fluid is given Section 13.1.2. As stated there, blood can be considered as an incompressible homogenous viscous fluid for flow in large blood vessels. Also, the laminar flow is dominant in the physiological flow environment. Therefore, the fundamental laws of physics, which include balance of mass and balance of linear momentum, are applicable here. These laws are expressed by continuity equation (3.3.3) and the Navier–Stokes equations (3.3.7) in Section 3.3.1.

If a *mass transport* by blood is considered, such as transport of gases (O_2, CO_2) or macromolecules (albumin, globumin), the process can be represented by diffusion with convection. Then, the fundamental law of mass balance is described by the diffusion–convection transport equation (3.3.9).

Continuum-based numerical methods, such as the finite element method (FE), described in Chapters 4 to 7, or the element-free Galerkin (EFG) method of Section 8.5, are applicable. We present the final form of these equations to emphasize some specifics related to blood flow. The incremental-iterative balance equation of a finite element for a time step 'n' and equilibrium iteration 'i' has a form (see equation (7.4.13))

$$\begin{bmatrix} \frac{1}{\Delta t}\mathbf{M} + {}^{n+1}\tilde{\mathbf{K}}_{vv}^{(i-1)} & \mathbf{K}_{vp} \\ \mathbf{K}_{vp}^T & 0 \end{bmatrix} \begin{Bmatrix} \Delta \mathbf{V}^{(i)} \\ \Delta \mathbf{P}^{(i)} \end{Bmatrix}_{blood}$$
$$= \begin{Bmatrix} {}^{n+1}\mathbf{F}_{ext}^{(i-1)} \\ 0 \end{Bmatrix} - \begin{bmatrix} \frac{1}{\Delta t}\mathbf{M} + {}^{n+1}\mathbf{K}_{vv}^{(i-1)} & \mathbf{K}_{vp} \\ \mathbf{K}_{vp}^T & 0 \end{bmatrix} \begin{Bmatrix} {}^{n+1}\mathbf{V}^{(i-1)} \\ {}^{n+1}\mathbf{P}^{(i-1)} \end{Bmatrix} + \begin{Bmatrix} \frac{1}{\Delta t}\mathbf{M}\, {}^{n}\mathbf{V} \\ 0 \end{Bmatrix}$$
(13.2.1)

where ${}^{n+1}\mathbf{V}^{(i-1)}$ and ${}^{n+1}\mathbf{P}^{(i-1)}$ are the nodal vectors of blood velocity and pressure, with the increments in time step $\Delta \mathbf{V}^{(i)}$ and $\Delta \mathbf{P}^{(i)}$ (the index 'blood' is used to emphasize that we are considering blood as the fluid); Δt is the time step size and the left upper indices 'n' and '$n+1$' denote start and end of time step; and the matrices and vectors are defined in (7.4.5)–(7.4.13). Note that the vector ${}^{n+1}\mathbf{F}_{ext}^{(i-1)}$ of external forces includes the volumetric and surface forces. In the assembling of these equations, as described in Section 4.2.2, the system of equations of the form (13.2.1) is obtained, with the volumetric external forces, and the surface forces acting only on the fluid domain boundary (the surface forces among the internal element boundaries cancel). The balance equations (13.2.1) can also be formed when the EFG method is employed, with the matrices and vectors corresponding to the 'free' points (see Section 8.5).

The specifics for the blood flow are that the matrix ${}^{n+1}\mathbf{K}^{(i-1)}$ may include *variability of the viscosity* if non-Newtonian behavior of blood is considered. According to (7.4.10) we have that

$$\left[K_{KJ}^{(i-1)} \right]_{kk} = \left[\hat{K}_{KJ}^{(i-1)} \right]_{kk} + \int_V \mu^{(i-1)} N_{K,j} N_{J,j} dV, \quad \text{sum on } j, \text{ no sum on } k \qquad (13.2.2)$$

where $\mu^{(i-1)}$ corresponds to the constitutive law for the last known conditions (at iteration '$i-1$'). In the case of use of the Cason relation (13.1.2), the second invariant of the strain rate $D_{II}^{(i-1)}$ is to be evaluated when computing $\mu^{(i-1)}$.

We note here that the penalty method can also be used, as well as the ALE formulation for large displacements of blood vessel walls (e.g. aneurism, or heart), see Sections 7.4 and 7.5.

Shear Stress Distribution

In addition to the velocity and pressure fields of the blood, the distribution of stresses within the blood can be evaluated. The stresses ${}^t\sigma_{ij}$ at time 't' follow from equations (3.3.4) and (3.3.5) of Section 3.3,

$$ {}^t\sigma_{ij} = -{}^tp\delta_{ij} + {}^t\sigma_{ij}^{\mu} \qquad (13.2.3) $$

(where we here use σ_{ij}^{μ} instead τ_{ij} since τ denotes the wall shear stress)

$$ {}^t\sigma_{ij}^{\mu} = {}^t\mu \left(v_{i,j} + v_{j,i} \right) \qquad (13.2.4) $$

is the viscous stress. Here, ${}^t\mu$ is the viscosity corresponding to the velocity vector ${}^t\mathbf{v}$ at a spatial point within the blood domain. The field of the viscous stresses is given by (13.2.4).

Further, the *wall shear stress* at the blood vessel wall is calculated as:

$$ {}^t\tau = {}^t\mu \frac{\partial^t v_t}{\partial n} \qquad (13.2.5) $$

where ${}^tv_t$ denotes the tangential velocity, and n is the normal direction at the vessel wall. Practically, we first calculate the tangential velocity at the integration points (see Section 7.4.1 for the interpolation of velocities) near the wall surface, and then numerically evaluate the velocity gradient $\partial^t v_t/\partial n$; finally, we determine the viscosity coefficient ${}^t\mu$ using the average velocity at these integration points. In essence, the wall shear stress is proportional to the shear rate γ at the wall, and the blood dynamic viscosity μ.

For a *pulsatile flow* the mean wall shear stress within a time interval T can be calculated as (Taylor *et al.* 1998)

$$ {}^T\tau_{mean} = \left| \frac{1}{T} \int_0^T {}^t\tau_n dt \right| \qquad (13.2.6) $$

Another scalar quantity is a *time-averaged magnitude* of the surface traction vector, calculated as

$$ {}^T\tau_{mag} = \frac{1}{T} \int_0^T |{}^t\mathbf{t}| dt \qquad (13.2.7) $$

where the stress vector ${}^t\mathbf{t}$ is given by the Cauchy formula (2.1.6). Also, a very important scalar in the quantification of unsteady blood flow is the *oscillatory shear index* (OSI) defined as (He & Ku 1996)

$$ {}^T OSI = \frac{1}{2} \left(1 - \frac{{}^T\tau_{mean}}{{}^T\tau_{mag}} \right) \qquad (13.2.8) $$

The shear stress action has very important effects in the circulatory system. For example, it was found that arteries adapt to long-term increases or decreases in wall shear stress. A decreased flow rate causes a thickening of the intimal layer to reestablish a normal wall shear stress. On the other hand, increased wall shear stress is a response to remodeling of arteries to large diameters.

Endothelial cells respond to shear stress. At the lumenal surface, the shear stress can be sensed directly as a force on an endothelial cell. Measurement of the velocity gradient near the wall, needed for the shear stress evaluation, is technically difficult. The velocity gradient depends highly on the shape of the velocity profile and on the measurement accuracy of distance from the wall (Zarins et al. 1983, Ku et al. 1985). Numerical calculation could be a good alternative to complex, invasive and expensive measurement techniques.

13.2.3 Modeling the deformation of blood vessels

In modeling blood flow in large blood vessels we will consider two distinct cases: (a) walls are rigid, and (b) walls are deformable. The assumption (a) will be mainly adopted in practical applications within Sections 13.3 to 13.7. Also, some of these examples will be solved assuming deformable walls, in order to get an insight into the stress–strain field within the wall tissue and to elucidate the coupling effects between the blood flow and wall deformation. In studying blood and blood vessel system, it is important to determine the stress-strain state in tissue, as well as the effects of the wall deformation on the blood flow characteristics and blood mechanical action (first of all the shear stress) on the epithelial layer.

As described in Section 13.1.3, blood vessel tissue has complex mechanical characteristics. The tissue can be modeled by using various material models, from linear elastic to nonlinear viscoelastic. We here summarize the governing finite element equations used in modeling wall tissue deformation with emphasis on implementation of nonlinear constitutive models.

The finite element equation of balance of linear momentum is derived from the fundamental differential equations (2.1.11) of balance of forces acting at an elementary material volume. In dynamic analysis we include the inertial forces in this equation according to (2.1.12). Then, by applying the principle of virtual work (see, for example, equation (4.2.20)) the differential equations of motion of a finite element are obtained as (see (5.3.7))

$$\mathbf{M}\ddot{\mathbf{U}} + \mathbf{B}^w \dot{\mathbf{U}} + \mathbf{K}\mathbf{U} = \mathbf{F}^{ext} \tag{13.2.9}$$

Here the element matrices are: $\mathbf{M}$ is mass matrix; $\mathbf{B}^w$ is the damping matrix, in case when the material has a viscous resistance; $\mathbf{K}$ is the stiffness matrix; and $\mathbf{F}^{ext}$ is the external nodal force vector which includes body and surface forces acting on the element. By the standard assembling procedure, described in Section 4.2.2, the dynamic differential equations of motion (5.2.7) are obtained. These differential equations can further be integrated in a way described in Section 5.3, with a selected time step size Δt. The nodal displacements $^{n+1}\mathbf{U}$ at end of time step are finally obtained according to equation (5.3.8):

$$\hat{\mathbf{K}}_{tissue}\,^{n+1}\mathbf{U} = {}^{n+1}\hat{\mathbf{F}} \tag{13.2.10}$$

where the tissue stiffness matrix $\hat{\mathbf{K}}_{tissue}$ and vector $^{n+1}\hat{\mathbf{F}}$ are expressed in terms of the matrices and vector in (13.2.9), according to equations (5.3.9) and (5.3.10). Note that this equation is obtained under the assumption that the problem is linear: displacements are small, the viscous resistance is constant, and the material is linear elastic.

BLOOD FLOW AND BLOOD VESSELS

In many circumstances of blood flow the wall displacements can be large, as in the case of aneurism, hence the problem becomes geometrically nonlinear, see Section 6.2.3. Also, the tissue of blood vessels has nonlinear constitutive laws, as described in Sections 2.4.2, 6.2, 11.1 and 13.1.3, leading to materially nonlinear FE formulation. Therefore, the approximations adopted to obtain equation (13.2.10) may not be appropriate. For a nonlinear problem, instead of (13.2.10) we have the incremental-iterative equation

$$^{n+1}\hat{\mathbf{K}}_{tissue}^{(i-1)} \Delta \mathbf{U}^{(i)} = {}^{n+1}\hat{\mathbf{F}}^{(i-1)} - {}^{n+1}\mathbf{F}^{\text{int}(i-1)} \quad (13.2.11)$$

where $\Delta \mathbf{U}^{(i)}$ are the nodal displacement increments for the iteration 'i', and the system matrix $^{n+1}\hat{\mathbf{K}}_{tissue}^{(i-1)}$, the force vector $^{n+1}\hat{\mathbf{F}}^{(i-1)}$ and the vector of internal forces $^{n+1}\mathbf{F}^{\text{int}(i-1)}$ correspond to the previous iteration (see Section 6.2.3 for details).

We here emphasize the material nonlinearity of blood vessels which is used in further applications. As presented in Section 6.2.3, the geometrically linear part of the stiffness matrix, $\left(^{n+1}\mathbf{K}_L\right)_{tissue}^{(i-1)}$, and nodal force vector, $^{n+1}\mathbf{F}^{\text{int}(i-1)}$, are defined in equation (6.2.21):

$$\left(^{n+1}\mathbf{K}_L\right)_{tissue}^{(i-1)} = \int_V \mathbf{B}_L^T {}^{n+1}\mathbf{C}_{tissue}^{(i-1)} \mathbf{B}_L dV, \quad \left(^{n+1}\mathbf{F}^{\text{int}}\right)^{(i-1)} = \int_V \mathbf{B}_L^T {}^{n+1}\boldsymbol{\sigma}^{(i-1)} dV \quad (13.2.12)$$

where the consistent tangent constitutive matrix $^{n+1}\mathbf{C}_{tissue}^{(i-1)}$ of tissue and the stresses at the end of time step $^{n+1}\boldsymbol{\sigma}^{(i-1)}$ depend on the material model used. Calculation of the matrix $^{n+1}\mathbf{C}_{tissue}^{(i-1)}$ and the stresses $^{n+1}\boldsymbol{\sigma}^{(i-1)}$ for the tissue material models used in further applications are given in Section 11.2. In each of the subsequent sections we will give the basic data about the models used in the analysis.

13.2.4 Blood–blood vessel interaction

In all models of blood flow where deformation of blood vessel walls is taken into account, we will implement the loose coupling approach for the fluid–structure interaction, see Section 7.6. The overall strategy adopted here consists of the following steps (see Table 7.6.1):

(a) For the current geometry of the blood vessel, determine blood flow (with use of the ALE formulation of Section 7.5 when the wall displacements are large). Wall velocities at the common blood–blood vessel surface are taken as the boundary condition for the fluid.

(b) Calculate the loads, arising from the blood, which act on the walls.

(c) Determine deformation of the walls taking the current loads from the blood.

(d) Check for the overall convergence which includes fluid and solid. If convergence is reached, go to the next time step. Otherwise go to step (a).

(e) Update blood domain geometry and velocities at the common solid-fluid boundary for the new calculation of the blood flow. In case of large wall displacements, update the FE mesh for the blood flow domain. Go to step (a).

13.3 Human aorta

Some basic data about the human aorta are introduced first and then the FE model for simulation of blood flow through a deformable and rigid aorta is described. At the end of this section we present some typical results.

13.3.1 Introduction

The aorta is large vessel originating from the heart that traverses the middle of the abdomen and bifurcates into two arteries supplying the legs with blood. The aorta arises from the left ventricle of the heart, forms an arch, and then extends down to the abdomen. Two smaller arteries, called renal arteries, which branch from aorta, have the low resistance so that two-thirds of the entering flow leaves the abdominal aorta through the arteries at the diaphragm. The aorta carries and distributes oxygen rich blood to all arteries. Most of the major arteries branch off from the aorta, with the exception of the main pulmonary artery (Vander *et al.* 1998).

The walls of the aorta consist of three layers as described in Section 13.1.3. They are the tunica adventitia, tunica media and tunica intima (Holzapfel & Weizsacker 1998). The structure of the aorta and coronary arteries is shown in Fig. 13.3.1 (see color plate). The blood is pumped from the left ventricle of heart directly to the aorta root for a short time period approximately 0.2–0.3 s. The *coronary arteries* are positioned immediately after the valves and they are the left main coronary artery, which consists of the left circumflex and the left anterior descending artery, and the right coronary artery. The *ascending aorta* goes to

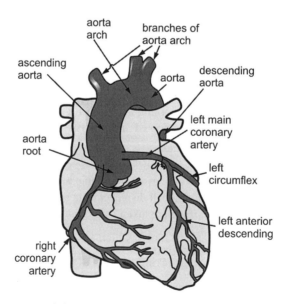

Fig. 13.3.1 The structure of the aorta and coronary arteries. The aorta consists of the aorta root, the aorta arch, the ascending and descending parts, and aortic branches. The coronary arteries exit from the aorta root and lead to a branching network of small arteries, arterioles, capillaries, venules and veins (see Plate 12)

the *aorta arch* and then to the *descending* aorta. Usually, under physiological normal human conditions, there are three aorta branches: brachiocephalic trunk, left common carotid artery, and subclavian artery.

Blood flow in the ascending aorta is complex pulsatile and three-dimensional. It also has rotational flow components at certain locations, as recorded by Segadal and Matre (1987). Motion and deformation of the aorta play an important role in the wall shear stress magnitudes, particularly at regions of stasis and flow reversals. A considerable number of numerical studies have been performed on blood flow in the aorta, initially assuming rigid walls and later with deformable walls (e.g. Liu *et al.* 2001).

In the presented solutions of blood flow within a human aorta, we consider the full fluid–structure interaction with walls deforming due to blood action (Kojić *et al.* 2003). The aorta model considered here assumes normal physiological healthy subjects.

13.3.2 Finite element model of the aorta

The finite element model of the aorta with arteries is shown in Fig. 13.3.2a. An overset grid approach is used, together with the body-fitted meshes, to generate separate meshes for the five distinct regions (objects). Blood flow domain is discretized by 3D eight-node finite elements for fluid, with velocities calculated at all nodes and pressure at the center (see Section 7.4). Blood is taken as an incompressible Newtonian fluid, appropriate for larger arteries. The blood density is $\rho = 1.05 \ g/cm^3$, and the kinematics viscosity is $\nu = 0.035 \ cm^2/s$.

The wall is modeled by the four-node shell elements (Slavković *et al.* 1994) assuming linear elastic isotropic material. A fluid–structure interaction algorithm is implemented (Filipović *et al.* 2006, Kojić *et al.* 1998; see Section 7.6). Boundary conditions for the blood are: prescribed inlet velocity profile which simulates cardiac cycle and zero surface traction at the outlet of the model. The flow rate time function for the input aorta profile is shown in Fig. 13.3.2b.

The geometric data used here correspond to a human aorta and are as follows: the inner aorta diameter is 2.4 *cm* and the arterial wall thickness is 0.2 *cm*; the inner diameters of

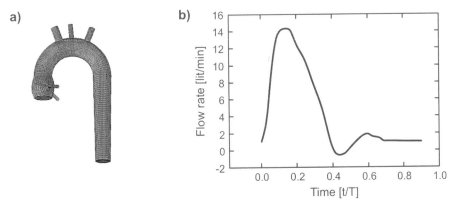

Fig. 13.3.2 FE model of the aorta with main artery branches. (a) Finite element mesh; (b) Input flow rate vs. time (pulsatile flow), T is the heart cycle

the brachiocephalic trunk, left common carotid artery, subclavian artery and left and right coronary artery, are (in *cm*) 1, 0.6, 0.8, 0.4, respectively.

Besides this example we provide on the web (see Software) a simplified model with rigid walls and without artery branches for which the finite element model can be generated parametrically.

13.3.3 Results and discussion

The contour slice of velocity magnitude and shear stress are shown in Figs. 13.3.3 (see color plate) and 13.3.4 (see color plate) for early systolic flow $t = 0.05$ s, in the case of deformable and rigid walls.

As expected, the region along the outer wall of the aorta arch distal from the branches has low velocities relative to those in the region in the ascending aorta and other parts of

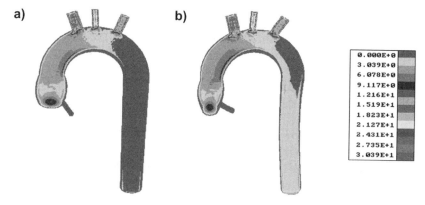

Fig. 13.3.3 The velocity magnitude field in the human aorta for early systolic flow $t = 0.05$ s. (a) Rigid walls; (b) Deformable walls (note that diameters are smaller with respect to rigid aorta shown in (a)) (see Plate 13)

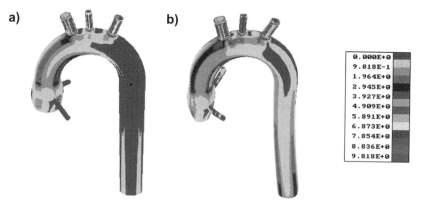

Fig. 13.3.4 The wall shear stress in the human aorta for early systolic flow $t = 0.05$ s; (a) Rigid walls; (b) Deformable walls (see Plate 14)

the aorta (the ascending aorta arch and the descending aorta). Also, the velocities are larger when deformation of the aorta wall is taken into account due to wall elasticity (see Fig.13.3 – color plate, where diameters of deformable aorta are smaller with respect to the rigid aorta due to inertial effect from the previous cycle).

The results for shear stress in Fig. 13.3.4 (see color plate) show generally lower shear stresses for deformable walls in comparison to rigid walls. This is probably due to more continuous changes of the velocities near the walls (smaller velocity gradient) under pulsatility conditions when the walls are considered deformable then rigid. The maximal shear stresses for the early systolic flow are observed in the aortic branches.

Comparing the solutions obtained using rigid and deformable walls, it is found that there is a difference of less than 2% in velocities, and about 22% in the wall shear stresses.

13.4 Abdominal aortic aneurysm (AAA)

An abdominal aortic aneurysm (AAA) is introduced in this section. Several types of AAA are modeled using rigid and deformable walls. Velocity and shear stress distribution inside the AAA are presented and discussed.

13.4.1 Introduction

In general, an aneurysm is a localized abnormal berry-like or gradual dilatation of any vessel, usually at or near a branch, which is caused by a localized damage or weakness of the vessel wall (Fig. 13.4.1 – see color plate). An abdominal aortic aneurysm (AAA) is an enlargement in the lining of the abdominal aorta which is the largest blood vessel in the body. A consensus definition of an aneurysm was defined in 1991 by the Society of Vascular Surgery and the International Society for Cardiovascular Surgery as: a permanent localized dilatation of an artery having at least 50% increase in diameter compared with

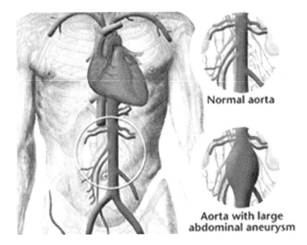

Fig. 13.4.1 Normal physiological abdominal aorta and aorta with large abdominal aneurysm (http://www.lifelinescreening.com/Disease/AAA/Pages/Index.aspx) (see Plate 15)

Table 13.4.1 Characteristics of true aneurysms

	Shape of dilatation	Diameter	Parts of vessel affected	Features
Berry-like	Small spherical	1.0–1.5 cm	Sidelong	Occurring in the brain arteries
Saccular	Spherical	5.0–20.0 cm	Sidelong	Often filled by thrombus
Fusiform	Gradual & progressive	Up to 20 cm	Circumferential	Frequently occurring in all parts of aorta
Dissecting	No dilatation	—	Blood dissecting between the layers of the vessel wall	Blood-filled channel within aortic wall

the expected normal diameter of the artery, or of the diameter of the segment proximal to the dilatation (Fukushima et al. 1989). It is considered that an aorta with a diameter of 3 cm and more is defined as aneurysmal, since the maximal diameter of a normal aorta is about 2 cm.

Aneurysms are classified as true and false. In a true aneurism, the blood remains in the circulatory system, while in a false aneurysm an expanding hematoma emerges from a hole on the artery wall. Hence, in the case of a false aneurysm the aorta wall is ruptured, not expanded. The true aneurysms are classified according to their shape: berry-like, saccular, fusiform and dissecting aneurysms. Characteristics of these aneurism shapes are given in Table 13.4.1 (Thompson & Bell, 2000).

The main causes of aneurysm are arteriosclerosis and cystic medial degeneration, but also genetic disorder, malfunction of the aorta, mycotic infections or arthritis (Cotran et al. 1994). Also, the cause of aneurysmal disorders is a loss of vessel distensibility, i.e. an increase of stiffness of the vessel wall. This stiffness change is due to the loss of elastin and increase in collagen content in the aortic wall, which in general occurs as a result of aging. Approximately 5% of men above the age of 60 develop an aneurysm. There is a risk that the aorta may rupture at the aneurism (Thompson & Bell 2000). The danger of rupture is directly related to the aneurysm size. The risk of rupture for a small AAA (with a diameter less than 4 cm) is about 2%, while an aneurysm larger than 5 cm has a risk of rupture of 5% to 10% per year (Cotran et al. 1994). Aneurysm has to be repaired because the death rate from ruptured AAA is almost 90%. About 80% of patients die before reaching the hospital and about 50% die during the rupture-repair surgery (Vliet & Boll 1997). Among those who died due to an aneurysm rupture are Albert Einstein and Lucille Ball.

We model here a few typical AAA with symmetric position related to the central axis. Deformable wall is modeled by applying fluid–structure interaction algorithm (see Section 7.6). Computer-generated simulations, and patient-specific computer simulations in particular, promise to become a very powerful tool in the clinical treatment of the AAA.

13.4.2 Modeling of blood flow within the AAA

The shape of AAA is very important. The severity of AAA is commonly estimated in clinical practice by considering the AAA maximal diameter. However, from the mechanical point of view, the hemodynamic effects and the mechanical stresses within the AAA tissue certinly are important in the process of the AAA rupture. Bulge diameter alone may not be a sufficient criterion for determination of rupure risk, therefore an insight into the hemodynamic effects and the stress–strain quantification and distribution within the vessel wall are of great significance even in medical practice.

Generation of the Finite Element Model
A simplified geometry of an aneurism is shown in Fig. 13.4.2. With the web software (Software – Section 13.4) a 3D finite element model for the blood flow domain can be parametrically generated. A transition smoothness between the surfaces is achieved by using Bezier's curves. Also, the results can be displayed with a user-friendly menu in a way suitable for an insight into medical aspects of the blood flow conditions. A detailed description of the software use is given on the web (within Tutorial of each example), while the description of geometric parameters is given in the caption of Fig. 13.4.2.

The theoretical background for the blood flow is given in Sections 3.3, 7.4 and 13.2.2.

Boundary Conditions
At the inflow aorta cross-section a fully developed parabolic flow is assumed, determined by a selected volume flux. The normal stress and tangential stress are set to be equal to zero (stress-free condition), or they are prescribed, at the outlet cross-section.

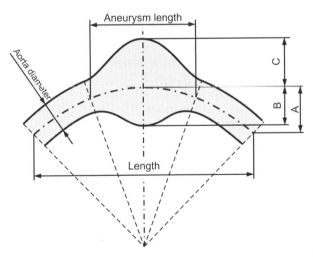

Fig. 13.4.2 Geometrical parameters of AAA: 'Length' is the parameter which defines the total horizontal projection of the generated aneurysm model; 'A' is the height of the arc of central line; 'Aorta diameter' is the abdominal aorta diameter; 'B' is the radius from the central line to the inner wall of the aneurysm; 'C' is the radius from the central line to the outer wall of the aneurysm; 'Aneurysm length' is an average length of the AAA

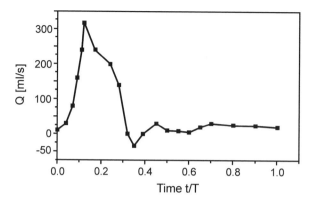

Fig. 13.4.3 A typical in-flow waveform at the aorta entry. Q is the volumetric in-flux and t/T is the relative time with respect to the cycle period T

It is assumed that the entering flow is pulsatile, with a typical waveform shown in Fig.13.4.3 (Ku 1997). As described in the software menu, the waveform can be changed.

13.4.3 Results

Results for two examples of the symmetric AAA are given here: (a) case with rigid walls, and (b) AAA with deformable walls. Results not shown here and solutions for other model parameters can be obtained using Software on the web.

Modeling of AAA Assuming Rigid Walls

We analyze an aneurism at the straight aorta domain, where aorta proximal and distal to the AAA bulge are idealized as straight rigid tube and branching arteries are excluded. The model has ratio $D/d = 2.75$ (Peattie et al. 2004) and geometry generated according to Fig. 13.4.2 (D and d are diameters of the bulge and aorta, respectively). The data are: blood density is $\rho = 1.05 \ g/cm^3$; kinematic viscosity (Newtonian fluid) $\nu = 0.035 \ cm^2/s$, $d = 12.7 \ mm$. The inflow velocity is defined by the flux function given in Fig. 13.4.3. The FE mesh consisted of approximately 8000 3D eight-node brick elements.

The results for the velocity and pressure at peak systole $t/T = 0.16$ are shown in Fig. 13.4.4 (see color plate). The velocity disturbance in the region of the aneurism is notable. Also, the region of maximum pressure is located inside AAA.

Modeling AAA with Deformable Walls

Here, an aneurysm of the straight aorta with deformable walls is modeled according to the fluid–structure interaction algorithm (Section 7.6). Blood flow is calculated using 2112 eight-node 3D elements, and 264 four-node shell elements used to model the aorta wall, with the wall thickness $\delta = 0.2 \ cm$. The material constants for blood are as in the previous example, while data for the vessel wall are: Young's modulus $E = 2.7 \ MPa$, Poisson's ratio

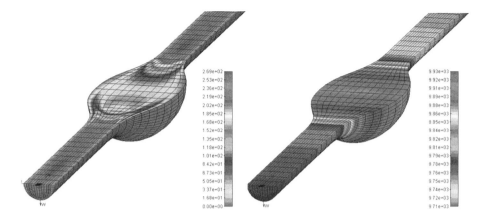

Fig. 13.4.4 Velocity field (left panel) and pressure distribution (right panel) for peak systole $t/T = 0.16$ of AAA for the model with $D/d = 2/75$, $d = 12.7$ mm (Peattie et al. 2004) (see Plate 16)

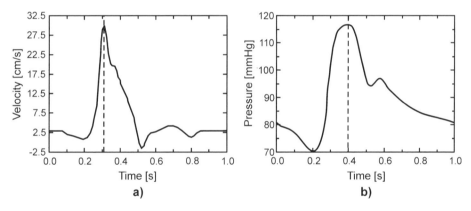

Fig. 13.4.5 Input velocity and output pressure profiles for the AAA on a straight vessel. Inlet peak systolic flow is at $t = 0.305$ s and outlet peak systolic pressure is at $t = 0.4$ s. (a) Velocity waveform; (b) Pressure waveform (Scotti et al. 2005)

$v = 0.45$, wall thickness $\delta = 0.2$ cm, tissue density $\rho = 1.1$ g/cm^3. Boundary conditions for the model are prescribed velocity profile (see Fig. 13.4.5a) and output pressure profile as given in Fig. 13.4.5b.

The results for velocity magnitude distribution at $t = 0.305$ s are shown in Fig. 13.4.6a (see color plate). The von Mises wall stress (see web – Theory, Chapter 2) distributions at $t = 0.4$ s is given in Fig. 13.4.6b (see color plate). It can be seen that the velocities are low in the domain of the aneurism, while the larger values of the wall stress are at the proximal and distal aneurism zones.

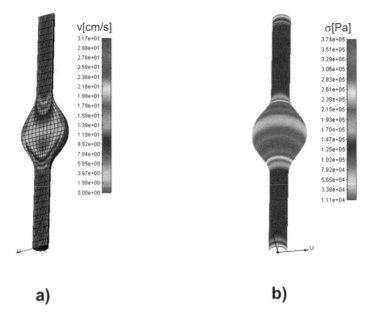

Fig. 13.4.6 Velocity magnitude field and von Mises wall stress distribution for symmetric AAA on a straight vessel. (a) Velocity field distribution for peak at $t = 0.305$ s; (b) von Mises wall stress distributions for blood pressure peak at $t = 0.4$ s (see Plate 17)

13.5 Blood flow through the carotid artery bifurcation

The topic addressed in this section is modeling of the carotid artery bifurcation which includes: (a) model generation: definition of geometry, automatic mesh generation, boundary and initial conditions, calculation of three-dimensional blood flow and deformable wall; (b) display of results for velocity, pressure and shear stress distribution for fluid domain, and von Mises stress distribution within the vessel wall.

13.5.1 Introduction

The common carotid arteries are two of several arteries that supply blood to the head. The right common carotid artery branches from the brachiocephalic artery and extends up the right side of the neck. The left common carotid artery branches from the aorta and extends up to the left side of the neck (Fig. 13.5.1a – see color plate). Each carotid artery branches into internal and external vessels near the top of the thyroid. A typical carotid artery bifurcation is shown in Fig. 13.5.1b (see color plate). The blood flows from the common carotid artery (CCA) into the internal carotid artery (ICA) and external carotid artery (ECA).

After heart disease and cancer, the third most common cause of death is stroke. Probably the most frequent stroke is of the embolic type with a heart disease as the source. The carotid bifurcation stenosis is also a significant cause of stroke, producing the infarction in the carotid region by embolization or thrombosis at the site of narrowing (Strandness & Eikelboom 1998). The thrombosis development and embolization is conditioned by the local hemodynamics which can be investigated experimentally and/or by computer modeling.

BLOOD FLOW AND BLOOD VESSELS

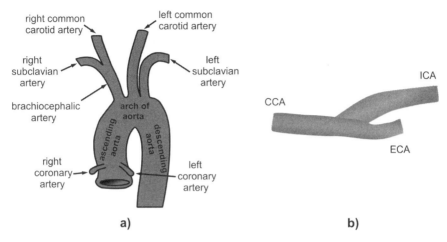

Fig. 13.5.1 Carotid artery bifurcation. (a) Position of carotid arteries in the arterial system; (b) Typical carotid artery bifurcation. CCA – common carotid artery, ICA – internal carotid artery, ECA – external carotid artery (see Plate 18)

The carotid artery stenosis has a number of risk factors in common with other atherosclerotic diseases. In general, increase of stroke risk is induced by many factors: age, systolic and diastolic hypertension, diabetes, cigarette smoking, etc. Changes of the geometrical vessel dimensions in the region of the carotid artery bifurcation certainly affect the blood flow and may lead to stenosis process. It has been shown that the vessel diameter at the carotid artery bifurcation changes considerably with age.

There is a number of references on experimental investigations of flow in the carotid artery, e.g. Ku *et al.* (1985), Perktold *et al.* (1991b), Zarins *et al.* (1983). We will use some of the results from these references to compare with our FE solutions.

13.5.2 Finite element model of the carotid artery bifurcation

Here we describe the basic concept of generation of the carotid artery bifurcation finite element (FE) models, with appropriate boundary conditions. A 3D finite element model with 3D fluid finite elements (eight-node isoparametric elements with velocity calculation at all nodes and pressure calculated at the element level, see Section 7.4) is generated for the carotid artery geometry as described below (see also web, Software). The wall is modeled with four-node shell elements (Section 4.5). The post-processing of results gives an insight into the local hemodynamics, as well as the blood mechanical action on the vessel walls, such as distributions of pressure and shear stress on the wall surfaces.

Model Geometry and FE Mesh for the Blood Flow Domain and Wall

A simplified carotid artery bifurcation geometry is shown in Fig. 13.5.2. The geometric parameters are used for the generation of the blood vessel internal surfaces, which are the boundaries for the blood flow domain. It is assumed that the cross-sections of all arteries are circular, with a transitional region at the branching. With the use of these geometric parameters, a 3D finite element model for the blood flow domain is generated; such an FE model is shown in Fig. 13.5.3a. It is assumed that the bifurcation

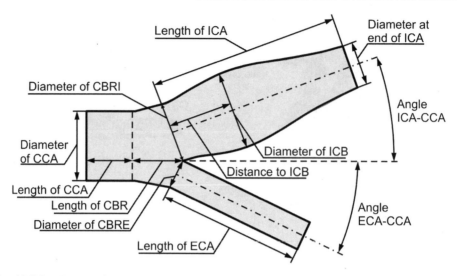

Fig. 13.5.2 Geometrical data for the carotid artery model. CCA – common carotid artery, CBR – carotid bifurcation region, CBRE – carotid bifurcation region external, ECA – external carotid artery, CBRI – carotid bifurcation region internal, ICA – internal carotid artery, ICB – internal carotid bulbus

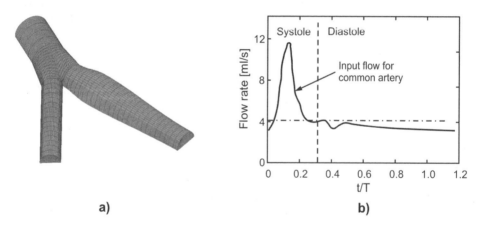

Fig. 13.5.3 An FE model of the carotid artery bifurcation. (a) FE mesh generated using the parameters shown in Fig. 13.5.2. The blood flow domain is modeled by 3D fluid finite elements; (b) Flow rate of the blood entering CCA in terms of the relative time t/T, where T is the cycle period (see web – Software, Section 13.5 Carotid Artery Bifurcation)

has the symmetry plane (the plane shown in Fig. 13.5.2), hence the FE model is generated for the half of the entire domain (part in front of the symmetry plane). A detailed description of the FE model generation, including material data for blood and range for all FE model parameters, is given on the web – Software (Section 13.5 Carotid Artery Bifurcation).

BLOOD FLOW AND BLOOD VESSELS 273

Boundary Conditions
At the inflow of CCA a fully developed flow is assumed. The inflow is pulsatile in character and is defined by volumetric blood flow rate Q (mL/s) at the common artery entering cross-section. The flow rate is specified by a waveform as shown in Fig. 13.5.3b (Perktold *et al.* 1991b). The waveform can be specified using the Software on the web.

At the fixed walls all velocity components are set to be zero. Also, the velocity components at the plane of symmetry in the direction normal to the plane are set to zero. It is assumed that the end cross-sections of the internal and external carotid arteries are stress-free, i.e. normal and tangential stresses are set to zero. This boundary condition does not represent the real flow conditions at the outflow boundaries, but, by varying the artery lengths, the physiological resistance to the blood flow in the region close to the bifurcation can be matched.

FE Model for Artery Wall
In the case of modeling deformation of the vessel wall, an FE mesh of four-node shell elements is generated to be compatible with the fluid 3D finite element surfaces bounding the blood flow domain. Details about this modeling are given in the menu of the Software on the web.

13.5.3 Example solutions

We first analyze blood flow through the carotid artery bifurcation with rigid walls and then with deformable walls. The first example, the case with the rigid walls, is also available on the web (see Software – Section 13.5 Carotid Artery Bifurcation).

Carotid Artery Bifurcation with Rigid Walls
The FE model is generated using the Software (see Fig. 13.5.2). Data are as follows: blood density is $\rho = 1.05$ g/cm^3; kinematic viscosity is $\nu = 0.035$ cm^2/s; (geometrical data – lengths in mm) diameter of the CCA = 6.2, length of the CCA = 7.44, diameter of the ICA = 6.5, length of the ICA = 26.04, diameter of the ECA = 3.658, length of the ECA = 18.6, diameter to the ICB = 6.5, distance to the ICB = 5.39, diameter at end of ICA = 4.34; angle between ICA and CCA = 25°, angle between ECA and CCA = 25°. The normal physiological pulsatile waveform form given in Fig. 13.5.3b is used for the input velocity profile.

The results for the velocity field at diastolic flow are shown in Fig. 13.5.4 (see color plate). A stagnation zone of flow is observed at the carotid bulbus distal to the bifurcation region along the internal carotid artery. The atherosclerosis appears in this region in more then 90% cases (Strandness & Eikelboom 1998), therefore determination of the stagnation regions within the flow field can be of use in clinical practice.

Carotid Artery Bifurcation with Deformable Walls
Here, we model blood flow within the bifurcation, including deformation of the vessel walls. The walls are modeled by the four-node shell finite elements (Section 4.5). Typical geometrical and material data, and flow conditions are used (Perktold & Rappitsch 1995). Geometrical data, with lengths in (mm) (according to Fig. 13.5.2), are: diameter of CCA = 6.0, length of the CCA = 7.0, diameter of ICA = 6.5, length of ICA = 26,

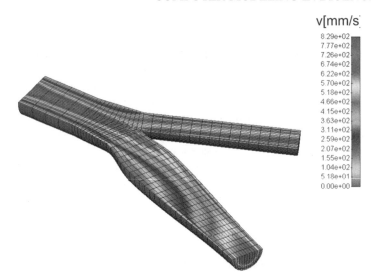

Fig. 13.5.4 3D field of velocity magnitude at the maximum systolic flow (at relative time $t/T = 0.11$ within the period T); see also web – Software for solutions for the entire period of the cycle (see Plate 19)

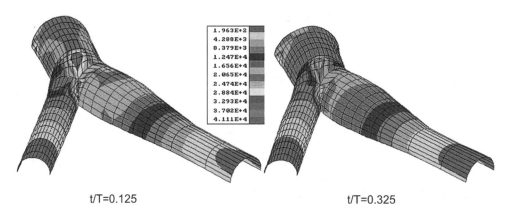

t/T=0.125   t/T=0.325

Fig. 13.5.5 Distribution of von Mises stress (Pa) within the artery walls of the carotid artery bifurcation due to action of blood, for systolic deceleration flow $(t/T = 0.125)$ and diastolic minimum flow $(t/T = 0.325)$ (see Plate 20)

diameter of ECA = 3.6, length of the ECA = 18.0, diameter to ICB = 6.5, distance to ICB = 5.3, diameter at end of ICA = 4.0; angle between ICA and CCA = 25°, angle between ECA and CCA = 25°. Material data for the blood are: density $\rho = 1.05 \ g/cm^3$ and dynamic viscosity $\mu = 0.0365 \, P$. Material data for the vessel walls are: Young's modulus $E = 0.361 \, MPa$, Poisson's coefficient $\nu = 0.45$, tissue density $\rho = 1.1 \ g/cm^3$. Wall thicknesses (in cm) of: CCA = 0.031, distal part (inner carotid wall) = 0.022, outer carotid = 0.02.

BLOOD FLOW AND BLOOD VESSELS

We here present some of the results, which are commonly given in the literature (e.g. Perktold *et al.* 1991b), and for certain pulse phase angles during the cycle. These phase angles (relative times) are: systolic peak flow, $t/T = 0.11$; systolic deceleration flow, $t/T = 0.125$; diastolic minimum flow, $t/T = 0.325$; diastolic flow, $t/T = 0.775$ (see Figs. 13.5.4–13.5.6).

The von Mises (effective) wall stress distribution at two relative times is shown in Fig. 13.5.5 – see color plate (Filipović 1999, Filipović & Kojić 2004, Filipović *et al.* 2006a, Kojić *et al.* 1998). It can be seen that the maximum stresses occur in the region of the bifurcation and are larger during systole.

The shear stress fields are shown in Fig. 13.5.6 (see color plate). As can be seen from the figure, the shear stresses are much higher at carotid branch than in the common, external and internal carotid artery. Also, shear stresses are much larger during systole than during diastole.

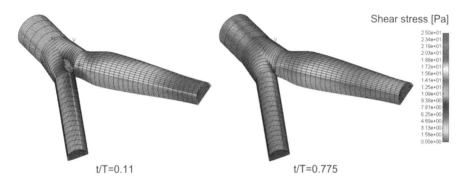

Fig. 13.5.6 Wall shear stress field at two relative times (left panel – systole; right panel – diastole). Systolic shear stresses are much larger than diastolic (see Plate 21)

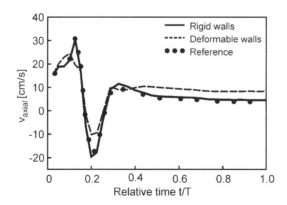

Fig. 13.5.7 Axial velocity component at the outer sinus wall position during a cycle period, with rigid and deformable walls (Filipović 1999, Perktold *et al.* 1991b, Perktold & Rappitsch 1995 – denoted as 'Reference' in the figure)

Change of the axial velocity component at the outer sinus wall position within the cycle is shown in Fig. 13.5.7. The solutions are obtained assuming rigid and deformable walls. It can be seen that wall deformation has an influence on velocity, but differences are form around 5% to 10%.

13.6 Femoral artery with stent

In this section the human femoral artery is analyzed. The basic anatomy and function of this artery is presented first, followed by the FE model description and results.

13.6.1 Femoral artery anatomical and physiological considerations and endovascular solutions

The purpose of this section is not to detail through the thorough network of the anatomical relationships, but to provide a short overview of the femoral artery anatomy with regard to important facts for its biomechanical behavior.

The common femoral artery (Fig. 13.6.1), after its pass beneath the inguinal ligament, courses through the so-called Scarpa's triangle (triangular space with muscles at the boundaries). A couple of centimeters below the inguinal ligament, the common femoral artery gives off its deep branch, responsible for almost all collateral supply, then continuing as

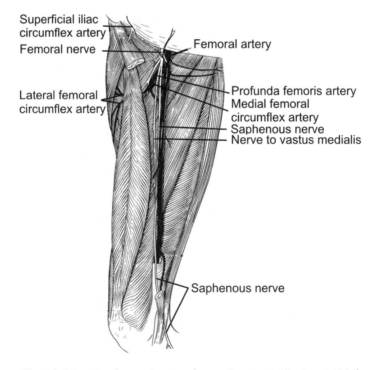

Fig. 13.6.1 The femoral artery (according to Hollinshead 1982)

a superficial femoral artery enters the aponeurotic canal (known as the adductor canal or Hunter's canal) at the caudal apex of the Scarpa's triangle. This canal is bounded by muscles at anterolateral and posterior surfaces, and by stiff aponeurosis between muscles at the anteromedial surface.

Symptomatic disease of the femoral, precisely the superficial femoral artery (SFA), usually arises at the so-called typical location, adductor canal. This part of the artery trajectory is important for several reasons. Passing through the tendon of the adductor magnus muscle, the artery is fixed in that region and is less viable than in its other segments. Multiple repetitive microtrauma to the vessel wall fixed to the canal could be stressed as a major potential stimulus for the atherosclerotic process. The artery is entrapped within the canal by the tendon itself. Therefore, the artery's ability to undergo the process of so-called positive remodeling, dilating and enlargement as a response to the plaque formation, is limited comparing to the rest of it (Blair 1990).

The SFA is a frequent target of atherosclerotic disease predominantly in the proximal section near the bifurcation into the deep femoral artery and in the distal section where the adductor muscles tend to compress the artery. In the past, the SFA revascularization was the domain of vascular surgery (femoropopliteal and femorodistal bypasses). However, with the development of endovascular treatment and advanced techniques as well as more sophisticated metallic endoprosthesis (stent) material, endovascular treatment is nowadays not just a treatment option but in most cases is preferable, at least as an initial revascularization procedure in the treatment of peripheral artery vascular disease.

Since the SFA is a unique vessel in terms of its anatomy, function and interventional requirements, it has no comparison with any other arterial vascular bed. It is a long conduction vessel with a high flow resistance underlying several different hemodynamic conditions.

The well-known factors influencing difficulties in any kind of treatment of femoropopliteal arterial segment include different external and internal forces, proximity of major flexing points, large muscle masses surrounding, lack of high diastolic flow, etc. Having relatively recent phylogenetic origin, there are also no collateral vessels along the entire course of the artery. From the very beginning of the endovascular approach to this territory, since Charles Dotter's pioneering angioplasty procedures (Dotter & Judkins 1964), there have been many controversies about the issue of their true clinical benefit. Shortly after the initial enthusiasm of every new device or procedure improvement, there followed discouraging facts that favored the opponents of endovascular therapies in this region. Even more extensive use of balloon expandable or self-expandable stainless steel stents did not approve itself as a method much better than angioplasty alone. Only recently, have clinical trials using nitinol self-expandable stents in femoropopliteal region showed a dramatic improvement in the mid- and long-term results (a description of the nitinol material is given on the web – Theory, Chapter 13).

However, stent fracture has emerged as a new problem in the percutaneous transluminal angioplasty of the superficial femoral artery. Walking more than 5000 steps per day was the strongest independent determinant variable associated with stent fracture by discriminant analysis ($p = 0.0027$, Osamu et al. 2006). Vigorous exercise adversely affected stent fracture in the group of 40 patients implanted with a nitinol stent in the SFA. Yet, the incidence of stent fracture is not equal for all nitinol stents currently in use for human superficial femoral arteries. This raises the question of possible causes and induces a search for explanation in biomechanical properties and different stent design. Computer modeling can be of help to gain insight into the mechanical working loads carried by the stent.

278   COMPUTER MODELING IN BIOENGINEERING

13.6.2 Analysis of the combined effects of the surrounding muscle tissue and inner blood pressure to the arterial wall with implanted stent

In an anatomical sense, the geometrical relationship between femoral artery and surrounding muscle tissue is very complex. In its spreading throughout the muscle structure of the thigh, the femoral artery is surrounded by adductor muscles. While walking, these muscles contract, inducing the pressure effect on the wall of the femoral artery. The goal of the analysis presented here is to reconstruct the appropriate geometrical and anatomical relationships of the femoral artery and muscle tissue in adductor canal, and then to reveal their interaction.

FE Model

In order to perform computer modeling of the combined effects of the surrounding muscle tissue and inner forces of blood against the arterial wall with implanted stent, a simplified finite element model (in the geometrical sense) is generated (Ranković 2007). We only consider part of the artery surrounded by the stent and segments within the region of 10 *mm* above and 10 *mm* below the stent. It is approximated that the arterial diameter and wall thickness are constant in the entire model and are taken to be 8 *mm* and 0.5 *mm*, respectively (Schmidt *et al.* 2000).

The FE model consists of the solid domain and the fluid domain (Fig. 13.6.2). The solid domain consists of stent, arterial wall and muscle bundle surrounding the arterial wall. The stent is modeled using 3D eight-node finite elements, arterial wall by using four-node shell elements, while muscle bundles are modeled using 3D eight-node elements with muscle fibers (see Sections 4.3, 4.5, 12.1 and 12.2 for the description of these elements). Fluid is modeled using 3D eight-node finite elements (Section 7.4). The thickness of the muscle tissue

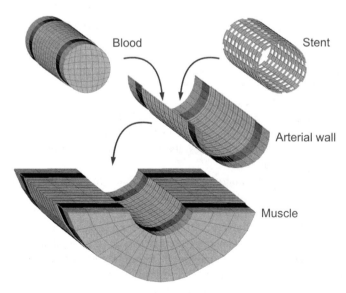

Fig. 13.6.2 Finite element model of femoral artery with stent, including the muscle surrounding (Ranković 2007)

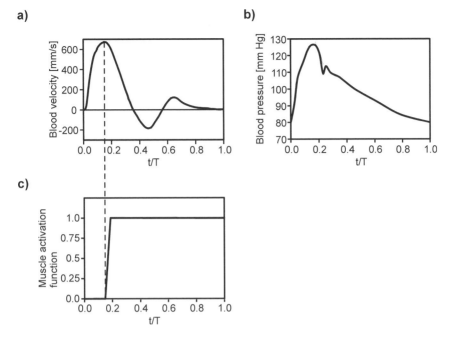

Fig. 13.6.3 Boundary conditions for blood flow and muscle activation within a heart cycle period T used in the femoral artery FE model. (a) Mean velocity at the inlet cross-section (according to Woodcock *et al.* 1975); (b) Blood pressure at the outlet (according to Patel *et al.* 1965); (c) Muscle activation function

is adopted to be 1 *cm*, with the muscle fibers in the circumferential direction. The results are presented for the extreme loading conditions of the artery wall arising from blood and muscle mechanical actions. They correspond to the moment of peak systole and maximum muscle activation.

The process of numerical analysis consists of two phases. In the first phase blood flow is computed until the systole peak is reached (maximal pressure and velocity values), imposing the boundary conditions shown in Figs. 13.6.3a,b. During this phase muscle tissue is assumed to be relaxed. In the second phase, starting immediately after the peak systole, activation of muscles with muscle contraction is assumed. The activation occurs in a short time period and muscles remain activated as shown in Fig. 13.6.3c. Therefore, the extreme mechanical loadings of the artery wall arising form the blood and muscle actions are modeled.

Boundary conditions for the solid surrounding the artery are as follows. It is assumed that the first and last cross-sections do not move axially, hence all FE element nodes in these cross-sections are axially restrained.

Material Characteristics

It is assumed that the wall material is orthotropic nonlinear elastic, and the Fung material model is adopted (Fung 1979). The strain energy function is defined in (11.2.32). The material parameters c, a_1, a_2, a_4 are determined using a data-fitting procedure, with data obtained experimentally (the experimental investigation of the arterial wall, Ranković 2007). Material parameters obtained from the fitting procedure are:

$$c = 0.7565 \; MPa, \; a_1 = 0.166, \; a_2 = 0.084, \; a_4 = 0.045 \tag{13.6.1}$$

For the stent material, the alloy of nitinol is adopted (for the definition of this material see web, Theory – Chapter 13). Material parameters characterizing this alloy are (Auricchio & Taylor 1997):

$$E = 60000 \; MPa \quad \nu = 0.3$$
$$\sigma_s^{AS} = 520 \quad \sigma_f^{AS} = 750 \quad \sigma_s^{SA} = 550 \quad \sigma_f^{SA} = 200 \tag{13.6.2}$$
$$\beta^{AS} = 250 \quad \beta^{SA} = 20 \quad \varepsilon_L = 7.5\% \quad C = 0$$

where all σ and β parameters are in (MPa). Material parameters of blood are: density $\rho = 1.05 \times 10^{-3} \; g/mm^3$ and dynamic viscosity $\mu = 3.675 \times 10^{-3} Pa \; s$.

Results

According to the boundary conditions and loads mentioned above, the numerical analysis of the material behavior of this complex model is performed. The extreme loading conditions are taken to be as described above, which correspond to the peak of systole and the muscle contraction activation according to diagram in Fig. 13.6.3. Transversal muscle contraction induces loading on the compound consisting of arterial wall and stent.

Figures 13.6.4a and 13.6.4b show the numerical solution for the hoop and longitudinal (axial) stresses in the arterial wall at a time corresponding to the maximal muscle contraction. The shell FE model of the stent and field of the effective stress within stent (see web, Theory – Chapter 2 for definition of the effective stress) at this time are shown in Fig. 13.6.5. In Fig. 13.6.6 we present the diagram of axial stress distribution in the arterial wall along the longitudinal line AB shown in Fig. 13.6.4b.

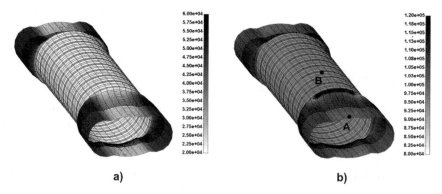

Fig. 13.6.4 Fields of hoop and axial stresses within arterial wall. (a) Hoop stress; (b) Axial stress (Pa)

BLOOD FLOW AND BLOOD VESSELS 281

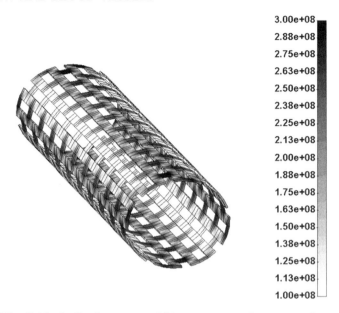

Fig. 13.6.5 The field of effective stress within stent at maximum muscle contraction (note that these stresses are much larger than within tissue)

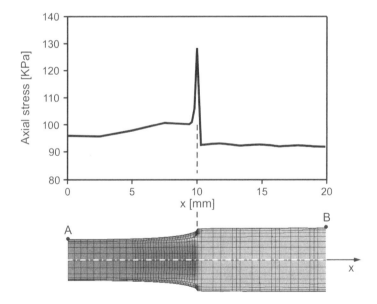

Fig. 13.6.6 Axial stress distribution in the arterial wall along the longitudinal line AB (see Fig. 13.6.4b) at maximum muscle contraction

It can be seen from the solutions shown above, that due to the pressure action of the surrounding muscle tissue, significant deformation occurs in the arterial wall regions localized distally with respect to stent. Because of large difference in stiffness between the arterial wall regions with and without stent, the concentration of axial stresses and deformations occurs in the zone around the stent boundary. In this zone the axial stress reaches the value between 120 and 130 *kPa*, which is around 30 *kPa* larger than within the rest of arterial wall. These large stress concentrations indicate a potential danger of the arterial wall endothelium injury in the zone of stent boundary.

Since an arterial wall endothelium injury leads to the endothelium prothrombotic behavior, the above result about the stress concentration suggests a possible cause of the repeated atherosclerotic changes.

13.7 Blood flow in venous system

In this section we present the basic information about the venous blood flow and introduce an FE model for flow in the straight lower leg vein without and with bandaging. Modeling of blood flow through the venous system in the leg could help the design compression therapy.

13.7.1 Introduction

In the human, the return of venous blood from the lower limbs to the heart requires the assistance of a pump structure equipped with nonreturn valves, since the force generated by the heart alone cannot overcome gravity to drive the blood from the toes to the brain. The pumping action in the deep veins is provided by the muscles. Muscle contractions, acting within the strict confines of the encircling deep facia, squeeze the blood, at high pressure, towards the heart. On the way to the heart, the blood also travels through the long and the short saphenous veins in the superficial system embedded in the fatty tissue surrounding the muscles and therefore at much lower pressure (Fig. 13.7.1) (Benbow *et al.* 1999). Provided

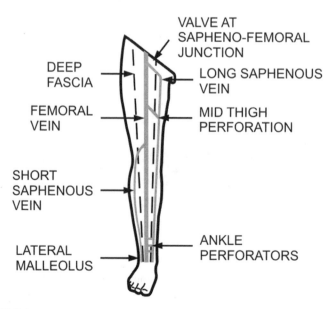

Fig. 13.7.1 Venous system in the leg (according to Benbow *et al.* 1999)

BLOOD FLOW AND BLOOD VESSELS

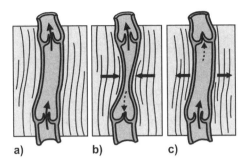

Fig. 13.7.2 Blood flow through the vein with the muscle interaction. (a) Resting condition; (b) Muscle contraction; (c) Muscle relaxation

the venous valves are working efficiently, and the muscle pump performs effectively, the return of the blood to the heart is assured (Fig. 13.7.2).

If the valves in the large veins become incompetent due to primary degeneration or post-thrombotic damage, blood will oscillate up and down in those segments lacking functional valves. The resulting retrograde flow in the veins of the lower leg leads to a reduced fall in venous pressure during walking (ambulatory venous hypertension). This causes fluid loss into the tissues and the formation of oedema. Compression of veins with incompetent valves produces an increase in orthograde (towards the heart) flow and a reduction in venous reflux.

13.7.2 Modeling blood flow through the veins

In order to help patients with problems in valves in large veins, the bilateral banding of the thighs and lower legs is implemented. This has the effect of reducing local blood volume, by redistributing blood towards central parts of the body (Nelzen *et al.* 1991), see Fig. 13.7.3.

We first analyze a simple example of an axisymmetric vein with the rigid walls and then a vein with compression therapy. The computational procedures include FE models of blood flow (Section 7.4) and solid–fluid interaction (Section 7.6).

Blood Flow in a Vein with Rigid Walls
Geometry of the straight axisymmetric vein and data are shown in Fig. 13.7.4a. The diameter D, length L, position and geometry of the valves define the basic geometry of the blood flow

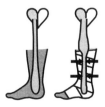

Fig. 13.7.3 Compression of the leg increases the blood flow in the preload of the heart

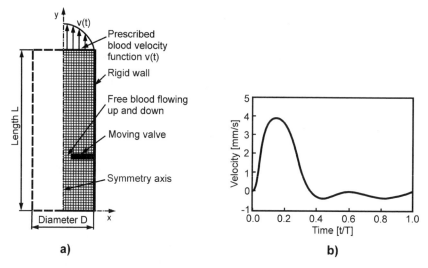

Fig. 13.7.4 Basic data for blood flow through a simple axisymmetric vein with rigid walls. The valves are modeled with a specific algorithm for the valves' position. The mean velocity calculated in the free region determines position of the valves: if the mean velocity is in the positive y direction the valves are open, otherwise the valves are closed according to the value of percentage parameter. (a) Geometrical data, FE mesh and boundary conditions (prescribed velocity on the top and traction free surface at the bottom); (b) Prescribed velocity during one cycle T

domain. Material data are blood density ρ as well as dynamics viscosity μ. The parabolic velocity profile is prescribed on the top of the model as a time function $v(t)$, see Fig. 13.7.4b.

Valves move within the cycles and represent a time-dependent boundary for the blood flow. A valve is completely open when the mean velocity near the valve is in the positive y direction, otherwise it is closed. Under normal physiological conditions the closure is complete (100%) and the blood cannot move down. In the case of disease the valves do not close completely and a gap which remains within the cross-section allows downward blood flow. In our model we introduce the percentage parameter which determines part of the closed cross-section when the valve is in the closed position.

The valve boundary condition is realized through a specific algorithm for the position of the valves. Also, in the 'Valve Data Dialog' of the Software on the web, the vertical position of the valve center, as well as its axial width, can be specified.

The velocity profiles which follow from the input waveform (Fig. 13.7.4b) is shown in Fig. 13.7.5 at times $t = 0.15\,s$, $t = 0.35\,s$, $t = 0.45\,s$ and $t = 0.60\,s$ (Kojić et al. 1998, Filipović et al. 2006c).

It can be seen from Fig. 13.7.5 that during systolic acceleration phase ($t = 0.15\,s$) all velocity profiles inside the vein are upward to the heart. The retrograde flow appears during systolic deceleration phase ($t = 0.35\,s$). In the early diastolic phase the valves are activated to block blood flow downward, but the failed valves (the percentage parameter of the closed cross-section by the valve = 50%) allow blood to move downwards ($t = 0.45\,s$). A small retrograde flow even appears during late diastolic phase ($t = 0.60\,s$) in the zone near the valves. At the end of cycle ($T = 1\,s$) the valves start to open and the blood velocities reach uniform distribution along the vein (not shown in the figure).

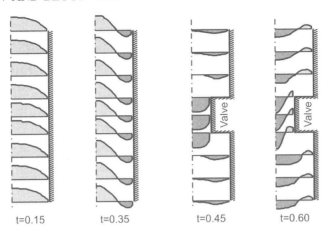

Fig. 13.7.5 Velocity profiles for blood flow through simple axisymmetric vein model with the rigid walls. Geometrical data (in *mm*): vein diameter $D = 5$, length $L = 100$. The percentage parameter of the closed cross-section by the valve $= 50\%$

Blood Flow through a Deformable Vein with Compression Therapy

We consider the same example as above, but now with deformable walls and compression therapy. Hence, the vein walls are subjected to mechanical action of blood and compression by bandages. Additional parameters (with respect to the model with rigid walls) to take into account vein wall deformability are the wall thickness δ, Young's modulus E, Poisson's ratio v and wall density ρ_w. The compression therapy is modeled by prescribing the force along the wall which is produced by the multilayer bandage system (Fig. 13.7.3). The compression forces vary along the domain of action.

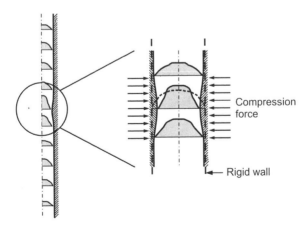

Fig. 13.7.6 Velocity distribution at $t = 0.15$ s (peak velocity) for blood flow through deformable vein with prescribed compression therapy force. Entering velocity profile is given in Fig. 13.7.4b

Fluid and solid are modeled by 2D axisymmetric finite elements and shell elements, respectively (see Sections 4.4 and 4.5). Tissue is assumed to be elastic and the fluid-structure interaction algorithm is implemented (see Section 7.6), with an incremental-iterative procedure.

The field of velocity magnitude at $t = 0.15\,s$ is shown in Fig. 13.7.6 for the following parameters: (lengths in mm) $D = 5, L = 100, \delta = 0.5; E = 3.61 N/mm^2, \nu = 0.49, \rho_w = 1.1 g/cm^3$; domain of force action $= 30-70\,mm$, compression forces $= 0-100\,N$.

Velocities are around 20% larger than in case when the compression therapy is not used (Fig. 13.7.5, $t = 0.15\,s$).

Both examples (rigid and deformable case) are available on the web Software where model parameters can be changed and influence of these changes can be explored.

13.8 Heart model

We first introduce some basic data about human heart functions, and then present solutions for an FE model of the left ventricular flow during filling phase.

13.8.1 Description of heart functioning

The heart is a potent biological pump consisting of two synchronized systems, the right and left heart (Fig. 13.8.1). The right heart receives blood from large veins and delivers it to the lung, and the left heart collects oxygenized blood from the lung and pumps it to a system circulation (Guyton 1991). The base of the heart structure is made of the fiber frame, which is composed of four fiber rings that are built of dense collagen fibers connected by conjunctive tissue to form *trigonum fibrosum*. Muscle tissues of atriums and ventricles make two anatomically and functionally independent complexes. Four fiber openings are attached with valves that along with left and right *trigonum* make the heart frame. A little bridge, which is built of specialized muscle tissue, is placed between atriums and ventricles. That

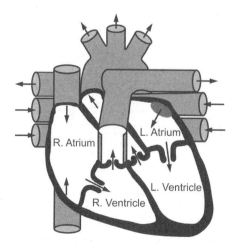

Fig. 13.8.1 Chambers of the heart (according to Mohr 2006)

specialized muscle tissue participates in conductivity of heart currents, known as Hiss's bundle, which is the only communication between heart cavities (Douglas *et al.* 2001). The functional tissue is composed of cardiocyte, the one-side densely packed contractile myofibrils arranged by groups.

The heart is reacting by synchronous contracting to any stimuli (electric impulse) that is above the threshold needed for depolarization, according to the rule 'all or nothing'. Between atria and ventricles are valves, needed to insure that blood flows only in one direction from atria to the ventricle. They are attached to the fibrous ring and are different anatomically. Between right atrium and ventricle is tricuspidal valve (with three cuspises) and in the left ventricle valve are bicuspidal (called mitral) valves. Retraction of the valves into the vantricule is prevented by contraction of papillary muscles in the initial part of the systole. Valves are shut passively because of increased pressure in the ventricules (Brandenburg *et al.* 1989).

Valves are also situated between ventricles and two major vessels: the aorta and pulmonary artery, preventing the regurgitation of the blood during diastole, when the pressure is higher in the arteries. Structurally, there are three symmetric, semilunar duplictures of the endothelia. Approximately, valve opening is repeated up to 100 000 times in a single day.

Energy for contraction is extracted by biochemical reactions, the predominant fuel is glucoses whose energy is transferred by the high energy compound adenosine triphosphate (ATP). Transformation of chemical into mechanical energy occurs during interactions of actin and myosin. Due to elastic fibers in the arteries, the energy is transferred in the systole part into potential energy needed to maintain continuous blood flow. In healthy individuals the heart approximately pumps out the blood 5–6 l/*min*. The heart cycle is divided into two phases:

(a) systole – during this phase blood is pumped out;

(b) diastole – during this phase blood is entering the heart.

In systole ventricles contract. Contraction of ventricles is separated in two phases:

(a) Isovolumetric contraction, with increasing pressure within ventricles and closed AV valves.

(b) Ejection phase – during which the blood is pushed out of the ventricle. During this phase, because of the higher pressure in the ventricles then in arteries, blood rushes through. Pressure reaches maximum values (normally 120 *mmHg* in the left and 25 *mmHg* in the right ventricle). After this period pressure drops and blood continues to move into large vessels due to inertia.

The pressure–volume (P–V) diagram of the left ventricle is given in Fig. 13.8.2. The starting point of the cycle is along the end-diastolic pressure, where a small preload of the resting heart exists. From this point the systole begins. After the mitral valve closes the isovolumic contraction (A) proceeds until the ventricle encounters its afterload, the aortic pressure.

The pressure first rises after the aortic valve opens and then drops during ejection (B). When the systole ends, the ventricular pressure and volume come to the end-systolic pressure–volume values. After the aortic valve closure removes the load (aortic pressure) from the ventricle, the relaxation begins under isovolumetric conditions (C) because blood can neither enter nor leave the ventricle. When the left ventricular pressure drops below that

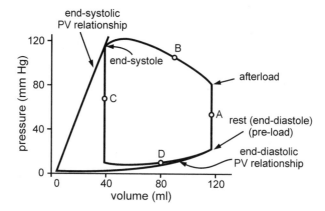

Fig. 13.8.2 Schematics of the pressure–volume loop of the left ventricle

in the left atrium, the mitral valve opens and the atrium empties into the ventricle during the phase of filling (D). The cycle is completed when ventricular relaxation is completed and ventricular pressure and volume again lie on the end-diastolic P–V values (Cheng *et al.* 2005).

Volume at the end of systole is called endsystolic (normally 60 mL), and the volume of the blood pumped out in a single cycle is the strike volume. Duration of systole is dependent on the heart frequency and with the frequency of 75/*min* it lasts 0.3 *s*. Amount of blood pumped out is dependent on the three factors: level of the stretching of heart, strength of the muscle contraction and the heart frequency.

To explain the conduction system of heart excitation we have to consider in more detail the following processes. First of all, there is the sinoatrial node (SAN) (see Fig. 13.8.3) (primary pacemaker cells) responsible for the initial electrical stimulus. These cells have the ability to depolarize and generate an action potential automatically. The electrical stimulus reaches the atrioventricular node and conductive tissue delays the excitation to the ventricles in order to provide temporal synchronization of contraction. After Tawara bundle and His

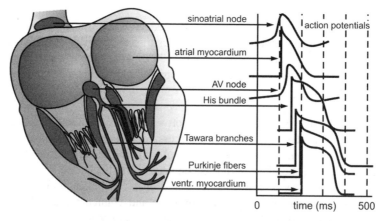

Fig. 13.8.3 Specialized excitation propagation tissues within the heart. Action potential curves are given at the right panel, with their temporal progression (adapted from Mohr 2006)

BLOOD FLOW AND BLOOD VESSELS

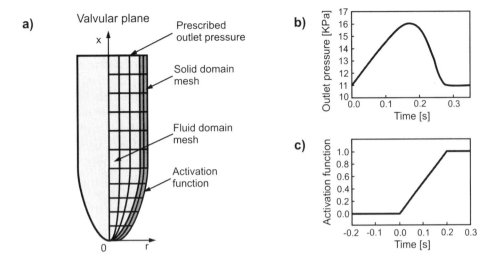

Fig. 13.8.4 FE model of the left ventricle. (a) Finite element mesh of the fluid and solid domains; (b) The outlet pressure at the valvular plane; (c) Activation function for the muscle fibers

bundle split, there are three branches. Two of these three branches conduct the excitation to the left, the third to the right ventricle. Further, the braches are splitting into a subendocardial network of Purkinje fibers which connects myocytes and activates muscle contraction. The atrioventricular node is called the secondary pacemaker. It also has the ability to self-depolarize, but with a lower frequency than the sinoatrial node. The action potential curves for each tissue type shown in Fig. 13.8.3 (right panel) vary in length and shape. This is due to the cell structure, amount of channels, pumps and exchangers.

13.8.2 Computational model

A simple FE model of the left ventricle is presented with some characteristic results.

Two-Dimensional Axisymmetric FE Model of Left Ventricle
The left ventricle is represented by an axisymmetric deformable body shown in Fig. 13.8.4. We model blood flow during filling phase by applying the fluid–solid interaction method of Section 7.6. Half of the radial plane is discretized using 2D axisymmetric four-node fluid elements with four velocites and pressure constant over the element (see Section 7.4) for the fluid domain. The ventricle wall is modeled by 2D axisymmetric four-node solid elements (Section 4.4), with fibers which have 3D direction. The finite element mesh of fluid and solid domain is shown in Fig. 13.8.4a.

The fluid mesh consists of 2000 elements with 2091 nodes. The Navier–Stokes equations (7.4.13) are solved using the ALE formulation for fluid with large displacements of the boundary (see Section 7.5). Also, a remeshing procedure is employed for the fluid domain in accordance with the motion of the ventricle wall. Boundary conditions for blood flow are: impermeable walls; no slip at the wall; valvular plane is not moving during the ejection phase; the aortic pressure is prescribed at the outlet section (Fig. 13.8.4b) according to the integrated lumped parameter model of systemic net (Redaelli & Montevecchi 1996). It is considered that blood behaves as a Newtonian fluid ($\rho = 1.05 \times 10^3$ kg/m^3, $\mu = 3.65 \times 10^{-3}$ kg/ms). The time step used in the calculation is $\Delta t = 0.005\ s$.

The ventricle wall model is simulated by muscle material model (see Section 12.2). Muscle fiber orientation is defined by direction vector in 3D space prescribed through input data. It has two components: in radial plane and circumferential direction. In this way we approximate in our axisymmetric FE model the real counter-rotating fibers within the heart wall (Redaelli & Montevecchi 1996). The inertial forces of the tissue are neglected.

It is assumed that initially the blood is at rest. The outlet blood pressure (Fig. 13.8.4b) is used as the boundary condition. At the same time the wall muscle fibers are activated according to the activation function shown in Fig. 13.8.4c (see Section 12.2).

Intraventricular pressure distribution pattern in systolic phase up to the pressure peak of 18 kPa (time $t = 0.12\ s$) is shown in Fig. 13.8.5 (see color plate) for five characteristic times. The field of blood velocity magnitude is shown in Fig. 13.8.6 (see color plate) for the same times. The velocity propagates from the valve toward the heart apex. This velocity propagation shows that the blood inertial effects are dominant in this time interval.

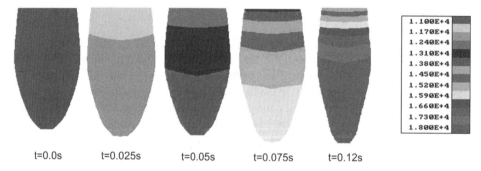

Fig. 13.8.5 Intraventricular pressure distribution pattern in systolic phase for five times. Intravascular pressure peak of 18 kPa is reached at $t = 0.12\ s$ (see Plate 22)

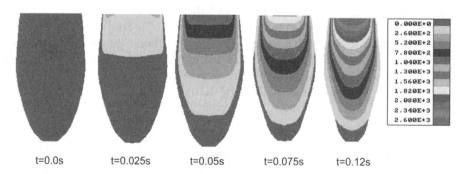

Fig. 13.8.6 Velocity field at the five characteristic times. The blood velocity propagates from the aortic valve to the heart apex, from 0 mm/s to $V_{max} = 2600\ mm/s$ at time $t = 0.12\ s$ (see Plate 23)

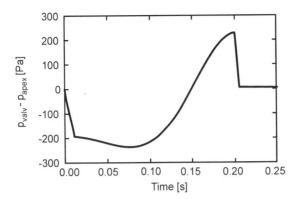

Fig. 13.8.7 Intraventricular pressure drop $p_{valv}(t) - p_{apex}(t)$ calculated during blood ejection phase (period between aortic valve opening and closure)

The intraventricular pressure drop between valvular (outlet) pressure p_{valv} and pressure at the apex p_{apex} inside the ventricle is given in Fig. 13.8.7. This result is in agreement with physiological data (Cheng *et al.* 2005).

The presented results show that even with a simple 2D axisymmetric model the realistic intraventricular pressure oscillations and blood velocity field within the ventricle can be obtained (see Software on the web).

References

Auricchio, F. & Taylor, R. (1997). Shape-memory alloys: modelling and numerical simulations of the finite-strain superelastic behavior, *Comp. Meth. Appl. Mech. Eng.*, **143**, 175–94.

Benbow, M., Burg, G., Camacho Martinez, F. *et al.* (eds.) (1999). *Compliance Network Physicians/HFL. Guidelines for the Outpatient Treatment of Chronic Wounds and Burns*, Blackwell Science, Berlin.

Blair, J.M., Glagov, S. & Zarins, C.K. (1990). Mechanism of superficial femoral artery adductor canal stenosis, *Surg. Forum*, **41**, 359–60.

Brandenburg, I., Robert, O., Valentini, F. & Giuliani, E.R. (1989). In D.C. McGoon (ed.), *Cardiology Fundamentals and Practice* (pp. 45–164), Year Book Medical Publishers, Inc.

Brooks, A.N. & Hughes, T.J.R. (1982). Streamline upwind/Petrov–Galerkin formulations for convection dominated flows with particular emphasis on the incompressible Navier–Stokes equations, *Comp. Meth. Appl. Mech. Eng.*, **32**, 199–259.

Cheng, Y., Oertel, H. & Schenkel, T. (2005). Fluid–structure coupled CFD simulation of the left ventricular flow during filling phase, *Ann. Biomed. Eng.*, **33**(5), 567–76.

Chien, S. (1970). Shear dependence of effective cell volume as a determinant of blood viscosity, *Science*, **168**, 977–9.

Collins, M.W., Pontrelli, G. & Atherton, M.A. (2004). *Wall–Fluid Interactions in Physiological Flows 6*, WIT Press.

Cotran, R.S., Kumar, V. & Robbins, S.L. (1994). *Robbins Pathologic Basis of Decease* (Chapter 11, pp. 499–504), Saunders, London.

Dean, W.R., (1928). The streamline motion of fluid in a curved pipe, *Philosoph. Mag.*, **5**, 673–95.

Dobrin, P.B. (1999). Distribution of lamellar deformation implications for properties of the arterial media, *Hypertension*, **33**, 806–10.

Doby, T. & Lowman, R.M. (1961). Demonstration of blood currents with radiopaque streamers, *Acta. Radio.*, **55**, 272–5.

Dotter, C.T. & Judkins, M.P. (1964). Transluminal treatment of arteriosclerotic obstruction. Description of a new technique and a preliminary report of its application. *Circulation*, **30**, 654–70.

Douglas, P.Z., Libby, P., Bonow, R.O. & Braunwald, E. (eds.) (2001). *Braunwald's Heart Disease: A Textbook of Cardiovascular Medicine*, 7th edition (pp. 360–94), FRCP Elsevier Saunders.

Fåhraeus, R. & Lindquist, T. (1931). The viscosity of the blood in narrow capillary tubes, *Am. J. Physiol.*, **96**, 562–8.

Filipović, N. (1999). *Numerical Analysis of Coupled Problem: Deformable Body and Fluid Flow*, Ph.D. Thesis, Faculty of Mechanical Engineering, University of Kragujevac, Serbia.

Filipović, N. & Kojić, M. (2004). Computer simulations of blood flow with mass transport through the carotid artery bifurcation, *Theoret. Appl. Mech.* (Serbian), **31**(1), 1–33.

Filipović, N., Kojić, M., Ivanović, M., Stojanović, B., Otasevic. L. & Ranković, V. (2006b). *MedCFD, Specialized CFD software for simulation of blood flow through arteries*, University of Kragujevac, Serbia.

Filipović, N., Kojić, M., Stojanović, B. & Ranković, V. (2006c). *VEINS, Specialized CFD software for simulation of blood flow through the veins*, University of Kragujevac, Serbia.

Filipović, N., Mijailović, S., Tsuda, A. & Kojić, M. (2006a). An implicit algorithm within the arbitrary Lagrangian–Eulerian formulation for solving incompressible fluid flow with large boundary motions, *Comp. Meth. Appl. Mech. Eng.*, **195**, 6347–61.

Fukushima, T., Matsuzawa, T. & Homma, T. (1989). Visualization and finite element analysis of pulsatile flow in models of the abdominal aortic aneurysm, *Biorheo.* **26**, 109–30.

Fung, Y.C., Fronek, K. & Patitucci, P. (1979). Pseudoelasticity of arteries and the choice of its mathematical expression, *Am. J. Physiol.*, **237**, 620–31.

Fung, Y.C. (1993). *Biomechanics: Mechanical Properties of Living Tissue*, 2nd edition, Springer-Verlag, New York.

Guyton, A.C. (1991). *Textbook of Medical Physiology*, 8th edition, W.B. Saunders Company.

Haber, S., Filipović, N., Kojić, M. & Tsuda, A. (2006). Dissipative particle dynamics simulation of flow generated by two rotating concentric cylinders, *Phys. Rev. E*, **74**, 1–8.

He, X. & Ku, D.N. (1996). Pulsatile flow in the human left coronary artery bifurcation: average conditions, *J. Biomech. Eng.*, **118**, 74–82.

Hollinshead, W.H. (1982). *Anatomy for Surgeons*, Volume 3, Harper & Row, Philadelphia.

Holzapfel, G.A. & Weizsacker, H.W. (1998). Biomechanical behavior of the arterial wall and its numerical characterization, *Comp. Biol. Med.*, **28**, 377–92.

Holzapfel, G.A., Gasser, C.T. & Ogden, R.W. (2000). A new constitutive framework for arterial wall mechanics and comparative study of material models, *J. Elasticity*, **61**, 1–48.

Kojić, M. & Bathe, K.J. (2005). *Inelastic Analysis of Solids and Structures*, Springer-Verlag, Berlin.

Kojić, M., Filipović, N., Vlastelica, I. & Živković, M. (2003). Modeling of blood flow in the human aorta with use of an orthotropic nonlinear material model for the walls, Second MIT Conference, Boston, USA.

Kojić, M., Filipović, N., Živković, M., Slavković, R. & Grujovic, N. (1998). *PAK-F Finite Element Program for Laminar Flow of Incompressible Fluid and Heat Transfer*, Faculty of Mechanical Engineering, University of Kragujevac, Serbia.

Ku, D.N. (1997). Blood flow in arteries, *Annu. Rev. Fluid Mech.*, **29**, 399–434.

Ku, D.N., Giddens, D.P., Zarins, C.Z. & Glagov, S. (1985). Pulsatile flow and arteriosclerosis in the human carotid bifurcation, *Atheroscler.*, **5**, 293–302.

Lei, M., Kleinstreuer, C. & Truskey, G.A. (1995). Numerical investigations and prediction of atherogenic sites in branching arteries, *J. Biomech. Eng.*, **117**, 350–7.

Liu, Y., Lai, Y., Nagaraj, A., Kane, B., Hamilton, A., Greene, R., McPherson, D.D. & Chandran, K.B. (2001). Pulsatile flow simulation in arterial vascular segments with intravascular ultrasound images, *Medical Eng. Phys.*, **23**, 583–95.

Loscalzo, J. & Schafer, A.I. (2003). *Thrombosis and Hemorrhage*, 3rd edition, Lippincott Williams & Wilkins, Philadelphia.

Milner, J.S., Moore, J.A., Rutt, B.K. & Steinman, D.A. (1998). Hemodynamics of human carotid artery bifurcations: computational studies with models reconstructed from magnetic resonance imaging of normal subjects, *J. Vasc. Surg.*, **28**, 143–56.

Mohr, M.B. (2006). *A hybrid deformable model of ventricular myocardium*, Ph.D. Thesis, University of Karlsruhe, Germany.

Moore, J.A., Steinman, D.A. & Ethier, C.R. (1998). Computational blood flow modeling: errors associated with reconstructing finite element models from MRI, *J. Biomech.*, **31**, 179–84.

Nelzen, O., Bergqvist, D. & Lindhagen, A. (1991). Leg ulcer etiology – a cross-sectional population study, *J. Vasc. Surg.*, **14**(4), 557–64.

Osamu, I. et al. (2006). Effect of exercise on frequency of stent fracture in the superficial femoral artery, *Am. J. Cardiology.*, **98**(2), 272–4.

Patel, D.J., Greenfield, J.C., Austen, W.G., Morrow, A.G. & Fry, D.L. (1965). Pressure-flow relationships in the ascending aorta and femoral artery of man, *J. Appl. Physiol.*, **20**(3), 459–63.

Peattie, R., Riehle, T. & Bluth, E. (2004). Pulsatile flow in fusiform models of abdominal aortic aneurysms: flow fields, velocity patterns and flow-induced wall stresses, *J. Biomech. Eng.*, **126**, 438–46.

Perktold, K., Hofer, M., Rappitsch, G., Loew, M., Kuban, B.D. & Friedman, M.H. (1998). Validated computation of physiologic flow in realistic coronary artery branch, *J. Biomech.*, **31**, 217–28.

Perktold, K., Peter, R.O., Resch, M. & Langs, G. (1991a). Pulsatile non-Newtonian blood flow in three-dimensional carotid bifurcation models: a numerical study of flow phenomena under diferent biurcation angles, *J. Biomech. Eng.*, **13**, 507–15.

Perktold, K. & Rappitsch, G. (1995). Computer simulation of local blood flow and vessel mechanics in a compliant carotid artery bifurcation model, *J. Biomech.*, **28**, 845–56.

Perktold, K., Resch, M. & Peter, R.O. (1991b). Three-dimensional numerical analysis of pulsatile flow and wall shear stress in the carotid artery bifurcation model, *J. Biomech.*, **24**, 409–20.

Pries, A.R. & Secomb, T.W. (2005). Microvascular blood viscosity and the endothelial surface layer, *Am. J. Physiol. Heart Crc. Physiol.*, **289**, 2657–64.

Ranković, V. (2007). *The correlation between characteristics of the stent and the biomechanical characteristics of the femoral artery in adduct canal*. Ph.D. Thesis, CIMSI, University of Kragujevac, Serbia.

Redaelli, A. & Montevecchi, F.M. (1996). Computational evaluation of intraventricular pressure gradients based on a fluid-structure approach, *J. Biomech. Eng.*, **118**, 529–37.

Rodkiewicz, C.M., Sinha, P. & Kennedy, J.S. (1990). On the application of a constitutive equation for whole blood, *J. Biomech Eng.*, **112**, 198–206.

Rosic, M., Pantovic, S., Rankovic, V., Obradovic, Z., Filipovic, N., Kojic, M. (2007). Evaluation of dynamic response and biomechanical properties of isolated blood vessels, *J. Biochem. Biophys. Methods*, in press.

Rushmer, R.F. (1976). *Cardiovascular Dynamics*, 4th editon. W.B. Saunders, Philadelphia.

Schmidt, A. et al. (2000). Arterial properties of the carotid and femoral artery in endurance-trained and paraplegic subjects, *J. Appl. Physiol.*, **89**, 1956–63.

Scotti, C.M., Shkolnik, A.D., Muluk, S.C. & Finol, E.A. (2005). Fluid–structure ineraction in abdominal aneurysm:effects of asymmetry and wall thickness, *Biomed. Eng. Online*, **4**, 64.

Segadal, L. & Matre, K. (1987). Blood velocity distribution in the human ascending aorta, *Circulation*, **36**, 90–100.

Slavković, R., Živković, M. & Kojić, M. (1994). Enhanced 8-node three-dimensional solid and 4-node shell elements with incompatible displacements, *Comm. J. Num. Meth. Eng.*, **10**, 699–709.

Strandness, D.E. & Eikelboom, B.C. (1998). Carotid artery stenosis – where do we go from here?, *European J. Ultrasound*, **7**, 17–26.

Taylor, C.A., Hughes, T.J.R. & Zarins, C.K. (1998). Finite element modeling of blood flow in arteries, *Comp. Meth. Appl. Mech. Eng.*, **158**, 155–96.

Thompson, M.M. & Bell, R.P.F. (2000). ABC of arterial and venous disease: arterial aneurysm, *B. Med. J.*, **320**, 1193–6.

Vander, A., Sherman, J. & Luciano, D. (1998). *Human Physiology: The Mechanism of Body Function*, 7th edition, WCB McGraw-Hill, New York.

Vliet, J.A.V.D. & Boll, A.P.M. (1997). Abdominal aortic aneurysm, *The Lancet*, **349**, 863–6.

Womersley, J. (1957). *An elastic tube theory of pulse transmission and oscillatory flow in mammalian arteries*, Tehnical Report WADC, Technical Report TR 56-614, Wright Air Development Center.

Woodcock, J.P., Morris, S.J. & Wells, P.N.T. (1975). Significance of the velocity impulse response and cross-correlation of the femoral and popliteal blood velocity time waveforms in disease of the superficial femoral artery, *Med. Biolog. Eng. Comp.*, **13**, 813–18.

Zarins, C.K., Giddens, D.P., Bharadvej, B.K., Sottiurai, V.S., Mabon, R.F. & Glagov, S. (1983). Carotid bifurcation ahterosclerosis: quantitative correlation of plaque localization with low velocity profiles and wall shear stress, *Circ. Res.*, 53, 502–14.

14

Modeling Mass Transport and Thrombosis in Arteries

In this chapter we first give an introduction to thrombosis process in large arteries. Then we present modeling the albumin and low density lipoprotein (LDL) transport as large molecules. A platelet mediated thrombosis is modeled using the continuum based approach (finite element method) as well as Dissipative Particle Dynamics (DPD) method. Numerical examples are given at the end of sections and the solutions are compared with experimental and analytical results from literature.

14.1 Introduction

Here, we present the basic information about platelet mediated thrombosis as the process which involves mass transport, platelet aggregation and accumulation on the vessel wall; while the description of the albumin and LDL transport is given in Section 14.2.1.

Atherosclerosis was thought to be a degenerative disease that was an inevitable consequence of aging. But research in the last two decades has shown that atherosclerosis is neither a degenerative disease nor inevitable. On the contrary, atherosclerosis seems to be a chronic inflammatory condition that is converted to an acute clinical event by the induction of plaque rupture, which in turn leads to thrombosis (Caro *et al.* 1971, Berliner *et al.* 1995). The basic mechanisms that induce the thrombosis are shown in Fig. 14.1.1, which depicts the sequence of changes in the artery wall that lead to a clinical event.

Platelets as blood constituent circulate through the blood, surveying the integrity of the vascular system (Ruggeri 2000). Platelets act in response to traumatic injuries in which the continuity of a vessel wall is interrupted and blood begins to pour outside (Hoff *et al.* 1975a,b, Ruggeri 2000). However, the process of platelet accumulation may lead to development of thrombus which narrows the vessel volume and produces severe constrictions in blood flow

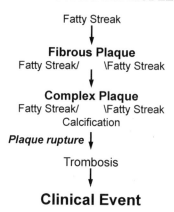

Fig. 14.1.1 Model showing the sequence of events from fatty streak to clinical event (according to Berliner *et al.* 1995)

and decease. The thrombus can even rupture, with embolization that can be of fatal outcome. A huge body of references is available about various aspects of thrombosis (Loscalzo & Schafer 2003).

Under normal conditions, platelets circulate in the blood flow with a disc shape and in a nonadherent state. If the circulating platelets contact the damaged wall they may be induced to begin an activation process in which the platelets change shape to a spherical spiny cell (see Fig. 13.1.3c), and then release chemicals into blood which can activate other platelets. Platelet activation can be described as the process of conversion of the smooth, nonadheret platelet into a sticky particle that releases and expresses chemicals with the ability to bind the plasma protein fibrinogen (Verstraete *et al.* 1998).

In a cylindrical vessel, the velocity profile of particles contained in circulating blood is parabolic; the shear rate decreases from the wall to the center of the lumen inversely to the flow velocity. In a flow field with high shear rate, only GP Ib α interaction with immobilized vWF multimers can initiate the tethering of circulating platelets to the vessel wall and to already adherent platelets (Ruggeri 2000). Each platelet has approximately 50 000 GPIIb/IIIa receptors which provide numerous possibilities for fibrinogen bridge connections (Fogelson & Guy 2004), Fig. 14.1.2.

It is common in modeling platelet-mediated thrombosis to adopt a continuum-based approach. Namely, the distribution of platelets within the blood is determined by calculating the field of platelet distribution within the blood as the carrying fluid. The fundamental equations rely on the convection–diffusion laws described in Section 3.3.1, and include the process of the platelet activation, as well as the boundary conditions specific for the thrombosis development. The continuum-based model of thrombosis is presented in Section 14.3. This method can be applied to stenotic flow in the coronary arteries, carotid bifurcation and other arteries.

Another approach of thrombosis modeling is to use a discrete particle method, such as dissipative particle dynamics (DPD) method described in Section 8.2. By employing this method, it is possible to track an individual platelet in the sequence of the process, including platelet activation, aggregation and adhesion to the wall. Application of the DPD method to model thrombosis within simple flowing conditions is given in Section 14.4.

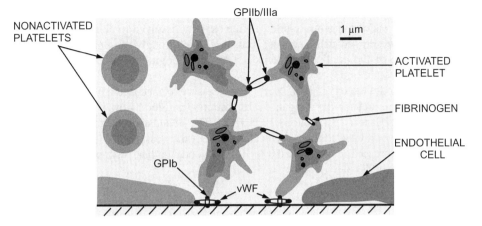

Fig. 14.1.2 Schematic representation of the mechanisms of platelet adhesion and aggregation in flowing blood (according to Fogelson & Guy 2004)

14.2 Modeling mass transport in arteries by continuum-based methods

In this section we give the fundamental relations of mass transport within a blood vessel, followed by the finite element formulation and solved examples.

14.2.1 The basic relations for mass transport in arteries

The metabolism of the artery wall is critically dependent upon its nutrient supply governed by transport processes within the blood. We here address two different mass transport processes in large arteries. One of them is the oxygen transport and the other is LDL transport. In Chapter 13 blood flow through the arteries is described as motion of a fluid-type continuum, with the wall surfaces treated as impermeable (hard) boundaries. However, transport of gases (e.g. O_2, CO_2) or macromolecules (albumin, globumin, LDL) represents a convection–diffusion physical process (see Section 3.3) with permeable boundaries through which the diffusion occurs. In the analysis presented further, the assumption is that the concentration of the transported matter does not affect the blood flow (i.e. a diluted mixture is considered). The mass transport process is governed by convection–diffusion equation (3.3.9),

$$\frac{\partial c}{\partial t} + v_x \frac{\partial c}{\partial x} + v_y \frac{\partial c}{\partial y} + v_z \frac{\partial c}{\partial z} = D \left(\frac{\partial^2 c}{\partial x^2} + \frac{\partial^2 c}{\partial y^2} + \frac{\partial^2 c}{\partial z^2} \right) \quad (14.2.1)$$

where c denotes the macromolecule or gas concentration; v_x, v_y and v_z are the blood velocity components in the coordinate system x, y, z; and D is the diffusion coefficient, assumed constant, of the transported material.

Transport of Oxygen
Oxygen as probably the most critical metabolite is supplied to the cells of the avascular tissue layer by diffusion. The arterial blood flowing within the vessel lumen and within the vasa vosurum are two sources of oxygen for the entire body.

Abnormalities in arterial wall oxygen implicate the formation of atherosclerotic lesions. There are two forms of oxygen in blood: (i) free oxygen dissolved in plasma, and (ii) bound to hemoglobin within red cells. Oxygen transport is a strongly nonlinear mass transport problem because of the nonlinear dependence of oxyhemoglobin concentration on plasma oxygen partial pressure.

In order to model the oxygen transport, it is necessary to specify the boundary conditions for the blood flow and for the oxygen concentration. The blood flow boundary conditions are described in Chapter 13 (Sections 13.2–13.8). The boundary condition specific for the oxygen transport is related to the oxygen flux through the vessel wall. Oxygen wall fluxes are frequently expressed in terms of the local Sherwood number, Sh_D, which is defined as (Moore & Etheir 1997)

$$Sh_D = \frac{q_w D_v}{D_b \left(PO_{2in} - PO_{2ref}\right)} \tag{14.2.2}$$

where q_w is the local wall oxygen flux, D_v is the arterial diameter, PO_{2in} and PO_{2ref} are the specified inlet and reference oxygen tensions (concentration of dissolved oxygen at which its partial pressure is in equilibrium with the solvent), respectively; and D_b is the oxygen diffusion coefficient. Relatively large Sherwood numbers are observed at the stenosis sites, consistent with high shear rates there.

Transport of the LDL

Another macromolecule directly responsible for the process of atherosclerosis is the low density lipoprotein (LDL) which is well known as atherogenic molecule. It is also known that LDL can go through the endothelium at least by three different mechanisms, namely, receptor-mediated endocytosis, pinocytotic vesicular transport, and phagocitosis (Goldstein et al. 1979). The permeability coefficient of an intact arterial wall to LDL has been reported to be of the order of $10^{-8}\,cm/s$ (Bratzler et al. 1977). The conversion of the LDL mass through a semipermeable wall, with the mass moving toward the vessel wall by a filtration flow and diffusing back to the mainstream at the wall, is described by the relation

$$c_w v_w - k \frac{\partial c}{\partial n} = K c_w \tag{14.2.3}$$

where c_w is the surface concentration of LDL, v_w is the filtration velocity of LDL transport through the wall, n is the coordinate normal to the wall, k is the diffusivity of LDL and K is the overall mass transfer coefficient of LDL at the vessel wall.

14.2.2 Finite element modeling of diffusion–transport equations

The basic FE equations for mass transport are presented in Section 7.4. In the case of blood flow with mass transport we have domination of the convection terms due to the low diffusion coefficient. Then it is necessary to have special stabilizing techniques in order to obtain a stable numerical solution. Here we have implemented the streamline upwind/Petrov–Galerkin stabilizing procedure (SUPG) (Brooks & Hughes 1982) within the numerical integration scheme described in Section 7.4. The incremental-iterative form of FE equations of balance are obtained through an extension of the system (7.4.9) by the diffusion equations (7.4.20), followed by the transformation of these additional equations into the incremental form (7.4.13). The final equations are

MODELING MASS TRANSPORT AND THROMBOSIS IN ARTERIES

$$\begin{bmatrix} \frac{1}{\Delta t}\mathbf{M}_v + {}^{n+1}\mathbf{K}_{vv}^{(i-1)} + {}^{n+1}\mathbf{K}_{\mu v}^{(i-1)} + {}^{n+1}\mathbf{J}_{vv}^{(i-1)} & {}^{n+1}\mathbf{K}_{vp}^{(i-1)} & 0 \\ \mathbf{K}_{vp}^T & 0 & 0 \\ {}^{n+1}\mathbf{K}_{cv}^{(i-1)} & 0 & \frac{1}{\Delta t}\mathbf{M}_c + {}^{n+1}\mathbf{K}_{cc}^{(i-1)} + {}^{n+1}\mathbf{J}_{cc}^{(i-1)} \end{bmatrix} \times$$
(14.2.4)

$$\begin{Bmatrix} \Delta \mathbf{V}^{(i)} \\ \Delta \mathbf{P}^{(i)} \\ \Delta \mathbf{C}^{(i)} \end{Bmatrix} = \begin{Bmatrix} {}^{n+1}\mathbf{F}_v^{(i-1)} \\ {}^{n+1}\mathbf{F}_p^{(i-1)} \\ {}^{n+1}\mathbf{F}_c^{(i-1)} \end{Bmatrix}$$

where the matrices are (no sum on j and sum on k; $j,k,l=1,2,3$)

$$(\mathbf{M}_v)_{jjKJ} = \int_V \rho N_K N_J dV \qquad (\mathbf{M}_c)_{jjKJ} = \int_V N_K N_J dV$$

$$\left({}^{n+1}\mathbf{K}_{cc}^{(i-1)}\right)_{jjKJ} = \int_V D N_{K,k} N_{J,k} dV \qquad \left({}^{n+1}\mathbf{K}_{\mu v}^{(i-1)}\right)_{jjKJ} = \int_V \mu N_{K,k} N_{J,k} dV$$

$$\left({}^{n+1}\mathbf{K}_{cv}^{(i-1)}\right)_{jjKJ} = \int_V N_K {}^{n+1}c_{,j}^{(i-1)} N_J dV \qquad \left({}^{n+1}\mathbf{K}_{vv}^{(i-1)}\right)_{jjKJ} = \int_V \rho N_K {}^{n+1}v_k^{(i-1)} N_{J,k} dV \quad (14.2.5)$$

$$\left({}^{n+1}\mathbf{J}_{vv}^{(i-1)}\right)_{jlKJ} = \int_V \rho N_K {}^{n+1}v_{j,l}^{(i-1)} N_J dV \qquad \left({}^{n+1}\mathbf{K}_{vp}^{(i-1)}\right)_{jjKJ} = \int_V \rho N_{K,j} \hat{N}_J dV$$

$$\left({}^{n+1}\mathbf{J}_{cc}^{(i-1)}\right)_{jjKJ} = \int_V \rho N_K {}^{n+1}v_k^{(i-1)} N_{J,k} dV$$

and the vectors are

$${}^{n+1}\mathbf{F}_c^{(i-1)} = {}^{n+1}\mathbf{F}_q + {}^{n+1}\mathbf{F}_{sc}^{(i-1)} - \frac{1}{\Delta t}\mathbf{M}_c\left\{{}^{n+1}\mathbf{C}^{(i-1)} - {}^n\mathbf{C}\right\} -$$
$${}^{n+1}\mathbf{K}_{cv}^{(i-1)}\left\{{}^{n+1}\mathbf{V}^{(i-1)}\right\} - {}^{n+1}\mathbf{K}_{cc}^{(i-1)}\left\{{}^{n+1}\mathbf{C}^{(i-1)}\right\} \qquad (14.2.6)$$

$$\left({}^{n+1}\mathbf{F}_q\right)_K = \int_V N_K q^B dV \quad {}^{n+1}\mathbf{F}_{sc}^{(i-1)} = \int_S D N_K \nabla^{n+1}\mathbf{c}^{(i-1)} \cdot \mathbf{n} dS$$

Here, the matrix $\mathbf{M}_v$ stands for $\mathbf{M}$ in (7.4.13); the matrices $\mathbf{M}_{cc}$ and $\mathbf{K}_{cc}$ are the 'mass' and convection matrices; $\mathbf{K}_{cv}$ and $\mathbf{J}_{cc}$ correspond to the convective terms of equation (14.2.1); and $\mathbf{F}_c$ is the force vector which follows from the convection-diffusion equation in (14.2.1). Also, equations (14.2.2) and (14.2.3) are used for the mass surface flux calculation at the wall boundary (see Theory, Chapter 14 on the web).

14.2.3 Examples

Solutions for two typical examples of mass transport are presented, for which also the Software on the web is provided.

Example 14.2-1. Modeling albumin transport in a large artery
The stenosed artery is shown in Fig. E14.2-1A. Due to axial symmetry, only a half is modeled and the blood is considered to be a Newtonian fluid. Two-dimensional axisymmetric

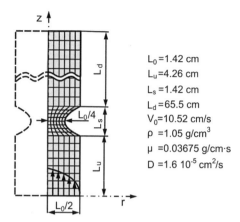

Fig. E14.2-1A Geometrical and material data for artery with stenosis

elements (see Section 4.4) are used. Data for blood density ρ, dynamic viscosity μ, diffusion coefficient for albumin transport D, the inflow mean velocity V_0, and geometry are given in the figure. It is assumed that the flow is steady.

Since the distance between the entrance and the stenosis position is large, it can be considered that the entering velocity profile is parabolic (see also Example (3.3-1):

$$v(r) = 2V_0 \left(1 - \left(\frac{2r}{L_0}\right)^2\right) \qquad (E14.2\text{-}1.1)$$

where r is the radial coordinate. At the wall velocities are equal to zero, while at the axis of symmetry only the radial velocity is equal to zero. At the outflow the zero traction stress is applied (see Example 7.4-2),

$$-p + \mu \frac{\partial v_z}{\partial z} = 0 \qquad (E14.2\text{-}1.2)$$

The boundary conditions for the concentration are: (a) at the inlet, $c = c_0 = 2.58 \times 10^{-3}\,mL/cm^3$; and (b) $\partial c/\partial z = 0$ at the axis of symmetry. Note that to mean velocity $V_0 = 10.52\,cm/s$ corresponds the Reynolds number $Re = 448$, while the Pecklet number is $P_e = 9.34 \times 10^5$ (defined as $P_e = L_0 V_0/D$).

The velocity field is shown in Fig. E14.2-1Ba (see color plate), where the field of disturbed flow is noticeable after the stenosis. Albumin concentration at the wall c_w normalized with respect to the inlet concentration c_0 is given in Fig. E14.2-1Bb (see color plate). A significant concentration increase in the domain of stenosis can be seen, with the peak at the distal region.

Other results, and dependence of the solutions on the geometric and material parameters, can be explored using Software on the web.

Example 14.2-2. Modeling the LDL transport through a straight artery with the filtration through the wall

The LDL transport through a straight artery is modeled in this example. The tube, representing the artery, has the diameter $d_0 = 0.6\,cm$ (Fig E14.2-2a). The filtration velocity

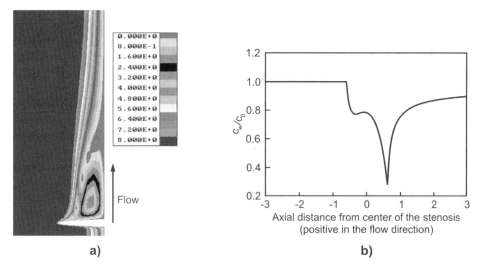

Fig. E14.2-1B Albumin transport in stenosed artery, the stenotic artery part: (a) Velocity field; (b) Normalized concentration at the wall c_w/c_0 (see Plate 24)

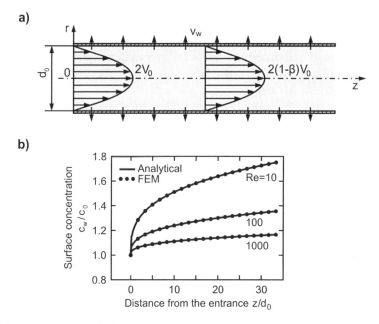

Fig. E14.2-2 Transport of the LDL through straight artery with semipermeable wall. (a) Schematic representation of velocity profiles (note that the profile changes in the flow direction since the wall is permeable), $\beta = 4v_w z/V_0 d_0$; (b) Normalized surface concentration of LDL, c_w/c_o, in terms of the normalized distance from the entrance z/d_0 (analytical and FE solutions)

through the vessel wall is $v_w = 4 \times 10^{-6} \, cm/s$ and the overall mass transfer coefficient of lipoproteins at the arterial wall, $K = 2 \times 10^{-8} \, cm/s$ (see (14.2.3)). Blood was modeled as a Newtonian fluid with density $\rho = 1.0 \, g/cm^3$ and viscosity $\mu = 0.0334 \, P$. The steady-state conditions for fluid flow and mass transport are assumed. The entering blood velocity is defined by the Reynolds number Re (calculated using the mean blood velocity and the artery diameter, see web – Theory, Chapter 3).

The 2D axisymmetric elements are used (see Section 4.4). The boundary conditions include prescribed parabolic velocity profile (see (E14.2-1.1)) and concentration c_0 at the inlet, zero stress at the outlet (see (E14.2-1.2)); and filtration at the walls according to (14.2.3).

The analytical solution for the axial and radial velocities, as well as for the concentration, is given in Yuan and Finkelstein (1956); details are given on the web – Examples, Chapter 14 (see also Software on the web).

Figure E14.2-2b shows distribution of the surface concentration of LDL along the axis of the artery for three Reynolds numbers. It can be seen that the concentration of LDL at the wall boundary layer increases with the axial distance from the entrance of the artery due the decrease of the fluid velocity (see the velocity profile in E14.2-2a, where $\beta = 4v_w z/V_0 d_0$).

Solutions for other parameters of the model can be obtained using Software (see web-Software).

14.3 Modeling thrombosis by continuum-based methods

The focus of this section is modeling mechanical aspects of thrombosis by continuum-based methods assuming activated platelets which are aggregating to the wall.

14.3.1 Model description

There are numerous computational approaches of modeling platelet transport and deposition relying on the convection–diffusion equation (14.2.1). These studies have typically required simplifying assumptions, such as diffusion-limited rates of platelet–surface adhesion, simple (Poiseuille or Couette) velocity fields, and constant platelet–surface reactivity (see e.g. Basmadjian 1990). Platelet activation, agonist transport, and bulk aggregation within a continuum model were also included in Fogelson (1984). Further, fictitious fluid-dynamic body forces were added to simulate flow disturbances due to platelet aggregation, with a discrete model of platelet–surface adhesion and bulk aggregation (Fogelson & Kuharsky 1998).

The simplest continuum-based model of thrombosis is to consider transport of platelets by convection and shear enhanced diffusion, treating platelets and blood as a dilute mixture with the governing equation (14.3.1) (Wootton et al. 2001, Bluestein et al. 1999). The diffusion coefficient is the effective diffusivity of platelets D_e. Usually, the effective diffusivity is expressed as a function of the maximum local shear rate, for example as

$$D_e = \alpha \dot{\gamma}_{max} + D_{th} \qquad (14.3.1)$$

where α is a coefficient fitted with experiments, D_{th} is the thermal diffusivity, and $\dot{\gamma}_{max}$ is the maximum shear strain rate.

The platelet flux represents the rate of platelet accumulation per unit surface area, which depends on the axial location x along the blood vessel,

$$j(x) = k_t c_w(x) \tag{14.3.2}$$

where $j(x)$ is the flux (volume per unit area), c_w is the concentration of free platelets at the surface, and k_t is the kinetic rate of aggregation of platelets to the wall. The rate of accumulation is coupled to the platelet transport through the boundary condition:

$$D \frac{\partial c}{\partial n}\bigg|_{wall} = j(x) \tag{14.3.3}$$

where n is the direction normal to the wall.

More complex continuum models of platelet mediated thrombosis were introduced by Fogelson (1984), Sorensen et al. (1999a,b) and Sorensen (2002). The thrombosis process is described by a coupled set of convection–diffusion–reaction equations similar to (14.2.1) (see also (3.3.9)),

$$\frac{\partial c_i}{\partial t} + v_x \frac{\partial c_i}{\partial x} + v_y \frac{\partial c_i}{\partial y} + v_z \frac{\partial c_i}{\partial z} = D_i \left(\frac{\partial^2 c_i}{\partial x^2} + \frac{\partial^2 c_i}{\partial y^2} + \frac{\partial^2 c_i}{\partial z^2} \right) + S_i \tag{14.3.4}$$

where D_i represents the diffusivity of species i in the blood, c_i is concentration and S_i is a source term for species i.

In Sorensen et al. (1999a) seven different species are considered: (i) normal, resting platelets; (ii) activated platelets; (iii) platelet-granule released; (iv) platelet-synthesized agonists; (v) prothrombin; (vi) thrombin; and (vii) ATIII. There is also a very complex list of relationships for the source terms S_i for each species of the model. The rate of activation of resting platelets k_{pa} and the amount λ_j of agonist j released per platelet are introduced. For example, $\lambda_j \cdot k_{pa} \cdot [RP]$ represents the rate at which agonist j is generated from newly activated platelets, where $[RP]$ is the concentration of resting platelets. Since very little quantitative information is available about the kinetics of platelet activation, a simplistic linear rate equation for k_{pa} with activation threshold is assumed:

$$k_{pa} = \begin{Bmatrix} 0, & \Omega < 1.0 \\ \dfrac{\Omega}{t_{act}}, & \Omega \geq 1.0 \end{Bmatrix} \tag{14.3.5}$$

where Ω is the activation function which depends on the concentration of the j-th agonist and t_{act} is a characteristic time for platelet activation. Usually this time is assumed to be $1s$ (Sorensen et al. 1999a). Although many of the coefficients in this model are not explicitly known (they were estimated), good results in comparison with simple experiments for platelet adhesion were obtained. There is also a model with four species (Fogelson 1992): (i) normal-resting platelets; (ii) active platelets; (iii) chemical agonist; (iv) concentration of platelet-platelet links.

We here describe in more detail a model with three species: (i) normal-resting platelets; (ii) active platelets; and (iii) chemical agonist. This model is used further in one example

(Example 14.3-2). The model represents the reduced Fogelson's model (Fogelson 1992) in which we neglected concentration of the platelet–platelet links. The governing equations of mass balance for this reduced model are:

$$\frac{\partial \phi_n}{\partial t} + \mathbf{v} \cdot \nabla \phi_n = D_n \Delta \phi_n - R(c)\phi_n$$

$$\frac{\partial \phi_a}{\partial t} + \mathbf{v} \cdot \nabla \phi_a = R(c)\phi_n \quad (14.3.6)$$

$$\frac{\partial c}{\partial t} + \mathbf{v} \cdot \nabla c = D_c \Delta c + AR(c)\phi_n$$

where ϕ_n is concentration of normal-resting platelets, $\mathbf{v}$ is blood flow velocity, D_n is diffusion coefficient for resting platelets, $R(c)$ is the conversion rate of the resting into activate platelets, ϕ_a is concentration of activated platelets, c is concentration of ADP, D_c is diffusion coefficient of signaling chemical ADP, and A is the rate of creation of ADP (assumed to be constant). It can be seen from (14.3.6) that the diffusion coefficient of activated platelets is assumed to be zero.

14.3.2 Examples

Example 14.3-1. Modeling platelet accumulation on collagen-coated wall of a tube with narrowing (stenosis)

We here model platelet deposition in a thrombogenic stenosis within a straight tube. The tube has a 75% reduction in diameter in the middle of the stenosis, as shown in Fig. E14.3-1A (Wootton et al. 2001). Blood is considered to be a Newtonian fluid with density 1.06 g/mL and viscosity 0.035 P. The Reynolds number based on the upstream diameter (see web-Theory, Chapter 3) is 160. Steady flow is assumed.

Platelet concentration is modeled using the governing equation for transport of a dilute mixture (14.3.1) which in the case of axisymmetric conditions is (see web – Theory, Chapter 3; u and v are axial and radial velocities of fluid, respectively):

$$u\frac{\partial c}{\partial x} + v\frac{\partial c}{\partial r} = \frac{\partial}{\partial x}\left(D\frac{\partial c}{\partial x}\right) + \frac{1}{r}\frac{\partial}{\partial r}\left(Dr\frac{\partial c}{\partial r}\right) \quad (E14.3\text{-}1.1)$$

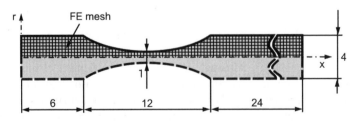

Fig. E14.3-1A Stenosis geometry for 75% reduction (lengths are in *mm*). One half of the radial plane x–r ($r \geq 0$) is modeled by axisymmetric finite elements. The finite element mesh is shown schematically

MODELING MASS TRANSPORT AND THROMBOSIS IN ARTERIES 305

It is assumed that the platelet concentration at the tube inlet is constant, with $c = c_0 = 5 \times 10^8$ platelet/mL for $x = 0$ as the boundary condition. The effective diffusivity D is taken to be dependent on the maximum local shear rate $\dot{\gamma}_{max}$ according to (14.3.1), with $\alpha = 7 \times 10^{-9}$ cm^2 and thermal diffusivity $D_{th} = 1.6 \times 10^{-9}$ cm^2/s. The platelet flux j is taken to be of the form (14.3.2), hence the relation (14.3.3) is applicable to the boundary condition at the wall surface.

Calculated velocity and concentration fields are shown in Fig. E14.3-1B (see color plate). The steady-state solution for the velocity (Fig. E14.3-1Ba – see color plate) shows that the flow disturbance occurs in the stenotic tube part. Due to the existence of a significant stagnation zone near the stenosis, the streamlines contours form a large vortex behind that narrowing, as expected (Fig. E14.3-1Bb).

The solution for wall shear strain rate is shown in Fig. E14.3-1Bc (see color plate). The shear rate at the wall has a peak just upstream of the stenosis throat. In Fig. E14.3-1Bd (see color plate) is shown the rate of platelet accumulation along the wall. It can be seen that a sharp decrease of platelet wall concentration rate occurs at the maximum narrowing region, which is consistent with results of Wootton et al. (2001).

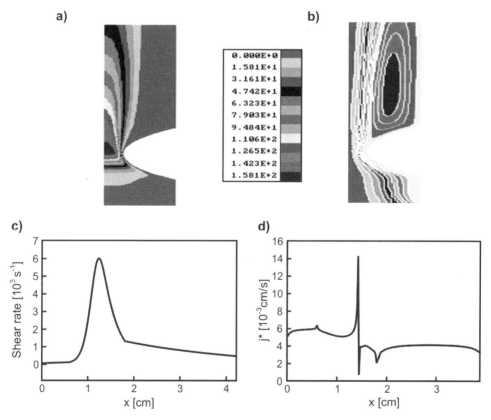

Fig. E14.3-1B A straight artery with 75% stenosis. (a) Intensity of blood velocity field (in cm/s); (b) Streamline contours; (c) Wall shear strain rate along the wall; (d) Platelet accumulation rate along the wall ($j^* = j(x)/c_0$, $j(x) = k_t c_w(x)$, see (14.3.2); $k_t = 5 \times 10^{-3}$ cm/s; Wootton et al. 2001) (see Plate 25)

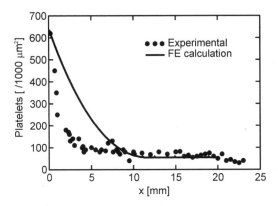

Fig. E14.3-2 Axial distribution of deposited platelets on collagen wall, as predicted by the three-species continuum model (see Section 14.3.1), and experimental results of Hubbell and McIntire (1986); after 120 s with wall shear rate = 500 s^{-1}

Example 14.3-2. Platelet aggregation in blood flow between two parallel plates
A simple experiment of determining platelet accumulation distribution along the collagen-coated plate after perfusion for two minutes at controlled wall shear rates is presented in Hubbell and McIntire (1986). A steady flow condition was maintained during the experiment. Red blood cells were disposed from the system. The lower surface exposed to the blood was coated with collagen, thus providing deposition along the entire lower plate. The gap between the two parallel plates was 200 μm. The flow is characterized by the wall shear rate of 500 s^{-1}, while the entrance platelet concentration was $2.0 \times 10^8 /mL$.

Our FE model of simple blood flow consists of 2D finite elements (see Sections 4.4 and 7.4). We used a three-species continuum model described in Section 14.3.1. The constants in (14.3.6) employed in this example are: $D_n = 1.6 \times 10^{-7}\ cm^2 s^{-1}$, $D_c = 8.0 \times 10^{-7}\ cm^2 s^{-1}$, $A = 1, R(c) = const = 1$.

The experimental and computed results for the adhered platelet distribution after 120 s for the shear rates of 500 s^{-1} are shown in Fig. E14.3-2; and they compare reasonably well.

14.4 Modeling of thrombosis by DPD

14.4.1 General considerations

To study the fundamental nature of platelet activation, aggregation and adhesion, it would be desirable to use the so-called 'Lagrangian' approach, tracking individual platelets in the sequence of the process. Although the obvious advantages of this discrete-type approach over the traditional continuum methods may have long been recognized, it was not technically feasible until recently. Facilitated by rapid increases in computer power, however, Lagrangian computational approaches have recently been the subject of intensive research with advances such that they are now becoming more applicable to real problems (see Chapter 8 and reference therein, e.g. Liu *et al.* 2004a, Espanol 1998).

The objective in this section is to apply one of these new computational methods, the dissipative particle dynamics (DPD) method, to simulate platelet-mediated thrombosis. In a simplified model, where presence of RBCs is neglected, blood is discretized into mesoscale particles representing plasma and platelets. Each platelet is modeled by one DPD particle. Besides the interaction repulsive, viscous and random forces among DPD particles, in a form described in Section 8.2, the attractive forces among activated platelets and with the wall, are included. These attractive forces represent the action of the proteins connecting the activated platelet, as schematically shown in Fig. 14.4.1. Namely, when a platelet is activated, its surface becomes sticky due to the expression of surface receptors (GP-IIb/IIIa) interacting with the plasma protein fibrinogen (see Fig. 14.1.2). Fibrinogen binds to these receptors of two activated platelets and forms a molecular bond between platelets.

Basic Equations

The basic equations of the DPD model of a fluid are presented in Section 8.2. These equations are also applicable to the modeling of platelet-mediated thrombosis. Taking each platelet

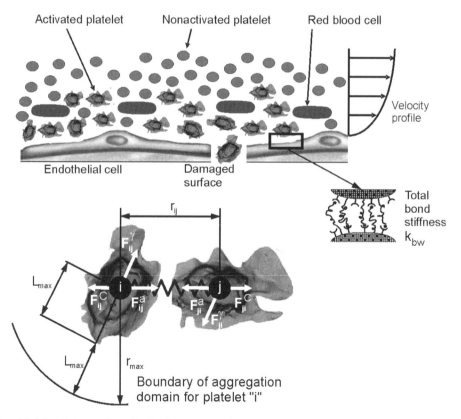

Fig. 14.4.1 Schematics of platelet aggregation and adhesion. Activated platelets in the vicinity of a injured wall epithelium and binding of platelets at the walls using springs. Interaction forces for two aggregated platelets (Filipović et al. 2008). The domain of the interaction between platelets is denoted by r_{max}

as a DPD particle, we have, besides the repulsive, dissipative and random forces, also the attractive force between the activated platelets. Then the equation (8.2.1) for a particle 'i' can be written as

$$m_i \dot{\mathbf{v}}_i = \sum_j (\mathbf{f}_{ij}^C + \mathbf{f}_{ij}^D + \mathbf{f}_{ij}^R + \mathbf{f}_{ij}^a) + \mathbf{f}_i^{ext} \qquad (14.4.1)$$

where m_i is the particle mass; $\dot{\mathbf{v}}_i$ is particle acceleration as the time derivative of velocity; $\mathbf{f}_{ij}^C$, $\mathbf{f}_{ij}^D$, $\mathbf{f}_{ij}^R$ and $\mathbf{f}_{ij}^a$ are the conservative (repulsive), dissipative, random and attractive interaction forces, that particle 'j' exerts on particle 'i', respectively, provided that particle 'j' is within the radius of influence r_c of particle 'i'; and $\mathbf{f}_i^{ext}$ is the external force acting on particle 'i'. Note that the attractive force exists between the activated platelet and the vessel wall, while it is equal to zero when either platelet 'i' or 'j' is not activated. Also, $\mathbf{f}_{ij}^a = \mathbf{0}$ between platelets and plasma DPD particles, as well as between the plasma particles.

When an activated platelet and a vessel wall are in proximity, they bind. However, when adhered platelets are exposed simultaneously to other forces stronger than the binding force, the bond breaks. To model platelet adhesion to vessel walls, we adopt as an approximation that the attractive forces are represented by springs, schematically shown in Fig. 14.4.1. The effective spring constant for platelet adhesion on the vessel wall, or to another stationary activated platelet, is denoted by k_{bw}.

An additional parameter involved in the model is the size of the domain from the collagen-coated wall (L_{max}^{wall}) in which the action of attractive force needs to be considered. Assuming a linear decrease of this force with the distance from the wall, we have that

$$f_w^a = k_{bw} \left(1 - L_w / L_{max}^{wall}\right) \qquad (14.4.2)$$

where L_w is the distance of the activated platelet from the wall.

In the examples given in next section we illustrate application of the DPD method to modeling of thrombosis.

14.4.2 Examples

In this section we first present modeling of blood flow in a microchannel with narrowing (stenosis) by two discrete particle methods, DPD and SPH (see Sections 8.2 and 8.4); and we also include the multiscale method (MBS) of Section 8.3. Then, computation of platelet accumulation on a collagen surface by using the DPD method is given as the second example.

Example 14.4-1. Blood flow in a microchannel with narrowing

Consider a steady blood flow between two parallel plates with narrowing, Fig. E14.4-1a. We solve this example using the DPD and SPH, as well as the multiscale (DPD+FE) method to demonstrate applicability of these methods to modeling of microcirculation.

Parameters used in the DPD and SPH models are the same as in Examples 8.2-1 and 8.4-1 (see also data in the figure caption). For comparison, the finite element (FE) solution is shown.

In the FE-DPD multiscale MBS model (see Section 8.3) the local DPD+FE domain, as well as the global FE domain are shown in Fig. E14.4-1b (Kojić 2008). At the common

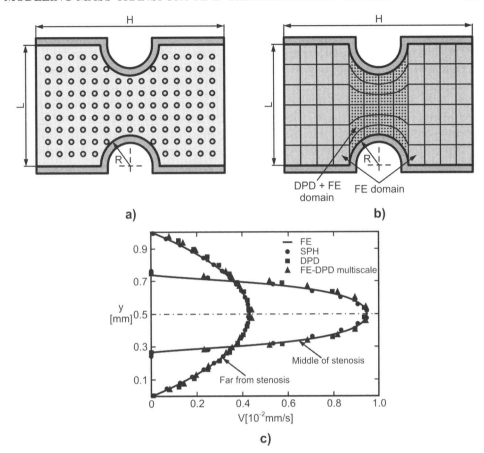

Fig. E14.4-1 Steady blood flow in a channel with narrowing. (a) Geometry of the channel and SPH initial particle positions; (b) Local (DPD+FE) and global (FE only) domains used for the multiscale MBS method; (c) Velocity profiles (FE, DPD, SPH and DPD-FE multiscale solutions). Data: kinematic viscosity $\nu = 10^{-6} \, m^2 s^{-1}$, fluid density $\rho = 10^3 \, kgm^{-3}$, lengths (m) $H = 2.14 \times 10^{-3} m$, $L = 1 \times 10^{-3} m$, $R = 2.5 \times 10^{-4} m$, acceleration $a = 10^{-3} \, ms^{-2}$, pressure gradient $\partial p / \partial x = \rho a = 1 \, kgm^{-2} s^{-2}$

boundary between the local and global domains the mesoscale DPD particle velocities are equal to the coarse scale FE velocities. The periodic boundary conditions are imposed at the common boundary to keep the number of particles constant.

Velocity profiles are shown in Fig. E14.4-1c, where a significant velocity increase in the domain of narrowing (stenosis) is notable. The solutions using the DPD, SPH and FE-DPD multiscale MBS methods compare well with the FE solution. Results for other model parameters can be obtained using the Software on the web.

Example 14.4-2. Platelet deposition in a perfusion chamber
To test application of the DPD method and the assumption about the wall attractive force, platelet deposition in a perfusion chamber is modeled. The model corresponds to the experiment of Hubbell and McIntire (1986), which is described in Example 14.3-2.

In the DPD model, a constant inflow flux of blood and the entering platelet concentration are imposed, as in the experiments. The flow domain is represented as a simple lattice with 10000×100 particles and periodic boundary conditions (see Section 8.2) along the direction of fluid flow. The external body force is used, necessary to reach the wall shear rates of the experiment. The conservative force parameter was taken as $a_{ij} = 25$, and the friction coefficient $\gamma = 4.5$ (see (8.2.3)). It is taken that L_{max}^{wall} be the whole chamber gap ($190 \, \mu m$) for calculation of the wall attractive force (14.4.2). The attractive forces between the activated moving platelets are neglected in the model. The solid walls are modeled by freezing the DPD particles at the wall surface, without possibility of breaking the bond. We also used the specular reflection boundary conditions where the velocity component tangential to the wall does not change while the normal component is reversed (see Section 8.2, eq. (8.2.8)).

The snapshot in Fig. E14.4-2Ab shows the platelet distribution within the blood and platelets adhered to the vessel wall. The experimental results and the computed results for the adhered platelet distribution after $120 \, s$ for the shear rates of $500 \, s^{-1}$ and $1500 \, s^{-1}$ are shown in Fig. E14.4-2B. It was found that the best fit of the numerical solution and experiment is achieved for the value of spring constant $k_{bw} = 50 \, N/m$ (Filipović et al. 2008).

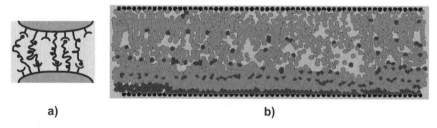

a) b)

Fig. E14.4-2A Platelet adhesion to the wall. (a) Schematics of the attraction between activated platelets and vessel wall; (b) Snapshot of DPD particles after $120 \, s$ of platelet deposition. Flow is from left to right

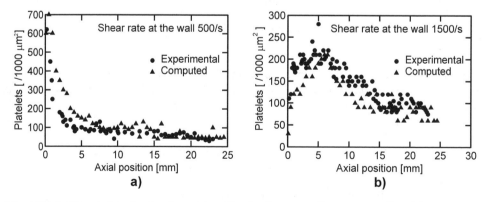

Fig. E14.4-2B Axial platelet deposition distribution on collagen as predicted by computer solution using the DPD method, and experimental results of Hubbell and McIntire (1986); after $120 \, s$. (a) Wall shear rate $= 500 \, s^{-1}$; (b) Wall shear rate $= 1500 \, s^{-1}$

It can be seen from the above that the computed results match the results experimentally recorded by Hubbell and McIntire. These results are also in agreement with the continuum-based solution of a diffusion-controlled transport of platelets over a reactive surface (Example 14.2-3).

References

Basmadjian, D. (1990). The effect of flow and mass transport in thrombogenesis, *Ann. Biomed. Eng.*, **18**, 685–709.

Berliner, J.A., Navab, M., Fogelman, A.M. et al. (1995). Atherosclerosis: basic mechanisms, oxidation, inflammation, and genetics, *Circulation*, **91**, 2488–96.

Bluestein, D., Gutierrez, C., Londono, M. & Schoephoerster, R.T. (1999). Vortex shedding in steady flow through a model of an arterial stenosis and its relevance to mural platelet deposition, *Ann. Biomed. Eng.*, **27**, 763–73.

Bratzler, R., Chislom, G. & Colton, C. (1977). The distribution of labeled low-density lipoproteins across the rabbit thoracic aorta in vivo, *Atherosclerosis*, **28**, 289–307.

Brooks, A.N. & Hughes, T.J.R. (1982). Streamline upwind/Petrov–Galerkin formulations for convection dominated flows with particular emphasis on the incompressible Navier–Stokes equations, *Comp. Meth. Appl. Mech. Eng.*, **32**, 199–259.

Caro, C.G., Fitz-Gerald, J.M. & Schroter, R.C. (1971). Atheroma and arterial wall shear. observation, correlation and proposal of a shear dependent mass transfer mechanism for atherogenesis, *Proc. R. Soc. London, Ser. B*, **177**, 109–59.

Filipović, N., Ravnic, D.J., Kojić, M., Mentzer, S.J., Haber, S. & Tsuda, A. (2008). Interactions of blood cell constituents: experimental investigation and computational modeling by discrete particle dynamics algorithm, *Microvasc. Res.*, **75**, 279–84.

Flekkoy, E.G., Coveney, P.V. & De Fabritiis, G. (2000). Foundations of dissipative particle dynamics, *Phys. Rev. E*, **62**, 2140–57.

Fogelson, A.L. (1984). A mathematical model and numerical method for studying platelet adhesion and aggregation during blood clotting, *J. Comp. Phys.*, **56**, 111–34.

Fogelson, A.L. (1992). Continuum models of platelet aggregation: formulation and mechanical properties, *Siam J. Appl. Math.*, **52**(4), 1089–110.

Fogelson, A.L. & Guy, R.G. (2004). Platelet–wall interaction in continuum models of platelet thrombosis: formulation and numerical solution, *Mathematical Med. Biol.*, **21**(4), 293–334.

Fogelson, A.L. & Kuharsky, A.L. (1998). Membrane binding-site density can modulate activation thresholds in enzyme systems, *J. Theor. Biol.*, **193**, 1–18.

Fuster V, Badimon L, Badimon JJ, Chesebro JH (1992a). The pathogenesis of coronary artery disease and the acute coronary syndromes, *N. Engl. J. Med.*, 326, 242–250.

Goldstein, J., Anderson, R. & Brown, M. (1979). Coated pits, coated vesicles, and receptor-mediated endocytosis, *Nature*, **279**, 679–84.

Groot, R.D. & Warren, P.B. (1997). Dissipative particle dynamics: bridging the gap between atomistic and mesoscopic simulation, *J. Chem. Phys.*, **107**(11), 4423–35.

Haber, S., Filipović, N., Kojić, M. & Tsuda, A. (2006). Dissipative particle dynamics simulation of flow generated by two rotating concentric cylinders, *Phys. Rev. E*, **74**, 1–8.

Hoff, H.F., Heideman, C.L., Jackson, R.L., Bayardo, R.J., Kim, H.S. &, Gotto, A.M. Jr. (1975a). Localization patterns of plasma apolipoproteins in human atherosclerotic lesions, *Circ. Res.*, **37**, 72–9.

Hoff, H.F., Titus, J.L., Bajardo, R.J., Jackson, R.L., Gotto, A.M., DeBakey, M.E. & Lie, J.T. (1975b). Lipoproteins in atherosclerotic lesions. localization by immunofluorescence of apo-low density lipoproteins in human atherosclerotic arteries from normal and hyperlipoproteinemics, *Arch. Pathol.*, **99**, 253–8.

Hoogerbrugge, P.J. & Koelman, J.M.V.A. (1992). Simulating microscopic hydrodynamic phenomena with dissipative particle dynamics, *Europhys. Lett.*, **19**, 155–160.

Hubbell, J.A. & McIntire, L.V. (1986). Technique for visualization and analysis of mural thrombogenesis, *Rev. Sci. Instrum.*, **57**(5), 892–7.

Kojić, M. (2008). On the application of dicrete particle methods and their coupling to the continuum-based methods within a multiscale scheme, *Advances in Nonlinear Sciences*, Monograph 2 – Yugoslavian Academy for Nonlinear Sciences (JANN), to appear.

Kojić, M., Filipović, N. & Tsuda, A. (2008). A mesoscopic bridging scale method for fluids and coupling dissipative particle dynamics with continuum finite element method, *Comp. Meth. Appl. Mech. Eng.* **197**, 821–33.

Loscalzo, J. & Schafer, A.I. (2003). *Thrombosis and Hemorrhage*, 3rd edition, Lippincott Williams & Wilkins, Philadelphia.

Moore, J.A. & Etheir, C.R. (1997). Oxygen mass transfer calculations in large arteries, *J. Biomech. Eng.*, **119**, 469–75.

Ruggeri, Z.M. (2000). Old concepts and new developments in the study of platelet aggregation, *J. Clin. Invest.*, **105**(6), 699–701.

Sorensen, E.N. (2002). *Computational simulation of platelet transport, activation and deposition*, Ph.D. Thesis, University of Pittsburgh.

Sorensen, E.N., Burgreen, G.W., Wagner, W.R. & Antaki, J.F. (1999a). Computational simulation of platelet deposition and activation: I. Model development and properties, *Ann. Biomed. Eng.*, **27**, 436–48.

Sorensen, E.N., Burgreen, G.W., Wagner, W.R. & Antaki, J.F. (1999b). Computational simulation of platelet deposition and activation: II. Results of Poiseulle flow over collagen, *Annals Biomed. Eng.*, **27**, 449–58.

Verstraete, M., Fuster, V. & Topol, E.J. (1998). *Cardiovascular Thrombosis: Thrombocardiology and Thromboneurology*, 2nd edition, Lippincot-Raven Publishers.

Wootton, D.M., Markou, C.P., Hanson, S.R. & Ku, D.N. (2001). A mechanistic model of acute platelet accumulation in thrombogenic stenosis, *Ann. Biomed. Eng.*, **29**, 321–9.

Yuan, S.W. & Finkelstein, A.B. (1956). Laminar pipe flow with injection and suction through a porous wall, *Trans. ASME*, **78**, 719–24.

15

Cartilage Mechanics

Cartilage is the main part of joints associated to protect bone rubbing on bone while our skeleton is moving. In this chapter we first describe cartilage and the spine. In order to develop a computational model, we present differential equations of balance in cartilage mechanics with additional effects of swelling pressure and electrokinetic coupling. Then we give finite element balance equations and some typical numerical examples for cartilage and spine deformation. Solved examples include electromechanical coupling of cartilage, the free-swelling problem, as well as one-dimensional and three-dimensional spinal motion segment modeling.

15.1 Introduction

General Properties of Cartilage
In general, joints are designed to prevent bone rubbing on bone in the moving parts of skeletons, and protect and cushion the bones from damage. Articular cartilage is the dense connective tissue that covers the bone (Fig. 15.1.1a).

The cartilage is a multiphasic, nonlinear permeable viscoelastic material, consisting of *two principal phases*: a solid and fluid. The *solid phase* is comprised primarily from *collagen* arranged in a specific fibrillar network (Ghadially 1978), proteoglycans, nonspecific glycoproteins and chondrocytes in the aggregated form (Mow *et al.* 1980). A movable *interstitial fluid phase*, which is predominantly water, contains approximately 78% of wet weight. Inside the solid phase we recognize the cellular–chondrocyte (2% of total volume), and acellular–extracellular matrix (20%).

The collagen fibres give tensile stiffness and proteoglycans from fluid phase give compressive stiffness to the cartilage. The exterior part of cartilage is covered by a dense fibrous membrane called the perichondrium. If cartilage is damaged, the healing is very difficult because there are no nerves or blood vessels in cartilage.

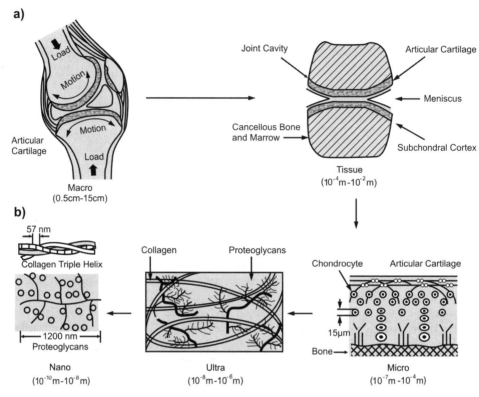

Fig. 15.1.1 The role of cartilage and cartilage structure. (a) Articular cartilage in a joint protects the bones from damage; (b) Cartilage structure (adapted from http://www.engin.umich.edu/class/bme456/cartilage/cart.htm)

Cartilage is classified histologically as elastic, fibrocartilagenous or hyaline which depends on its molecular composition. The ear and the larynx are composed of elastic cartilage, while fibrocartilage is associated with the menisci of the knee and the intervertebral discs. Hyaline cartilage is the predominant form of cartilage, and is most commonly associated with the skeletal system. Figure 15.1.1a represents the composite structure of joints which consists of bone, articular cartilage, ligaments, tendons, muscle and the joint capsule. The more detailed actual bearing surface of the joint is shown in Fig. 15.1.1b.

Both solid and fluid phases are considered to be incompressible (Mow *et al.* 1980). A fundamental observation in cartilage is *fluid exuding* when cartilage is compressed. It is known that *interstitial fluid pressure* supports almost 95% of applied load and the rest is supported by a solid matrix. There is also a phenomenon addressed as *creep* where due to fluid exudation, fluid loss increases the swelling pressure enough to support the applied external pressure. During compression fluid exudation causes the stress to rise above the elastic equilibrium value which is associated with *stress relaxation*. It is assumed that the stress relaxation is caused by fluid redistribution within the extracellular matrix (Armstrong *et al.* 1984).

Experimental investigations show that *swelling pressure* is generated (or electrical potential is created) within the cartilage during cartilage deformation, as a consequence of chemical

CARTILAGE MECHANICS

diffusion processes. These specific effects are very important for cartilage mechanical behavior and will be described in a mathematical form in Section 15.2 for further use in computer modeling.

Today, the two main diseases caused by damage of the collagen network within cartilage are arthritis and osteoporosis. These diseases are directly related to change in cartilage permeability as the mechanical property. If permeability is increased fluid can be easily exuded and consequently the fluid pressure becomes lower. This significantly decreases the load supported by the fluid phase, leading to higher stresses in the solid phase and its damage.

The Spine Anatomy

Numerical solutions for deformation of cartilage between vertebrae of the spine will be presented in Section 15.4, and we here give a brief description of the spine anatomy. The spine is one of the most important parts of the body because it gives the body structure and support. A column of nerves that connects the brain to the rest of the body, allowing the control of the movements, is called the *spinal cord*. Ligaments and muscles as addition to the spinal column also give stability of the body. The spinal column has three main sections: the *cervical spine*, the *thoracic spine* and the *lumbar spine*. The first seven vertebrae is called the cervical spine. The mid-back, called the thoracic spine, consists of 12 vertebrae. The lower portion of the spine, the lumbar spine, is usually made up of five vertebrae (Yoganandan *et al.* 1987). Cartilage situated between two more rigid body vertebrae (Fig. 15.1.2 – see color plate) represents the *spinal motion segment* (SMS) – a repeating unit from which the spine is composed (Simon *et al.* 1985).

The *vertebrae* protect and support the spinal cord. They also bear the majority of the weight carried by the spine. Vertebrae, like all bones, have an outer shell, called *cortical bone*, which is hard and strong. The inside is made of a soft, spongy type of bone, called *cancellous* bone. Interverterbal disc has a strong outer ring of fibers named the *annulus*,

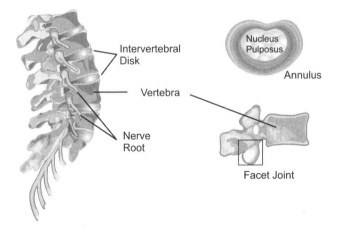

Fig. 15.1.2 Spine anatomy. There are L1 to L5 spinal motion segments (SMS) which consist of vertebrae and intervertebral discs. The annulus and nucleus pulposus are two main materials of disc, while vertebrae have inside a soft spongy type of bone, called cancellous bone, and an outer shell called cortical bone which is much stronger to support the spinal cord (see Plate 26)

and a soft, jelly-like center called the nucleus pulposus (Fig. 15.1.2 – see color plate). The annulus as the strongest area of the disc keeps the disc's center intact. The facet joints are the real joints inside the spinal column. They link the vertebrae together and give them the flexibility to move against each other (Bogduk *et al.* 1992).

15.2 Differential equations of balance in cartilage mechanics

As described in Section 15.1, cartilage represents a porous medium containing fluid that fills the pore space. Therefore, we have a mixture of two phases: solid as a supporting matrix, and fluid. Under loading the solid matrix deforms changing its size and shape. The pore space is also changing. Deformation of the solid induces motion of the fluid within the pore space. On the other hand, the fluid acts on the solid by shear due to relative motion with respect to the microstructural solid architecture and by the pressure. Considering this mixture of the two phases we see that these two phases are mechanically coupled when subjected to mechanical and/or biochemical action.

We further describe the basic quantities used in the cartilage mechanical model, which includes additional effects expressed by the swelling pressure or electrokinetic coupling. Then, the governing equations for this model are derived (Kojić *et al.* 2001): the balance of linear momentum and mass balance equations.

15.2.1 Basic physical quantities, swelling pressure and electrokinetic coupling

Definition of Stress, Strain and Fluid Velocity

These quantities are defined in Section 3.4 for a solid–fluid mixture. A short summary of these definitions is given here for the completeness of this section.

The total stress $\boldsymbol{\sigma}$, as a force per unit surface of the solid–fluid mixture, can be expressed in terms of the stress carried by the solid, $\boldsymbol{\sigma}_s$, and the fluid pressure, p, as (3.4.6): $\boldsymbol{\sigma} = (1-n)\boldsymbol{\sigma}_s - n\mathbf{m}p$, where $\mathbf{m}$ is a constant vector defined as $\mathbf{m} = [1\ 1\ 1\ 0\ 0\ 0]$, and '$n$' is porosity. The effective stress, $\boldsymbol{\sigma}'$, is given in (3.4.9), $\boldsymbol{\sigma}' = \boldsymbol{\sigma} + \mathbf{m}p$. The one-index notation is used as in Section 3.4.

The cartilage is considered as a deformable continuum (deformable mixture), hence the strains $\mathbf{e}$ are calculated from displacements $\mathbf{u}$. The displacements and strains of the mixture refer to the solid phase. In the case of small deformations, we use the small strains defined in (2.1.25), $e_{ij} = 0.5\left(\partial u_i/\partial x_j + \partial u_j/\partial x_i\right)$. When the cartilage undergoes large deformations, various strain measures may be employed, as given in (2.4.15)–(2.4.17). For example, the logarithmic strains for the configuration are given in (2.4.17), ${}^t e_{ij} = \sum_{\alpha=1}^{3} \ln {}_0^t\lambda_\alpha \,({}^t\overline{\mathbf{p}}_\alpha)_i \,({}^t\overline{\mathbf{p}}_\alpha)_j$ where the ${}_0^t\lambda_\alpha$ are the stretches in the principal directions ${}^t\overline{\mathbf{p}}_\alpha$ of the left Cauchy–Green deformation tensor ${}_0^t\mathbf{B}$.

There are two types of fluid velocities: microvelocity and macrovelocity. The microvelocity $\mathbf{v}_f$ is the true velocity of fluid particles passing through the pores. On the other hand, the macrovelocity $\mathbf{q}$, called Darcy's velocity in general flows through porous media, is defined as the volume of fluid passing in a unit time through a unit area of the mixture.

CARTILAGE MECHANICS

The relationship between the fluid velocities $\mathbf{v}_f$ and $\mathbf{q}$ is given in (3.4.3): $\mathbf{q} = n(\mathbf{v}_f - \dot{\mathbf{u}})$, where $\dot{\mathbf{u}}$ is the velocity of a point of the mixture. Therefore, Darcy's velocity is the relative velocity of fluid with respect to the solid.

We will employ a simple isotropic elastic material model for the solid and the effective stress principle, therefore the constitutive relations are given in (3.4.7): $\boldsymbol{\sigma}' = \mathbf{C}^E (\mathbf{e} - \mathbf{e}^p)$, where $\mathbf{C}^E$ is the elastic constitutive matrix of the solid material, given in (2.2.5); and $\mathbf{e}^p$ is the strain in the solid due to pressure, $\mathbf{e}^p = -\mathbf{m}p/(3K_s)$, where K_s is the bulk modulus of the solid material, see (2.2.11).

Swelling Pressure

Two approaches in the mathematical description of the swelling pressure are presented. First, according to Laible *et al.* (1993), the total pressure at a point, p_{tot}, can be written as the sum of the fluid pressure p and the swelling pressure p_c,

$$p_{tot} = p + p_c \tag{15.2.1}$$

Further, the swelling pressure can be expressed as a nonlinear function of change in the water content ζ,

$$p_c = p_{co} + k_c(\zeta)\zeta \tag{15.2.2}$$

where p_{co} is the initial swelling pressure, and $k_c(\zeta)$ is a function determined empirically. The variable ζ can be expressed by the divergence of the relative displacement of the fluid $\mathbf{u}_f$, i.e.

$$\zeta = \nabla^T \mathbf{u}_f \tag{15.2.3}$$

The fluid displacement, on the other hand, can be obtained from Darcy's velocity $\mathbf{q} = \dot{\mathbf{u}}_f$, hence the parameter ζ is related to the fluid compressibility.

Another approach for defining the swelling pressure is given in Simon and Gaballa (1988). There, a concentration strain $\mathbf{e}^c$ is introduced,

$$\mathbf{e}^c = \mathbf{m}\alpha_c c \tag{15.2.4}$$

where c is the local ion concentration, and α_c is the coefficient of chemical contraction. This strain enters the constitutive relation (3.4.7) for solid, hence we have

$$\boldsymbol{\sigma}' = \mathbf{C}^E (\mathbf{e} - \mathbf{e}^p - \mathbf{e}^c) \tag{15.2.5}$$

On the other hand, the concentration is governed by Fick's law (see (3.2.5)),

$$\mathbf{q}^c = -\beta_c \nabla c \tag{15.2.6}$$

where $\mathbf{q}^c$ is the ionic flux, and β_c is the ionic diffusion coefficient. The governing equation for the ion concentration follows from the mass balance (see (3.2.8)),

$$\nabla^T (\beta_c \nabla c) - \frac{\partial c}{\partial t} = 0 \tag{15.2.7}$$

The ionic diffusion rate is large when compared with the relative fluid velocity, so that the field of concentration can be considered stationary in the analysis. Then, taking that β_c can be considered constant, the last equation reduces to the Laplace equation $\nabla^2 c = 0$.

Electrokinetic Coupling

The swelling pressure effects can be interpreted through the electrokinetic coupling (Frank & Grodzinsky 1987a,b; Sachs & Grodzinsky 1989). It was experimentally found that the following relation can be established (which encompasses Ohm's and Darcy's laws as special cases):

$$\begin{Bmatrix} \mathbf{q} \\ \mathbf{j} \end{Bmatrix} = \begin{bmatrix} -k_{11} & k_{12} \\ k_{21} & -k_{22} \end{bmatrix} \begin{Bmatrix} \nabla p \\ \nabla \phi \end{Bmatrix} \qquad (15.2.8)$$

where $\mathbf{j}$ is the current density, ϕ is the electrical potential, k_{11} is (short-circuit) Darcy's hydraulic permeability, k_{22} is the electrical conductivity, and k_{12} and k_{21} are the electrokinetic coupling coefficients that are mutually equal according to the Onsager reciprocity.

15.2.2 Equations of balance

After the above definitions of the basic quantities, we here present the governing equations for the cartilage model. As in Section 3.4, the cartilage is considered as a continuum schematically shown in Fig. 15.2.1. According to the description of the cartilage model in Section 15.2.1, the variables at a material point P of the mixture are: displacement of the mixture (displacement of the solid), $\mathbf{u}$; relative fluid velocity with respect to the solid (Darcy's velocity), $\mathbf{q}$; the fluid pressure, p; swelling pressure p_c – used when the water content approach according to Laible *et al.* (1993) is used; electrical potential, ϕ, if the swelling pressure effects are defined by the electrokinetic coupling.

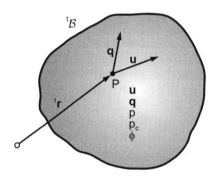

Fig. 15.2.1 Configuration $^t\mathcal{B}$ of cartilage at time t, considered as a continuous medium (mixture), and variables at a material point P whose position vector is $^t\mathbf{r}$. The variables are: $\mathbf{u}$ – displacement of the mixture, $\mathbf{q}$ – relative fluid velocity with respect to the solid (Darcy's velocity), p – fluid pressure, p_c – swelling pressure used according to Laible *et al.* (1993); ϕ – electrical potential, if the swelling pressure effects are interpreted by electrokinetic coupling

CARTILAGE MECHANICS

Differential Equations of Motion

First, we have that the differential equation of motion (3.4.5) remains the same,

$$\nabla^T \boldsymbol{\sigma} + \rho \mathbf{b} - \rho \ddot{\mathbf{u}} + \rho_f \dot{\mathbf{q}} = \mathbf{0} \tag{15.2.9}$$

where $\rho = (1-n)\rho_s + n\rho_f$ is the mixture density, and **b** is body force per unit mass. Derivation of this equation of balance of linear momentum is presented in detail in Section 3.4. Next, the equation of balance of linear momentum for the fluid is given in (3.4.4) if the swelling effects are neglected,

$$-\nabla p + \rho_f \mathbf{b} - \mathbf{k}^{-1}\mathbf{q} - \rho_f \ddot{\mathbf{u}} - \frac{\rho_f}{n}\dot{\mathbf{q}} = \mathbf{0} \tag{15.2.10}$$

However, if the swelling pressure according to Laible *et al.* (1993) is used, then the total pressure p_{tot} given in (15.2.1) must be used in (15.2.10) and in the expression (3.4.9): $\boldsymbol{\sigma}' = \boldsymbol{\sigma} + \mathbf{m}p$, instead of the pressure p.

If the electrokinetic coupling is employed, then the governing equation (15.2.10) is modified as follows (Kojić *et al.* 2001). First, we see that the resistance force per unit volume of the mixture, $\mathbf{F}_w$, acting on the fluid, follows from the first equation of the system (15.2.8),

$$\mathbf{F}_w = -k_{11}^{-1}\mathbf{q} + k_{11}^{-1}k_{12}\nabla\phi \tag{15.2.11}$$

Then, (15.2.10) changes to

$$-\nabla p + \rho_f \mathbf{b} - k_{11}^{-1}\mathbf{q} + k_{11}^{-1}k_{12}\nabla\phi - \rho_f \ddot{\mathbf{u}} - \frac{\rho_f}{n}\dot{\mathbf{q}} = \mathbf{0} \tag{15.2.12}$$

Further, the continuity equation for the current density must be satisfied,

$$\nabla^T \mathbf{j} = 0 \tag{15.2.13}$$

Substituting the current density **j** from the second equation of the system (15.2.8) into (15.2.13), the following equation is obtained:

$$k_{21}\nabla^T\nabla p - k_{22}\nabla^T\nabla\phi = 0 \tag{15.2.14}$$

This equation is solved together with other governing equations of the model.

Continuity Equation

The continuity equation for the fluid is given in (3.4.10), according to Lewis and Schrefler (1987):

$$\nabla^T \mathbf{q} + \left(\mathbf{m}^T - \frac{\mathbf{m}^T \mathbf{C}^E}{3K_s}\right)\dot{\mathbf{e}} + \left(\frac{1-n}{K_s} + \frac{n}{K_f} - \frac{\mathbf{m}^T \mathbf{C}^E \mathbf{m}}{9K_s^2}\right)\dot{p} = 0 \tag{15.2.15}$$

when the swelling pressure effects are neglected, or when these effects are expressed by electrokinetic coupling. However, when these effects are expressed according to (15.2.1) or by ionic concentration, additional terms appear in (15.2.15), as presented on the web – Theory, Chapter 15. Detailed derivation of (15.2.15) is given in web – Theory, Chapter 3.

Change of porosity is described by (3.4.12) and remains the same in the case when swelling pressure effects are taken into account.

15.3 Finite element modeling of cartilage deformation

In this section we present the equations of balance of a finite element which are obtained by transforming the fundamental relations given in Section 15.2 into the finite element equations. The derivation is analogous to that given in Section 7.7. Part of the derivations which is the same as in Section 7.7 is omitted, while details are given for the swelling effects. It is assumed that the displacements and strains of the solid matrix are small and that the material is elastic with the constitutive relations given in Section 15.2.1.

15.3.1 Finite element balance equations

Model with Electrokinetic Coupling

In this case the nodal variables are: displacements of solid, $\mathbf{U}$; fluid pressure, $\mathbf{P}$; Darcy's velocity, $\mathbf{Q}$; and electrical potential, $\mathbf{\Phi}$. The equations corresponding to $\mathbf{U}$ and $\mathbf{P}$ are the same as in the system of equations (7.7.4) and are obtained from (15.2.9) and (15.2.15) as described in Section 7.7: applying the principle of virtual work and the Galerkin method.

Further, as in Section 7.7, we multiply (15.2.12) by the interpolation matrix $\mathbf{N}_q^T$ for the relative velocity of fluid $\mathbf{q}$, and integrate over the finite element volume V. The resulting equation is

$$-\int_V \mathbf{N}_q^T \nabla p \, dV - k_{11}^{-1} k_{12} \int_V \mathbf{N}_q^T \nabla \phi \, dV + \int_V \mathbf{N}_q^T \rho_f \mathbf{b} \, dV - k_{11}^{-1} \int_V \mathbf{N}_q^T \mathbf{q} \, dV - \int_V \mathbf{N}_q^T \rho_f \ddot{\mathbf{u}} \, dV - \int_V \mathbf{N}_q^T \frac{\rho_f}{n} \dot{\mathbf{q}} \, dV = \mathbf{0} \quad (15.3.1)$$

The second system of equations follows from (15.2.15) and is given by (7.7.3).

Finally, we multiply the continuity equation (15.2.14) by the interpolation matrix $\mathbf{N}_\phi^T$ for the electrical potential and integrate over the volume, V

$$k_{21} \int_V \mathbf{N}_\phi^T \nabla^T \nabla p \, dV - k_{22} \int_V \mathbf{N}_\phi^T \nabla^T \nabla \phi \, dV = 0 \quad (15.3.2)$$

The interpolation matrices $\mathbf{N}_q$ and $\mathbf{N}_\phi$ are the $N \times 1$ vector-column matrices, where N is the number of finite element nodes used in the interpolation. Note that (usually) in practical applications the interpolation matrices for the displacements $\mathbf{N}_u$ and for the relative velocities $\mathbf{N}_q$ are quadratic, with respect to the natural coordinates, while $\mathbf{N}_p$ for the pressure and $\mathbf{N}_\phi$ for the electrical potential are linear.

A standard procedure of integration over the element volume is performed and the Gauss theorem is employed (see Section 7.1). An implicit time integration scheme is adopted, hence

CARTILAGE MECHANICS

the condition that the balance equations are satisfied at the end of each time step is imposed. The system of differential equations for a finite element is:

$$\begin{bmatrix} \mathbf{M}_{uu} & 0 & 0 & 0 \\ 0 & 0 & 0 & 0 \\ \mathbf{M}_{qu} & 0 & 0 & 0 \\ 0 & 0 & 0 & 0 \end{bmatrix} \begin{Bmatrix} {}^{n+1}\ddot{\mathbf{U}} \\ {}^{n+1}\mathbf{P} \\ {}^{n+1}\ddot{\mathbf{Q}} \\ {}^{n+1}\ddot{\mathbf{\Phi}} \end{Bmatrix} + \begin{bmatrix} 0 & 0 & \mathbf{C}_{uq} & 0 \\ \mathbf{C}_{pu} & \mathbf{C}_{pp} & 0 & 0 \\ 0 & 0 & \mathbf{C}_{qq} & 0 \\ 0 & 0 & 0 & 0 \end{bmatrix} \begin{Bmatrix} {}^{n+1}\dot{\mathbf{U}} \\ {}^{n+1}\dot{\mathbf{P}} \\ {}^{n+1}\dot{\mathbf{Q}} \\ {}^{n+1}\dot{\mathbf{\Phi}} \end{Bmatrix}$$

$$+ \begin{bmatrix} \mathbf{K}_{uu} & \mathbf{K}_{up} & 0 & 0 \\ 0 & 0 & \mathbf{K}_{pq} & 0 \\ 0 & \mathbf{K}_{qp} & \mathbf{K}_{qq} & \mathbf{K}_{q\phi} \\ 0 & \mathbf{K}_{\phi p} & 0 & \mathbf{K}_{\phi\phi} \end{bmatrix} \begin{Bmatrix} \Delta \mathbf{U} \\ \Delta \mathbf{P} \\ \Delta \mathbf{Q} \\ \Delta \mathbf{\Phi} \end{Bmatrix} = \begin{Bmatrix} {}^{n+1}\mathbf{F}_u \\ {}^{n+1}\mathbf{F}_p \\ {}^{n+1}\mathbf{F}_q \\ {}^{n+1}\mathbf{F}_\phi \end{Bmatrix} \quad (15.3.3)$$

The matrices and vectors in this equation, which do not refer to the electrokinetc coupling, are given in (7.7.5). The matrices and vectors not given in (7.7.5) are:

$$\mathbf{K}_{q\phi} = -k_{11}^{-1} k_{12} \int_V \mathbf{N}_q^T \mathbf{N}_{\phi,x} dV \quad \mathbf{K}_{\phi p} = k_{21} \int_V \mathbf{N}_{\phi,x}^T \mathbf{N}_{p,x} dV$$

$$\mathbf{K}_{\phi\phi} = -k_{22} \int_V \mathbf{N}_{\phi,x}^T \mathbf{N}_{\phi,x} dV \quad (15.3.4)$$

$${}^{n+1}\mathbf{F}_q = \int_V \mathbf{N}_q^T \rho_f {}^{n+1}\mathbf{b} dV - \mathbf{K}_{qp}{}^n\mathbf{P} - \mathbf{K}_{qq}{}^n\mathbf{Q} - \mathbf{K}_{q\phi}{}^n\mathbf{\Phi}$$

$${}^{n+1}\mathbf{F}_\phi = \int_A \mathbf{N}_\phi^T \mathbf{n}^T \mathbf{j} dA - \mathbf{K}_{\phi p}{}^n\mathbf{P} - \mathbf{K}_{\phi\phi}{}^n\mathbf{\Phi}$$

In these expressions $\mathbf{n}$ is the normal vector to the boundary, and A is the boundary area.

The above equations are further assembled (see Section 4.2) and the resulting FE system of equations is integrated incrementally, with a time step Δt, transforming this system into a system of algebraic equations. A Newmark integration method, such as described in Section 5.3, may be implemented for the time integration. The unknowns in this algebraic system of equations are increments of the nodal variables.

Note that in the case of nonlinear behavior of the solid material, and/or large displacements of solid, an incremental-iterative scheme must be used as given according to (7.1.11). Also, change of porosity can be taken into account according to (7.7.6). Details about the procedure for solving nonlinear problems are given on the web – Theory, Chapter 7, and Chapter 15.

Model with Water Content

If the swelling pressure effects are taken through the equations (15.2.1)–(15.2.3), the system of equations (15.3.3) will have $\mathbf{U}$, $\mathbf{P}$ and $\mathbf{Q}$ as the nodal point variables, and the force ${}^{n+1}\mathbf{F}_q$ (with omission of the term $-\mathbf{K}_{q\phi}{}^n\mathbf{\Phi}$) will have an additional term ${}^n\mathbf{F}_q^c$,

$${}^n\mathbf{F}_q^c = -\int_V \mathbf{N}_q^T \nabla {}^n p_c dV \quad (15.3.5)$$

where ${}^n p_c$ is evaluated according to (15.2.2), and ${}^n\zeta$ follows from (15.2.3),

$${}^n\zeta = \nabla^T \, {}^n\mathbf{u}_f \quad (15.3.6)$$

Model with Ion Concentration

Finally, if the swelling pressure effect is described by (15.2.4), there is no electrical potential terms in the system (15.3.3) and in the expressions (15.3.4), but then the force $^{n+1}\mathbf{F}_u$ in (7.7.5) is

$$^{n+1}\mathbf{F}_u = \int_V \mathbf{N}_u^T \rho^{n+1}\mathbf{b}dV + \int_A \mathbf{N}_u^T {}^{n+1}\mathbf{t}dA - \int_V \mathbf{B}^{T\,n}\boldsymbol{\sigma}dV - \mathbf{K}_{up}{}^n\mathbf{P} + \frac{\alpha_c E}{1-2\nu}\int_V \mathbf{Bm}\,{}^n c dV \quad (15.3.7)$$

Here, the last term corresponds to swelling pressure effects expressed by ion concentration.

15.4 Examples

We here give five examples with selected set of parameters. Solutions for a range of the parameters can be obtained using the Software on the web.

Example 15.4-1. One-dimensional mechanical-to-electrical transduction of soft biological tissue

We analyze the electrokinetic transduction in charged, homogenous, isotropic, hydrated material with a platen on the top (Fig. E15.4-1a). An imposed displacement of amplitude u_0 elicits a stress and electrokinetic response. The sinusoidal dynamic stiffness and streaming potential are calculated in response to a sinusoidal displacement under open circuit conditions ($\mathbf{j} = \mathbf{0}$), and the predictions of the analytical and FEM (finite element method) are compared with the experimental results (Frank & Grodzinsky 1987a,b).

The displacement of the top surface is given as

$$u(t) = u_0 \cos \omega t \quad (E15.4\text{-}1.1)$$

where w is the circular frequency, $w = 2\pi f$, with f being the frequency. The conditions at $z = 0$ are: fluid can flow freely through the porous platen so that fluid pressure $p = 0$, and the displacement is given in (E15.4-1.1).

The analytical solution for the cartilage stiffness is (Frank & Grodzinsky 1987a,b)

$$\Lambda_c = E\gamma L \coth(\gamma L) \quad (E15.4\text{-}1.2)$$

where $\gamma^2 = jw/(Ek)$, E is the aggregate modulus of elasticity, k is permeability coefficient, and j is the imaginary unit. If the platen spring stiffness is Λ_s, which is in series with the cartilage stiffness, then the total dynamic stiffness of the system, Λ, is

$$\Lambda = \frac{\Lambda_s E\gamma L \coth(\gamma L)}{\Lambda_s + E\gamma L \coth(\gamma L)} \quad (E15.4\text{-}1.3)$$

The analytical solution for the streaming potential is

$$V = k_e \left(\frac{\Lambda_s E\gamma L \tanh\dfrac{\gamma L}{2}}{\Lambda_s + E\gamma L \coth(\gamma L)} \right) \frac{u_0}{L} \quad (E15.4\text{-}1.4)$$

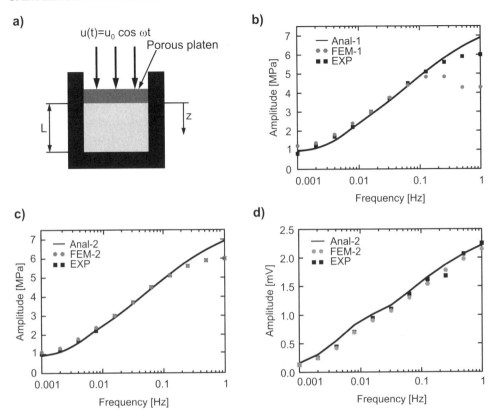

Fig. E15.4-1 Mechanical-to-electrical transduction. (a) Schematics of the model; (b) Amplitude of the dynamic stiffness in terms of the excitation frequency, with material constants given in Frank and Grodzinsky (1987a,b), (solutions: ANAL-1, FEM-1 are analytical and by FEM; EXP – experimental); (c) Amplitude of the dynamic stiffness with adjusted material constants (solutions: ANAL-2, FEM-2, Kojić et al. 2001); (d) Amplitude of the streaming potential under open circuit conditions (constants as in ANAL-2 and FEM-2), Data common for all solutions: $\nu = 0.1$, $L = 680 \times 10^{-6} m$, $k_e = -2.18 \times 10^{-8}$ V/Pa, $k_i = -2.07 \times 10^{-8}$ V/Pa, $u_0 = 10 \times 10^{-6} m$

where k_e is the material constant $k_e = k_{21}/k_{22}$ defined in (15.2.8). The material constants, corresponding to the cartilage and common for all solutions, are given in Fig. 15.4.1, according to Frank and Grodzinsky (1987a,b).

Amplitudes of the dynamic stiffness obtained analytically, experimentally and by FE analysis are shown in Fig. E15.4-1b. The FE model assumes the plane strain conditions, small displacement formulation (geometrical linearity, Kojić et al. 2001), with material constants given in the figure caption and $E = 1$ MPa, $k = 3 \times 10^{-15} m^4/Ns$. The FEM and analytical solutions are denoted in the figure as FEM-1 and ANAL-1. A notable deviation of these solutions from the experimental results (EXP) can be seen in the domain of higher frequencies. On the other hand, this deviation is reduced if the values $E = 0.91$ MPa and $k = 2.6 \times 10^{-15} m^4/Ns$ are used (Kojić et al. 2001), as shown in Fig. E15.4-1c (numerical solution denoted as FEM-2). The analytical and FEM (corresponding to constants used in FEM-2) solutions for the streaming potential are given in Fig. E15.4-1d and agree well with experiments.

Example 15.4-2. One-dimensional electrical-to-mechancial transduction of soft biological tissue

Here, Example 15.4-1 is considered, but the tissue is now subjected to electrical excitation on the top in order to analyze the electrical-to-mechanical transduction. In experimental investigation, mechanical stresses within the tissue are generated by a sinusoidal current applied to the electrodes (Frank & Grodzinsky 1987a). The boundary conditions are as shown in Fig. E15.4-2a. At $z = 0$ and $z = L$: $u = 0$ (restrained displacements), $q = 0$ (impermeable walls); at $z = 0$: $J = J_0 \cos(\omega t)$, where J_0 and ω are amplitude of the current density and circular frequency, respectively. The same FE model is used as in the previous example.

The analytical solution for mechanical stress σ (in the complex domain) is given by (Frank & Grodzinsky, 1987a)

$$\sigma = -\frac{k_i J_0}{jwL} \left(\frac{\Lambda_s E\gamma L \tanh \frac{\gamma L}{2}}{\Lambda_s + E\gamma L \coth(\gamma L)} \right) \quad (E15.4\text{-}2.1)$$

where the definition of all quantities is given in the previous example.

Amplitudes of the mechanical stress are obtained analytically (ANAL-2) and by FEM analysis (FEM-2) (constants corresponding to the FEM-2, geometrical linearity and constant porosity) are shown in Fig. E15.4-2b for a range of the current frequency. The FEM solution compares well with the experimentally recorded stress. Averaging of the stress along the tissue depth is performed for the FEM solution representation.

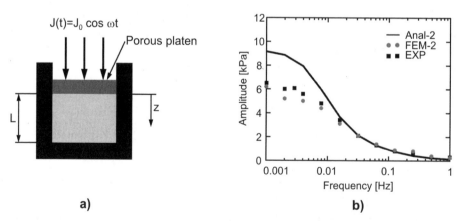

Fig. E15.4-2 Electrical-to-mechanical transduction. (a) Schematics of the model; (b) Dependence of amplitude of the mechanical stress on the excitation frequency generated by the electric current (material constants correspond to FEM-2 solution in Example 15.4-1). Data: $E = 0.91 MPa$, $\nu = 0.1$, $L = 680 \times 10^{-6} m$, $k = 2.6 \times 10^{-15} m^4/Ns$, $k_e = -2.18 \times 10^{-8}$ V/Pa, $k_i = -2.07 \times 10^{-8}$ V/Pa, $J_0 = 3.8 A/m^2$

CARTILAGE MECHANICS

Example 15.4-3. One-dimensional free-swelling problem

The model consists of a vertical column of cartilage tissue open at the top and closed at the base, with geometrical and material data given in Fig. E15.4-3A (Laible et al. 1993). Free swelling is analyzed in time period from $t = 0$ to $t = 20\,000$ s. All three approaches for the description of swelling pressure effects (see Section 15.2) are used.

At time $t = 0$ the swelling pressure is $300\,kPa$. As the material swells, the fluid is imbibed and the swelling pressure drops. During this process a nonzero stress develops in the fluid and the opposite stress develops in the solid, until at the equilibrium stage these stresses become equal and opposite, with the value of the swelling pressure. At the equilibrium, the state represents the end of the free swelling process (here at time $t = 2 \times 10^4$ s), when the fluid velocity becomes equal to zero.

In the first approach (change in water content), the swelling pressure is discribed by a function of time (Laible et al. 1993):

$$p_c = \beta (fcd_{total})^2 \left(\frac{twc}{twc - (A + e^{Bp_c})} \right)^2 \tag{E15.4-3.1}$$

where $\beta = 20.65$, $A = 0.8$, $B = 0.3$, $fcd_{total} = 0.12$ are empirical coefficients; twc is total water content, $twc = \nabla^T \mathbf{u}_f$, $\mathbf{u}_f$ is the fluid 'displacement' – with increment in time step Δt equal to $\Delta \mathbf{u}_f = \Delta t \mathbf{q}$. Figure 15.4-3Ba shows the history of swelling pressure, obtained by using the expression (E15.4-3.1) and the FE model; the initial condition $p_{tot} = p_c = 300\,kPa$ was used.

In the ionic diffusion approach, the coefficient of chemical contraction is taken as $\alpha_c = 0.093$ (1/molar) (Simon and Gaballa 1988). To determine the same response obtained by the change in water content approach we fitted ionic concentration c, and found that $c = 2.1$ (Filipović 1999, Kojić et al. 2001).

In the third approach the electrical potential as boundary condition on the top of the column is used. The material constants in (15.2.8) are used from Sachs and Grodzinsky (1989),

$$\mathbf{k}^{-1} = \mathbf{b} = \begin{pmatrix} b_{11} & b_{12} \\ b_{21} & b_{22} \end{pmatrix} = \begin{pmatrix} 10^{15}\,\frac{Ns}{m^4} & 10^7\,\frac{Vs}{m^2} \\ 10^7\,\frac{N}{Am} & 1\,\Omega m \end{pmatrix} \tag{E15.4-3.2}$$

and the change of electrical potential with time is obtained as shown in Fig. E15.4-3Bb.

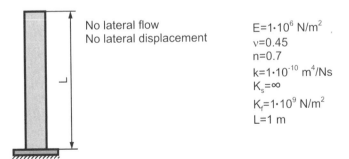

No lateral flow
No lateral displacement

$E = 1 \cdot 10^6$ N/m²
$v = 0.45$
$n = 0.7$
$k = 1 \cdot 10^{-10}$ m⁴/Ns
$K_s = \infty$
$K_f = 1 \cdot 10^9$ N/m²
$L = 1$ m

Fig. E15.4-3A Geometrical and material data for free swelling of a column (E is Young's modulus, v is Poisson's ratio, n is porosity, k is permeability, and K_s and K_f are the bulk moduli of solid and fluid)

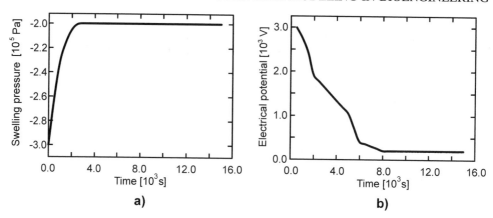

Fig. E15.4-3B Swelling pressure and electrical potential versus time which give the same column response when using the water content approach or electrokinetic coupling. (a) Swelling pressure; (b) Electric potential at the top of the column

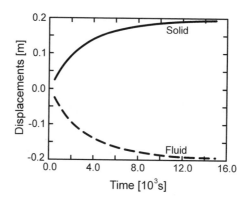

Fig. E15.4-3C Solid and fluid displacements versus time during free swelling (all three approaches)

Solutions for the displacement of solid and fluid, obtained by the three approaches are shown in Fig. E15.4-3C. It can be seen that the solid expands (positive displacement) while the water moves in the opposite direction and displacements of both phases diminish over time.

The swelling pressure p_c (obtained from (E15.4-3.1)) and fluid pressure p changes with time are shown in Fig. E15.4-3D. Both pressures approach to limiting values with time. Note that at the final stage of free swelling the total pressure $p_{tot} = p + p_c$ (see (15.2.1)) tends to zero. Also, when using the ionic diffusion approach, we cannot distinguish the swelling from the fluid pressure; which represents a shortcoming of this approach. This example also demonstrates that the electrokinetic coupling and swelling pressure effects are equivalent.

CARTILAGE MECHANICS

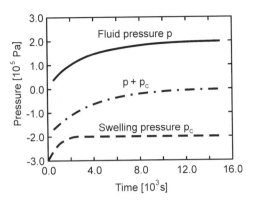

Fig. E15.4-3D Swelling pressure (p_c) and fluid pressure (p) versus time during free swelling. Total pressure $p_{tot} = p + p_c$ tends to zero with time

Example 15.4-4. One-dimensional model of creep response of human spinal motion segment (SMS)

Various FE procedures have been developed for the study of SMS. These procedures are based on the elastic or viscoelastic material models (Simon *et al.* 1985, Laible *et al.* 1993). Also, a simplified one-dimensional analytical model (Simon *et al.* 1985) was proposed to interpret the experimental results (Kazarian 1975).

Here, a one-dimensional model of the SMS is represented by the cylindrical column, constrained laterally and under the condition that no fluid flow is allowed through the sides and the bottom of the cylinder (see also Examples 3.4-1 and 7.7-1). The column is subjected to a step load and free drainage is allowed at the top surface of the cylinder (Fig. E15.4-4a). The applied total stress at the free top surface is $p_0 = 1$ *MPa*.

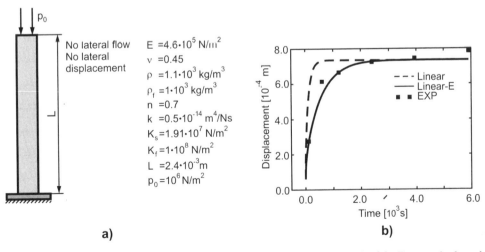

Fig. E15.4-4 One-dimensional model for creep response of SMS. (a) Geometrical and material data; (b) Displacement of the top surface without (Linear) and with (Linear-E) electrokinetic coupling (geometrically linear model), EXP – experimental

From a fitting procedure with respect to the analytical solution, the following values for material constants are obtained: $E = 4.6 \times 10^5 \ N/m^2$ and $k = 0.5 \times 10^{-14} \ m^4/Ns$. The constants for electrokinetic coupling are as in Example 15.4-3. The FE solutions using the plane strain finite elements are obtained employing these constants and others given in Fig. E15.4-4a and assuming geometric linearity (change of column height is neglected). The FE solutions for the column settlement, without and with electrokinetic coupling, are shown in Fig. E15.4-4b. It can be seen that the electrokinetic coupling plays a significant role in the mechanical response, and that solution for the displacement (when the electrokinetic coupling is taken into account) is in a good agreement with experimental results (Kazarian 1975).

Example 15.4-5. Static response of human spinal motion segment (SMS)

In this example we analyze the static response of human SMS based on the poroelastic model, subjected to axial loading, $F = 400 \ N$. The geometry of half of the SMS is shown in Fig. E15.4-5a (see color plate). Three-dimensional finite elements are employed and half of

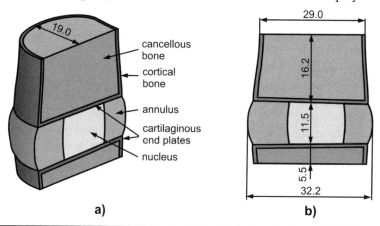

Property	Annulus	Nucleus	Cortical bone	Cartilaginous end plate	Cancellous bone
E [N/m²]	4.55 × 10⁵	4.55 × 10⁵	2.41 × 10⁷	2.41 × 10⁸	2.41 × 10⁷
ν	0.45	0.45	0.25	0.25	0.25
ρ [kg/m³]	1.061 × 10³	1.342 × 10³	4.184 × 10²	4.184 × 10²	4.184 × 10²
ρ_f [kg/m³]	1 × 10³	1 × 10³	–	–	–
n	0.7	0.7	1.0	1.0	1.0
k [m⁴/Ns]	1 × 10⁻¹⁴	1 × 10⁻¹⁴	–	–	–
K_s [N/m²]	1.01 × 10⁷	1.01 × 10⁵	1 × 10³⁵	1 × 10³⁵	1 × 10³⁵
K_f [N/m²]	2.21 × 10⁹	2.21 × 10⁹	–	–	–

Fig. E15.4-5 Dynamic response of human spinal motion segment (SMS). (a) Human SMS (intervertebral disk) and one-half of the model; (b) Geometrical data (length in *mm*) and material properties used for the FE model; (c) von Mises stress distribution (*MPa*); (d) Relative fluid velocity distribution (*m/s*) (see Plate 27) (*Continued on page 329*)

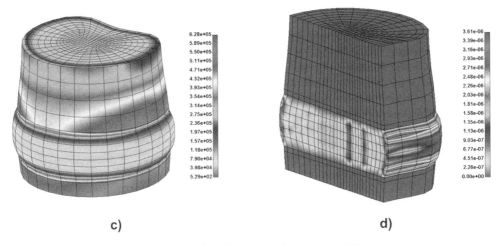

Fig. E15.4-5 (*continued from page 328*)

the SMS (shown in Fig. E15.4-5a, left panel – see color plate) is modeled due to symmetry, with the appropriate boundary conditions at the symmetry plane (no displacements and no velocities through the symmetry plane). Five different materials are employed, for the annulus, nucleus, cortical bones, cartilaginous end plates and cancellous bone. The elastic material constants are given in Fig. E15.4-5b (see Plate 27). The constants are taken from Argoubi and Shirazi-Adl (1996).

The fields of the effective (von Mises) stress and fluid velocity magnitude are shown in Figs. E15.4-5c,d, respectively (see color plate). The locations of maximum von Mises stress and relative velocity between annulus and cancellous bone are in agreement with experimental results (Argoubi & Shirazi-Adl 1996).

References

Argoubi, M. & Shirazi-Adl, A. (1996). Poroelastic creep response analysis of a lumbar motion segment in compression, *J. Biomech.*, **29**, 1331–9.

Armstrong, C.G., Lai, W.M. & Mow, V.C. (1984). An analysis of the unconfined compression of articular cartilage, *J. Biomech. Eng. Trans. ASME*, **106**, 165–73.

Bogduk, N., Macintosh, J.E. & Pearcy, M.J. (1992). A universal model of the lumbar back muscles in the upright position, *Spine*, **17**, 897–913.

Filipović, N. (1999). *Numerical Analysis of Coupled Problem: Deformable Body and Fluid Flow*, Ph.D. Thesis, Faculty of Mechanical Engineering, University of Kragujevac, Serbia.

Frank, E.H. & Grodzinsky, A.J. (1987a). Cartilage electromechanics – I. Electrokinetic transduction and the effects of electrolyte pH and ionic strength, *J. Biomech.*, **20**, 615–27.

Frank, E.H. & Grodzinsky, A.J. (1987b). Cartilage electromechanics – II. A continuum model of cartilage electrokinetics and correlation with experiments, *J. Biomech.*, **20**, 629–39.

Ghadially, F.N. (1978). Fine structure of joints. In L. Sokoloff (ed.), *The Joints and Synovial Fluid, 1* (pp. 105–76). Academic Press, New York.

Kazarian, L.E. (1975). Creep characteristics of human spinal column, *Orth. Clin. N. Am.*, **6**, 3–18.

Kojić, M., Filipović, N. & Mijailović, S. (2001). A large strain finite element analysis of cartilage deformation with electrokinetic coupling, *Comp. Meth. Appl. Mech. Eng.*, **190**, 2447–64.

Laible, J.P., Pflaster, D.S., Krag, M.H., Simon, B.R. & Haugh, L.D. (1993). A poroelastic-swelling finite element model with application to the intervertebral disc, *Spine*, **18**, 659–70.

Lewis, R.W. & Schrefler, B.A. (1987). *The Finite Element Method in the Deformation and Consolidation of Porous Media*, John Wiley& Sons, Ltd, Chichester, England.

Mow, V.C., Kuei, S.C., Lai, W.M. & Armstrong, C.G. (1980). Biphasic creep and stress relaxation of articular cartilage: theory and Experiments, *J. Biomech. Eng. Trans. ASME*, **102**, 73–84.

Sachs, J.R. & Grodzinsky, A.J. (1989). An electromechanically coupled poroelastic medium driven by an applied electric current: surface detection of bulk material properties, *PCH*, **11**, 585–614.

Simon, B.R. & Gaballa, M.A. (1988). Poroelastic element models for the spinal motion segment including ionic swelling. In R.L. Spilker & B.R. Simon (eds.), *Computational Methods in Bioengineering* (pp. 93–9), American Society of Mechanical Engineers, New York.

Simon, B.R., Wu, J.S.S., Carlton, M.W., Evans, J.H. & Kazarian, L.E. (1985). Structural models for human spinal motion segments based on a poroelastic view of the intervertebral disc, *J. Biomech. Eng.*, **107**, 327–35.

Yoganandan, N., Myklebust, J.B., Ray, G. & Sances, A. (1987). Mathematical and finite element analysis of spine injuries, *CRC Crit. Rev. Biomed. Eng.*, **15**(1), 29–93.

16

Cell Mechanics

In this chapter we consider mechanical behavior of cells. In order to introduce mechanical models of cells, we first describe the architectural organization and microstructural mechanics of the cell cytoskeleton (CSK) and how they determine mechanical responses and functions of adherent cells. In particular, the cellular tensegrity model is defined which has emerged as a leading model of cytoskeletal mechanics. A simple tensegrity model illustrates key mechanisms by which the CSK develops resistance to mechanical distortion and governs cell deformability.

The biphasic cell model is further introduced, where the cell is considered as a deformable continuum composed of an elastic cortical membrane surrounding a cytoplasmic solid–fluid mixture. Within the mixture, the solid phase represents the CSK and the fluid phase is the cytosol.

Using these two models, several examples illustrate mechanical behavior of cells. We also provide the software on the web for further exercises in modeling mechanical response of cells.

16.1 Introduction to mechanics of cells

Mechanical stress, including gravity, pressure, tension, compression, hemodynamic stress and motions, plays a critical role in living tissue development and extends to the cell level. Cell differentiation, growth, secretion, gene expression and signal transduction – all can be modified due to mechanical stress acting on living cells. Yet, little is known about how cells convert these mechanical signals into biochemical and biological responses. In order to elucidate this process, it is necessary to understand mechanisms by which cells develop and transmit mechanical stress that oppose cell deformation. A growing body of evidence shows that deformability of adherent cells is governed by the cytoskeleton (CSK), an intracellular molecular network that mechanically stabilizes the cell and actively generates mechanical stress.

Mechanical distortion of cell shape can impact many cell behaviors, including motility, contractility, growth, differentiation and apoptosis (e.g. Folkman & Moscona 1978). Mechanical forces produce changes in cell function by inducing restructuring of the intracellular CSK and thereby impacting cellular biochemistry (e.g. Ingber 1993). Through largely unknown mechanisms, mechanical signals are transduced into biochemical signals that lead to changes in gene expression and protein synthesis (e.g. Ingber 1997). Motile cells also can sense the mechanical stiffness of their extracellular matrix (ECM), and preferentially move toward areas of greater rigidity (Lo et al. 2000). The changes in cell extension and movement that make cell migration possible depend on the ability of the substrate to resist cell tractional forces. Changes in the cellular force balance (e.g. Sheetz 2001), in turn, alters cellular biochemistry which further strengthens the cell's ability to resist to applied loads (e.g. Sheetz 2001). Even at the nuclear and cytoplasmic levels, a key role for mechanical distension is evident in the control of subcellular structure and function (e.g. Ingber 1993).

The above-cited observations suggest that regulation of many vital cellular behaviors is centered around cell adhesion, spreading and a mechanical distending stress borne by the CSK. We must therefore search for a model of the cell that will allow us to relate cytoskeletal mechanics to biochemistry at the molecular level, and to translate this description into quantitative, mathematical terms. The former will permit us to define how specific molecular components contribute to cell behaviors. The latter will allow development of computational approaches to address levels of complexity and multicomponent interactions that exist in cells.

Mechanical Properties and Role of Biopolymers of the Cytoskeleton

Major stress-bearing components of the CSK are filamentous actin (F-actin), microtubules and intermediate filaments, and each of them has its specific mechanical role within the CSK (Ingber 2003). Actin filaments are semi-flexible polymers of high tensile stiffness (elastic modulus $E \sim 10^0$ GPa) and persistence length $L_p \sim 10^1$ μm (Gittes et al. 1993); L_p is a measure of filament flexibility and is roughly the minimum length at which the filament ends become uncorrelated to Brownian motion ($L_p = EI/kT$, where E is the elastic modulus of the filament, I is its cross-sectional moment of inertia, k is Botzmann's constant and T is absolute temperature; see Table 16.1.1). However, within the CSK, actin filaments are much shorter (< 1 μm) then their L_p and thus they appear as straight-line segments (Fig. 16.1.1 – see color plate).

Table 16.1.1 Mechanical properties of actin filaments, stress fibers, microtubules and intermediate filaments obtained from *in vitro* measurements: d is filament diameter, L_p is persistence length, E is elastic modulus, F_{max} is tensile force at which the filament breaks, E_{max} is elastic modulus corresponding to F_{max}, and ε_{max} is strain corresponding to F_{max}. Data for microtubules include the inner and outer diameters

Filament type	d (nm)	L_p (μm)	E (MPa)	F_{max} (nN)	E_{max} (MPa)	ε_{max} $(\%)$
Actin filaments	5–10	10–20	2600	0.4	—	0.9
Stress fibers	200–500	—	1.45	380	100	275
Microtubules	12/25	5000	1200	—	—	—
Intermediate filaments	10	1	6.4	14	80	220

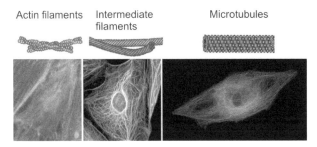

Fig. 16.1.1 The immuno-fluorescent images of the principal stress-bearing components of the cytoskeleton: actin filaments, intermediate filaments and microtubules. The blue oval in the left panel is the nucleus. The artistic depiction of molecular structure of each filament is shown above the corresponding image (from Ingber 1998) (see Plate 28)

Cytoskeletal actin filaments are often crosslinked with myosin crossbridges. Myosins are molecular motors which are capable of generating tensile force in the actin filaments (see Chapter 12 for details). As a result of this action, the CSK becomes prestressed. Actin filaments are also grouped together with myosin and other actin-binding proteins to form 200–500 nm diameter bundles known as the actin stress fibers.

Stress fibers are less stiff ($E \sim 10^0\ MPa$) and much more extensible than individual actin filaments (Deguchi *et al.* 2006). However, while actin filaments appear linearly elastic, stress fibers exhibit stiffening under sustained tension (Table 16.1.1). Within the CSK, stress fibers carry both tensile and compressive loads.

Isolated microtubules appear straight, as rigid tubes of nearly the same stiffness as actin filaments; but of much greater persistence length, $L_p \sim 10^3\ \mu m$, see Table 16.1.1 (Gittes *et al.* 1993). Based on this high L_p, microtubules should appear straight on the whole cell level if they were not mechanically loaded. However, immunofluorescent images of the CSK of living cells show that microtubules appear curved (e.g. Ingber 2003), see Fig. 16.1.1. It follows therefore that some type of mechanical force must act on microtubules; conceivably compression in microtubules lead to their buckling.

Intermediate filaments are much more flexible ($L_p \sim 10^0\ \mu m$) and less stiff ($E \sim 10^0$–$10^1\ MPa$) than actin filaments and microtubules. They are extensible and exhibit stiffening similar to stress fibers, see Table 16.1.1 (Fudge *et al.* 2003). Since within the CSK the typical length of intermediate filaments (10–20 μm) is much greater than their L_p, they appear to provide very soft elasticity to the cell. Within the CSK of living cells, intermediate filaments appear slack (Fig. 16.1.1); however, during large deformation of the cell they become tensed and provide structural stability to the CSK at high strains (Wang & Stamenović 2000). Intermediate filaments also provide lateral elastic support to microtubules and thereby effectively increase the critical buckling force of microtubules (Brodland & Gordon 1990).

While all three filamentous biopolymer systems are important for mechanical function of the cell, experimental studies in which these systems were selectively disrupted show that the cytoskeletal actin has the major contribution to the overall mechanical response; whereas the contributions of microtubules and intermediate filaments are relatively smaller (Wang *et al.* 1993). However, this contribution may change. For example, with increasing of cell spreading on the ECM, the fraction of the cytoskeletal prestress that is balanced by the ECM increases at the expense of microtubules (Hu *et al.* 2004). With increasing distortion of the cell, the contribution of intermediate filaments becomes more important (Wang & Stamenović 2000).

Besides the described complexity of the load bearing within the cell, mechanical models have been introduced which to a certain extent can describe the mechanical behavior of cells.

16.2 Cell mechanical models

In this section we present two basic models commonly used in cell mechanics: tensegrity model of cytoskeleton, and biphasic continuum model.

16.2.1 Stabilizing influence of CSK prestress – cellular tensegrity model

As described in Section 16.1, the cytoskeleton is a molecular network composed of filamentous biopolymers including actin microfilaments, microtubules, intermediate filaments and a number of crosslinking proteins (Fig. 16.1.1). Experiments on cultured living cells have shown that the CSK is a prestressed structure and that the cytoskeletal prestress appears to play a central role in determining and possibly regulating cell mechanical properties. These properties impact cell biochemical activities, including movement and growth as well as contractility (Wang *et al.* 2001). Cytoskeletal prestress is generated: (a) actively by the cell's contractile apparatus (molecular myosin motors); and (b) passively by mechanical distension of the cell as it adheres to the ECM, as well as by swelling pressure (turgor) of the cytoplasm.

Tensegrity Model
The idea that mechanical prestress may determine cell shape stability was initially explored in a model that depicts the cell as a tensed membrane surrounding a viscous cytoplasm (e.g. Evans & Yeung 1989). This idea was further advanced by the hypothesis that prestress in the tensed intracellular cytoskeletal lattice, rather than the cortical membrane, is primarily responsible for shape stability in adherent mammalian cells (e.g. Ingber 2003). A special class of reticulated mechanical structures, known as 'tensegrity' structures, describe these mechanical conditions within the CSK (Wang *et al.* 1993, Fig. 16.2.1).

Tensegrity architecture represents a class of prestressed structures which maintain their structural integrity, even before application of external loading, because of prestress in their cable-like structural members.[1] A hallmark property that stems from this feature is that structural stiffness of the network is proportional to the level of the prestress that it supports (e.g. Volokh & Vilnay 1997). In tensegrity architecture the prestress in the cable network is balanced by compression of internal elements that are called struts (Fig. 16.2.1).

According to the cellular tensegrity model, actin filaments and intermediate filaments are envisioned as tensile elements, whereas microtubules (stabilized by lateral guy wire-like connections) and thick crosslinked actin bundles (e.g. within filopodia) act as compression

---

[1] Tensegrity architecture is a building principle introduced by Fuller (1961) and is defined as a system through which structures are stabilized by continuous tension (i.e., prestress) carried by the structural members. Fuller referred to this architecture as 'tensional integrity', or 'tensegrity'.

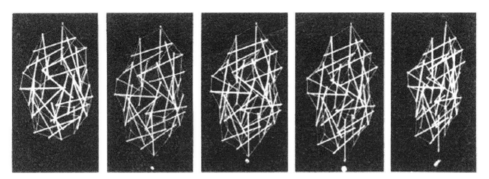

Fig. 16.2.1 A tensegrity cell model under different mechanical loads. The model consists of struts (stiff elements, thick lines) and elastic cables (thin lines). The model is suspended from above and loaded by increased force downward, from zero at the left panel to a maximal force at the right panel (Wang *et al.* 1993, with permission)

elements. In addition to these compression elements within the CSK, the cell's tethers to the ECM (which is physically connected to the CSK and critical for cell shape stability), known as focal adhesions, also balance a portion of the prestress.

The central mechanism by which prestressed structures, including tensegrities, develop restoring stress in the presence of external loading is primarily by geometrical rearrangement of their pre-tensed members. The greater the pre-tension carried by these elements, the less geometrical rearrangement they undergo under an applied load, and thus, the less deformable (more rigid) the structure will be. This explains why structural stiffness increases in proportion with the level of the prestress. This property is independent of whether this prestress is balanced by internal compression-bearing struts, by the ECM or by cytoplasmic swelling pressure. Results obtained from mechanical measurements of cultured cells are consistent with this *a priori* prediction for prestressed structures (Fig. 16.2.2). Namely, at steady state, cell stiffness (elastic modulus) increases in proportion with increasing cytoskeletal prestress (Wang *et al.* 2001, 2002).

In summary, in the cellular tensegrity model, the CSK and ECM are assumed to form a single, synergetic, mechanically stabilized system. Two key premises of the model are: (i) the prestress, carried by the actin network and intermediate filaments, confers shape stability to the cell; and (ii) this prestress is partly balanced by CSK-based microtubules and partly by the ECM (Stamenović 2006). Thus, a disturbance of this complementary force balance would cause load transfer between these three distinct systems that would, in turn, affect cell deformability and alter stress-sensitive biochemical activities at the molecular level.

Alternative Prestressed Models

There are different prestressed structural models of the CSK in the literature, most notably models based on a cortical membrane (Discher *et al.* 1998) and tensed cable networks (Coughlin & Stamenović 2003). These models have been successful at explaining *some* aspects of cellular mechanics, but lack the ability to describe many other mechanical behaviors that are important for cell function. In particular, the cortical membrane model ignores the contribution of the ECM to cellular mechanics, and it cannot explain the observed transmission of mechanical signals from cell surface to the nucleus, as well as to basal focal

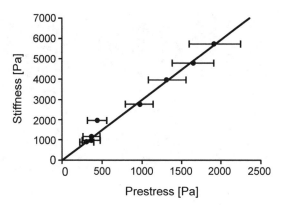

Fig. 16.2.2 Cytoskeletal stiffness of human airway smooth muscle cells increases linearly with increasing cytoskeletal prestress, consistent with the *a priori* prediction of the tensegrity model. Data are obtained from magnetic cytometry measurements (stiffness) and traction microscopy measurements (prestress) in cultured human airway smooth muscle cells. Dots are data ±SE; line is linear regression (according to Wang *et al.* 2002)

adhesions ('action at a distance') (Wang *et al.* 2001). On the other hand, all of these features (and many others) can be explained by models that depict the CSK as a prestressed tensegrity structure (e.g. Stamenović & Ingber 2002). Moreover, other models cannot explain how mechanical stresses applied to the cell surface result in force-dependent changes in biochemistry at discrete sites inside the cell (e.g. focal adhesions, microtubules), whereas the tensegrity can (Ingber 1997).

The cortical membrane model and the tensed cable network model also fall into the category of stress-supported structures. In fact, according to the definition based on structural stability (Connelly & Back 1998), all prestressed structures are tensegrity structures. They differ from each other only in the manner by which they balance the prestress. However, in the structural mechanics literature, distinction is made between tensed cable nets, and tensegrity structures with cables and internal struts (Volokh & Vilnay 1997). A key distinction is that in cable-and-strut structures at each free node one compression strut balances tensile forces in the remaining cables, whereas in other prestressed cable structures at each free node force balance only includes cable tensile forces.

16.2.2 Mathematical model of a six-strut tensegrity structure

A six-strut tensegrity structure (Fig. 16.2.3a) has been used in the past as a conceptual model of the CSK (e.g. Stamenović *et al.* 1996, Volokh *et al.* 2000, Wang & Stamenović 2000, Wendling *et al.* 1999). It is composed of six compression-bearing struts interconnected by 24 tension bearing cables. At the reference state, before application of an external load, the cables are under tension. This pre-tension confers the shape stability to the structure and modulates its structural rigidity (e.g. Stamenović *et al.* 1996).

Geometrical Description
At the reference configuration, all the struts are of length L_0. It is shown below that the corresponding length of the cable segment $l_0 = \sqrt{3/8}\ L_0$ and the corresponding distance

CELL MECHANICS

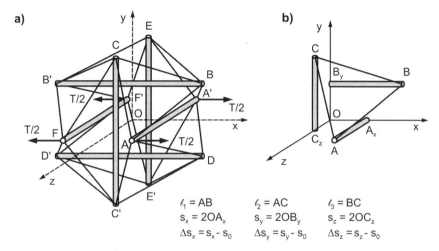

$l_1 = AB$
$s_x = 2OA_x$
$\Delta s_x = s_x - s_0$

$l_2 = AC$
$s_y = 2OB_y$
$\Delta s_y = s_y - s_0$

$l_3 = BC$
$s_z = 2OC_z$
$\Delta s_z = s_z - s_0$

Fig. 16.2.3 Six-strut tensegrity model. (a) Model geometry. Stretching force of magnitude $T/2$ is applied at the nodes A, A′, F and F′; (b) Portion of the model inside the first quadrant of the x, y, z coordinate system (A_x, B_y, C_z are points of struts at the coordinate axes x, y, z)

of the pairs of parallel struts is $s_0 = L_0/2$. The coordinate system xyz is placed at the geometrical center of the structure such that the coordinate axes are parallel with the pairs of parallel struts (Fig. 16.2.3a). The structure is stretched uniaxially, in the x-direction, by forces of magnitude $T/2$ applied at the endpoints of the struts AA′ and FF′. This causes: (a) changes in the strut length from L_0 to L_I (struts AA′ and FF′), L_{II} (struts BB′ and DD′), and L_{III} (struts CC′ and EE′); (b) changes in the distance between the pairs of parallel struts from s_0 to s_x (struts AA′ and FF′), s_y (struts BB′ and DD′), and s_z (struts CC′ and EE′); and (c) changes in length of cable segments from l_0 to l_1 (segments AB, A′B, AD, A′D, FB′, F′B′, FD′, F′D′), l_2 (AC, AC′, A′E, A′E′, FC, FC′, F′E, F′E′) and l_3 (BC, B′C, DC′, D′C′, BE, B′E, DE′, D′E′). Changes in the distances between a pair of parallel strut, $\Delta s_\alpha \equiv s_\alpha - s_0$ ($\alpha = x, y, z$), are referred to as extensions. Coordinates of the nodal points at the reference state are: A = $(L_0/4, 0, L_0/2)$, A′ = $(L_0/4, 0, -L_0/2)$, B = $(L_0/2, L_0/4, 0)$, B′ = $(-L_0/2, L_0/4, 0)$, C = $(0, L_0/2, L_0/4)$, C′ = $(0, -L_0/2, L_0/4)$, D = $(L_0/2, -L_0/4, 0)$, D′ = $(-L_0/2, -L_0/4, 0)$, E = $(0, L_0/2, -L_0/4)$, E′ = $(0, -L_0/2, -L_0/4)$, F = $(-L_0/4, 0, L_0/2)$, F′ = $(-L_0/4, 0, -L_0/2)$. Relationships between L_I, L_{II}, L_{III}, s_x, s_y, s_z, l_1, l_2 and l_3 are derived below.

Consider the portion of the structure in the first quadrant of the x, y, z coordinate system (Fig. 16.2.3b). Let A_x and B_y denote middle points of AA′ and BB′ struts, respectively. Then, AB = l_1, $OA_x = s_x/2$, $OB_y = s_y/2$, $AA_x = L_I/2$, and $BB_y = L_{II}/2$. Thus,

$$l_1 = \sqrt{(BB_y - OA_x)^2 + OB_y^2 + AA_x^2} = \frac{1}{2}\sqrt{(L_{II} - s_x)^2 + s_y^2 + L_I^2}. \quad (16.2.1)$$

Expressions for l_2 and l_3 are obtained in a similar manner

$$l_2 = \frac{1}{2}\sqrt{s_x^2 + L_{III}^2 + (L_I - s_z)^2}, \quad l_3 = \frac{1}{2}\sqrt{L_{II}^2 + (L_{III} - s_y)^2 + s_z^2}, \quad (16.2.2)$$

Equilibrium Equations

Equilibrium equations are obtained by considering balance of forces in x-, y- and z-directions at each node. These include forces F_1, F_2 and F_3 in the cables of corresponding lengths l_1, l_2 and l_3, respectively; the compression in the struts P_I, P_{II} and P_{III} corresponding to lengths L_I, L_{II} and L_{III}, respectively; and the external force $T/2$. The following relationships are obtained

$$T = 2F_1 \frac{s_x - L_{II}}{l_1} + 2F_2 \frac{s_x}{l_2}, \qquad F_1 \frac{s_y}{l_1} = F_3 \frac{L_{III} - s_y}{l_3}, \qquad (16.2.3\text{a,b})$$

$$F_2 \frac{L_I - s_z}{l_2} = F_3 \frac{s_z}{l_3}, \qquad P_I = F_1 \frac{L_I}{l_1} + F_2 \frac{L_I - s_z}{l_2}, \qquad (16.2.4\text{a,b})$$

$$P_{II} = F_1 \frac{L_{II} - s_x}{l_1} + F_3 \frac{L_{II}}{l_3}, \qquad P_{III} = F_2 \frac{L_{III}}{l_2} + F_3 \frac{L_{III} - s_y}{l_3} \qquad (16.2.5\text{a,b})$$

Equations (16.2.3a) and (16.2.4b) represent the balance of forces at nodes A and F in the x- and z-directions, respectively; equations (16.2.3b) and (16.2.5a) reperesent balance of forces at B and D in the y- and x-directions; and equations (16.2.4a) and (16.2.5b) represent balance of forces at C and E in the z- and y-directions. Balance of forces in other nodes is satisfied by the symmetry of the structure.

There are altogether 15 unknown variables (l_1, l_2, l_3, L_I, L_{II}, L_{III}, s_x, s_y, s_z, F_1, F_2, F_3, P_I, P_{II}, P_{III}) and nine equations. Six additional equations are obtained from the constitutive equations for cables and struts. We assume that cables and struts are two-force members and that cables support only tension. In past studies, the cables were usually viewed as linear elastic (Stamenović et al. 1996, Volokh et al. 2000, Wang & Stamenović 2000, Wendling et al. 1999), whereas the struts were considered either rigid (Stamenović et al. 1996, Wang & Stamenović 2000, Wendling et al. 1999) or elastic such that they buckle under compression (Volokh et al. 2000).[2] Since the models with rigid struts yielded qualitatively similar results as models with buckling strut, for simplicity we here consider the case with rigid struts. In that case, L_I, L_{II} and L_{III} are all equal to L_0 and equations (16.2.4b)–(16.2.5b) become redundant. Thus the number of unknowns is nine and the number of equations is six. Three additional equations represent the constitutive equations of linearly elastic cables and are given as:

$$F_i = \begin{cases} EA \dfrac{l_i - l_R}{l_R} & \forall \; l_i > l_R \\ 0 & \forall \; l_i \leq l_R \end{cases} \qquad i = 1, 2, 3 \qquad (16.2.6)$$

where E and A are Young's modulus and cross-sectional area of the cable, respectively; and l_R is the resting length. Values of E and A are selected based on experimental data (Table 16.1.1), whereas l_R is selected according to a desired pre-tension in the cables.

The response of the model under the uniaxial load, shown in Fig. 16.2.3, is given in Example 16.3-1.

---

[2] It is noteworthy that the six-strut tensegrity model also can be modified for studying cell viscoelasticity by assuming that the cables are linear viscoelastic (Voigt) elements (e.g., Cañadas et al., 2002; Sultan et al., 2004).

16.2.3 Biphasic models

It is observed experimentally that cells have elastic and viscous response when subjected to loading. This response can be modeled by a solid continuum with viscoelastic constitutive laws. For a recent review of these models see Mofrad and Kamm. (2006). On the other hand, from the description of living cell in Section 16.1 it can be seen that cells are very complex structures, with water, charged or uncharged micromolecules, ions and other molecules, and it is hard to find a phenomenological constitutive relationships which can represent cell behavior under complex mechanical and other actions (as osmotic or electric). It is natural to investigate models which are close to the real physical composition of the cell, and biphasic, fluid–solid mixture, models are along this line.

According to biphasic models, a cell can be considered as a solid–fluid mixture continuum, with cytoplasm as the fluid and cytoskeleton as the solid. In the solid–fluid mixture formulation the viscous response comes from the solid–fluid interaction. Furthermore, the viscoelasticity can also be included in the solid-phase constitutive law. These models can further be extended to three-phasic (fluid–solid–ion) models to include coupling of mechanical, chemical and electrical events. A review of multiphasic cell models is given by Guilak *et al.* (2006).

We use here the cartilage model of Section 15.2 as the basis for the biphasic model of a cell. For simplicity, the additional effects arising due to action of osmotic pressure will be neglected. But, in order to model biochemical processes within the cell which result in mechanical internal mechanical stresses, we include activation within the continuum model (Kojić *et al.* 2007), as in the case of muscle modeling (Chapter 12).

Schematics of the biphasic model is shown in Fig. 16.2.4. The field variables at a material point of the continuum are: displacement of solid **u**, relative fluid velocity with respect to solid (Darcy's velocity) **q**, and fluid pressure p. The total stress within the solid includes the

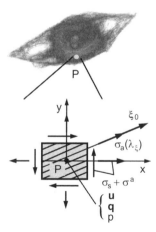

Fig. 16.2.4 Biphasic model of cell. Stresses at a material point P within the solid phase include the passive and active parts $\boldsymbol{\sigma}_s$ and $\boldsymbol{\sigma}^a$ (2D representation of stresses in the figure). The stress $\boldsymbol{\sigma}^a$ is acting along the fibers (direction $\boldsymbol{\xi}_0$) and depend on the fiber stretch λ_ξ. The field variables of the model are: displacement of solid **u**, relative fluid velocity with respect to solid (Darcy's velocity) **q**, and fluid pressure p

passive part $\boldsymbol{\sigma}_s$; and active part $\boldsymbol{\sigma}^a$, acting along the skeleton fiber direction $\boldsymbol{\xi}_0$ at a material point. The stress $\boldsymbol{\sigma}_s$ can be determined from the constitutive law (e.g. elastic as for cartilage model in Section 15.2). On the other hand, the stress $\boldsymbol{\sigma}^a$ can be expressed in terms of the fiber stretch λ_ξ and the activation level (expressed for example by an activation function $\alpha_a(t)$, see (12.2.11) in Section 12.2).

The governing equations of the model have the same form as in Sections 3.4 and 15.2. The balance of linear momentum (3.4.5) remains the same,

$$\nabla^T \boldsymbol{\sigma} + \rho \mathbf{b} - \rho \ddot{\mathbf{u}} + \rho_f \dot{\mathbf{q}} = 0 \tag{16.2.7}$$

where $\boldsymbol{\sigma}$ is the total stress which can be expressed in terms of $\boldsymbol{\sigma}_s$, $\boldsymbol{\sigma}^a$ and p, as

$$\boldsymbol{\sigma} = (1-n)(\boldsymbol{\sigma}_s + \boldsymbol{\sigma}^a) - n\mathbf{m}p \tag{16.2.8}$$

n is porosity; $\rho = (1-n)\rho_s + n\rho_f$ is the mixture density; and $\mathbf{m}$ is a constant vector defined as $\mathbf{m}^T = [1\ 1\ 1\ 0\ 0\ 0]$ which provides that the pressure component contributes to the normal stresses only. Assuming elastic behavior of the solid skeleton, the continuity equation can be written as (see (3.4.10))

$$\nabla^T \mathbf{q} + \left(\mathbf{m}^T - \frac{\mathbf{m}^T \mathbf{C}^E}{3K_s}\right)\dot{\mathbf{e}} + \left(\frac{1-n}{K_s} + \frac{n}{K_f} - \frac{\mathbf{m}^T \mathbf{C}^E \mathbf{m}}{9K_s^2}\right)\dot{p} = 0 \tag{16.2.9}$$

where $\mathbf{C}^E$ is the elastic constitutive matrix, and K_s and K_f are bulk moduli for the solid skeleton material and fluid, respectively.

The finite element equations are as (see (7.7.4))

$$\begin{bmatrix} \mathbf{M}_{uu} & 0 & 0 \\ 0 & 0 & 0 \\ \mathbf{M}_{qu} & 0 & 0 \end{bmatrix} \begin{Bmatrix} ^{n+1}\ddot{\mathbf{U}} \\ ^{n+1}\ddot{\mathbf{P}} \\ ^{n+1}\ddot{\mathbf{Q}} \end{Bmatrix} + \begin{bmatrix} 0 & 0 & ^n\mathbf{C}_{uq} \\ ^n\mathbf{C}_{pu} & \mathbf{C}_{pp} & 0 \\ 0 & 0 & \mathbf{C}_{qq} \end{bmatrix} \begin{Bmatrix} ^{n+1}\dot{\mathbf{U}} \\ ^{n+1}\dot{\mathbf{P}} \\ ^{n+1}\dot{\mathbf{Q}} \end{Bmatrix}$$

$$+ \begin{bmatrix} ^n\mathbf{K}_{uu} & ^n\mathbf{K}_{up} & 0 \\ 0 & 0 & \mathbf{K}_{pq} \\ 0 & \mathbf{K}_{qp} & \mathbf{K}_{qq} \end{bmatrix} \begin{Bmatrix} \Delta \mathbf{U} \\ \Delta \mathbf{P} \\ \Delta \mathbf{Q} \end{Bmatrix} = \begin{Bmatrix} ^{n+1}\mathbf{F}_u \\ ^{n+1}\mathbf{F}_p \\ ^{n+1}\mathbf{F}_q \end{Bmatrix} \tag{16.2.10}$$

The matrices and vectors are given in (7.7.5). We note here that the nodal force $^{n+1}\mathbf{F}_u$ includes the total stress $^{n+1}\boldsymbol{\sigma}$ given in (16.2.8). These equations can be further written in incremental-iterative form (see web – Theory, Chapter 7). Other computational details are presented in Section 7.7.1, such as change of the porosity n and others.

In several examples we illustrate application of the above equations (see also web, Software).

16.3 Examples: modeling of cell in various mechanical conditions

Examples presented here illustrate responses of the cell subjected to external loading and internal excitation. The calculated responses give some of the characteristic mechanical

CELL MECHANICS

properties of cells in a qualitative sense, when the mechanical models of Section 16.2 are used. Additional information about how these models may help to elucidate mechanical behavior of the cell can be gained using the Software (see web – Software).

Example 16.3-1. Modeling the cytoskeleton using tensegrity architecture
In this example, some of the cell mechanical characteristics are obtained from the six-strut tensegrity model of the cytoskeleton, shown in Fig. 16.2.3.

It is found that the force–extension behavior of the model is nonlinear and that it depends on the pre-strain (and thereby pre-stress) in the cables (Fig. E16.3-1a). It is important that this nonlinearity is the result of geometrical rearrangements of the cables and not of material properties of individual cables which are assumed to be linearly elastic.

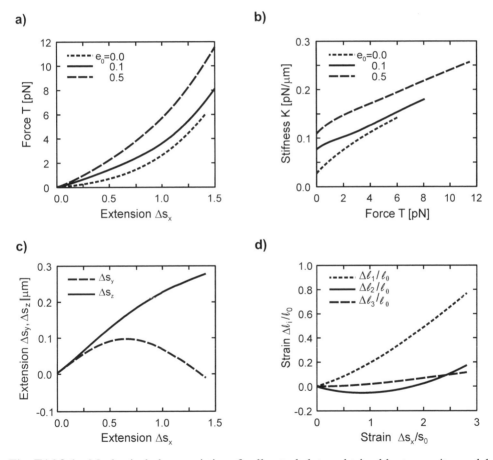

Fig. E16.3-1 Mechanical characteristics of cell cytoskeleton obtained by tensegrity model (FE model of cables and struts, see Fig. 16.2.3 for the model description). (a) Dependence of axial force T on the extension Δs_x for the pre-strain of cables $e_0 \equiv l_0/l_R - 1 = 0.0, 0.1$ and 0.5; (b) Stiffness $K = T/\Delta s_x$ vs. Δs_x for $e_0 = 0.0, 0.1$ and 0.5; (c) Dependence of lateral extensions Δs_y and Δs_z on the axial extension Δs_x for $e_0 = 0.1$; (d) Cable strains $\Delta l_i/l_0 = l_i/l_0 - 1$ ($i = 1, 2, 3$) vs. structural axial strain $\Delta s_x/s_0$ for $e_0 = 0.1$

Structural stiffness increases with the pre-strain, moreover, the structure exhibits stiffening with increasing extension. Taken together, these results are consistent with experimental data obtained from measurements on living cells (Wang et al., 1993; Wang & Ingber, 1994). Another interesting feature is that for small axial extensions Δs_x, lateral extensions in the y- and z-directions Δs_y and Δs_z, respectively, increase and are of similar magnitude (Fig. E16.3-1c). However, for large values of Δs_x, we have that Δs_z continues to increase while Δs_y peaks and decreases (Fig. E16.3-1c). This asymmetry indicates that the model is anisotropic. Cells also exhibit anisotropic behavior (Hu et al. 2004) and fractional change in the axial displacement $\Delta s_x/s_0$ (i.e. axial strain, see Fig. 16.2.3 and the model description) is greater than fractional changes of length of each cable (Fig. E16.3-1d). In other words, the entire structure deforms much more than its individual cables. This is consistent with behavior of cells where the cells undergo much larger strains than the individual components of the cytoskeleton. Note that these results obtained by the FE method also agree with results obtained previously by directly solving the governing equations numerically (using Mathematica software, Stamenović et al. 1996).

Example 16.3-2. Deformation of red blood cells subjected to action of optical tweezers
The deformation of red blood cells (RBCs) has been the subject of many investigations. The RBC deformation is particularly very large during blood flow through capillaries. There, an RBC of a biconcave shape with diameter of around $8\,\mu m$ passes through capillaries of diameter as small as of $3\,\mu m$, changing to a bullet shape with large strains, and then recovers its initial shape after leaving the capillaries. This deformability is necessary for mass exchange and normal function of blood. Loss of deformability occurs in severe diseases such as malaria.

Various experimental techniques have been introduced to investigate mechanical characteristics of RBCs, one of which is extension by optical tweezers (Dao et al. 2003, Mills et al. 2004). Also, a number of mechanical models have been used for calculating the RBC mechanical response when subjected to loading. One of them consists of a shell for the membrane, with neo-Hookean material model, and fluid for cytosol surrounded by the membrane (Dao et al. 2003, Mills et al. 2004).

In this example we use elastic material for the membrane and the biphasic model described in Section 16.2.3 for the cytosol. Isoparametric 3D finite elements for biphasic medium (Section 7.7) are used for both the membrane and the cytosol, with the zero-porosity for the membrane (Kojić et al. 2007). It is assumed that the RBC has initially a biconcave shape (Fig. E16.3-2Aa – see color plate). We model one-eighth part of the cell (in the first coordinate quadrant) loaded by $1/4F$, due to symmetry in geometry and loading. The force is distributed on a part of the surface, as it is in the experiment, and increases slowly so that the quasi-static deformation is assumed. The data used in the model are: membrane thickness $\delta = 90\,nm$; Young's moduli for membrane and biphasic model $(pN/\mu m^2)$: 1.772×10^2 and 4×10^1; porosities: $n = 0$ and $n = 0.7$; bulk moduli (for cytosol) $(pN/\mu m^2)$: $K_s = 1.54 \times 10^1$, $K_f = 1.0 \times 10^9$; permeability $k = 1 \times 10^2\,\mu m^4/pN s$.

When the RBC is subjected to axial forces, it deforms into the direction of force action and contracts in the direction normal to the action of forces. The deformed configuration with

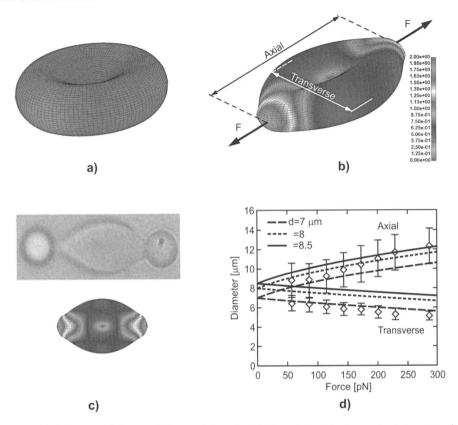

Fig. E16.3-2A A biphasic FE model of RBC subjected to uniaxial extension forces. One-eighth of the cell is modeled due to symmetry (3D biphasic finite elements). (a) Initial biconcave shape; (b) Deformed shape at force of 300 pN; (c) Deformed shape (top view) experimentally recorded and computed; (d) Change of axial and transverse diameters in terms of extensional force for three initial diameters d (computed results are represented by lines, and experimental by bars) (see Plate 29)

the displacement field is shown in Fig. E16.3-2Ab (see color plate). Experimentally recorded and computed shapes (top view) of the deformed cell agree reasonably well (Fig. E16.3-2Ac – see color plate).

Change of the cell diameters in terms of the extension force F (computed for three values of initial diameters, and experimentally recorded – Dao et al. 2003) in the direction of force action (axial) and in direction orthogonal to this one (Transverse in the figure) in terms of the axial force F is shown in Fig. E16.3-2Ad. The axial diameter increases and the transverse diameter decreases nonlinearly with the force increase.

Deformation of RBCs assuming cylindrical shape and change of the axial and transverse diameters with the force increase is shown in Fig. E16.3-2B. It can be seen that, as

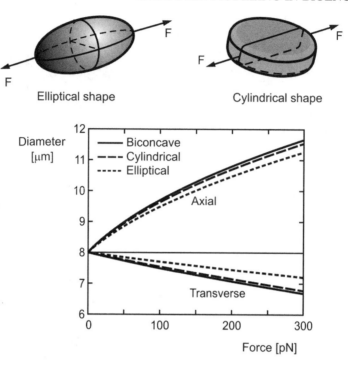

Fig. E16.3-2B Computed change of RBC diameters for initial biconcave, cylindrical and elliptical shape. The softest is the biconcave RBC

expected, the RBCs of cylindrical and elliptical shape are stiffer with respect to the RBCs of biconcave shape.

The computational results agree reasonably well with experiments and show that the biphasic model can be used for modeling the RBC mechanical response.

Example 16.3-3. Cell deformation induced by magnetic bead twisting

The method of magnetic twisting cytometry is used to probe mechanical properties of adherent cells (e.g. Maksym *et al.* 2000). A study of cell deformation during this experiment using a finite element model is given in Mijailović *et al.* (2002). Here, we model this experiment by representing the cell as a 2D biphasic continuum of Section 16.2.3. The material data for this cell model are as for cytosol in Example 16.3-2. The radius of the bead is $R = 4.5\,\mu m$, and the depth of bead indentation is $h = 0.45\,\mu m$.

The field of effective stress (see web – Theory, Chapter 2) within the cell is shown in Fig. E16.3-3a (see color plate) when the bead is subjected to the force $F = 800\,pN$ parallel to the cell surface. The maximum stresses are in the domains around the edges of the contact between the cell and bead. Distribution of the tangential and normal stresses on the cell contact surface with the bead is shown in Fig. E16.3-3b (see color plate). These results agree with those reported in Mijailović *et al.* (2002) obtained by FE modeling with the cell represented by an isotropic elastic body.

Detailed analysis of cell deformation under the force or moment and for various model parameters can be performed using Software on the web.

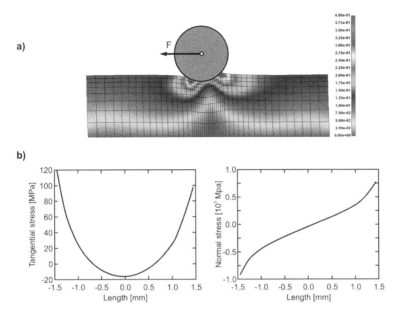

Fig. E16.3-3 Cell deformation due to action of bead tightly bound to the cell surface. The bead is subjected to a force. (a) Field of effective stress within the cell represented by a 2D biphasic continuum; (b) Tangential and normal tractions along the contact surface between the cell and bead (see Plate 30)

Example 16.3-4. Modeling of cell crawling

In this example we model motion of a cell over a plane surface. The cell consists of membrane, interior, nucleus and skeleton (Kojić et al. 2007). The plane strain 2D model is considered. Initial dimensions, shape and the FE mesh of the cell are shown in Fig. E16.3-4b (position 1) – see color plate. Data for the membrane and the biphasic medium characteristics are given in the caption of Fig. E16.3-4. Cytoskeleton is modeled by a set of fibers (truss finite elements) connecting the nodes parallel to the surface and around the nucleus. It is assumed that the nucleus is stiffer than the cell cytoplasm, with no fibers. We use the constitutive law of the fibers shown in Fig. E16.3-4a (right panel) – see color plate, with the activation function shown in Fig. E16.3-4a (left panel). Computational procedure is analogous to that presented in Chapter 12.

It is assumed that the cell has the protrusion and that it is attached at the front part to the surface (McGrath & Dewey 2006) at the position 1. Then, due to activation the cell deforms and slides over the surface. It is assumed that the activation function is linear (Fig. E16.3-4a left panel – see color plate) reaching maximum at time equal to 5 s. The deformed cell shape and the displacement field within the cell are shown in the position 2 of the figure. It is then assumed that the cell attaches to the surface at its rear region, with detachment of the protrusion end. Activation decreases to zero and the cell further moves due to relaxation, reaching the initial shape at the position 3.

Modeling of the cell crawling with changing the model parameters can be performed using the Software on the web.

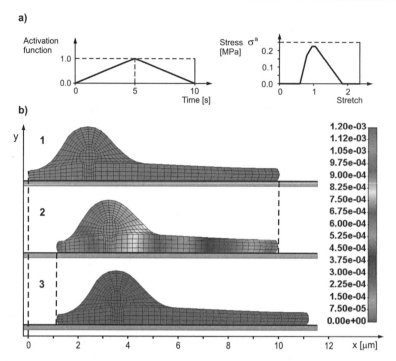

Fig. E16.3-4 Crawling of cell over a flat surface (2D plane strain conditions in plane x–y). Biphasic model includes: cytoplasm with cytoskeleton and nucleus, and membrane. (a) Activation function of skeleton structure (left panel) and constitutive law for the active stress σ^a (right panel); (b) Three positions of the cell during crawling (1 – initial, after first step; 2 – middle, when detachment of the front and attachment of the rear part occur; 3-after relaxation) with the displacement field. Data: Young's moduli (MPa) for solid within solid-fluid mixture, and within nucleus $E = 0.1$ and $E = 0.3$, respectively; initial porosity $n = 0.7$; permeability $k = 10^2$ $\mu m^4/pNs$; solid and fluid density $\rho = 10^{-9}$ $mg/\mu m^3$; bulk moduli of solid and fluid $K_s = 8.3333 \times 10^{-2}$ MPa, $K_f = 10^9 pN/\mu m^2$ (see Plate 31)

References

Brodland, G.W. & Gordon, R. (1990). Intermediate filaments may prevent buckling of compressively loaded microtubules, *ASME J. Biomech. Eng.*, **112**, 319–21.

Cañadas, P., Laurent, V.M., Oddou, C. & Isabey, D. (2002). A cellular tensegrity model to analyze the structural viscoelasticity of the cytoskeleton, *J. Theor. Biol.*, **218**, 155–73.

Connelly, R. & Back, A. (1998). Mathematics and tensegrity, *Am. Sci.*, **86**, 142–51.

Coughlin, M.F. & Stamenović, D. (2003). A prestressed cable network model of adherent cell cytoskeleton, *Biophysic. J.*, **84**, 1328–36.

Dao, M., Lim, C.T. & Suresh, S. (2003). Mechanics of the human red blood cell deformed by optical tweezers, *J. Mech. Phys. Solids*, **51**, 2259–80.

Deguchi, S., Ohashi, S. & Sato, M. (2006). Tensile properties of single stress fibers isolated from cultured vascular smooth muscle cells, *J. Biomech.*, **39**, 2603–10.

Discher, D.E., Boal, D.H. & Boey, S.K. (1998). Stimulations of the erythrocyte cytoskeleton at large deformation. II. Micropipette aspiration, *Biophysic. J.*, **75**, 1584–97.

Evans, E. & Yeung, A. (1989). Apparent viscosity and cortical tension of blood granulocytes determined by micropipet aspiration, *Biophysic. J.*, **56**, 151–60.

Folkman, J. & Moscona, A. (1978). Role of cell shape in growth and control, *Nature*, **273**, 345–9.

Fudge, D.S., Gardner, K.H., Forsyth, V.T., Riekel, C. & Gosline, J.M. (2003). The mechanical properties of hydrated intermediate filaments: insights from hagfish slime threads, *Biophysic. J.*, **85**, 2015–27.

Fuller, B. (1961). Tensegrity, *Portfolio Artnews Annu.*, **4**, 112–27.

Gittes, F., Mickey, B., Nettleton, J. & Howard, J. (1993). Flexural rigidity of microtubules and actin filaments measured from thermal fluctuations in shape, *J. Cell Biol.*, **120**, 923–34.

Guilak, F., Haider, M.A., Setton, L.A., Laursen, T.A. & Baaijens, F.P.T. (2006). Multiphasic models of cell mechanics. In M.R.K. Mofrad & R.D. Kamm (eds.), *Cytoskeletal Mechanics* (Chapter 5), Cambridge University Press, Cambridge, England.

Hu, S., Chen, J. & Wang, N. (2004). Cell spreading controls balance of prestress by microtubules and extracellular matrix, *Front. Biosci.*, **9**, 2177–82.

Ingber, D.E. (1993). The riddle of morphogenesis: a question of solution chemistry or molecular cell engineering?, *Cell*, **75**, 1249–52.

Ingber, D.E. (1997). Tensegrity: the architectural basis of cellular mechanotransduction, *Annu. Rev. Physiol.*, **59**, 575–99.

Ingber, D.E. (1998). The architecture of life, *Sci. Am.*, **278**, 48–57.

Ingber, D.E. (2003). Cellular tensegrity revisited I. Cell structure and hierarchical systems biology, *J. Cell Sci.*, **116**, 1157–73.

Kojić, M., Isailović, V., Stojanović, B. & Filipović, N. (2007). Modeling of cell mechanical response by biphasic models with activation, *J. Serbian Soc. Comp. Mech.*, **1**(1), 135–43.

Lo, C.-M., Wang, H.-B., Dembo, M. & Wang, Y.-L. (2000). Cell movement is guided by the rigidity of the substrate, *Biophys. J.*, **79**, 144–52.

Maksym, G.N., Fabry, B., Butler, J.P. *et al.* (2000). Mechanical properties of cultured human airway smooth muscle cells from 0.05 to 0.4 Hz, *J. Appl. Physiol.*, **89**, 1619–32.

McGrath, J.L. & Dewey, C.F. Jr. (2006). Cell dynamics and the actin cytoskeleton. In M.R.K. Mofrad & R.D. Kamm (eds.), *Cytoskeletal Mechanics* (Chapter 9), Cambridge University Press, Cambridge, England

Mijailović, S.M., Kojić, M., Živković, M., Fabry, B. & Fredberg, J.J. (2002). A finite element model of cell deformation during magnetic bead twisting, *J. Appl. Physiol.*, **93**, 1429–36.

Mills, J.P., Qie, L., Dao, M., Lim, C.T. & Suresh, S. (2004). Nonlinear elastic and viscoelastic deformation of the human red blood cell with optical tweezers, *MCB*, **1**(3), 169–80.

Mofrad, M.R.K. & Kamm, R.D. (2006). Continuum elastic or viscoelastic models of the cell. In M.R.K. Mofrad & R.D. Kamm (eds.), *Cytoskeletal Mechanics* (Chapter 4), Cambridge University Press, Cambridge, England.

Sheetz, M.P. (2001). Cell control by membrane-cytoskeleton adhesion, *Nature Rev. Mol. Cell Biol.*, **2**, 392–6.

Stamenović, D. (2006). Models of cytoskeletal mechanics based on tensegrity. In M.R.K. Mofrad & R.D. Kamm (eds.), *Cytoskeletal Mechanics* (Chapter 6), Cambridge University Press, Cambridge, England.

Stamenović, D., Fredberg, J.J., Wang, N., Butler, J.P. & Ingber, D.E. (1996). A microstructural approach to cytoskeletal mechanics based on tensegrity, *J. Theor. Biol.*, **181**, 125–36.

Stamenović, D. & Ingber, D.E. (2002). Models of cytoskeletal mechanics of adherent cells, *Biomech. Model. Mechanobiol.*, **1**, 95–108.

Sultan, C., Stamenović, D. & Ingber, D.E. (2004). A computational tensegrity model predicts dynamic rheological behaviors in living cells, *Ann. Biomed. Eng.*, **32**, 520–30.

Volokh, K.Y. & Vilnay, O. (1997). New cases of reticulated underconstrained structures, *Int. J. Solids Struct.*, **34**, 1093–104.

Volokh, K.Y., Vilnay, O. & Belsky, M. (2000). Tensegrity architecture explains linear stiffening and predicts softening of living cells, *J. Biomech.*, **33**, 1543–9.

Wang, N., Butler, J.P. & Ingber, D.E. (1993). Mechanotransduction across cell surface and through the cytoskeleton, *Science*, **260**(5111), 1124–7.

Wang N, Ingber DE (1994). Control of the cytoskeletal mechanics by extracellular matrix cell shape and mechanical tension, *Biophys. J.*, 66, 2181-2189.

Wang, N., Naruse, K., Stamenović, D. *et al.* (2001). Mechanical behavior in living cells consistent with the tensegrity model, *Proc. Natl. Acad. Sci. USA*, **98**, 7765–70.

Wang, N. & Stamenović, D. (2000). Contribution of intermediate filaments to cell stiffness, stiffening and growth, *Am. J. Physiol. Cell Physiol.*, **279**, C188–C194.

Wang, N., Tolić-Nørrelykke, I.M., Chen, J. *et al.* (2002). Cell prestress. I. Stiffness and prestress are closely associated in adherent contractile cells, *Am. J. Physiol. Cell. Physiol.*, **282**, C606–C616.

Wendling, S., Oddou, C. & Isabey, D. (1999). Stiffening response of a cellular tensegrity model, *J. Theor. Biol.*, **196**, 309–25.

17

Extracellular Mechanotransduction: Modeling Ligand Concentration Dynamics in the Lateral Intercellular Space of Compressed Airway Epithelial Cells

Cellular mechanotransduction, i.e. the transduction of mechanical stimuli into cellular signals, can occur through autocrine signaling in a dynamically changing extracellular space. We developed computational models to analyze how alterations in the geometry of an epithelial lateral intercellular space (LIS) affect the concentrations of constitutively shed ligands inside and below the LIS of cultured airway epithelial cells. The two presented models, based on pure diffusion and diffusion–convection, utilize the finite element method to solve for the concentration of ligands inside and outside of the LIS. Using these models, we examined the temporal relationship between geometric changes and ligand concentration, and the dependence of this relationship on system characteristics such as ligand diffusivity, shedding rate and rate of deformation. Our results reveal how the kinetics of mechanical deformation can be translated into varying rates of ligand accumulation, a potentially important mechanism for cellular discrimination of varying rate-mechanical processes. Furthermore, our results demonstrate that rapid changes in LIS geometry can transiently increase ligand concentrations in underlying media or tissues, suggesting a mechanism for communication of mechanical state between epithelial and subepithelial cells. All of these results underscore the insight gained by employing numerical modeling to explore the complex process of extracellular mechanotransduction.

Most of the text and figures presented in this chapter are based on references Kojić (2007) and Kojić et al. (2006) (adapted with permission from *Biophysical Journal* 2006).

17.1 Autocrine signaling in airway epithelial cells

17.1.1 Introduction

The Role of the Airway Epithelium in Asthma
The epithelial cells that line the human airway serve as the body's natural barrier against a wide range of unwanted gases and particulate matter. But they are more than a simple barrier. In disease states, such as asthma, the structural organization of the airway epithelium may change. For example, depending on the severity of their disease, an asthmatic can have pronouncedly thickened epithelium, thickening of the collagen layer just below the true basement membrane, smooth muscle cell proliferation, and other structural changes that together decrease the airway luminal area. Furthermore, mast cells in the asthmatic epithelium can serve as a sensor for allergens, thus recruiting an immune response that can ultimately lead to progression of the disease.

Recently, another role for the epithelium has emerged whereby the epithelium is involved in mechanotransduction of forces that accompany airway constriction. Elucidating the mechanism by which the epithelium transduces mechanical stimuli, such as compressive stress, could lead to a better understanding of the role played by the epithelium in a complex disease such as asthma. Ultimately one could envision new drugs that could help break the vicious cycle involving the epithelium, the immune system and the subepithelial tissue that contributes to the progression of asthma. Also, it is likely that the mechanotransduction mechanism involved in airway epithelial cells could be applicable to other epithelial cells or tissues in the body, increasing the need to uncover the underlying processes.

The Lateral Intercellular Space (LIS) of Airway Epithelial Cells
Neighboring airway epithelial cells are connected at the apical surface via tight junctions that are impermeable to larger molecules, such as proteins (Tschumperlin *et al.* 2004, Willumsen 1994). These tight junctions serve as the natural barrier from the outside world and their intactness is crucial for normal function. Below the tight junctions, on the lateral side of neighboring cells, lies an extracellular area termed the lateral intercellular space (LIS). The LIS is thus defined on the apical side by the tight junction, the lateral surfaces of neighboring cells, and basally by the basement membrane *in vivo*, or porous substrate on which cells are grown *in vitro* (see Fig. 17.1.1).

On the lateral surface of the cells there is a host of different membrane-bound ligands and their corresponding receptors. The bound ligands are released from the cell surface into the LIS via the action of a sheddase that cleaves the connecting bond (Harris *et al.* 2003). If a ligand is released into a collapsing LIS, its local concentration will increase. If the process occurs uniformly along the lateral surface, this establishes a concentration gradient in the

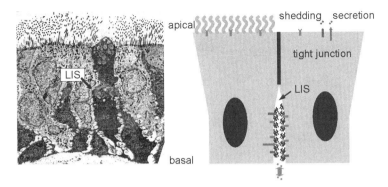

Fig. 17.1.1 Scanning electron micrograph (left panel) of airway epithelium (according to Evans *et al.* 2001) and schematic (right panel) of the LIS separating two neighboring epithelial cells

apico-basal direction of the LIS, whereby the highest concentration is near the tight junction and the lowest, essentially zero, concentration is at the basal surface.

In the LIS, glycoprotein projections from the lateral cell surface form a mesh called the glycocalyx. The dense network of the glycocalyx could play a crucial role in hindering the diffusion of large molecules (Kovbasnjuk *et al.* 2000). Furthermore, being charged the glycocalyx could also provide significant chemical impedance to positively charged ligands thus hindering movement down their concentration gradient to the area below the basal surface of the cell where their concentration is likely to be quite small. The effect of the glycocalyx on the diffusion coefficient of large molecules has been studied in the LIS of canine epithelial cells, showing a several fold decrease in diffusion coefficient and thus effectively a corresponding increase in resistance to diffusion, since diffusion coefficient $\sim 1/$(diffusion resistance) (Kovbasnjuk *et al.* 2000). Although the apical surface is also rich in ligand secretion activity and glycocalyx, the presence of a large reservoir of apical fluid will keep effective concentrations low, making the most important changes that occur during prolonged bronchoconstriction in the LIS.

17.1.2 The EGF–receptor autocrine loop in the LIS

As mentioned above, numerous receptors extend into the LIS from the cellular surface. One type of receptor that plays a crucial role in mechanotransduction is the epidermal growth factor (EGF) receptor (Tschumperlin *et al.* 2002). We have previously shown that the EGF receptor (EGFR) functions as part of an autocrine loop in normal human bronchial epithelial (NHBE) cells (see Fig. 17.1.2). An activating signal can be an increase in the number of ligands bound through an increase in local ligand concentration.

There are at least four known members of the EGF receptor family (erbB1-4) and our main focus will be on erbB1, also known as the classical EGF receptor (referred to as EGFR) (Yarden & Sliwkowski 2001). The EGFR contains three domains: an intrinsic kinase domain in the cytoplasm, a membrane-spanning domain, and an ectodomain that binds EGFR ligands. When the ligand binds to the receptor, the monomeric receptors dimerize and phosphorylate specific cytoplasmic tyrosine residues. The phosphorylated residues then serve as a site of attachment for effector molecules that further activate signal transduction

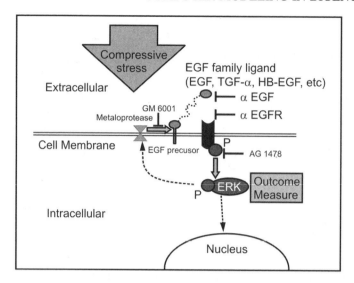

Fig. 17.1.2 Schematic of the epidermal growth factor receptor (EGFR) autocrine loop. Light gray indicates antibodies that block various sites (GM6001 blocks metalloprotease activity, ligand and receptor antibodies are labeled with an α, AG1478 blocks EGFR phosphorylation). Phosphorylated mitogen-activated protein kinase ERK was the outcome measure

pathways (Raab & Klagsbrun 1997). Downstream from receptor activation phosphorylation of the mitogen-activated protein kinase ERK occurs (Tschumperlin *et al.* 2002). The next step in the pathway is translocation of p-ERK to the nucleus (see Fig. 17.1.2) where it helps activate various transcription factors such as *cfos* and *jun*, hence forming the loop: **ligand–EGFR–ERK–transcription-ligand**.

Phosphorylated ERK can also signal to metalloproteases to cleave the membrane bound ligand (Harris *et al.* 2003) forming another autocrine loop: **ligand–EGFR–ERK–metalloprotease–ligand**. It should be noted that the two loops mentioned above could be interconnected and part of other, broader feedback mechanisms utilized by the cell to sense and appropriately respond to changes in the local environment.

17.1.3 Modeling the effects of compressive stress on epithelial cells *in vitro*

During bronchoconstriction, because the basement membrane is relatively inelastic, the epithelium buckles under the influence of contracting smooth muscle. The pattern of buckling depends on the mechanical properties of the airway components as well as the thickness of the epithelium. Previous numerical modeling studies and experiments have shown that a thicker epithelium needs smaller luminal pressures to cause airway closure (Hrousis et al. 2002, Wiggs *et al.* 1997). Depending on the 'state' of the epithelium the cells will thus fold in different patterns (by state we mean the degree of airway remodeling that has occurred as a result of disease).

In an asthmatic airway the effect of airway remodeling is the classic rosette folding pattern (see Fig. 17.1.3) in which neighboring cells located in the cleft areas get squeezed

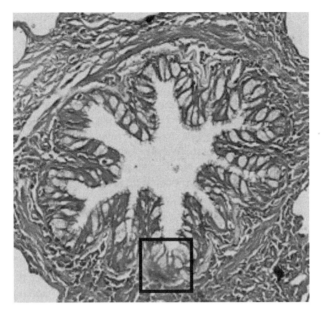

Fig. 17.1.3 A constricted remodeled airway. The box indicates an area where cells get pushed against each other

against each other during bronchoconstriction. Thus they experience a compressive stress, whose magnitude is $\sim 30\,cm$ H_2O (Wiggs *et al.* 1997).

To test the effects of this compressive stress *in vitro* we use a system that can apply a pneumatic pressure gradient across cultured normal human bronchial epithelial (NHBE) cells (Ressler *et al.* 2000, Swartz *et al.* 2001). The cells (obtained from Clonetics) are grown on microporous polyester substrates (pore size 0.4 microns, Transwell-Clear). Initially, for the first week, the cells are fed from the top and bottom. After this, for the next two weeks the cells are only fed through the bottom pores, and a cell–air interface is established at the top. This interface is a signal for the cells to polarize and differentiate into the NHBE phenotype. Once the cells are fully differentiated, the pressure device is used to apply compressive stress (see Fig. 17.1.4). The device consists of a plug for the transwell and a pressure tank that supplies compressed air (with 5% CO_2) through a tube going through the plug to the apical surface of the cells. The media bath below the substrate remains at atmospheric pressure, hence a pressure gradient can be established across the cells.

Previous Experimental Findings
Earlier experiments with this pressure device provided several key findings about the effects of compressive stress (Tschumperlin *et al.* 2004):

1. The EGFR via signaling through one of its ligands, heparin-binding epidermal growth factor (HB-EGF), plays a key role in transducing the mechanical stimulus.

2. ERK phosphorylation occurs downstream of the EGFR.

3. Metalloproteases involved in shedding of ligand into the LIS are an integral part of the mechanotransduction pathway.
4. The LIS collapses, decreasing its volume substantially due to compression.

In order to have a more accurate picture of how the collapsing LIS results in increased signaling through the EGFR a 1D LIS model based on diffusion has been put forth (Tschumperlin et al. 2004). The proposed mechanism functions as follows. Under compression, the LIS width (and thus volume) decreases pushing out the intercellular fluid basally through the porous substrate (since tight junctions form an impermeable barrier on the apical surface). HB-EGF, a large, charged protein, known to be released into the LIS remains there due to its size and charge, hence the LIS concentration of HB-EGF increases. In other words, the HB-EGF in the LIS finds itself in a smaller volume and thus the concentration increases. This increase in turn causes a greater number of HB-EGF molecules to be bound to EGFR than before the LIS collapse, effectively activating the mechanotransduction loop described above.

The model can be expressed in mathematical form assuming that diffusion of HB-EGF occurs in the apical-to-basal direction down its concentration gradient, where the highest concentration is immediately below the impermeable tight junctions and the lowest (essentially zero) concentration is in the media bath. Hence, the balance equation of diffusion (3.2.8) is now

$$\frac{\partial C}{\partial t} = D\frac{\partial^2 C}{\partial x^2} + \frac{2q}{w} \qquad (17.1.1)$$

where C is the ligand (HB-EGF) concentration, t is time, x is the apico-basal direction, D is the diffusion coefficient of ligand through the LIS/glycocalyx matrix, q is the rate of shedding of ligand from the cell surface into the LIS and w is the LIS width. Here we assume that the LIS collapse can be regarded as two parallel plates coming closer together. From (17.1.1), the steady-state solution

$$C = \frac{q}{wD}(h^2 - x^2) \qquad (17.1.2)$$

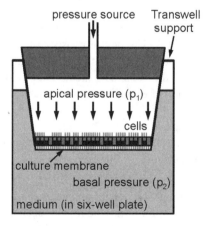

Fig. 17.1.4 The pressure device supplies a pressure gradient p_1-p_2 across the cells

EXTRACELLULAR MECHANOTRANSDUCTION

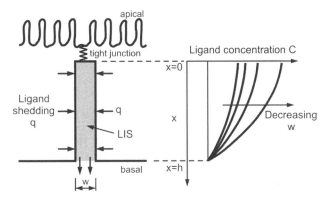

Fig. 17.1.5 Schematic of the initial LIS model (left panel) and the corresponding steady-state solutions (right panel)

indicates that the concentration is inversely proportional to the width, thus C increases as w decreases; here h is the LIS height (see Fig. 17.1.5). Such a deduction could be viewed as satisfactory from a qualitative standpoint, but for deeper engineering insight the LIS dynamics must be taken into account.

Rationale for a Focus on HB-EGF

Previous experiments (Tschumperlin *et al.* 2004, Tschumperlin & Drazen 2001, Tschumperlin *et al.* 2002) have shown that HB-EGF is the key EGFR ligand in the mechanotransduction pathway. These experiments revealed that if antibodies were used against other common EGFR ligands, such as EGF or TGF-α, there was no measurable deterioration of ERK phosphorylation, whereas an HB-EGF antibody substantially downgraded the signal. This begs the question why HB-EGF, and not one or all of the other ligands?

We believe that two major characteristics distinguish HB-EGF, namely its size and charge. For example, HB-EGF has a molecular weight of about $22\,kDa$, whereas TGF-α and EGF are both only about $5.5\,kDa$. In experiments performed in Spring's laboratory (Kovbasnjuk *et al.* 2000), the diffusion coefficient in a similar LIS architecture decreased by an order of magnitude as the size of the molecule probed increased from $3\,kDa$ to $10\,kDa$. Thus it is conceivable that just due to size alone HB-EGF would have a much smaller diffusion coefficient and would be 'left behind' relative to the other ligands during the 'washout' phase of an LIS collapse where fluid is squeezed out. Furthermore, HB-EGF is more likely to get trapped or hindered in the glycocalyx due to its positively charged domains (Raab & Klagsbrun 1997) that could interact with negatively charged areas of the glycocalyx.

The end result would be a relatively higher increase in concentration of HB-EGF compared to other ligands in the collapsing LIS. In other words, the concentration of HB-EGF would increase drastically (while the concentration of other ligands would be mostly unchanged) and hence would provide a new input signal to the cell, initiating the mechanotransduction pathway. Further experimental evidence for such a mechanism comes from exogenously adding a bolus of HB-EGF and seeing a similar response in ERK phosphorylation as in the case of compressive stress (Tschumperlin *et al.* 2004). Therefore, since an increase in HB-EGF can cause the same cellular response as compressive stress, it is feasible that the collapsing LIS does just that: increases local HB-EGF concentration via the mechanism described above. In following chapters we explore these possibilities in more detail through the use of computer models.

Role of Computer Modeling
The established system of epithelial stress by pneumatic pressure was used to examine mechanotransduction in cultured human airway epithelial cells (Tschumperlin *et al.* 2004). However, the experiments performed can only provide a partial view of the system. To have a more complete and general understanding, mathematical and computer models are necessary (Kojić 2007, Kojić *et al.* 2006). Specifically, not every quantity can be measured experimentally in real time, and even if such measures were possible, one would still have to determine the relative importance of various ligand–receptor pairs in signal transduction. Computer models therefore become a valuable tool in this selection process and can also provide insight that was not initially apparent. When coupled to experiments, these models allow for a more complete understanding of observed phenomena.

Different generations of the computer model presented in subsequent sections reflect higher levels of complexity, all aimed at not only having a deeper understanding of the experimental results, but also elucidating the mechanisms involved in mechanotransduction of compressed airway epithelial cells.

17.2 The dynamic diffusion model

17.2.1 Introduction

In the previous section, we described a 1D steady-state (i.e. time independent) model of LIS ligand concentration, developed by Tschumperlin and colleagues (Tschumperlin *et al.* 2004). This model yielded the parabolic concentration profiles (see Fig. 17.1.5) in the LIS for HB-EGF ligand for only two time points: first for the steady-state pre-collapse condition, and then, depending on the change in LIS width w, the final post-collapse steady-state condition. The implicit assumption was that after the collapse occurs, the LIS width stays constant and equal to the width at the end of the collapse.

The key result of the steady-state model was that the ligand concentration was inversely proportional to the LIS width. Thus, during a pressure-induced collapse the LIS width would decrease and the final, steady-state concentration would proportionally increase. This was an important first step in establishing the plausibility of the mechanotransduction mechanism via an increase in LIS concentration. However, the steady-state model could not address how the concentration changes during a collapse because all of the time-dependent terms were set to zero. Since ligand dynamics can dominate signal transduction (Sasagawa *et al.* 2005), we present in this chapter a dynamic diffusion model that outputs LIS ligand concentration as a function of both time and space. We then explore the parameter space of the model by varying the rate of width decrease, the diffusion coefficient and the shedding rate of ligand into the LIS.

The Dynamic Diffusion Equation
The steady-state model was based on removing the time dependence from the concentration, effectively making the concentration a function of only the spatial coordinate x. Hence the basic diffusion equation becomes (17.1.1)

EXTRACELLULAR MECHANOTRANSDUCTION

$$\frac{\partial C}{\partial t} = 0 = D\frac{\partial^2 C}{\partial x^2} + \frac{2q}{w} \quad (17.2.1)$$

where C is the ligand (HB-EGF) concentration, x is the apico-basal direction, D is the diffusion coefficient of ligand through the LIS/glycocalyx matrix, q is the rate of shedding of ligand from the cell surface into the LIS and w is the LIS width. Solving (17.2.1), the steady-state concentration profile becomes

$$C(x) = \frac{q}{wD}(h^2 - x^2) \quad (17.2.2)$$

Here h is the total apical-basal height of the LIS. From this equation we see how for a given w we can obtain a steady-state concentration profile. Obviously, a smaller width results in a greater concentration.

Although this was an important initial step in linking geometric changes of the LIS to extracellular ligand concentration, many important issues could not be addressed from this simple model. One such issue is the time dependence of changes in ligand concentration as the LIS collapses. In other words, the key question becomes: how does the LIS ligand concentration change over time during and after the collapse?

To answer this question we needed to employ numerical methods to solve the diffusion equation in which C and w are considered to be a function of time:

$$\frac{\partial C(x,t)}{\partial t} = D\frac{\partial^2 C}{\partial x^2} + \frac{2q}{w(t)} \quad (17.2.3)$$

17.2.2 Finite element model of dynamic diffusion

In order to solve (17.2.3) we first discretize the LIS space into 1D finite elements (see Fig. 17.2.1). Then, using the Galerkin weighted method (Huebner 1975) we transform (17.2.3) into the following form (see equation (7.1.2)):

$$\int_L \psi(x) \left[\frac{\partial C}{\partial t} + \frac{\partial}{\partial x}\left(D\frac{\partial C}{\partial x}\right) + q_V = 0 \right] dL \quad (17.2.4)$$

where $\psi(x)$ is a weighted function, $q_V = 2q/w$ is the source term due to the shedding, and L is the selected domain. Equation (17.2.4) represents the so-called *weak form* (Bathe 1996, Hughes 1987) of the differential equation (17.2.3) since (17.2.3) is not necessarily satisfied at each point of the domain.

In the isoparametric formulation of a 1D finite element of the length L we use interpolation functions N_K as the weighted functions (where the subscript K refers to the node number) (Huebner 1975). The nodes of the finite elements are represented with circles in Fig. 17.2.1. Therefore, we obtain N weak-form equations for a finite element,

$$\int_L N_K \left[-\frac{\partial C}{\partial t} + \frac{\partial}{\partial x}\left(D\frac{\partial C}{\partial x}\right) + q_V = 0 \right] dL \quad K = 1, 2, \ldots, N \quad (17.2.5)$$

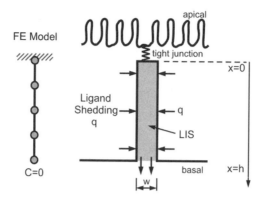

Fig. 17.2.1 Schematic of the 1D LIS finite element (FE) model. At the apical surface of the LIS ($x = 0$) there is an impermeable wall (tight junction) and at the basal surface ($x = h$) we assume the ligand concentration is zero. Each finite element consists of two nodes (depicted as circles); note: FE model not drawn to scale

where N is the number of nodes per element. In our case there are two nodes per element, thus $K = 1, 2$. Applying Gauss' theorem (see (1.4.11) in Section 1.4) to the diffusion term within the integral we obtain

$$\int N_K \frac{\partial}{\partial x}\left(D\frac{\partial C}{\partial x}\right) dL = \int_S N_K D \frac{\partial C}{\partial x} dS - \int_L D \frac{\partial N_K}{\partial x} \frac{\partial C}{\partial x} dL \qquad (17.2.6)$$

where S is the element surface and the term $D\frac{\partial C}{\partial x}$ represents the flux q_s of C through the element surface, which in our case are the two cross-sectional areas (one at each of the element nodes and equal to $w \times (unit\ depth)$). The surface integral can then be written as

$$Q_K^S = \int_S N_K D \frac{\partial C}{\partial x} dS = \int_S N_K\, q_s\, dS \qquad (17.2.7)$$

where $Q_K^S(t)$ is the surface flux corresponding to the node 'K'.

Next we perform the interpolation of the variable C within the element as

$$C = N_K C^K = N_1 C^1 + N_2 C^2 + \ldots + N_N C^N \qquad (17.2.8)$$

where C^K are the nodal values of C. Now, we substitute (17.2.6) and (17.2.8) into (17.2.5) and also use (17.2.7), to obtain

$$\begin{aligned}&\mathbf{M\dot{C}} + \mathbf{KC}^K = \mathbf{Q}^S + \mathbf{Q}^V, \quad \text{or} \\ & M_{KJ}\dot{C}^J + K_{KJ} C^J = Q_K^S + Q_K^V, \quad K, J = 1, 2, \ldots, N\end{aligned} \qquad (17.2.9)$$

Here $\mathbf{C}$ is the vector of nodal values C^K, $\dot{\mathbf{C}}$ is the vector of time derivatives at nodal points, and the matrices and vectors are:

EXTRACELLULAR MECHANOTRANSDUCTION

$$M_{KJ} = \int_L N_K N_J \, dL, \quad K_{KJ} = \int_L D \frac{\partial N_K}{\partial x} \frac{\partial N_J}{\partial x} dL,$$

$$Q_K^S = \int_S N_K q_S \, dS, \quad Q_K^V = \int_S N_K q_V \, dS$$

(17.2.10)

The system of equations (17.2.9) represents the equations of balance for a finite element. Assemblage of the element equations is then performed and appropriate boundary conditions must be implemented prior to solving the equations of the whole system (see Section 4.1). The equation of balance for the finite element assemblage retains the form of (17.2.9). Note that the surface fluxes Q_K^S cancel over the internal surfaces of the finite elements, i.e. the flux leaving one node of the finite element is equal to the flux going into the same node of the neighboring finite element which shares that node. Therefore, the contribution to the system surface vector $\mathbf{Q}^S$ comes only from the flux through the surface of the whole domain.

The incremental form for time integration of (17.2.9) can be written as

$$\left(\frac{1}{\Delta t}\mathbf{M} + \mathbf{K}\right)^{n+1}\mathbf{C} = {}^{n+1}\mathbf{Q}^S + {}^{n+1}\mathbf{Q}^V + \frac{1}{\Delta t}\mathbf{M}\,{}^n\mathbf{C}$$

(17.2.11)

where Δt is the time step, and the upper left indices n and $n+1$ denote values at the start and end of the n-th time step (see (7.1.9)).

17.2.3 Exploring the parameter space of the diffusion equation

Based on the diffusion equation (17.2.3), the ligand concentration C depends on three parameters: the diffusion coefficient D, the shedding rate q, and the LIS width w. To investigate the effect of these parameters on the changes in concentration we employed our finite element (FE) model to see the interdependence of the three parameters and thus get a sense of the relative importance of experimentally determining each/all of them.

The LIS geometry was regarded as a 1D rod (see Fig. 17.2.1), consisting of 50 finite elements. The boundary conditions were: no flux through the top, apical surface; and zero concentration at the bottom, basal surface. The diffusion equation (17.2.3) was then transformed to the appropriate 1D finite element form (see (17.2.11)) and solved using the PAK software package (Kojić et al. 1998). We define x as the depth coordinate, being 0 at the most apical (tight junction) surface (Fig. 17.2.1).

To explore the effect on concentration, the parameters D, w and q were changed. We first assume constant D and q, while the width w changes. We further assume that the LIS width w decreases linearly during the prescribed time of collapse. For example, in previous work (Tschumperlin et al. 2004) it was established that the LIS width shrinks to close to 90% in a time that is less than 20 minutes. Exactly how fast the LIS shrinks was unclear since the imaging technique only allowed measurements at 20-minute intervals. Thus, we first solved the ligand concentration profiles for a very rapid (1 s) LIS collapse from $w = 2$ to 0.6 microns (see Fig. 17.2.2a). The diffusion coefficient was taken from the literature and estimated to be $18\,\mu m^2/s$ (Kovbasnjuk et al. 2000), whereas the shedding rate was roughly approximated to be 12.6 $(ng.\mu m)/(mL.s)$ (based on personal communication with Dr. Ivan Maly, unpublished results). The height h of the LIS was taken from previous experiments (Tschumperlin et al. 2004) to be $h = 17\,\mu m$. The circles (Fig. 17.2.2a)

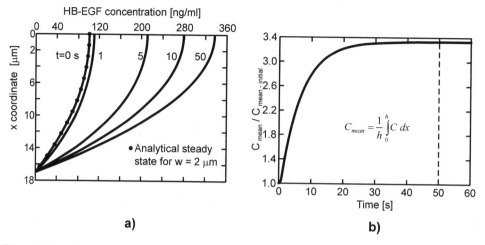

Fig. 17.2.2 Case for $w = 2 \rightarrow 0.6\,\mu m$ in $1\,s$, $D = 18\,\mu m^2/s$, $q = 12.6\,(ng.\mu m)/(mL.s)$. (a) Numerical solution for concentration profiles along the LIS depth; (b) Corresponding fold increase in mean LIS concentration

represent the analytical solution (see (17.1.2)) for $w = 2\,\mu m$, which is in agreement with our numerical solution. The curves to the right of initial $t = 0$ curve represent the time evolution of concentration increase as the LIS collapses from $w = 2$ to $0.6\,\mu m$ in $1\,s$. The curve at $t = 50\,s$ is the new steady-state concentration profile corresponding to $w = 0.6\,\mu m$.

In addition to looking at the concentration profile along the depth of the LIS it is worthwhile to examine the relative fold increase of the mean LIS concentration defined as:

$$C_{mean} = \frac{1}{h} \int_0^h C\,dx \qquad (17.2.12)$$

The ratio of C_{mean} over the initial C_{mean}, i.e. the fold increase in mean LIS ligand concentration, for the same conditions of LIS collapse as described above, indicates an initial rapid increase followed by a steady-state plateau (see Fig. 17.2.2b).

Previously we have stated that we observed a change in LIS volume of about 90% in $< 20\,min$ and we show the effects of a rapid 1 second collapse in Fig. 17.2.2. Next we show how the LIS concentration dynamics change when the time of collapse increases. Specifically, we examine four cases: $t_{collapse} = 1, 100, 500, 1000$ seconds, and assume that LIS width changes from $w = 2$ to $0.2\,\mu m$ during the prescribed time of collapse, with all other parameters being the same as above in Fig. 17.2.2. The results for these four cases are shown in Fig. 17.2.3. As the figure indicates, the varying time-of-collapse has a profound effect on the dynamics, where the longer $t_{collapse}$ results in a slower approach to steady state.

In all of the above calculations, both the diffusion coefficient and the shedding rate were assumed to be constant during and after the LIS collapse. We next explore what are the effects of variable diffusion coefficient D on the diffusion process in the LIS. Namely, it is likely that the ligand diffusion coefficient through the LIS can change as a result of structural changes in the LIS glycocalyx. In effect, the glycocalyx matrix becomes 'denser' and further hinders diffusion of a large molecule like HB-EGF (molecular weight of 22 kDa)

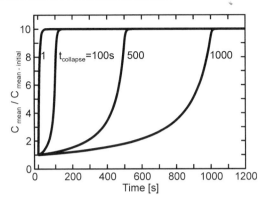

Fig. 17.2.3 Variable $t_{collapse}$ (number next to each curve indicates duration of collapse) for $w = 2 \to 0.2$ μm, $D = 18$ $\mu m^2/s$, $q = 12.6$ $(ng.\mu m)/(mL.s)$

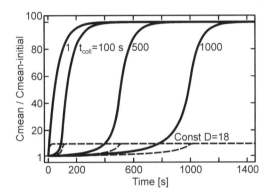

Fig. 17.2.4 Variable diffusion coefficient and variable $t_{collapse}$ for: $w = 2 \to 0.2$ μm, $D = 18 \to 1.8$ $\mu m^2/s$, $q = 12.6$ $(ng.\mu m)/(mL.s)$. Dashed line represents Fig. 17.2.3, with constant $D = 18$ $\mu m^2/s$

(Kovbasnjuk et al. 2000, Xia et al. 1998). To see how a changing diffusion coefficient affects the mean concentration, we assume that the diffusion coefficient D decreases linearly with LIS collapse. In other words, D is a linear function of w: $D = 18$ $\mu m^2/s$ for $w = 2$ μm and $D = 1.8$ $\mu m^2/s$ for $w = 0.2$ μm. The results are displayed in Fig. 17.2.4. For comparison, we also plot the previously determined case of constant $D = 18$ $\mu m^2/s$ (dashed line). The decreasing diffusion coefficient has a retarding effect on the concentration increase toward the new steady state for each of the four $t_{collapse}$ cases.

The third, final parameter that could be variable (and was until now assumed constant) is the shedding rate q. It is conceivable that the LIS autocrine loop involving the EGFR-ERK-metalloprotease mechanism (see Fig. 17.1.2) could change due to increased ligand binding to the EGFR. We examine two cases of changing q: one being a linear and sustained five-fold increase (top dashed line in Fig. 17.2.5); and the other being a transient five-fold increase followed by a symmetric, linear decrease to the initial value of q (solid line in Fig. 17.2.5). For clarity, we only consider one time of collapse: $t_{collapse} = 1 s$. The resulting

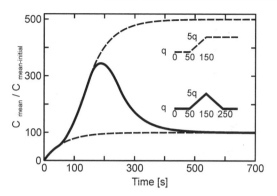

Fig. 17.2.5 Variable q, w, D: $\dot{w} = 2 \to 0.2\,\mu m$ in 1 second, $D = 18 \to 1.8\,\mu m^2/s$. Top dashed line: change in $q = 12.6 \to 63$ $(ng\,\mu m)/(mL\,s)$ during $50 \to 150$ s. Solid line: $q = 12.6 \to 63$ during $50 \to 150$ s followed by $q = 63 \to 12.6$ during $150 \to 250$ s. Bottom dashed line represents Fig.17.2.2b where $q = \text{const} = 12.6$ $(ng\,\mu m)/(mL\,s)$

concentration curves differ in shape and character, prompting the need to further investigate and experimentally determine not only how q might change, but also how all of the parameters in the diffusion equation behave in the *in vitro* pressure system.

The modeling results presented in this section were based on dynamic diffusion and hence yielded LIS ligand concentrations as a function of time and space. By varying parameters in the diffusion equation we were able to gain insight into ligand dynamics for a range of conditions (e.g. fast vs. slow change in LIS geometry). The dynamic diffusion model was built upon the steady-state model and represented an important step forward in understanding how LIS collapse affects ligand concentration. However, the effect of convection, i.e. fluid flow, was completely neglected. The assumption of a purely diffusive process meant that the fluid squeezed out of the LIS did not carry any concentration with it (or it carried such a small amount that it was negligible). Such an assumption may not be valid, especially for a rapid collapse, where most of the fluid in LIS flows into the underlying reservoir in a short period of time. Another issue that the pure-diffusion model could not address was what happens in the underlying reservoir close to the LIS boundary. To adequately examine these and other convection-related issues, we had to develop a more complete transport model, which coupled diffusion and convection, and also included the underlying media reservoir. We present this diffusion–convection model next.

17.3 The dynamic diffusion and convection model

17.3.1 Introduction

In the previous section we examined dynamic diffusion in the lateral intercellular space of two neighboring airway epithelial cells. The model presented above established only the steady-state LIS ligand concentration profiles. Furthermore, the dynamic diffusion model was based on pure diffusion and neglected potentially important effects of convection.

In this section we develop a generalized finite element solution of the coupled 1D diffusion–convection equation to evaluate the temporal changes in ligand concentration

EXTRACELLULAR MECHANOTRANSDUCTION

occurring in a dynamically collapsing interstitial space between epithelial cells (Kojić *et al.* 2006). We introduce a new geometry for the model that accommodates diffusion and convection of ligands that are shed into a lateral intercellular space that is continuous with an underlying media reservoir (Kojić *et al.* 2006). Employing the model, we explore the parameter space of the governing equations, examining the effect of ligand diffusivity, shedding rate and rate of extracellular space change on the kinetics of ligand accumulation. The new model geometry reveals the transient effect of convection on ligand concentration changes in the underlying space (e.g media for the *in vitro* case or tissues *in vivo*), suggesting a potential mechanism for communication of a change in the mechanical state of the epithelium to underlying tissues. Moreover, the diffusion–convection model offers a novel explanation for how cells could discriminate between mechanical processes occurring over a range of rates in different physiological scenarios. We use insights gained from the model to propose two explanations for a selective contribution of the EGF family-ligand heparin-binding EGF (HB-EGF) to the transduction of mechanical stress via autocrine signaling in a collapsing extracellular space.

17.3.2 Finite element model of coupled diffusion and convection

We modeled the lateral intercellular space (LIS) separating neighboring cells as idealized parallel plates and assumed free diffusion of ligand into the media reservoir below the LIS. Boundary conditions to represent the special case of an epithelial layer were imposed: impermeable tight junction at the apical surface; open to a large reservoir (e.g. the underlying media) such that sufficiently far below the basal surface the ligand concentration is assumed to be zero (see Fig. 17.3.1). Previously we solved the 1D diffusion equation (Fick's law) analytically, with a source term included to account for the constitutive shedding of ligand into the LIS, to obtain the steady-state ligand concentration profile within the LIS:

$$\frac{\partial C}{\partial t} = D_{LIS}\frac{\partial^2 C}{\partial x^2} + \frac{2q}{w} \qquad (17.3.1)$$

Here the ligand shedding rate q (distributed uniformly along the lateral cell boundary) and ligand diffusion coefficient in the LIS D_{LIS} are assumed to be constants, and w is the LIS width. Solving for the steady-state ligand concentration $C(x)$ yields:

$$C(x) = C_h + \frac{q}{wD}\left(h^2 - x^2\right) \qquad (17.3.2)$$

where C_h is the ligand concentration at the LIS boundary $x = h$.

To account for convective effects, as well as to determine how the concentration at and below the LIS boundary changes during a collapse, we now introduce an extended model geometry with three domains: LIS, transitional and radial (see Fig. 17.3.1). The LIS domain includes the LIS space, from the tight junction to the basal boundary. The transitional domain corresponds to the space between LIS and radial domains, where we numerically switch from a Cartesian to a cylindrical coordinate system. The radial domain represents the outside space (i.e. underlying media or tissues) and allows for radial diffusion of ligand once it leaves the LIS. The governing transport equations for each domain are:

$$\text{LIS:} \quad \frac{\partial C}{\partial t} = D_{LIS}\frac{\partial^2 C}{\partial x^2} + \frac{2q}{w} - V_x\frac{\partial C}{\partial x} \qquad (17.3.3a)$$

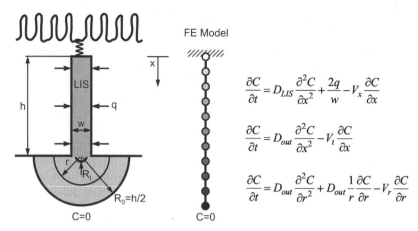

Fig. 17.3.1 Schematic of LIS finite element (FE) model. Neighboring cells are separated by the LIS. Ligands are constitutively shed into LIS from the cell surface at a rate q. In the space below the LIS it is assumed that at a radial distance $R_0 = h/2$ (where h is the LIS height) the ligand concentration is zero. The second boundary condition is an impermeable wall at the top due to the tight junctions (no flux at $x = 0$). Using 1D isoparametric finite elements we discretize the space into three domains: LIS (for $0 < x < h$), transitional (for $h < x < h + R_t$) and radial (for $R_t < r < R_0$). Note that the finite elements are not to scale. Each of the domains has its own governing diffusion–convection equation. In the three domains the corresponding bulk fluid velocities are V_x in the LIS, V_t in the transitional, and V_r in the radial domain. The diffusivities D_{LIS} and D_{out} (inside and outside the LIS, respectively) may be different. The transitional domain extends $R_t = w/\pi$ below the LIS

$$\text{Transitional:} \quad \frac{\partial C}{\partial t} = D_{out}\frac{\partial^2 C}{\partial x^2} - V_t \frac{\partial C}{\partial x} \qquad (17.3.3b)$$

$$\text{Radial:} \quad \frac{\partial C}{\partial t} = D_{out}\frac{\partial^2 C}{\partial r^2} + D_{out}\frac{1}{r}\frac{\partial C}{\partial r} - V_r \frac{\partial C}{\partial r} \qquad (17.3.3c)$$

where D_{LIS} and D_{out} are the ligand diffusivities in the LIS and outside space, respectively, V_x is the bulk fluid velocity in the LIS caused by changes in LIS dimensions, V_t is the fluid velocity in the transitional domain (assumed to be uniform), and V_r is the radial fluid velocity at a radius r measured from the LIS boundary. Note that in (17.3.3b,c) there is no q (ligand shedding) term because shedding is assumed to occur only from the lateral surfaces of the LIS.

By conservation of mass and fluid incompressibility it can be shown that in the LIS:

$$V_x = \frac{\dot{w}}{w}x \qquad (17.3.4a)$$

where $\dot{w} = dw/dt$ is the rate of change of LIS width, while in the radial domain:

$$V_r \pi r = const \qquad (17.3.4b)$$

The transitional regime was included to avoid numerical difficulties that can occur when switching from Cartesian to cylindrical coordinate systems. The transitional region begins

EXTRACELLULAR MECHANOTRANSDUCTION 365

at the LIS boundary and extends to a distance $R_t = w/\pi$ below the LIS. This distance was determined by matching the fluxes corresponding to Cartesian (w) and radial (πR_t) lengths, through which the flux passes. We further approximate the velocity field in this domain as uniform, being equal to the bulk velocity at the LIS exit $V_t = V_x(x = h)$. The approximations made in this domain have little impact on the overall concentration profile inside and outside of the LIS.

The radial domain encompasses the region between R_t (end of the transitional domain) and $R_0 = h/2$ (where we assume the ligand concentration to be zero). Mathematically, the zero-concentration boundary would be infinitely far away from the LIS (i.e. $R_0 \to \infty$), but for efficient numerical simulations we determined that for an LIS height $h = 15\ \mu m$ (Tschumperlin et al. 2004), $R_0 = 7.5\ \mu m$ is sufficiently far away from the LIS boundary such that further increasing R_0 had little effect on the overall concentration profile (data not shown). Hence, for all of the simulations we fixed the value of $R_0 = 7.5\ \mu m$ to be half of the previously measured LIS height $h = 15\ \mu m$ (Tschumperlin et al. 2004).

The diffusion–convection equations, along with the boundary conditions of no flux at the most apical point (impermeable tight junction) and zero concentration at R_0, were solved using the PAK finite element method software package (Kojić et al. 1998). The LIS and outside space were discretized by 1D isoparametric finite elements (see Fig. 17.3.1). The governing differential equations (17.3.3a–c) were first converted to the appropriate finite element system of first-order nonlinear differential equations, which were further linearized and integrated in time using a time step Δt. A Newton–Raphson iterative scheme was employed for each time step Δt. The final system of incremental-iterative equilibrium equations for a time step n and iteration i is (see (7.1.11))

$$\hat{\mathbf{K}} \Delta \mathbf{C}^{(i)} = {}^{n+1}\mathbf{Q}^{(i-1)} - {}^{n+1}\mathbf{F}^{(i-1)} \qquad (17.3.5)$$

where $\Delta \mathbf{C}^{(i)}$ is the vector of concentration increments at the finite element nodal points, $\hat{\mathbf{K}}$ is the system matrix, ${}^{n+1}\mathbf{Q}^{(i-1)}$ is the convection and shedding vector, and ${}^{n+1}\mathbf{F}^{(i-1)}$ is the out-of-balance vector. In component form, the terms of (17.3.5) are

$$\hat{K}_{KS} = M_{KS} + K_{KS} = \frac{1}{\Delta t} \int_L N_K N_S dx + D \int_L \frac{\partial N_K}{\partial x} \frac{\partial N_S}{\partial x} dx \qquad (17.3.6a)$$

$${}^{n+1}Q_K^{(i-1)} = Q_K^V + Q_K^{Vel} = \int_L N_K \frac{2q}{w} dx - \int_L N_K {}^{t+\Delta t}V \frac{\partial^{t+\Delta t}C^{(i-1)}}{\partial x} dx \qquad (17.3.6b)$$

$${}^{n+1}F_K^{(i-1)} = K_{KS} {}^{n+1}C_S^{(i-1)} + \frac{1}{\Delta t} M_{KS} ({}^{n+1}C_S^{(i-1)} - {}^n C_S) \qquad (17.3.6c)$$

where N_K are interpolation functions, L is the length of the finite element, C_S is concentration at node S of the finite element, while V and C are the velocity and concentration within the finite element.

In the radial domain below the LIS axially symmetric 1D finite elements were used. From the equation of balance in cylindrical coordinates (see (17.3.3c)) and integration over rdr instead of dx we obtain the finite element equations for the axially symmetric 1D elements in the form of (17.3.5).

A time series of the concentration profiles $C(x, t)$ during a linear decrease in w by 85%, from 1.5 to 0.225 μm (based on previous experimental results (Tschumperlin et al. 2004))

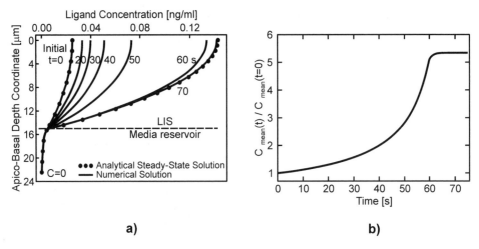

Fig. 17.3.2 Solutions obtained by solving for the ligand concentration from the governing diffusion–convection equations. (a) Evolution of the time-dependent concentration profile during an LIS collapse to 15% of its original width (from 1.5 to 0.225 μm) over 60 seconds. Concentration is plotted vs. the apico-basal depth coordinate, where $x = 0$ at the apical tight junction, $x = 15$ at the LIS boundary, and at a depth of 22.5 μm the concentration is zero. For comparison the open circles represent the analytical steady-state solution prior to LIS collapse and upon reaching the new, post-collapse steady state. The numbers next to the curves indicate seconds after onset of collapse; (b) Concentration profiles from (a) plotted as fold-mean concentrations: $C_{mean}(t)/C_{mean}(t=0)$. Here, $C_{mean}(t) = \frac{1}{h}\int_0^h C(x,t)\,dx$, where h is the LIS depth equal to 15 μm, and $C_{mean}(t=0)$ is the mean LIS ligand concentration just prior to the change in LIS width

over a 60-second duration is shown in Fig. 17.3.2a. The circles correspond to the analytical, pre- and post-collapse steady-state solutions (see (17.3.2)). Each of the solid-line curves represents a solution of (17.3.3a–c) at different time points. The height of the LIS was chosen to be 15 μm and the LIS width w to be initially 1.5 μm (Tschumperlin et al. 2004) (thus the outside space extended to $R_0 = h/2 = 5w_0 = 7.5\ \mu m$ below the LIS). For this example the ligand diffusivity and shedding rate were arbitrarily selected ($D_{LIS} = D_{out} = 75\ \mu m^2/s$ and $q = 10$ molecules/cell/minute).

Another way to represent the same time series of ligand concentration profiles is to calculate the fold change in the mean ligand concentration: $C_{mean}(t)/C_{mean}(t=0)$ (Fig. 17.3.2b). Here, $C_{mean}(t) = 1/h \int_0^h C(x,t)\,dx$ and $C_{mean}(t=0)$ is the mean LIS ligand concentration just prior to the change in LIS width.

17.3.3 Exploring the parameter space of the governing equations

Cells express a variety of autocrine mediators with a range of diffusivities dependent on molecular size and charge characteristics (Kovbasnjuk et al. 2000, Xia et al. 1998). These mediators are likely released at various rates, further influencing their kinetics within the LIS. We therefore explored the parameter space of the governing equations to characterize the relative importance of ligand properties by varying the diffusion coefficient D and the shedding rate q over several orders of magnitude. In all these simulations, the LIS width was decreased to 15% of its initial, pre-collapse value (from 1.5 to 0.225 μm) linearly over 60 seconds. The magnitude of this change was selected to correspond to previous experimental results in bronchial epithelial cells compressed by an apical to basal pressure gradient of 30 cm H_2O (Tschumperlin et al. 2004). The rate of LIS width change was arbitrarily selected; we address the importance of this parameter later.

The diffusion–convection equations (17.3.3a–c) were first solved for a constant q of 10 molecules per cell per minute (DeWitt et al. 2001) (evenly distributed along the cell boundary), with diffusion coefficients of 100, 10, 1 and 0.1 $\mu m^2/s$ (Fig. 17.3.3). Diffusivities on the order of 100 $\mu m^2/s$ characterize free diffusion of smaller molecules whose molecular weight ranges from about 0.1–10 kDa, whereas D on the order of 10 $\mu m^2/s$ corresponds to free diffusion of larger molecules 10–1000 kDa. The cases of $D = 0.1$ and 1 $\mu m^2/s$ represent hindered diffusion of large molecules (Kovbasnjuk et al. 2000). We further assumed equal diffusivities inside and outside of the LIS, $D_{LIS} = D_{out}$ (this assumption will be addressed later). The results of these simulations are plotted alongside each other in Fig. 17.3.3a. The case of $D = 0.1 \mu m^2/s$ is shown in Fig. 17.3.3b due to the difference in the time scale. In general, the smaller the ligand diffusivity (or conversely the larger the ligand), the slower the increase in the normalized mean concentration during LIS width change.

A similar order-of-magnitude analysis was performed for shedding rates from 0.1 to 1000 molecules/cell/min (DeWitt et al. 2001) (data not shown). While the shedding rate affected the absolute value of ligand concentration, it did not alter the normalized fold change in concentration induced by LIS collapse; this was true both at steady state and during the dynamic changes in LIS dimensions. Hence, the influence of shedding rate was limited to effects on the absolute ligand concentrations in our system.

Estimating the HB-EGF Diffusion Coefficient in the LIS

In the LIS of MDCK epithelial cells, which share a generally similar architecture with human bronchial epithelial cells, large molecules experience hindered diffusion due to the protrusion

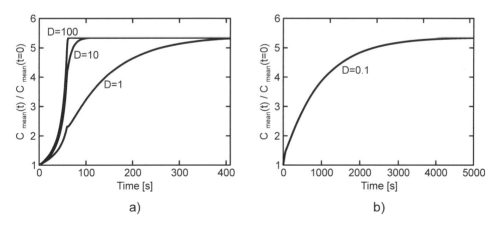

Fig. 17.3.3 Effect of diffusion coefficient on concentration. (a) A 60-second collapse was examined for cases of $D = 1$, 10, 100 $\mu m^2/s$, where the LIS width decreases linearly to 15% of its original value (from 1.5 to 0.225 μm) over a 60-second interval. Diffusion coefficient units are $\mu m^2/s$. (b) The case of $D = 0.1 \mu m^2/s$ (time scale goes to 5000 s). All cases assume a constant shedding rate of $q = 10$ molecules/cell/min

of a glycocalyx into the LIS (Kovbasnjuk *et al.* 2000). Specifically, while a 3 kDa molecule diffuses in the LIS of MDCK cells as if in free solution, a significant decrease in diffusion coefficient is observed for a 10 kDa molecule relative to that for free diffusion (Fig. 17.3.4). Because our previous work in human bronchial epithelial cells suggested a key role for HB-EGF in mechanotransduction (Tschumperlin *et al.* 2004, Tschumperlin *et al.* 2002), we used these existing diffusion data to estimate the diffusivity of HB-EGF in a typical LIS. HB-EGF that is proteolytically processed and shed into the LIS has a molecular weight of about 22 kDa (Harris *et al.* 2003, Raab & Klagsbrun 1997) and is heavily charged (Raab & Klagsbrun 1997). Previous studies have shown that interactions between a charged molecule and the extracellular glycocalyx can hinder diffusion (Dowd *et al.* 1999). Therefore, we assumed that HB-EGF diffusion in the LIS would be significantly hindered, due to both charge interactions and given that this effect is readily apparent for a smaller 10 kDa molecule (Kovbasnjuk *et al.* 2000) (Fig. 17.3.4). Based on these data we approximated the HB-EGF diffusion coefficient in the LIS as $D_{LIS} = 1.8 \,\mu m^2/s$, while outside the LIS it was assumed to be an unhindered (free solution) value of $D_{out} = 75 \,\mu m^2/s$. The choice of the hindered LIS diffusion coefficient is only an order-of-magnitude estimate based on hindered diffusion of large molecules in the LIS.

17.3.4 Rate sensitivity of extracellular mechanotransduction

Biomechanical forces develop on a range of time scales, from milliseconds for traumatic injury, to days to weeks or months for cellular proliferation and tissue morphogenesis. We have previously shown that the relevant time scale for compression of the LIS of

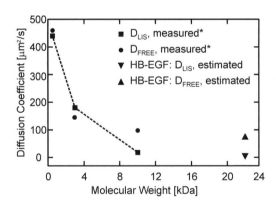

Fig. 17.3.4 Diffusion coefficients for 524 kDa HPTS, and for 3 and 10 kDa dextran molecules measured by Kovbasnjuk *et al.* (*indicates Kovbasnjuk *et al.* 2000) in the LIS of MDCK cells (squares connected by dashed line), and in free solution (circles). For a 10 kDa dextran with $D_{out} = D_{FREE} = 98 \,\mu m^2/s$, hindered diffusion in the LIS was observed ($D_{LIS} = 18 \,\mu m^2/s$). For HB-EGF of size 22 kDa we estimated a LIS diffusion coefficient of $D_{LIS} = 1.8 \,\mu m^2/s$ (triangle pointing down), an order of magnitude less than that measured for a 10 kDa dextran; outside the LIS the HB-EGF free solution value was estimated as $D_{out} = D_{FREE} = 75 \,\mu m^2/s$ (triangle pointing up), based on extrapolation of the measured free solution values assuming $D_{FREE} \propto (mol.wt.)^{-1/3}$

EXTRACELLULAR MECHANOTRANSDUCTION

airway epithelial cells and subsequent cellular signaling is on the order of seconds to minutes (Tschumperlin et al. 2004). Specifically, we know that at 20 minutes after the onset of 30 cm H$_2$O of transcellular compressive stress a new steady state in LIS geometry is established, whereby the LIS width shrinks to 15% of its original value (Tschumperlin et al. 2004). However, due to limitations in the system available for imaging a few years ago we did not know the temporal behavior of the LIS width during this 20-minute span. We thus modeled a range of different rates of collapse of the LIS width occurring linearly over durations from 1 to 1200 seconds (Figs. 17.3.5a,b). We incorporated our estimates for hindered diffusion of HB-EGF inside the LIS, and free diffusion outside of the LIS. For clarity, the 60-second case is illustrated in both panels. As the figures indicate, the rate of change in LIS dimension plays a dominant role in defining both the rate of ligand accumulation, and the shape of relationship between ligand concentration and time.

In order to illustrate the localized variation in concentration induced by dynamic changes in LIS geometry, we calculated the HB-EGF concentration profiles in the LIS at several times during (solid lines) and after (dashed lines) LIS collapse for the 1 and 10 s cases (Figs. 17.3.6a,b). These two cases are identical to those shown in Fig. 17.3.5b, but now represent concentration profiles as functions of depth and time.

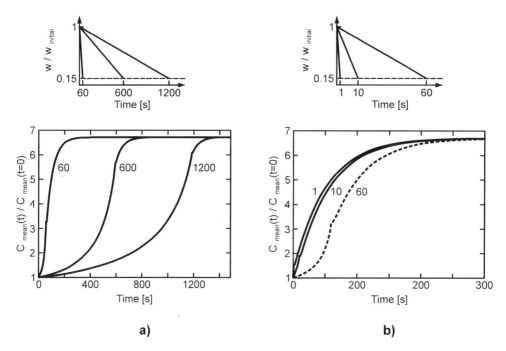

Fig. 17.3.5 (a) Incorporating the estimated HB-EGF diffusion coefficients (see Fig. 17.3.4), we examined three cases of LIS collapse when the LIS width decreased linearly to 15% of its initial, pre-collapse value (from 1.5 to 0.225 μm) over $t_{collapse} = 60, 600, 1200\,s$. (b) Additional cases, where the same LIS collapse (to 15% of the initial width) occurred in $t_{collapse} = 1$ and $10\,s$ (solid lines). The 60-second collapse case from (a) (dashed line) is shown for comparison. For all cases we assumed that the shedding rate is constant and equal to $q = 10$ molecules/cell/min

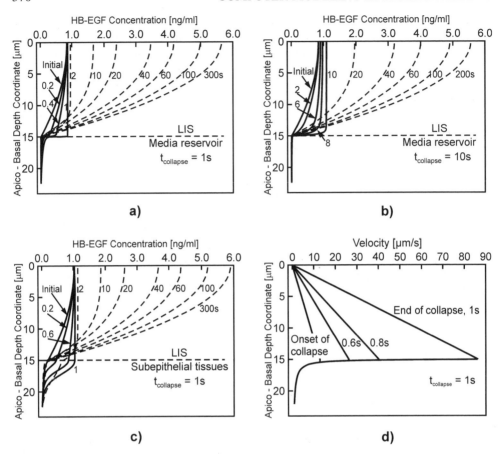

Fig. 17.3.6 (a) Evolution of concentration profiles for a 1 s linear collapse (see Fig. 17.3.5b) for the *in vitro* case where: $D_{LIS} = 1.8\,\mu m^2/s$ and $D_{out} = 75\,\mu m^2/s$, LIS height $h = 15\,\mu m$ (boundary between LIS and underlying media), and concentration is assumed zero at 7.5 μm below the LIS boundary. Solid lines indicate times when the LIS width is changing (number next to curve represents the time in seconds corresponding to the profile), broken lines represent times after the LIS has reached its new steady-state geometry; (b) Concentration profiles for a 10 s linear collapse (see Fig. 17.3.5b) for the *in vitro* case with the same parameters as in part (a); (c) Evolution of concentration profiles for a 1 s linear collapse for the *in vivo* case where: $D_{LIS} = D_{out} = 1.8\,\mu m^2/s$. The geometry is the same as that described in part (a); (d) Velocity profiles corresponding to case (a)

Considering first the case where the LIS collapse occurs over 1 s, we see that due to the rapid decrease in LIS width the concentration within the LIS becomes uniform, reaching a level equal to the most apical concentration (Fig. 17.3.6a). In the space immediately below the LIS the concentration tends to increase during the collapse due to convection, since velocities increase with width decrease (see (17.3.4a)). This increase in extra-LIS concentration (equaling about one-quarter of the most apical LIS concentration at the end of the collapse) permeates several microns into the underlying media. After the collapse has ended, convection stops and diffusion alone becomes the governing process. This transition from a diffusive–convective to a purely diffusive regime results in a change in slope (at the end of the collapse) of the normalized mean concentration curve (Fig. 17.3.5b). After the conclusion of the LIS collapse the concentration profile transitions to a parabolic shape (Fig. 17.3.6a). Soon thereafter the effect of shedding ligand into a now much smaller space causes the concentration to increase until the new steady state is reached after $300\,s$ (Fig. 17.3.6a).

In the case of a $10\,s$ collapse (Fig. 17.3.6b) the concentration at the end of the collapse also tends to be uniform throughout the LIS and equal to the concentration at position $x = 0$, however the $x = 0$ concentration is greater at $10\,s$ than at the start of collapse. Another difference between the 1 and $10\,s$ cases is that much less increase in concentration below the LIS boundary is observed for the $10\,s$ collapse (diminished convective effects for the $10\,s$ case). The new steady state is reached some $200\,s$ after the onset of collapse (Fig. 17.3.6b).

The previous two cases were intended to approximate the *in vitro* situation, where airway epithelial cells were grown on a porous substrate below which lies an essentially infinite reservoir of media. The HB-EGF diffusion coefficients inside and outside the LIS were assumed to be different based on the hindered diffusion in the LIS and free diffusion outside of the LIS. We modified these assumptions to simulate a scenario potentially encountered *in vivo*: instead of media below the cells, we assumed that sub-LIS tissues would hinder diffusion by the same amount as seen in the LIS ($D_{LIS} = D_{out} = 1.8\,\mu m^2/s$). For a very rapid $1\,s$ collapse, where again the LIS width decreases linearly to 15% of its initial, pre-collapse value, we determined the evolution of the concentration profiles (Fig. 17.3.6c). As in the $1\,s$ *in vitro* case there is a tendency toward uniform $x = 0$ concentration throughout the LIS during the collapse. Here though, the concentration changes permeate much deeper below the cells. For instance, just prior to the end of collapse the ligand concentration $3\,\mu m$ below the LIS reaches 40% of the LIS value, representing a 10-fold increase from an initial pre-collapse value of 0.04 to $0.4\,ng/ml$ in one second. These results highlight the fact that a rapid *in vivo* LIS collapse could transiently signal to underlying cells via a convective increase in ligand concentration that permeates into the surrounding tissues. This suggests a potential mechanism for communicating events that affect the epithelium to subepithelial tissues.

To see how the LIS collapse affects bulk velocity profiles inside and outside of the LIS, we examined the $1\,s$ *in vitro* case from above (see Figs. 17.3.6a,d). The bulk velocity profile inside the LIS is linear (starting from zero at the impermeable tight junction $x = 0$), whereas outside the LIS the velocity decreases proportionally to the inverse of the radius (Fig. 17.3.6d and Fig. 17.3.1). Both the linear and $1/r$ dependence follow from conservation of mass (see (17.3.4a,b)). For the $1\,s$ *in vitro* case the corresponding local Peclet numbers along the depth of the LIS ($Pe = V_x h/D_{LIS}$) can be calculated to range from 0 at $x = 0$ to > 700 at the LIS boundary. Thus, since $Pe \gg 1$ for most of the LIS, convection dominates during rapid collapse. Furthermore, a Peclet number can be obtained across the LIS width w, ranging from $Pe = \dot{w}(w/2)/D_{LIS} = 0.5$ at the LIS wall to 0 at a distance $w/2$ from the wall. Here $\dot{w}$ represents the rate of change of the width w, i.e. the velocity of the LIS wall. The small values of the Peclet number over the LIS width, combined with the large height-to-width ratio of the LIS geometry, justify our use of a 1D model in which we assume uniform concentrations across the LIS width.

Determining Maximum Rate of Ligand Concentration Change During LIS Collapse

Computational and experimental studies have demonstrated that receptor activation and downstream signaling are influenced not only by the magnitude, but also by the rate of ligand concentration change in the cellular microenvironment (Sasagawa *et al.* 2005). To explore this facet of transduction in our model, we first differentiated the normalized C_{mean} curves shown in Figs. 17.3.5a,b with respect to time and then found the maximum rate of

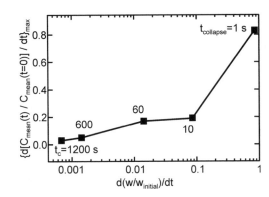

Fig. 17.3.7 Maximum rate of change of the fold-mean concentration as a function of the corresponding rate of LIS collapse for five cases from Figs. 17.3.5a,b. Individual squares represent the maximum rate of concentration change for a given rate of collapse. For example, in the case of $t_{collapse} = 10\,s$ the LIS collapses by 85% linearly over $10\,s$ and thus the rate of collapse is constant and equal to $0.085\,s^{-1}$. For this $10\,s$ collapse, the maximum rate of change of concentration was obtained by finding the largest slope of the $C_{mean}(t)/C_{mean}(t=0)$ curve (see $10\,s$ curve in Fig. 17.3.5b)

concentration change. In Fig. 17.3.7, the maximum rate of concentration change (i.e. the maximum slope of the fold-mean curves of Figs. 17.3.5a,b) is plotted versus the time derivative of the corresponding collapse of LIS width (see Figs. 17.3.5a,b $w/w_{initial}$ linear relationships). In our simulations the LIS width decreased linearly over time and the resulting time derivatives (i.e. rate of collapse) were constant for each case. The largest rate of concentration change was for the fastest collapsing LIS, i.e. the $1\,s$ collapse. A four-fold decrease in the maximum rate of ligand accumulation was observed when comparing the $10\,s$ case to the $1\,s$ case, with a small further decrement of 10% occurring between the $10\,s$ and $60\,s$ cases. The slower collapsing LIS cases (such as $600\,s$ and $1200\,s$) exhibited maximum rates of ligand concentration change that were lower by more than an order of magnitude when compared to the $1\,s$ case. Thus, the rate of LIS geometry change profoundly affects the peak rate of LIS ligand concentration change.

The results of our analysis demonstrate that while the magnitude of ligand concentration change depends on the change in w, the kinetics of ligand accumulation depend predominately on $\dot{w}$ and D. Strikingly, these results suggest that all other parameters being equal, the fastest change in ligand concentration will occur for the highest diffusivity (and hence, smallest) molecules. How then can we explain the selective role for HB-EGF in transducing mechanical stress in human airway epithelial cells exposed to compressive stress (Tschumperlin et al. 2004, Tschumperlin & Drazen 2001, Tschumperlin et al. 2002) when it is known that these cells can shed other ligands that bind to the same receptor (e.g. TGF-alpha) and exhibit higher diffusivities?

17.3.5 HB-EGF vs. TGF-alpha concentration dynamics

While proteolytically processed and shed HB-EGF is $\sim 22\,kDa$ in size, shed TGF-alpha (and EGF) is about four times smaller, being $\sim 5.5\,kDa$ (Harris et al. 2003, Raab & Klagsbrun 1997). If we were to assume free diffusion of each ligand, the difference in ligand size would predict a ~ 35% difference in the expected free diffusion coefficient for these two ligands (Fig. 17.3.4). However, returning to the diffusivity measurements made in the LIS of MDCK cells, it was found that these two ligands straddle the molecular size range over which diffusion becomes significantly hindered in the LIS (Kovbasnjuk et al. 2000). Thus, the diffusivity in the LIS can be approximated as $D_{LIS} = 1.8\,\mu m^2/s$ for HB-EGF (22 kDa), while for TGF-alpha (5.5 kDa) the $D_{LIS} = 120\,\mu m^2/s$ is the same as for free diffusion (see Fig. 17.3.4). If we further assume both ligands are shed at the same rate $q = 10$ molecules/cell/min, the solution of the governing diffusion–convection equations during a 60-second collapse yields the absolute mean concentration curves (not normalized) for HB-EGF and TGF-alpha shown in Fig. 17.3.8a. For comparison, the case assuming free diffusion for HB-EGF both inside and outside the LIS ($D_{LIS} = D_{out} = 75\,\mu m^2/s$) is also shown (dashed line in Fig. 17.3.8a). Note that the units here are picoM; thus while the mass concentration of free-diffusing HB-EGF is higher than for TGF-alpha, its molar concentration is lower due to its larger molecular weight.

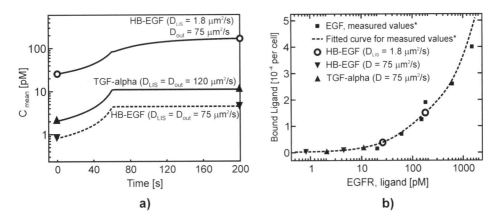

Fig. 17.3.8 (a) Mean molar LIS concentration for HB-EGF (top solid and bottom dashed curves) and TGF-alpha (solid middle curve) for a $60\,s$ LIS collapse when the LIS width decreases to 15% of its initial, pre-collapse value (from 1.5 to $0.225\,\mu m$). For HB-EGF, we examined two cases: free solution inside and outside of LIS with $D_{LIS} = D_{out} = 75\,\mu m^2/s$ (bottom dashed curve), and hindered diffusion in the LIS with $D_{LIS} = 1.8\,\mu m^2/s$ and free diffusion outside of the LIS with $D_{out} = 75\,\mu m^2/s$ (top solid curve). For TGF-alpha, free diffusion was assumed both inside and outside of LIS with $D_{LIS} = D_{out} = 120\,\mu m^2/s$ (middle solid curve). In all cases shedding rate was constant and equal to $q = 10$ molecules/cell/min. The open circles represent hindered HB-EGF mean concentration at the start and after collapse, while crosses and triangles represent free diffusion of TGF-alpha and HB-EGF mean concentrations, respectively pre- and post-collapse. (b) EGFR–ligand binding curve (dashed line) based on fitting the experimental values (solid squares) (* indicates Lauffenburger et al. 1998). A portion of the curve (relevant to our concentration ranges) was used. The final point on the actual curve was $20 \times 10^3\,pM$ and the corresponding bound ligand was 11×10^4 per cell (Lauffenburger et al. 1998). The mean concentrations for HB-EGF and TGF-alpha for pre- and after-collapse are represented by the symbols shown in panel (a). TGF-alpha concentration increases due to the collapse, but (like free-diffusion HB-EGF) remains on the flat part of the curve, whereas hindered HB-EGF concentration increases along the steep part of the curve, rendering it a more effective molecule for receptor activation

We observe that for hindered HB-EGF in the LIS, the mean absolute concentrations are an order of magnitude higher those of TGF-alpha (Fig. 17.3.8a). A corollary to this result is that in order to have a similar LIS concentration for both ligands, with the assumption of hindered diffusion in the LIS for HB-EGF, the cell must shed TGF-alpha at a rate ~ 10 times higher than that of HB-EGF. Furthermore, this result reveals two potential explanations for a selective role for HB-EGF in extracellular mechanotransduction. In the first case, the different mean concentrations that arise in the LIS as a consequence of different ligand diffusivities could place HB-EGF and TGF-alpha on different parts of an EGFR–ligand binding curve (Lauffenburger *et al.* 1998) (see Fig. 17.3.8b). Assuming a constant and equivalent shedding rate for each ligand, and equivalent ligand–receptor affinities (Jones *et al.* 1999) and signaling properties, the absolute concentrations of each of the ligands at the start and end of collapse correspond to the values of Fig. 17.3.8a. Therefore, the differences in HB-EGF and TGF-alpha concentrations could result in HB-EGF shifting up the EGFR–ligand binding curve. On the other hand, the low concentration of TGF-alpha (as well as free-diffusing HB-EGF) could place it on the flat portion of the curve, rendering it an ineffective activator of the EGFR in response to mechanical deformation (Fig. 17.3.8b).

A second potential explanation is that the molecular sieving properties of the LIS might become amplified by the geometric decrease in LIS space (Fig. 17.3.9). While we have thus

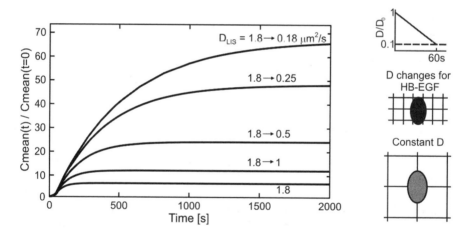

Fig. 17.3.9 Fold change in mean concentrations for various geometry-dependent changes in diffusion coefficients during LIS collapse. All cases are for a 60 s LIS collapse where the LIS width decreases to 15% of its initial, pre-collapse value (from 1.5 to 0.225 μm, Fig. 17.3.5b). The case of constant HB-EGF diffusion coefficient $D_{LIS} = 1.8\,\mu m^2/s$ is shown as the bottom curve. In the other four cases we assume that the collapsing LIS causes a linear decrease (following LIS geometry) in HB-EGF diffusion coefficient during the 60 s collapse. For example, the top curve $D_{LIS} = 1.8 \to 0.18\,\mu m^2/s$ indicates the case where the diffusion coefficient changes linearly over 60 seconds from the initial value of 1.8 $\mu m^2/s$ to the final value of 0.18 $\mu m^2/s$ (see side panel). In the other three cases of decreasing D_{LIS}, the diffusion coefficient linearly decreases over 60 seconds from 1.8 to 0.25, 0.5, and 1 $\mu m^2/s$. The side schematics illustrate how the shrinking volume of the LIS could considerably amplify the effect of hindered diffusion

far assumed that during the LIS collapse the shrinking of the intercellular space does not affect ligand diffusivity, the decrease in LIS width could form a more tightly packed space and a greater barrier to diffusion, especially for large, highly charged molecules like HB-EGF, while leaving smaller ligands like TGF-alpha relatively unaffected. We modeled this putative effect by assuming that the size/charge interactions (Dowd *et al.* 1999, Kovbasnjuk *et al.* 2000) would decrease the HB-EGF diffusion coefficient during the course of an LIS collapse. In Fig. 17.3.9 we illustrate several scenarios in which the HB-EGF diffusivity decreases linearly along with the linear LIS width decrease over $60\,s$ (see Figs 17.3.5 and 17.3.9). A decrease in D_{LIS} during collapse could amplify the increase in HB-EGF concentration, potentially mediating or magnifying cellular mechanotransduction.

17.3.6 Discussion

In this chapter we developed a computational framework to help understand how the concentration of constitutively shed ligands changes as a result of simple geometric changes in the spaces separating cells. Our computational model includes both diffusive and convective effects, allowing us to study the temporal relationship between deformation and ligand accumulation, and the dependence of this relationship on system characteristics such as ligand diffusivity, shedding rate and rate of deformation. The model geometry is expanded over previous efforts (Tschumperlin *et al.* 2004) to include both the LIS and the underlying space, thereby also providing an assessment of the effect of convection and diffusion on ligand concentration in the basal space underlying the LIS.

The modeling results reveal several key facets of extracellular mechanotransduction. How fast the local ligand concentration changes depends primarily on the rate of change of the extracellular geometry (Figs. 17.3.5a,b); on the other hand, the magnitude of the change in concentration (at steady state) is entirely determined by the magnitude of the geometry change. While the fold-change in ligand concentration that occurs with LIS collapse is independent of the ligand shedding rate, the absolute concentration of ligand is not (Fig. 17.3.8a). Thus, ligands with different shedding rates could occupy different regimes on a receptor dose–response curve (Fig. 17.3.8b). Similarly, the absolute concentration of a ligand depends on its diffusivity; low diffusivity molecules accumulate at higher baseline concentrations when shed into the LIS, and vice versa (see (17.3.2) and Fig. 17.3.8a). We used these system properties to propose two explanations for the selective role of HB-EGF as a key mechanotransduction ligand in bronchial epithelial cells (Tschumperlin *et al.* 2004, Tschumperlin *et al.* 2002); both mechanisms are based on the large size of HB-EGF relative to other EGF-family ligands (Harris *et al.* 2003), and the assumption that HB-EGF diffusion will be hindered in the LIS (Kovbasnjuk *et al.* 2000).

By including convection and expanding the model geometry, we were able to examine how dynamic changes in LIS geometry alter the ligand concentration in the underlying space (which we chose to be either a media reservoir or subepithelial tissues, Figs. 17.3.6a–c). We showed that for low diffusivity molecules and fast geometric changes, convection leads to large but transient increases in ligand concentration that permeate several microns below the cellular layer. This convective effect could allow nearly immediate communication of the mechanical state of epithelial cells to underlying cells, which frequently share responsibility for management of tissue architecture (Swartz *et al.* 2001).

The modeling results demonstrate how the varying kinetics of geometric changes in the extracellular space are translated into varying rates of change of ligand concentration

(Fig. 17.3.6). Recent experimental and computational studies have clearly demonstrated that the rate of ligand concentration change encodes important signaling information (Sasagawa et al. 2005, Schoeberl et al. 2002). Together these observations raise the possibility that cellular mechanotransduction through the proposed extracellular mechanism could discriminate between different rate processes, based on the velocity of ligand accumulation and subsequent receptor activation.

Our results raise other questions. For example, what are the effects of varying the magnitude or rate of loading on geometric changes, and how do these loading conditions relate to various physiological scenarios? While not available when this model was first developed, dynamic measures of the geometric response of the interstitium to loading, as detailed in Kojić (2007), can be coupled to the model described here to predict the overall relationship between mechanical loading and local autocrine ligand concentration (Kojić 2007).

Computational modeling of the EGFR system, from autocrine activity (Maheshwari et al. 2001, Monine et al. 2005, Shvartsman et al. 2002a, Shvartsman et al. 2002b, Shvartsman et al. 2001) to receptor trafficking (Lauffenburger & Linderman 1993, Resat et al. 2003, Wiley et al. 2003) and downstream signal pathways (Kholodenko et al. 1999, Sasagawa et al. 2005, Schoeberl et al. 2002, Wiley et al. 2003), has been essential to our understanding of this important biological pathway. The model described here explores a previously ignored idea in which changes in the concentration of shed ligands in an extracellular compartment, occur based solely on geometric changes. If linked together with previously developed cell membrane and intracellular compartmental models of ligand kinetics, receptor trafficking and intracellular signaling, the combined models could provide a comprehensive framework for understanding how mechanical or architectural changes in cells and tissues that modulate extracellular geometry are converted into biological signaling responses.

References

Bathe, K.J. (1996). *Finite Element Procedures*, Prentice-Hall, Englewood Cliffs, NJ.

DeWitt, A.E., Dong, J.Y., Wiley, H.S. & Lauffenburger, D.A. (2001). Quantitative analysis of the EGF receptor autocrine system reveals cryptic regulation of cell response by ligand capture, *J. Cell Sci.*, **114**(12), 2301–13.

Dowd, C.J., Cooney, C.L. & Nugent, M.A. (1999). Heparan sulfate mediates bFGF transport through basement membrane by diffusion with rapid reversible binding, *J. Biol. Chem.*, **274**(8), 5236–44.

Evans, M.J., Van Winkle, L.S., Fanucchi, M.V. & Plopper, C.G. (2001). Cellular and molecular characteristics of basal cells in airway epithelium, *Exp. Lung Res.*, 27(5), 401–15.

Harris, R.C., Chung, E. & Coffey, R.J. (2003). EGF receptor ligands, *Exp. Cell Res.*, **284**(1), 2–13.

Hrousis, C.A., Wiggs, B.J., Drazen, J.M., Parks, D.M. & Kamm, R.D. (2002). Mucosal folding in biologic vessels, *J. Biomech. Eng.*, **124**(4), 334–41.

Huebner, K.H. (1975). *The Finite Element method for Engineers*, John Wiley & Sons, Inc., New York.

Hughes, T.J.R. (1987). *The Finite Element Method. Linear Static and Dynamic Finite Element Analysis*, Prentice-Hall, Englewood Cliffs, NJ.

Jones, J.T., Akita, R.W. & Sliwkowski, M.X. (1999). Binding specificities and affinities of egf domains for ErbB receptors, *FEBS Lett.*, **447**(2–3), 227–31.

Kholodenko, B.N., Demin, O.V., Moehren, G. & Hoek, J.B. (1999). Quantification of short term signaling by the epidermal growth factor receptor, *J. Biol. Chem.*, 274(42), 30169–81.

Kojić, M., Slavković, R., Živković, M. & Grujovic, N. (1998). *PAK-T Finite Element Program for Linear and Nonlinear Heat Conduction*, Faculty of Mechanical Engineering, University of Kragujevac, Serbia.

Kojić, N. (2007). *Mechanotransduction via airway epithelial cells: the effect of compressive stress*, Ph.D. Thesis, Harvard-MIT Division of Health Sciences and Technology, MIT.

Kojić, N., Kojić, M. & Tschumperlin, D.J. (2006). Computational modeling of extracellular mechanotransduction, *Biophys. J.*, **90**(11), 4261–70.

Kovbasnjuk, O.N., Bungay, P.M. & Spring, K.R. (2000). Diffusion of small solutes in the lateral intercellular spaces of MDCK cell epithelium grown on permeable supports, *J. Membr. Biol.*, **175**(1), 9–16.

Lauffenburger, D.A. & Linderman, J.J. (1993). *Receptors: Models for Binding, Trafficking, and Signaling*, Oxford University Press, Oxford, England.

Lauffenburger, D.A., Oehrtman, G.T., Walker, L. & Wiley, H.S. (1998). Real-time quantitative measurement of autocrine ligand binding indicates that autocrine loops are spatially localized, *Proc. Natl. Acad. Sci. USA*, **95**(26), 15368–73.

Maheshwari, G., Wiley, H.S. & Lauffenburger, D.A. (2001). Autocrine epidermal growth factor signaling stimulates directionally persistent mammary epithelial cell migration, *J. Cell Biol.*, **155**(7), 1123–8.

Monine, M.I., Berezhkovskii, A.M., Joslin, E.J., Wiley, H.S., Lauffenburger, D.A. & Shvartsman, S.Y. (2005). Ligand accumulation in autocrine cell cultures, *Biophys. J.*, **88**(4), 2384–90.

Raab, G. & Klagsbrun, M. (1997). Heparin-binding EGF-like growth factor, *Biochim. Biophys. Acta*, **1333**, 179–99.

Resat, H., Ewald, J.A., Dixon, D.A. & Wiley, H.S. (2003). An integrated model of epidermal growth factor receptor trafficking and signal transduction, *Biophys. J.*, **85**(2), 730–43.

Ressler, B., Lee, R.T., Randell, S.H., Drazen, J.M. & Kamm, R.D. (2000). Molecular responses of rat tracheal epithelial cells to transmembrane pressure, *Am. J. Physiol. Lung Cell Mol. Physiol.*, **278**(6), 1264–72.

Sasagawa, S., Ozaki, Y., Fujita, K. & Kuroda, S. (2005). Prediction and validation of the distinct dynamics of transient and sustained ERK activation, *Nat. Cell Biol.*, **7**(4), 365–73.

Schoeberl, B., Eichler-Jonsson, C., Gilles, E.D. & Muller, G. (2002). Computational modeling of the dynamics of the MAP kinase cascade activated by surface and internalized EGF receptors, *Nat. Biotechnol.*, **20**(4), 370–5.

Shvartsman, S.Y., Hagan, M.P., Yacoub, A., Dent, P., Wiley, H.S. & Lauffenburger, D.A. (2002a). Autocrine loops with positive feedback enable context-dependent cell signaling, *Am. J. Physiol. Cell Physiol.*, **282**(3), C545–9.

Shvartsman, S.Y., Muratov, C.B. & Lauffenburger, D.A. (2002b). Modeling and computational analysis of EGF receptor-mediated cell communication in *Drosophila oogenesis*, *Development*, **129**(11), 2577–89.

Shvartsman, S.Y., Wiley, H.S., Deen, W.M. & Lauffenburger, D.A. (2001). Spatial range of autocrine signaling: modeling and computational analysis, *Biophys. J.*, **81**(4), 1854–67.

Swartz, M.A., Tschumperlin, D.J., Kamm, R.D. & Drazen, J.M. (2001). Mechanical stress is communicated between different cell types to elicit matrix remodeling, *Proc. Natl. Acad. Sci. USA*, **98**(11), 6180–5.

Tschumperlin, D.J., Dai, G., Maly, I.V. et al. (2004). Mechanotransduction through growth-factor shedding into the extracellular space, *Nature*, **429**(6987), 83–6.

Tschumperlin, D.J. & Drazen, J.M. (2001). Mechanical stimuli to airway remodeling, *Am. J. Respir. Crit. Care Med.*, **164**(10Pt2), 90–4.

Tschumperlin, D.J., Shively, J.D., Swartz, M.A. et al. (2002). Bronchial epithelial compression regulates MAP kinase signaling and HB-EGF-like growth factor expression, *Am. J. Physiol Lung Cell Mol. Physiol.*, **282**(5), 904–11.

Wiggs, B.R., Hrousis, C.A., Drazen, J.M. & Kamm, R.D. (1997). On the mechanism of mucosal folding in normal and asthmatic airways, *J. Appl. Physiol.*, **83**(6), 1814–21.

Wiley, H.S., Shvartsman, S.Y. & Lauffenburger, D.A. (2003). Computational modeling of the EGF-receptor system: a paradigm for systems biology, *Trends Cell Biol.*, **13**(1), 43–50.

Willumsen, N.J., Davis, C.W. & Boucher, R.C. (1994). Selective response of human airway epithelia to luminal but not serosal solution hypertonicity, *J. Clin. Invest.*, 94, 779–87.

Xia, P., Bungay, P.M., Gibson, C.C., Kovbasnjuk, O.N. & Spring, K.R. (1998). Diffusion coefficients in the lateral intercellular spaces of Madin–Darby canine kidney cell epithelium determined with caged compounds, *Biophys. J.*, **74**(6), 3302–12.

Yarden, Y. & Sliwkowski, M.X. (2001). Untangling the ErbB signalling network, *Nature Rev. Mol. Cell Biol.*, **2**(2), 127–37.

18

Spider Silk: Modeling Solvent Removal during Synthetic and *Nephila clavipes* Fiber Spinning

The process by which spiders make their mechanically superior fiber involves removal of solvent (water) from a concentrated protein solution while the solution flows through a progressively smaller diameter spinning canal. To probe the effects of solvent removal during elongational flow, which is exhibited in the spinning canal of the spider, on fiber mechanical properties a study using synthetic materials was first conducted (Section 18.1). The study establishes in Section 18.2 a model for solvent removal during dry spinning of synthetic fibers. Central to the synthetic model is the determination of the dependence of the solvent diffusion coefficient on the solvent concentration. The procedures used to obtain the variable synthetic diffusion coefficient, and the subsequent model of solvent removal during synthetic fiber spinning, were then applied (Section 18.3) to the *Nephila clavipes* (golden orb) spider. As for the synthetic case, it was assumed that internal diffusion governs solvent (water) removal in the spinning canal of the spider during fiber formation. The modeling results provide the key spinning parameters which suggest that simple diffusion, along with the dry wall boundary condition, is a viable mechanism for water removal during typical *Nephila* fiber spinning.

Most of the text and figures presented in this chapter are based on references Kojić et al. (2006) and Kojić et al. (2004) (adapted with permission from *Biomacromolecules* **2004**, 5, 1698–1707. Copyright 2004 American Society).

18.1 Determination of the solvent diffusion coefficient in a concentrated polymer solution

18.1.1 Introduction

Diffusion is one of the governing processes in many chemical engineering applications, such as solvent removal during dry spinning of fibers out of a concentrated polymer solution. Neglecting non-Fickian effects (El Afif & Grmela 2002), the internal diffusion of solvent molecules through the polymer solution can be modeled by Fick's law, in which the diffusion coefficient is the material parameter (Deen 1998, Incropera & DeWitt 1996, Middleman 1998, Mills 1995). When polymer solutions experience significant changes in solvent concentration due to mass transfer, the diffusion coefficient can vary considerably as a function of solvent concentration (Kobuchi & Arai 2002, Mills 1995). Therefore, the essential problem becomes establishing the dependence of the diffusion coefficient on solvent concentration.

In this section we describe how to obtain such a dependence by finite element modeling of a simple evaporation experiment (Kojić et al. 2006). A small amount of the solution was placed in a pan and allowed to evaporate into air, while measuring mass loss over time. The governing process occurring within the polymer solution is diffusion, while convection dominates on the vapor (air) side. We model this process by imposing the appropriate boundary conditions and present a computational procedure to numerically determine the diffusion coefficient of solvent through the polymer solution as a function of solvent concentration. Subsequently, the dependence of the diffusion coefficient on concentration enables the calculation of the time evolution of solvent concentration profiles along the depth of the pan.

The results for the diffusion coefficient obtained in this section were then directly applied to the practical application of spinning synthetic silk-like fibers (see Section 18.2), where solvent removal during spinning becomes important. Furthermore, the general experimental and numerical procedures described in this section were also utilized in Section 18.3 in order to determine the diffusion coefficient of water through the native spider silk spinning material.

Experimental Procedure

The polymer solution described in this chapter consists of 35% polymer and 65% solvent by weight (Table 18.1.1). We will further refer to the THF/DMAc system as solvent, and the Elasthane/PTMO system as polymer. The solution was placed in a Seiko TG/DTA–320

Table 18.1.1 Composition of polymer solution

35% Polymer*	65% Solvent
20% Elasthane 80A**	90% THF
15% PTMO-2900	10% DMAc

Notes: *All percentages in weight percent. **Elasthane™ 80A polyurethane is a thermoplastic elastomer formed as the reaction product of a polyol, an aromatic diisocyanate, and a low molecular weight glycol used as a chain extender. Polytetramethylene oxide (PTMO) is reacted in the bulk with aromatic isocyanate, 4,4′-methylene bisphenyl diisocyanate (MDI), and chain extended with 1,4-butanediol. The Polymer Technology Group, Berkeley, California. http://www.polymertech.com/

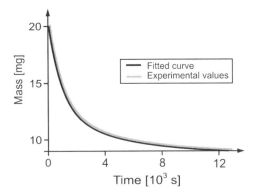

Fig. 18.1.1 Mass change over time of the polymer solution used in the pan experiment

(Thermogravimetric and Differential Thermal Analyzer) machine, which was used to record mass as a function of time for 12240 seconds (3.4 hours). A 5 mm diameter aluminum pan was used to hold an initial amount of 18.912 mg of the solution. The TG/DTA provided a closed environment at a temperature of $T = 25\,°C$, while blowing air at a rate of $150\,mL/min$. Mass loss was recorded on the computer using the standard Seiko TG/DTA software and is shown in Fig. 18.1.1.

Determining the Governing Process
In order to leave the polymer solution, the solvent molecules must first migrate to the surface and then evaporate into the air. The migration to the surface corresponds to internal diffusion, while the evaporation from the surface is related to convective mass transfer. For rapid convective removal, as is the case in our experiment, there is a much greater resistance encountered on the internal (solution) side than on the convective (vapor/air) side. The ratio of the resistances is commonly referred to as the mass transfer Biot number (Middleman 1998):

$$Bi_m = \frac{h_m \alpha L}{D_{ss}} \qquad (18.1.1)$$

where D_{SS} is the internal solvent diffusion coefficient mm^2/s, h_m is the convective mass transfer coefficient [mm/s], L is the characteristic length mm, and α is the partition coefficient between the two phases at the interface. For our case, an order of magnitude estimate gives:

$$D_{ss} \sim 10^{-5} \frac{mm^2}{s}, \quad h_m \sim 1 \frac{mm}{s} \qquad (18.1.2)$$
$$L \sim 1\,mm, \quad \alpha \sim 10^{-3}$$

The corresponding Biot number then becomes $Bi_m \sim 100$. This high value of the Biot number implies that the governing process is internal diffusion (Middleman 1998, Mills 1995).

18.1.2 Numerical procedure

Here, we first give the fundamental diffusion equation for the pan experiment conditions, then summarize the relevant finite element equations, and present the computational procedure.

The diffusion of solvent through a polymer solution is governed by the general form of Fick's law (Mills 1995) (see also (3.2.7) in Section 3.2)

$$\frac{\partial \rho_s}{\partial t} = \nabla \left[\rho_T D_{ss} \nabla \left(\frac{\rho_s}{\rho_T} \right) \right] \qquad (18.1.3)$$

where ρ_s is the partial mass density (further referred to as concentration) of the solvent, ρ_T is the total mass density (mass per unit volume of solution),

$$\rho_T = \rho_s + \rho_p \qquad (18.1.4)$$

with ρ_p being the concentration (partial mass density) of the polymer. As stated above, it is assumed that

$$D_{ss} = D_{ss}(\rho_s) \qquad (18.1.5)$$

The diffusion of solvent in the pan experiment can be considered a one-dimensional process, for which the model is shown in Fig. 18.1.2. The spatial derivatives in (18.1.3) then reduce to the partial derivatives with respect the x-coordinate, hence

$$\frac{\partial \rho_s}{\partial t} - \frac{\partial}{\partial x} \left[\rho_T D_{ss} \frac{\partial}{\partial x} \left(\frac{\rho_s}{\rho_T} \right) \right] = 0 \qquad (18.1.6)$$

In order to transform the expression in brackets to a more suitable form, we use the following relation

$$\rho_T = \left(1 - \frac{\bar{\rho}_p}{\bar{\rho}_s} \right) \rho_s + \bar{\rho}_p = a_1 \rho_s + a_2 \qquad (18.1.7)$$

where $\bar{\rho}_s$ and $\bar{\rho}_p$ are the material densities (mass per unit volume of pure substance) of the solvent and polymer, respectively. The material densities are constant, hence the coefficients

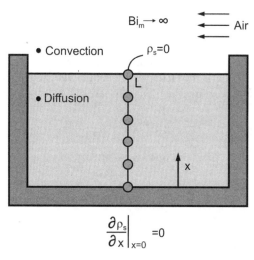

Fig. 18.1.2 Finite element model of the pan experiment

SPIDER SILK: MODELING SOLVENT REMOVAL

a_1 and a_2 are two constants depending on the material characteristics. Therefore, (18.1.6) takes on the form

$$\frac{\partial \rho_s}{\partial t} - \frac{\partial}{\partial x}\left(\alpha_m \frac{\partial \rho_s}{\partial x}\right) = 0 \qquad (18.1.8)$$

with no heat (here mass) volumetric source, where

$$\alpha_m = a_2 \frac{D_{ss}}{a_1 \rho_s + a_2} \qquad (18.1.9)$$

is the coefficient depending on ρ_s.

The usual Galerkin procedure (Huebner 1975, Hughes 1987, Kojić et al. 1998) (see Section 7.1) is employed to transform (18.1.8) into the finite element balance equations. Thus, integration over finite element volume V of (18.1.8) gives

$$\int_V N_K \frac{\partial \rho_s}{\partial t} dV - \int_V N_K \frac{\partial}{\partial x}\left(\alpha_m \frac{\partial \rho_s}{\partial x}\right) dV = 0 \qquad (18.1.10)$$

where N_K are finite element interpolation functions ($K = 1, 2, \ldots, N$, where N is number of element nodes), and $dV = A dx$ with A being the element cross-section area. By linearization around the time t (start of time step) a system of algebraic incremental equations is obtained. An incremental-iterative form of these equations, assuming an implicit integration scheme (i.e. the equilibrium is sought iteratively for the end of the time step), can be written as (Bathe 1996, Kojić & Bathe 2005, Kojić et al. 1998) (see (7.1.11) and (17.3.5))

$$^{n+1}\hat{\mathbf{K}}^{(i-1)} \Delta \boldsymbol{\rho}_s^{(i)} = {}^{n+1}\mathbf{F}^{(i-1)} - {}^{n+1}\hat{\mathbf{K}}^{(i-1)\,n+1}\boldsymbol{\rho}_s^{(i-1)} + \frac{1}{\Delta t}{}^{n+1}\mathbf{M}^{(i-1)\,n}\boldsymbol{\rho}_s \qquad (18.1.11)$$

where i denotes the iteration counter, and the index $n+1$ indicates that the evaluation is performed at the end of the time step n. The vectors $^{n+1}\boldsymbol{\rho}_s^{(i-1)}$ and $\Delta\boldsymbol{\rho}_s^{(i)}$ are the nodal vectors for the concentrations and concentration increments. The matrix $^{n+1}\hat{\mathbf{K}}^{(i-1)}$ is

$$^{n+1}\hat{\mathbf{K}}^{(i-1)} = \frac{1}{\Delta t}{}^{n+1}\mathbf{M}^{(i-1)} + {}^{n+1}\mathbf{K}^{(i-1)} \qquad (18.1.12)$$

where the components of the finite element matrices are

$$^{n+1}M_{KL}^{e(i-1)} = A^{(i-1)} \int_{^{n+1}L^{(i-1)}} N_K N_L dx \qquad (18.1.13)$$

$$^{n+1}K_{KL}^{e(i-1)} = A^{(i-1)} \int_{^{n+1}L^{(i-1)}} \frac{\bar{\rho}_p}{\rho_T^{(i-1)}} D_{ss}^{(i-1)} \frac{\partial N_K}{\partial x} \frac{\partial N_L}{\partial x} dx \qquad (18.1.14)$$

The vector $^{n+1}\mathbf{F}^{(i-1)}$ in (18.1.11) is the mass flux through the boundary, i.e.

$$F_K^{(i-1)} = \int_{A^{(i-1)}} N_K q_A dA \qquad (18.1.15)$$

384 COMPUTER MODELING IN BIOENGINEERING

where q_A is the mass flux through current area $A^{(i-1)}$. In practical application of (18.1.11) we impose zero concentration boundary condition at the free solvent surface, and the flux terms $^{n+1}\mathbf{F}^{(i-1)}$ cancel at the internal nodes in the FE assemblage process (see Section 4.1). In our case the element area A is constant, equal to the cross-sectional area of the pan, while the element lengths changes. Hence, the line integrals are evaluated over the last known element length, calculated as

$$^{n+1}L^{(i-1)} = L_p + \frac{1}{\bar{\rho}_s} \int_{t+\Delta t L^{(i-2)}} {}^{t+\Delta t}\rho_s^{(i-1)} dx \qquad (18.1.16)$$

where $L_p = \left({}^{0}\rho_p/\bar{\rho}_p\right)^0 L$ is the length of the finite element occupied by the polymer, which does not change; ${}^{0}\rho_p$ and ${}^{0}L$ are the initial polymer concentration and element length, respectively. The iterations in (18.1.11) continue until a selected numerical tolerance is reached, e.g. $\|\Delta\boldsymbol{\rho}_s^{(i)}\| \leq \varepsilon$, where ε is a small number.

In order to obtain the relationship between the solvent diffusion coefficient and solvent concentration we propose the following two procedures:

(a) calculation of the dependence $(D_{ss})_{mean}$ on $(\rho_s)_{mean}$

(b) determination of $D_{ss}(\rho_s)$ that matches the experimental results.

Computational steps used in procedure (a) are given in Fig. 18.1.3. As shown in the figure, a value of D_{ss}, denoted here as $(D_{ss})_{mean}$, is assumed for the current time step and this value is the same for all material points. We then iterate on $(D_{ss})_{mean}$ until the total mass of the polymer solution $^{t+\Delta t}m_{calc}$ matches the experimental value $^{t+\Delta t}m_{exp}$ at the end of the time step. The corresponding mean solvent concentration $(\rho_s)_{mean}$ is defined as

$$(\rho_s)_{mean} = \frac{\sum_{elements} \int_{L_{element}} \rho_s dx}{\sum_{elements} L_{element}} \qquad (18.1.17)$$

The iterations continue until the equation

$$^{n+1}f = {}^{n+1}m_{calc} - {}^{n+1}m_{exp} = 0 \qquad (18.1.18)$$

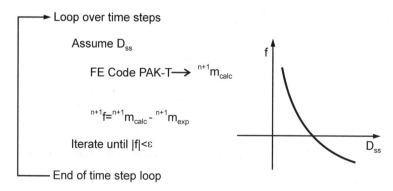

Fig. 18.1.3 Computational steps to determine dependence $(D_{ss})_{mean}$ on $(\rho_s)_{mean}$

is satisfied, where

$$^{n+1}m_{calc} = A \sum_{elements} \int_{^{n+1}L} {}^{n+1}\rho_s dx + m_p \qquad (18.1.19)$$

Here, m_p is the total mass of polymer. The iterations according to (18.1.11) are performed for each trial value of $(D_{ss})_{mean}$.

In order to find the $D_{ss}(\rho_s)$ relationship, denoted above as the procedure (b), we perform the following computational procedure. We form the error function as the difference between the calculated and experimentally recorded mass,

$$^{n+1}e(D_1, D_2, \ldots, D_m; t_{n+1}) = A \int_0^{^{n+1}L_{tot}} {}^{n+1}\rho_s(D_1, D_2, \ldots, D_m; x) dx + m_p - {}^{n+1}m_{exp} \qquad (18.1.20)$$

where $D_1, D_2, \ldots, D_m$ are values of the diffusion coefficient D_{ss} on a multilinear curve $D_{ss}(\rho_s)$ with pairs $(\rho_{s1}, D_1), (\rho_{s2}, D_2), \ldots, (\rho_{sm}, D_m)$; t_{n+1} is the time at end of n-th time step, i.e. $t_{n+1} = n\Delta t$; $^{n+1}\rho(D_k, x)$ and $^n m_{exp}$ are the solvent concentration and experimentally determined mass at time t_{n+1}; m_p is mass of polymer; and $^{n+1}L_{tot}$ is the total height of mass within the pan at time t_{n+1}. We assume an initial multilinear curve based on the relationship $(D_{ss})_{mean} - (\rho_s)_{mean}$ in Fig. 18.1.4 (see Example in Section 18.1.3).

In an iteration scheme we then calculate new values of the parameters $D_k^{(j)}$ as

$$D_k^{(j)} = D_k^{(j-1)} - \left(\frac{\partial^{n+1}e}{\partial D_k}\right)^{(j-1)} \Delta D_k^{(j-1)}, \qquad k = 1, 2, \ldots, m \qquad (18.1.21)$$

where j is the iteration counter in iterations on the parameters D_k. Since there is no analytical form for dependence $^{n+1}e(D_k)$, we calculate derivatives $\partial^{n+1}e/\partial D_k$ numerically by a perturbation procedure. In evaluating the vector $\partial^{n+1}e/\partial D_k$ we perform a weighting of $\partial^{n+1}e/\partial D_k$ with weighting coefficients being proportional to the gradients $\partial \rho_s/\partial x$ for each parameter D_k. Also, we normalize the vector $\partial^{n+1}e/\partial D_k$. Increments $\Delta D_k^{(j-1)}$ in (18.1.21) are taken as $\Delta D_k^{(j-1)} = 0.05 D_k^{(j-1)}$. Further, we impose the condition that the slope $\partial D_{ss}/\partial \rho_s$ increases for a concentration range $0 \leq \rho_s \leq (\rho_s)_{incr}$, and then decreases for $(\rho_s)_{incr} \leq \rho_s \leq {}^0\rho_s$, where ${}^0\rho_s$ is the initial concentration in the pan experiment. These conditions are imposed in order to obtain an 'S'-curve for $D_{ss}(\rho_s)$ observed experimentally for some polymer solutions (Kobuchi & Arai 2002). In our analysis we found that error was not significantly sensitive on the value $(\rho_s)_{incr}$.

For current values of parameters $D_k^{(j)}$ obtained within the iteration scheme (18.1.21), we calculate the total error per time step as

$$e_{tot} = \frac{1}{n_{tot}} \sum_{n=1}^{n_{tot}} |^{n+1}e| \qquad (18.1.22)$$

where n_{tot} is the total number of steps in the analysis. We stop iterations when we reach a value $(e_{tot})_{min}$ after which there was no further decrease of e_{tot}.

18.1.3 Example

We show the following example as an illustration of the proposed computational scheme. The basic FE code PAK-T (Kojić et al. 1998) for heat conduction, with necessary modifications, is employed. A simple bisection method with an acceleration scheme is used to determine the trial values of $(D_{ss})_{mean}$ for solving (18.1.18). The materials used in the experiment are described in Section 18.1.1. The data are as follows (densities in g/cm^3 and length in cm)

$$\bar{\rho}_s = 0.894 \quad \bar{\rho}_p = 1$$
$$^0\rho_s = 0.609 \quad ^0\rho_p = 0.325 \quad ^0L_{tot} = 0.112 \tag{18.1.23}$$

The experimental mass loss $m_{exp}(t)$ is shown in Fig. 18.1.1. The finite element model, along with the appropriate boundary conditions, is depicted in Fig. 18.1.2. The boundary conditions are:

$$1)\ (\rho_S)_{surface} = 0, \quad 2)\ \left.\frac{\partial \rho_S}{\partial x}\right|_{x=0} = 0 \tag{18.1.24}$$

The first boundary condition follows from the high value of the Biot number, while the second boundary condition comes from the impermeability of the pan.

Figure 18.1.4 displays the dependence $(D_{ss})_{mean}$ on $(\rho_s)_{mean}$ as the result of the computational procedure (a). The decrease of the diffusion coefficient with decreasing solvent concentration implies that the resistance to solvent diffusion increases as the polymer becomes more concentrated and experiences conformational changes. The dependence seems almost linear for a wide range of solvent concentrations. This linear relationship is then used as a starting relationship for the procedure (b), i.e. determination of $D_{ss}(\rho_s)$.

In calculation of $D_{ss}(\rho_s)$ we used $n_{tot} = 10^4$ with $\Delta t = 1$ s to cover the whole time interval of our experiments. We found that the best convergence $e_{tot} \to 0$ was obtained when the error ^{n+1}e and derivatives $\partial^{n+1}e/\partial D_k$ were calculated for time t_n in (18.1.20) and (18.1.21) which corresponds to the maximum error ^{n+1}e in the whole time interval. We mention here that we have performed evaluation of the coefficients D_k using the error according to (18.1.22) and found difficulties in convergence to the experimental curve $m_{exp}(t)$. If we

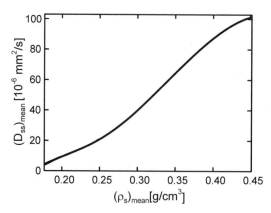

Fig. 18.1.4 Dependence of $(D_{ss})_{mean}$ on $(\rho_s)_{mean}$ according to procedure (a)

used error measure (18.1.22), but without the absolute values of ^{n+1}e, we found that we could obtain convergence of the error to a small value. In our final procedure we assumed an initial multilinear curve with number of points $m = 6$ (dashed line Fig. 18.1.5), based on the relationship $(D_{ss})_{mean} - (\rho_s)_{mean}$ in Fig. 18.1.4, with 20 iterations. Then we used 20 iterations with 41 parameters to obtain the final $D_{ss}(\rho_s)$ relationship (solid line Fig. 18.1.5) that matched the experimental results (see Fig. 18.1.6). We stopped calculation of parameters D_k after 20 iterations for both curves since further iterations on D_k did not lessen the deviation of the calculated curve $m(t)$ from the experimental curve $m_{exp}(t)$ (measured also by error (18.1.22)). The division of the interval of $(\rho_s)_{max}$ along the ρ_s-axis was always into equal segments.

Figure 18.1.6 shows the calculated curves $m(t)$ for three constant values of diffusion coefficient, and two curves corresponding to the initial and final relationships $D_{ss}(\rho_s)$ (see Fig. 18.1.5). From the results shown in Figs. 18.1.5 and 18.1.6 we see that: (i) diffusion coefficients depends on concentration, since constant values of D_{ss} give very large deviation of calculated mass $m(t)$ with respect to the experimentally measured mass change $m_{exp}(t)$; (ii) the initial relationship $D_{ss}(\rho_s)$ leads to a significant difference between the calculated and measured mass over time; the final relationship $D_{ss}(\rho_s)$, which is notably different from the initial one, gives the solution $m(t)$ close to the experimental curve $m_{exp}(t)$.

Finally, Fig. 18.1.7 displays several profiles of the solvent concentration ρ_s in the pan for different times. These profiles were calculated by using the $D_{ss}(\rho_s)$ relationship of Fig. 18.1.5. The x-coordinate corresponds to Fig. 18.1.2 and is essentially the pan depth coordinate, where $x = 0$ is the bottom of the pan. As Fig. 18.1.7 indicates, during the process of diffusion ρ_s decreases from an initial $(t = 0)$ uniform distribution, defined as 100% initial, to smaller and smaller values as more solvent is lost. For example, after $t = 10^4$ seconds the concentration of solvent varies from 41% at the bottom of the pan to zero at the free surface. The shaded area under the $t = 1.3 \times 10^6 s$ curve represents the mass of the solvent (per unit pan cross-sectional area) left in the pan at that time relative to initial. It is also worth noting that the gradient $\partial \rho_s / \partial x$ has a high value in the vicinity of the free surface in the initial period, and then decreases over time.

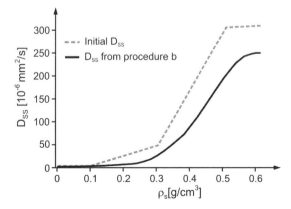

Fig. 18.1.5 Solvent diffusion coefficient D_{ss} vs. solvent concentration ρ_s obtained using procedure (b) (solid line). Dashed line represents the initial multilinear D_{ss} curve used in procedure (b)

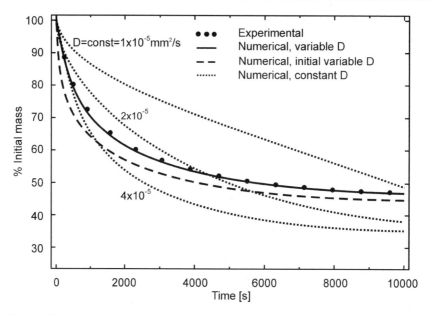

Fig. 18.1.6 Computed mass curves obtained with constant and variable (from Fig. 18.1.5) diffusion coefficients, and experimental curve

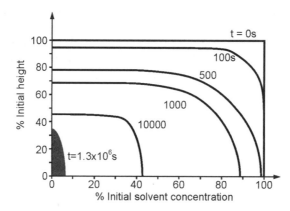

Fig. 18.1.7 Distribution of the solvent concentration (relative to initial) along the depth of the pan ($x = 0$ is the bottom of the pan, Fig. 18.1.2) for several times, with variable diffusion coefficient obtained from procedure (b) (Fig. 18.1.5)

18.2 Modeling solvent removal during synthetic fiber spinning

18.2.1 Introduction

In the previous section we proposed a computational procedure for the determination of $D_{ss}(\rho_s)$ based on a simple pan-weighing experiment. This dependence of solvent diffusion

SPIDER SILK: MODELING SOLVENT REMOVAL

coefficient on concentration is a material characteristic of the polymer solution. The proposed procedure can be implemented in chemical engineering practice where diffusion is the governing process. One such application involves modeling of solvent removal during spinning of a synthetic silk-like fiber (Kojić et al. 2004) from the polymer solution described in Table 18.1.1. The obtained relationship $D_{ss}(\rho_s)$ facilitates the determination of the correlation between the solvent removal process and the mechanical properties of the spun fiber (Kojić et al. 2004).

The impetus for this kind of synthetic study was to probe the effects of solvent removal during elongational flow, which is exhibited in the spinning canal of the spider, on fiber mechanical properties. Spiders produce fibers that are mechanically superior to essentially any known material, but the specifics of the spinning process still remain a mystery. The modeling of solvent removal during spinning of synthetic fibers presented in this section was then applied to an actual spinning canal of a *Nephila clavipes* spider (Section 18.3).

Synthetic Materials and Experimental Procedure

A synthetic solution out of which fibers were spun consisted of three major components: polymer, resin and solvent. The polymer used was commercial polyurethane Elasthane 80A, while the resin was PTMO-2900 (see Table 18.1.1). The resin's main function was to act as a plasticizer, giving a more robust spinning solution. For the solvent component, a combination of THF and DMAc was used. The recipe for the Elasthane solution is given in Table 18.1.1. Percentages of polymer and solvent reflect those of spider protein (spidroin) and water in the spider spinning dope (Chen et al. 2002). The final recipe was obtained based on 'spinnability' of small diameter fibers in an effort to approach those of spider silk, which are $\sim 1\,\mu m$.

A schematic diagram of the experimental setup is given in Fig. 18.2.1. The Elasthane solution was placed into a Becton–Dickinson plastic syringe (fitted with a 32-gage ¼ inch EFD dispensing needle tip) and pushed out by a KD Scientific (KDS100) syringe

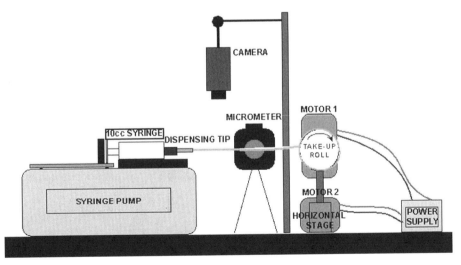

Fig. 18.2.1 Schematic diagram of the experimental setup used for fiber spinning. The horizontal stage moves in a direction perpendicular to the page in order to ensure that the fiber does not wind on top of itself on the take-up roll (spool)

pump. The viscous solution was then scooped from the needle tip, stretched and placed onto the rotating collection spool. The spool consisted of a Pakon plastic slide mount (34.5 mm × 23 mm open area) connected to a rotating motor shaft. The spool system stands on a mobile stage, whose movement prevents fibers from folding on top of each other. In other words, the stage moves the spool in a direction perpendicular to the fiber and thus enables collection of a continuous, single fiber.

In any one experiment, the key adjustable parameters are: the exit velocity of the spinning solution at the needle dispensing tip (the flow rate of the solution out of the syringe divided by the cross-sectional area of the needle); the velocity of the take-up roll (spool); and the length of the spin line. We define the spin line as the path taken by the solution from its exit at the needle tip to the first point of contact with the rotating spool. The spin line represents the axial coordinate of the fiber where a velocity gradient exists, since the velocity at the spool point of contact is always greater than the needle exit velocity. Thus, the spinning solution experiences an elongational flow while on the spin line.

18.2.2 Governing process during synthetic solvent removal

In order for the solvent to evaporate into the surrounding air, the solvent molecules must first move through the solution and then evaporate at the air/fiber interface. The movement through the solution is a diffusive process, further referred to as internal diffusion, while the evaporation into air corresponds to external convection. To determine the relative importance of each of these two processes it is useful to look at resistances, a standard procedure in heat and mass transfer analysis (Mills 1995). A ratio of the internal diffusion resistance to external convection resistance is defined as the mass transfer Biot number (Deen 1998, Middleman 1998) (see Section 18.1):

$$Bi_m = \frac{h_m d_{fiber} \alpha}{D_{ss}} \qquad (18.2.1)$$

where D_{ss} is the internal solvent diffusion coefficient mm^2/s, h_m is the convective mass transfer coefficient mm/s, d_{fiber} is the fiber diameter mm, and α is the partition coefficient between the two phases at the interface.

In order to determine an order of magnitude estimate of the Biot number all of the four quantities in (18.2.1) must be approximated. We first focus on the convective (vapor) side, and then examine the diffusion part.

The Convective Mass Transfer Coefficient
On the vapor side, for flow across a cylinder the following empirical relation holds (Mills 1995):

$$Sh_{Re \to 0} = \frac{h_m d_{fiber}}{D_{sair}} \approx 0.3 \Rightarrow h_m = \frac{0.3 D_{sair}}{d_{fiber}} \qquad (18.2.2)$$

where the dimensionless Sherwood number, Sh, is the mass transfer equivalent of the Nusselt number for heat transfer, h_m is the convective mass transfer coefficient mm/s, d_{fib} is the fiber diameter mm, and D_{sair} is the solvent vapor diffusion coefficient through the air mm^2/s. Also the assumption of a very small Reynolds number, Re, is justified since the air velocity (V_{air}) is small ($< 1\,m/s$) and the diameter of the fiber is on the order of microns. Hence,

$$Re = \frac{V_{air}\, d_{fiber}}{\nu_{air}} \sim \frac{(1m/s)\,(10^{-6}\,m)}{1.6 \times 10^{-5}\,m^2/s} = 0.06 \qquad (18.2.3)$$

where ν_{air} is the kinematic viscosity for air at room temperature of $25\,°C$.

To find the diffusion coefficient of solvent through the air, D_{sair}, in (18.2.2), we employ the kinetic theory of gases, which gives (Poling et al. 2001):

$$D_{sair} = \frac{0.00143\, T^{1.75}}{P M_{sair}^{0.5} \left[(\Sigma_v)_{solvent}^{1/3} + (\Sigma_v)_{air}^{1/3} \right]^2} \qquad (18.2.4)$$

Here, D_{sair} is in cm/s^2, P is the pressure in bars, T is the temperature in Kelvin, M_{sair} is a combination of the molecular weights g/mol of the solvent and air $M_{sair} = \dfrac{2}{(1/M_{solvent}) + (1/M_{air})}$, and $(\Sigma_v)_{solvent}$ and $(\Sigma_v)_{air}$ are the sum of the atomic diffusion volumes according to Fuller (Fuller et al. 1969, Poling et al. 2001) for the solvent and air, respectively. For our case, after substitution of the appropriate values, (18.2.4) gives $D_{sair} = 10\,mm^2/s$. This value can then be substituted into (18.2.2) to give the convective mass transfer coefficient of $h_m = \dfrac{0.3\, D_{sair}}{d_{fiber}} \sim 1000\,mm/s$, since the diameter of the fiber is on the order of a micron.

The Internal Solvent Diffusion Coefficient

To determine the diffusion coefficient of the solvent through the solution (on the fiber side) we applied the procedures described in Section 18.1. Briefly, a small amount of the spinning solution was placed in a pan surrounded by a controlled environment, and the corresponding mass loss due to solvent evaporation was recorded. This mass loss was then modeled by numerical methods, which yielded the dependence of the solvent diffusion coefficient on solvent concentration (Fig. 18.2.2) (Kojić et al. 2006).

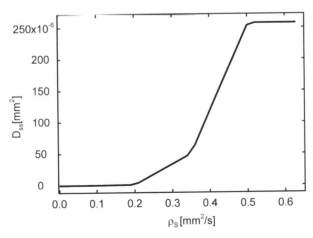

Fig. 18.2.2 Solvent diffusion coefficient D_{ss} vs. solvent concentration ρ_s used in synthetic fiber spinning calculations

Returning to the Biot number, all of the quantities in (18.2.1) can now be approximated, with the partition coefficient estimated as the ratio of the partial densities of the solvent on the liquid and vapor side ($\alpha \sim 10^{-3}$). Hence,

$$Bi_m = \frac{h_m d_{fiber} \alpha}{D_{ss}} \sim \frac{10^3 \cdot 10^{-3} \cdot 10^{-3}}{10^{-5}} = 100 \qquad (18.2.5)$$

Thus, the high value of the Biot number indicates that the internal diffusive resistance is much greater than the convective resistance. This, in turn, implies that internal diffusion of solvent through the fiber solution is the governing process in solvent removal on the spin line and the convective (vapor) side can be neglected.

18.2.3 Numerical modeling of synthetic internal solvent diffusion

The diffusion of solvent in the case of fiber spinning can be considered an axially symmetric process. Namely, there is axially symmetric diffusion within a fiber cross-section, with axial motion of the cross-section. The size of the cross-section decreases with time due to the solvent evaporation and due to the axial velocity gradient, as described below. We further use the notation and general equations for the polymer solution depicted in Section 18.1 (see equations (18.1.3–10)).

In the case of axially symmetric in-plane diffusion, the system of differential equations reduces to the differential equation (see Example 1.5-5)

$$\frac{\partial \rho_s}{\partial t} - \frac{1}{r}\frac{\partial}{\partial r}\left[r\rho_T D_{ss}\frac{\partial}{\partial r}\left(\frac{\rho_s}{\rho_T}\right)\right] = 0 \qquad (18.2.6)$$

where r is the radial coordinate of the fiber. This equation can also be written in the form

$$\frac{\partial \rho_s}{\partial t} - \frac{1}{r}\frac{\partial}{\partial r}\left(r\alpha_m \frac{\partial \rho_s}{\partial r}\right) = 0 \qquad (18.2.7)$$

where

$$\alpha_m = \bar{\rho}_p \frac{D_{ss}}{a_1 \rho_s + \bar{\rho}_p} \qquad (18.2.8)$$

with

$$a_1 = (1 - \frac{\bar{\rho}_p}{\bar{\rho}_s}) \qquad (18.2.9)$$

and $\bar{\rho}_s$ and $\bar{\rho}_p$ are the material densities (mass per unit volume of pure substance) of the solvent and polymer, respectively. The material densities are constant and represent the material characteristics.

The finite element equilibrium equations are given in (18.1.11), with the components of the finite element matrices, comprising the matrix $^{t+\Delta t}\hat{\mathbf{K}}^{(i-1)}$ (see equations (18.1.12)), are

$$^{n+1}M_{KL}^{e(i-1)} = L_x \int_{n+1 L^{(i-1)}} N_K N_L r dr \qquad (18.2.10)$$

$$^{n+1}K_{KL}^{e(i-1)} = L_x \int_{n+1 L^{(i-1)}} \frac{\bar{\rho}_p}{\rho_T^{(i-1)}} D_{ss}^{(i-1)} \frac{\partial N_K}{\partial r}\frac{\partial N_L}{\partial r} r dr \qquad (18.2.11)$$

SPIDER SILK: MODELING SOLVENT REMOVAL

where $^{n+1}L^{(i-1)}$ is the element length in the radial direction corresponding to the iteration $i-1$ (for the first iteration, $i=1$, we use $^{n+1}L^{(0)} = {}^0L$, the initial radial length), and L_x is the element length in the axial direction x. The vector $^{n+1}\mathbf{F}^{(i-1)}$ is the mass flux through the element boundary at the node K,

$$F_K^{(i-1)} = R_K \int_{L_x} q_A dx \qquad (18.2.12)$$

where q_A is the mass flux through the current area, and R_K is the radial coordinate corresponding to the finite element node K. Since only in-plane diffusion is considered, then $L_x = 1$. The iterations in the incremental-iterative system of algebraic equations for the assemblage of the finite elements (see (18.1.11)) continue until a selected numerical tolerance is reached, e.g. $\|\Delta\boldsymbol{\rho}_s^{(i)}\| \leq \varepsilon$, where ε is a small number.

The line integrals in (18.2.10) and (18.2.11) are evaluated over the last known element length $^{t+\Delta t}L^{(i-1)}$. This length is calculated from the element volume $^{n+1}V^{(i-1)}$ (with unit axial length L_x) corresponding to the solvent concentration $^{n+1}\rho_s^{(i-1)}$,

$$^{n+1}V^{(i-1)} = {}^nV_p + \frac{2\pi}{\bar{\rho}_s} \int_{n+1 L^{(i-2)}} {}^{n+1}\rho_s^{(i-1)} r dr \qquad (18.2.13)$$

where ^nV_p is volume of the polymer that corresponds to the start of n-th time step, calculated as $^nV_p = {}^nV - {}^nV_s$, with ^nV and ^nV_s being the total volume of the finite element and the volume occupied by the solvent, respectively. It is assumed that V_p does not change within the time step Δt.

Now we consider the element lengths change, and hence the change of the fiber radius due to axial motion. In the case when the axial velocity is uniform along the fiber axis, the axial motion has no effect on the fiber radius change and the radius decrease occurs only due to solvent evaporation. However, if at a considered time and at a considered axial position of the cross-section, the axial velocity has a non-zero axial gradient $k_{ax} = \partial v/\partial x \neq 0$ ($k_{ax} > 0$ for the spin line), the cross-section size changes also due to the axial motion. The fiber material can be considered incompressible, and in the case of no evaporation the mass continuity equation for two cross-sections at a distance Δx can be written as

$$v_1 A_1 - (v_1 + k_{ax}\Delta x)A_2 = 0 \qquad (18.2.14)$$

where v_1 and A_1 are velocity and size of the first cross-section (at position x) and A_2 is the area of the cross-section 2, at position $x + \Delta x$. The radius R_2 follows from this equation,

$$R_2 = R_1 \exp\left(-\frac{k_{ax}\Delta x}{2v_1}\right) \qquad (18.2.15)$$

and the increment Δx is

$$\Delta x = \frac{v_1}{k}(1 - \exp(-k_{ax}\Delta t)) \qquad (18.2.16)$$

The evaporation and the axial motion occur at the same time, and the diffusion depends on the current fiber radius, therefore both incompressibility and diffusion must be accounted for in a time step. We adopt the following procedure:

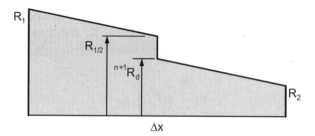

Fig. 18.2.3 Graphical representation of the numerical procedure that takes into account both diffusion and incompressibility effects during fiber spinning. The middle vertical from $R_{1/2}$ to $^{n+1}R_d$ indicates diffusive loss, while the parallel diagonal lines from R_1 to $R_{1/2}$ and from $^{n+1}R_d$ to R_2 depict changes due to incompressibility

1. Determine the radius $R_{1/2}$ from (18.2.15) by using $\Delta x/2$ for Δx, discretize this cross-section into the finite elements, and calculate the polymer volumes ^nV_p for each element.

2. Calculate mass loss due to diffusion in the time step, starting from the radius $R_{1/2}$ and using the polymer volumes ^nV_p. Determine the radius $^{n+1}R_d$ corresponding to the end of the diffusion calculations.

3. Apply the incompressibility condition (18.2.14) for the second half of the time step (i.e. in (18.2.15) substitute $^{n+1}R_d$ for R_1 and $\Delta x/2$ for Δx).

A graphical representation of the above steps is shown in Fig. 18.2.3. Note that the fiber radius at the equilibrium iteration i is

$$^{n+1}R^{(i)} = \sum_{elements} {}^{n+1}L^{(i)} \qquad (18.2.17)$$

where $^{n+1}L^{(i)}$ correspond to the current element volume $^{n+1}V^{(i)}$ according to (18.2.13). The radius $^{n+1}R_d$ represents the final value of $^{n+1}R^{(i)}$.

18.2.4 Example: Synthetic fiber spinning

The general numerical procedures described in the previous section were applied to two cases of fiber spinning from the same material (Table 18.1.1), and then used to predict their corresponding mechanical properties. The material and partial densities g/cm^3 used in (18.2.10–11) are

$$\bar{\rho}_s = 0.894, \quad \bar{\rho}_p = 1, \quad {}^0\rho_s = 0.603, \quad {}^0\rho_p = 0.325 \qquad (18.2.18)$$

where the bar indicates the pure substance density, and the zero indicates the initial partial density (concentration).

The first case examined was that of relatively thick fiber and shorter spin line, while the second case models a thin fiber with a longer spin line. Specifically, the relevant data was:

$$\text{Case 1) } v_{exit} = 4.7\,cm/s, \quad {}^0R_{fiber} = 40\,\mu m, \quad L_{spin} = 7.5\,cm, \quad v_{spool} = 10\,cm/s$$
$$\text{Case 2) } v_{exit} = 4.7\,cm/s, \quad {}^0R_{fiber} = 15\,\mu m, \quad L_{spin} = 9.5\,cm, \quad v_{spool} = 68\,cm/s \qquad (18.2.19)$$

SPIDER SILK: MODELING SOLVENT REMOVAL

Here, v_{exit} is the exit velocity out of the syringe tip, $^0R_{fiber}$ is the initial fiber diameter at a short distance from the tip ($< 1\,mm$), L_{spin} is the length of the spin line (the distance from the syringe tip to the spool), and v_{spool} is the spool speed at the point of contact of the spool and fiber (the end of the spin line).

In order to solve for the radius and concentration profiles along the spin line, an assumption that the fiber velocity increases linearly from the syringe tip to the spool was made. Thus, in (18.2.14) $k_{ax} = \dfrac{\partial v}{\partial x} = \dfrac{v_{spool} - v_{exit}}{L_{spin}} = const$. Also needed was the solvent concentration boundary condition on the fiber surface $(\rho_S)_{surface} = 0$, based on the large Biot number (see (18.2.5)).

Following the numerical procedure described in the previous section, the modified finite element code PAK-T (Kojić et al. 1997) for nonlinear heat conduction was employed for the two fiber spinning cases. The main goal was to determine how the radius and solvent concentration profiles along the spin line change under different spinning conditions. Figures 18.2.4 and 18.2.5 show the results for the two cases, which differ in the length of the spin line and spool velocity.

The radius change and the concentration profiles (shaded) at several axial positions are shown for the two cases. Also, the ratio of ρ_{mean}/ρ_{s_0} is given for each profile as an indication of how much solvent is left, where ρ_{s_0} is the initial solvent concentration from (18.42), and

$$\rho_{mean} = \dfrac{\sum\limits_{elements} \int\limits_{L_{element}} \rho_s r\,dr}{R^2 \pi} \qquad (18.2.20)$$

Here R is the current radius (at the end of the considered time step) of the fiber cross-section.

As Figs. 18.2.4 and 18.2.5 indicate, at the end of the spin line the thicker fiber (case 1 has relatively almost twice as much solvent than the thinner fiber (case 2)). Thus, we expect significant differences in the mechanical properties of these two fibers, since in

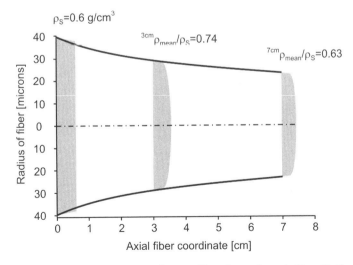

Fig. 18.2.4 Case 1: radius and concentration profiles along the spin line. Ratios of ρ_{mean}/ρ_{s_0} are given at $x = 3\,cm$, and at the end of the spin line, $x = 7\,cm$

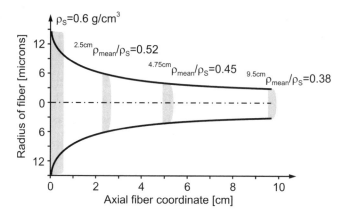

Fig. 18.2.5 Case 2: radius and concentration profiles along the spin line. Ratios of ρ_{mean}/ρ_{s_0} are given at $x = 2.5\,cm$, $x = 4.75\,cm$, and at the end of the spin line, $x = 9.5\,cm$

the thinner fiber case more solvent was removed during the elongational flow experienced on the spin line. While on the spin line, the individual chains were stretched along the fiber axis and, with the simultaneous removal of solvent, able to interact with each other. The end result is an ordered structure along the fiber axis (as with a spider silk fiber) that should give good mechanical properties. It should be noted that once on the spool the fiber does not experience any change in velocity. Therefore, in the seconds subsequent to attaching to the spool the remaining solvent evaporates and essentially 'freezes' the configuration formed at the end of the spin line. In other words, since there is no more elongational flow on the spool, no further stretching of the fiber polymer chains occurs.

The fibers spun under the spinning conditions for case 2 described above are shown in Fig. 18.2.6, where a human hair is shown for size comparison. The thicker fibers of case 1 looked the same and are omitted for clarity.

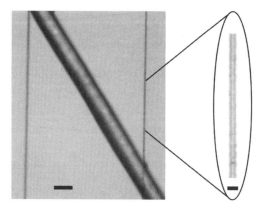

Fig. 18.2.6 Left: a human hair between spun fibers. Bar is 80 microns. Right: enlarged image of the spun fiber. Bar is 7 microns

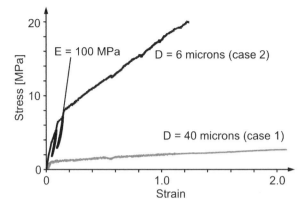

Fig. 18.2.7 Engineering stress–strain curves for the two cases of spun fibers. The black line drawn in for case 2 corresponds to unloading the sample in the linear regime and the slope of this line is the elastic modulus E. The same graphic modulus representation for case 1 was omitted for clarity

Mechanical Properties of Spun Synthetic Fibers

To determine the mechanical properties of the fibers, a uniaxial testing machine developed by Sauri Gudlavalleti and Lallit Anand of MIT was used (Gudlavalleti *et al.* 2005). This device is capable of testing small diameter fibers and as such was used for our purposes. The engineering stress–strain curves for the two different fibers are shown in Fig. 18.2.7.

The smaller diameter fibers had considerably better mechanical properties, with an elastic modulus (E) of $100\,MPa$ and a toughness (area under the stress–strain curve, equivalent to the energy to break) of $15\,MJ/m^3$. In contrast, the thicker fibers had an elastic modulus of $20\,MPa$ and a toughness of only $3\,MJ/m^3$. For comparison purposes, the native spider silk dragline fiber (from which the spider hangs from) has an elastic modulus of $10\,GPa$ and toughness of $150\,MJ/m^3$ (Gosline *et al.* 1999), and thus is vastly superior to the synthetic fibers spun in this experiment.

Nonetheless, the fiber of case 2 showed five-fold better mechanical properties than the thicker case 1 fiber, which had roughly twice as much solvent at the end of the spin line. Therefore, the mechanical tests verify that more solvent removal on the spin line along with a correspondingly smaller diameter leads to a fiber with better mechanical properties.

A similar principle is employed by the spider, whereby most of the water is removed as the spinning solution flows through the spinning canal. The numerical modeling of this water removal process is presented in the next section, starting with obtaining the water diffusion coefficient through the spinning dope.

18.3 Modeling solvent removal during *Nephila clavipes* fiber spinning

18.3.1 Introduction

The practice of spinning a solid protein fiber out of a concentrated water solution has been utilized by spiders for millions of years (Selden 1989, Shear & Palmer 1989). The resulting spider-silk fiber has mechanical properties that are superior to any known material (Gosline *et al.* 1999). Although the chemical composition and genetic basis of the proteins have been established (Vollrath & Knight 2001), the spinning process by which the fiber is formed remains a mystery (Kaplan *et al.* 1994, Vollrath & Knight 1999). During this process most of the solvent (water) is removed and simultaneously the stretched protein chains are aligned due to the elongational flow. The flow, i.e. movement of the silk solution through the progressively tapered spinning canal, forces the protein chains to extend along the long axis of the canal. The end result is a fiber with exceptional mechanical properties along the fiber axis (Gosline *et al.* 1999).

By learning how the spider makes its fiber, one could conceive new processing techniques (Kaplan *et al.* 1994) that would yield novel materials, such as a synthetic spider silk analog. In particular, the spinning conditions experienced in the spider's spinning canal, i.e. velocity and water concentration profiles, could give new insight into the spider spinning process. Here we apply the procedures presented in the previous sections of this chapter to examine the spinning conditions in the canal of the *Nephila clavipes* (golden orb) spider.

18.3.2 *Nephila* water diffusion coefficient

In order to explore the spinning conditions within the *Nephila* canal, the major ampullate gland was dissected out of a female *Nephila clavipes* (golden orb) spider provided by the Miami MetroZoo (Fig. 18.3.1 – see color plate). The major ampullate gland contains a reservoir used for dope storage and a progressively narrowing spinning canal, or S-duct. This gland is used for spinning dragline (from which the spider hangs from) and spoke-like web-frame fibers. While the dope flows through the canal it loses water, and modeling of this process requires obtaining the diffusion coefficient of water through the dope.

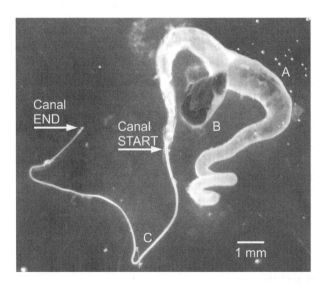

Fig. 18.3.1 Major ampullate gland of a female *Nephila clavipes* spider. A: Ampulla, dope reservoir; B: A blob of dope used in the pan weighing experiment; C: Spinning canal, S-duct (see Plate 32)

To this end the pan-weighing experiment was employed (Kojić et al. 2006), where a small amount, such as the protruding blob of dope in Fig. 18.3.1B (see color plate), was collected and placed in a pan in a controlled environment. The evaporation of water was monitored and this mass–time curve was modeled by using numerical methods, as described in Section 18.2.3. Specifically, 12.3 mg of *Nephila* dope was placed in a 5-*mm*-diameter aluminum pan that sat in a in a Seiko TG/DTA-320 machine. The TG/DTA provided a closed environment at a temperature of $T = 25\,°C$, while blowing air at a rate of $150\,mL/min$. Mass loss was recorded for 9480 seconds (2.6 hours) using the standard Seiko TG/DTA software (dashed curve in Fig. 18.3.2). The experimental mass–time *Nephila* curve indicates that the dope is an aqueous protein solution consisting of about 70 wt. % of water, which is in agreement with measurements made by Chen et al. (2002).

After performing the numerical procedure described previously in Section 18.3.2, where internal diffusion of water was assumed to be the governing process, a diffusion coefficient was obtained which gave a numerical mass–time curve that best matched the experimental one (solid line in Fig. 18.3.2). This water diffusion coefficient through the *Nephila* dope was determined to be

$$D_{water-dope} = 2.15 \times 10^{-5}\,mm^2/s \tag{18.3.1}$$

further abbreviated as D_{wd}. Unlike the synthetic material (Fig. 18.2.2), the water diffusion coefficient was a constant, i.e. it was independent of the water concentration. A possible benefit to the spider of the constant diffusion coefficient could be seen in the constant relative resistance seen by the water as it travels from the inside of the fiber to the surface. In other words, as the fiber becomes more and more dry the resistance seen by the escaping water remains constant and does not increase as in the synthetic case.

Once the key material parameter, i.e. the water diffusion coefficient, was obtained, the spider spinning process could be more closely examined. Thus, the rest of this section focuses on the spinning canal, starting with the governing process for water removal.

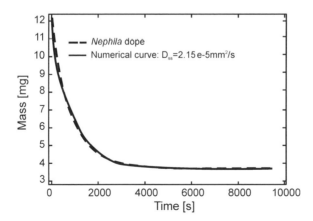

Fig. 18.3.2 Results of the pan-weighing experiment. The dashed curve represents the experimentally measured mass change over time for a blob of *Nephila* dope (Fig. 18.3.1B). The solid line indicates the numerical curve obtained by using a constant water diffusion coefficient of $D_{water-dope} = 2.15 \times 10^{-5}\,mm^2/s$

Governing Process in the Spinning Canal: Internal Water Diffusion

The spinning canal can be approximated as a progressively narrowing tapered tube through which dope (stored in the gland ampulla) flows (Fig. 18.3.1). A measurement of the canal geometry is shown in Fig. 18.3.3. The measurements obtained were in agreement with those previously published by Knight and Vollrath (1999). An assumption of rapid convective removal of water from the wall, yielded a high value of the Biot number (see (18.2.5)). Thus, the governing process of water removal is diffusion of water through the fiber (termed internal diffusion). The concentration on the canal wall surface was therefore assumed to be zero (i.e. dry-wall boundary condition). A possible mechanism by which the spider could achieve this includes specialized epithelial cells acting as an ion pump (Vollrath & Knight 2001). Regardless of the specific mechanism, the analyses presented in this chapter assume a diffusion dominated scenario, by which internal diffusion is the limiting factor for water removal. In other words, water must 'fight' its way through the fiber and is quickly removed upon reaching the wall. The rationale for the above assumptions was to probe whether simple diffusion (along with rapid removal from the wall) could be a viable mechanism for water loss during fiber formation in the spinning canal. Other possible ways of water removal could be osmotic pressure effects that could drive water to leave the fiber.

With the assumptions about diffusion being the key process, a general numerical scheme was developed in order to model water removal in the spinning canal and is presented in the following section.

18.3.3 Modeling of internal water diffusion

Here we present the numerical procedure for modeling water diffusion, assuming that the internal diffusion of water is the governing process. The diffusion occurs while the material is traveling through the canal, which can be considered as an axisymmetric tube with variable cross-section. The model which will be introduced relies on the following physical conditions (assumptions):

1. A set of material particles lying initially at the entering tube cross-section travels to the exit, remaining in the plane orthogonal to the canal axis and with a decrease of the radius according to the canal size (the canal is filled with the spinning material).
2. The diffusion is an axisymmetric process and occurs in radial directions.

The diffusion of a solvent through a concentrated solution is governed by Fick's law (see 3.2.3 in Section 3.2), which in the case of in-plane axisymmetric conditions is expressed by the differential equation of the form (18.2.6) (Mills 1995),

$$\frac{\partial \rho_w}{\partial t} - \frac{1}{r}\frac{\partial}{\partial r}\left[r\rho_T D_{wd} \frac{\partial}{\partial r}\left(\frac{\rho_w}{\rho_T}\right)\right] = 0 \quad (18.3.2)$$

where ρ_w is the water partial mass density (concentration), r is the radial coordinate of the canal, ρ_T is the total mass concentration (total mass per unit volume of solution),

$$\rho_T = \rho_w + \rho_p \quad (18.3.3)$$

where ρ_p is the concentration of protein.

SPIDER SILK: MODELING SOLVENT REMOVAL

Comparing to (18.3.2) and (18.3.3) to (18.2.6) and (18.1.3), we have that here the diffusion coefficient of water through the dope, D_{wd}, is constant as opposed to being variable for the synthetic case (Sections 18.1 and 18.2). It should be noted that we assume that the constant diffusion coefficient does not change in the canal when the dope experiences elongational and shear flow.

The finite element method was used to model the diffusion as in Section 18.2.3 since here we also have an axisymmertic diffusion–convection problem. The differential equation (18.3.2) is transformed into the finite element balance equations. The cross-section size changes in time and the water content changes in each finite element, therefore the problem is nonlinear. The finite element equations of mass balance are (see (18.1.11))

$$^{n+1}\hat{\mathbf{K}}^{(i-1)}\Delta\boldsymbol{\rho}_w^{(i)} = {}^{n+1}\mathbf{F}^{(i-1)} - {}^{n+1}\hat{\mathbf{K}}^{(i-1)}\,{}^{n+1}\boldsymbol{\rho}_w^{(i-1)} \tag{18.3.4}$$

The vectors $^{n+1}\boldsymbol{\rho}_w^{(i-1)}$ and $\Delta\boldsymbol{\rho}_w^{(i)}$ are the nodal vectors for the water concentrations and concentration increments. The matrix $^{n+1}\hat{\mathbf{K}}^{(i-1)}$ and the vector $^{n+1}\mathbf{F}^{(i-1)}$ correspond to the equilibrium iteration $i-1$, and are evaluated over the last known volumes of the elements.

The boundary conditions used in the finite element model include the axial symmetry condition and the assumption that the water concentration at the external boundary of the fiber is equal to zero. The condition that the canal is filled with the spinning material is achieved in the model by using the continuity equation. Namely, considering two cross-sections (1 and 2) at the axial distance Δx, the following equation can be written

$$\Delta t v_1 A_1 - \Delta t v_2 A_2 - \Delta x \frac{\Delta m_d}{\bar{\rho}_w} = 0 \tag{18.3.5}$$

where v_1 and v_2, and A_1 and A_2 are the velocities and cross-sectional areas at the first and second cross-section (Fig. 18.3.3), respectively; Δm_d is mass of water left the silk in time period Δt for which the particles move from the section 1 to section 2; and $\bar{\rho}_w$ is the pure water mass density ($\bar{\rho}_w = 1$). The sizes of the cross-sections are known from the canal profile (Fig. 18.3.3).

In the computational procedure we follow 'a cross-section' from the entering position, assuming an entering velocity $v_{entering}$ and calculate the water diffusion for that cross-section using the finite element model, with satisfying the continuity equation (18.3.5) for each time step. Practically, with the known velocity $v_1 = v_t$ (time at start of time step) we calculate diffusion (the mass loss Δm_d) using the iteration scheme in (18.3.4), and repeat the calculation until the continuity equation (18.3.5) is satisfied. The computational steps are as follows:

1. Assume velocity at end of time step $v_{t+\Delta t}$

2. Calculate the distance Δx and the mean radius R_{mean} to model the diffusion

3. Solve for the diffusion using the iterative scheme, (18.3.4). Repeat the computational steps 1–3 until the continuity equation (18.3.5) is satisfied

The mean radius is calculated from the canal geometry and the value Δx using the velocities v_t and $v_{t+\Delta t}$. When the iterations on the velocity $v_{t+\Delta t}$ are completed, we start with

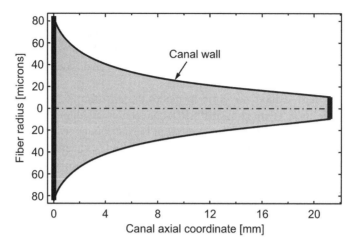

Fig. 18.3.3 Measured radius profile of the *Nephila* spinning canal (Fig. 18.3.1). A dry-wall boundary condition is assumed along the canal wall. The gray symbolizes the spinning material in the canal, whereas the bounding thick vertical lines indicate cross-sections at the canal entrance and exit

the new cross-section position, with the velocity $v_{t+\Delta t}$ as the starting velocity for the next time step, and so on.

The *Nephila* analysis differs from the synthetic due to the prescribed geometry of the canal and unknown velocity profile. Thus, similar basic equations, such as Fick's law, were solved for both synthetic and *Nephila* fiber spinning, taking into account the different prescribed conditions and the finite element code was adapted accordingly for both cases. The results for *Nephila* obtained from the procedure described above are presented next.

18.3.4 Example: The *Nephila* spinning canal

To examine how water is removed during a typical *Nephila* spinning process, the following assumptions were made:

1. The fiber is being pulled out at a speed of 10–20 *mm/s*.

2. The fiber exiting the canal is nearly dry.

The first assumption represents a typical exit velocity during normal web making (Shao & Vollrath 2002, Vollrath & Knight 2001), while the second attempts to evaluate the complete diffusive water loss.

With the above assumptions in place, the numerical procedure of Section 18.3.3 was employed in order to determine the velocity and concentration profiles of cross-section of spinning material moving from the beginning to the end of the spinning canal.

The material densities used were (in g/cm^3)

$$\bar{\rho}_W = 1, \quad \bar{\rho}_p = 1,$$
$$^0\rho_W = 0.7, \quad ^0\rho_p = 0.3 \tag{18.3.6}$$

where the bar represents pure substance density and the superscript zero indicates the initial partial density (concentration) of water and protein.

The adjustable parameter in the model was the entering velocity, and for $v_{entering} = 0.2\ mm/s$ the obtained velocity profile of the cross-section as it traverses down the spinning canal is shown in Fig.18.3.4. As the figure indicates the velocity slowly increases in the first half of the canal, reaching only $1.6\ mm/s$ at the halfway point, and then drastically increases as the fiber diameter further narrows. Such a velocity profile is well suited for optimizing water loss, since in the first sections of the canal the fiber is thicker and thus more time is needed for water to reach the surface of the fiber. Therefore, a relatively small velocity assures longer diffusion times for the large diameter cross-sections. As the fiber cross-section becomes smaller the water has a shorter distance to overcome and thus shorter diffusion times enable greater cross-sectional velocities. The fiber exit velocity (at a distance of $21\ mm$ from the canal entrance) of $16\ mm/s$ was within the range of assumption 1.

For the above velocity conditions, the corresponding cross-section position in the canal over time was determined (Fig. 18.3.5). The graph indicates that a cross-section traveling from the canal entrance needs to flow 20 seconds in the canal if most of the water is to be removed. In other words, in 20 seconds the cross-section has traveled $21\ mm$ through the

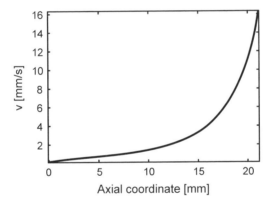

Fig. 18.3.4 Velocity profile of a cross-section that travels through the spinning canal

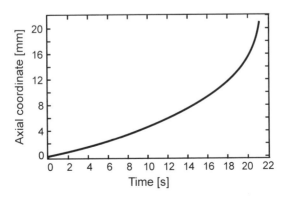

Fig. 18.3.5 Position in the canal of a traveling cross-section as function of spinning time

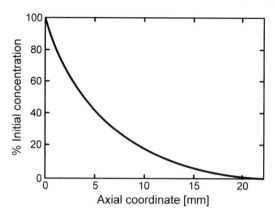

Fig. 18.3.6 Relative water concentration (expressed as % of initial) of a cross section traversing the spinning canal. The relative concentration is obtained by dividing the mean concentration (see equation (18.3.7)), by the initial water concentration (see equation (18.3.6))

entire length of the canal. Also, the moving cross-section spends more than 50% of the total spinning time in the first 25% of the canal, for reasons described above. Interestingly, the final 5 mm of the canal are traversed in less than one second.

The corresponding water concentration profile of the cross-section as it travels through the canal is shown in Fig. 18.3.6. As a relative measure of the water content in the cross-section a ratio of mean concentration to initial concentration, expressed in percent, is used.

$$\rho_{mean} = \frac{\sum_{elements} \int_{L_{element}} \rho_w r \, dr}{R^2 \pi} \qquad (18.3.7)$$

A gradual water loss in the cross-section is observed, suggesting that the spinning conditions, i.e. velocity and canal geometry, are optimized in such a way to provide continuous water removal. Our model shows that most of the water (about 60%) is removed in the first quarter of the canal. Furthermore, the relative concentration ratio already drops to 15% at the halfway point, suggesting that most of the water has already been removed in the first half of the spinning canal. At the canal exit, the ratio of the mean to initial water concentration is only 2%, which implies an essentially dry fiber, in accordance to assumption 2.

The results indicate that a cross-section traveling from the entrance to the canal exit must spend 20 seconds in the canal, while its velocity increases from 0.2 to 16 mm/s, if nearly all of the water is to be removed. These findings thus suggest that simple diffusion, along with the dry wall boundary condition, is a viable mechanism for water removal during *Nephila* fiber spinning.

The exact boundary condition at the wall, presently unknown, may vary due to the different types of epithelium encountered in the spinning canal (Vollrath & Knight 2001). These variable boundary conditions could then alter water removal. Once determined, these new parameters could be incorporated into the numerical model presented above. The results of the model would thus yield a more accurate description of water loss during fiber spinning and provide further insight into the 400 million-year-old mechanism by which the spider makes its mechanically superior fiber.

References

Bathe, K.J. (1996). *Finite Element Procedures*, Prentice-Hall, Englewood Cliffs, NJ.

Chen, X., Knight, D.P. & Vollrath, F. (2002). Rheological characterization of nephila spidroin solution, *Biomacromolecules*, **3**(4), 644–8.

Deen, W.M. (1998). *Analysis of Transport Phenomena*, Oxford University Press, New York.

El Afif, A. & Grmela, M. (2002). Non-Fickian mass transport in polymers, *J. Rheo.*, **46**, 591–628.

Fuller, E.N., Ensley, K. & Giddings, J.C. (1969). Diffusion of halogenated hydrocarbons in Helium: the effect of structure on collision cross sections, *J. Physical Chem.*, **73**, 3679–85.

Gosline, J.M., Guerette, P.A., Ortlepp, C.S. & Savage, K.N. (1999). The mechanical design of spider silks: from fibroin sequence to mechanical function, *J. Exp. Biol.*, **202**, 3295–303.

Gudlavalleti, S. Gearing, B. & Anand, L. (2005). Flexure-based micromechanical testing machines, *Experimental Mech.*, **45**, 412–19.

Huebner, K.H. (1975). *The Finite Element method for Engineers*, John Wiley & Sons, Inc., New York.

Hughes, T.J.R. (1987). *The Finite Element Method. Linear Static and Dynamic Finite Element Analysis*, Prentice-Hall, Englewood Cliffs, NJ.

Incropera, F.P. & DeWitt, D.P. (1996). *Fundamentals of Heat and Mass Transfer*, John Wiley & Sons, Inc., New York.

Kaplan, D., Adams, W.W., Farmer, B. & Viney, C. (1994). *Silk Polymers: Materials Science and Biotechnology*, ACS, Washington, DC.

Knight, D.P. & Vollrath, F. (1999). Liquid crystals and flow elongation in a spider's silk production line, *Proc. Roy. Soc. Lond. B*, **266**, 519–23.

Kobuchi, S. & Arai, Y. (2002). Prediction of mutual diffusion coefficients for acrylic polymer + organic solvent systems using free volume theory, *Prog. Polymer Sci.*, **27**, 811–14.

Kojić, M. & Bathe, K.J. (2005). *Inelastic Analysis of Solids and Structures*, Springer-Verlag, Berlin.

Kojić, M., Slavković, R., Živković, M. & Grujovic, N. (1997). *PAK-T Finite Element Program for Linear and Nonlinear Heat Conduction*, Faculty of Mechanical Engineering, University of Kragujevac, Serbia.

Kojić, M., Slavković, R., Živković, M. & Grujovic, N. (1998). *The Finite Element Method* (in Serbian), Faculty of Mechanical Engineering, University of Kragujevac, Serbia.

Kojić, N., Kojić, M., Gudlavalleti, S. & McKinley, G.H. (2004). Solvent removal during synthetic and *Nephila* fiber spinning, *Biomacromolecules*, **5**, 1698–707.

Kojić, N., Kojić, A. & Kojić, M. (2006). Numerical determination of the solvent diffusion coefficient in a concentrated polymer solution, *Commun. Numer. Meth. Eng.*, **22**, 1003–13.

Middleman, S. (1998). *An Introduction to Mass and Heat Transfer: Principles of Analysis and Design*, John Wiley & Sons, Inc., New York.

Mills, A.F. (1995). *Heat and Mass Transfer*, Richard D. Irwin, Chicago.

Poling, B.E., Prausnitz, J.M. & O'Connell, J.P. (2001). *The Properties of Gases and Liquids*, McGraw-Hill, New York.

Selden, P.A. (1989). Orb-weaving spiders in the early Cretaceous, *Nature*, **340**, 711.

Shao, Z. & Vollrath, F. (2002). Surprising strength of silkworm silk, *Nature*, **418**(6899), 741.

Shear, W.A. & Palmer, J.M. (1989). A Devonian spinneret: early evidence of spiders and silk use, *Science*, **246**, 479–81.

Vollrath, F. & Knight, D.P. (1999). Structure and function of the silk production pathway in the spider *Nephila edulis*, *Int. J. Biol. Macromolec.*, **24**, 243–9.

Vollrath, F. & Knight, D.P. (2001). Liquid crystalline spinning of spider silk, *Nature*, **410**(6828), 541–8.

19

Modeling in Cancer Nanotechnology

The strategy currently followed to deliver nano-sized particulates to solid tumors is based on the well-known enhanced permeability and retention effect where particles sufficiently small to spontaneously extravasate through the fenestrated tumor vessels can be transported from the vascular compartment to the inner region of the tumor mass and from there release their payload. An alternative active strategy is gaining consensus and is based on the targeting of the tumor vasculature through ligand–receptor specific interactions exploiting the biological and biophysical differences between normal and tumor vessel walls. Such an active strategy requires a detailed analysis of the transport and adhesive interaction of nano-sized particulate systems within the tumor vasculature which is characterized by permeable walls, high interstitial fluid pressure, and expression of specific receptor molecules.

In this chapter, the analysis of the transport of solute molecules resembling nano-sized particles under laminar flow is presented solving the classical diffusion–advection equation in a straight capillary. The effect of vessel wall permeability as well as the complex rheological behavior of blood are considered explicitly keeping the formulation tractable. Possible future directions of research are then presented in the closing paragraph where a finite element approach is described to treat the transport of nonconventional particulate systems having a nonspherical shape.

19.1 Introduction

Small-molecule agents and monoclonal antibodies (mAbs) as well as particulate formulations have been developed, subjected to clinical trials and some are already employed in the clinics to cure cancer by targeting receptor molecules expressed specifically within the

---

Computer Modeling in Bioengineering   Edited by M. Kojić, N. Filipović, B. Stojanović, N. Kojić
© 2008 John Wiley & Sons, Ltd

diseased cells. Despite this, the vast majority of malignancies have proven to be resistant to such interventions, partially due to the requisite dose limitations for preventing adverse effects on normal tissues but largely due to the barriers of different nature that these systemically administered agents should avoid before reaching their biological target (tumor microenvironment) (Minchinton & Tannock 2006, Ferrari, 2005a,b).

It is reported that small molecules and therapeutic antibodies reach the desired biological target only in one part per 10 000–100 000 molecules (Li *et al.* 2004). Most of these molecules are lost within the body, in that for their small size (< 1–$5\,nm$) they can easily cross the endothelial barrier and diffuse through the extracellular matrix of almost any normal tissue. Others are eliminated from the blood pool through the action of the immune system. And those that reach the tumor vasculature are prevented from penetrating deep in the tumor mass by the adverse interstitial fluid pressure and by the composition and highly intricate structure of the extracellular matrix of tumors (Heldin *et al.* 2004). In addition to this, it is now well accepted that the progression of a tumor mass cannot be retarded or inhibited by targeting a single molecule due to the great heterogeneity of most tumors. Instead multi-targeted therapies should be employed with a combination of agents targeted to several distinct molecules, that for instance could be delivered simultaneously at the same site through the use of particulate systems (Imai & Takaoka 2006).

In the delivery of nano-sized particulates to solid tumors, the strategy that has been traditionally considered and is currently employed by the few particulate formulations available in the clinics is based on the well-known effect of enhanced permeability and retention effect (EPR): nanoparticles sufficiently small to spontaneously extravasate through fenestrations found in the tumor vasculature can be entrapped in the extracellular matrix and transported from the vascular compartment to the inner region of the tumor mass. However, the high interstitial fluid pressure and the composition and structure of the extracellular matrix adverse even more the extravasation and subsequent transport of these nanometer particles towards the tumor cells.

As an alternative to the passive strategy based on the EPR effect, an active delivery strategy based on the targeting of the tumor vasculature through ligand–receptor specific interactions is currently gaining more and more consensus. There are striking biological differences between normal and tumor endotheliums, which provide a scientific rationale for vascular targeting (Neri & Bicknell 2005). The use of microarrays, phage display and SAGE libraries (Trepel *et al.* 2002) has led to the identification of several tumor endothelial markers (TEMs) that are almost exclusively expressed on abnormal rather then normal endothelial cells. And all these molecules constitute good candidates as 'docking sites' for the circulating nanoparticles. In vascular targeting, nanoparticles should be able to recognize specific molecular markers expressed over the cells lining the diseased blood vessels (tumor endothelium) and adhere firmly to the vessels withstanding the dislodging hydrodynamic forces. From there, the nanoparticles can release their payload, drug molecules or smaller particulate formulations specifically designed to efficiently be transported within the tumor mass. Consequently, the transport of nano-sized molecules and particulate systems within capillaries with permeable walls, as is the case of tumors, and in the presence of blood cells altering significantly the rheological properties of the hosting fluid is of vital importance in the design and development of such particulate systems, and this is one of the most active field of research in cancer nanotechnology.

19.2 The transport of particulates in capillaries

The longitudinal dispersion of a bolus of solute in a solvent flowing in a channel is of broad interest within several fields, including chemical and biomedical engineering, fluid dynamics and environmental sciences.

General

Taylor (1953) first studied the effect of shear on axial dispersion in fully developed laminar flow of a Newtonian fluid in a circular tube. A similar solution is readily obtained for the flow between plates. To introduce the concept of shear-augmented dispersion, consider a bolus of a passive species in a fully developed incompressible laminar Newtonian flow in a straight channel (Fig. 19.2.1). The bolus is carried downstream by the Poiseuille flow. At the leading edge of the bolus, the solute diffuses from the high concentration region near the center of the tube toward the low concentration region at the wall. In doing so, the amount of material traveling at a speed greater than the average is reduced, thereby reducing the rate of axial spread of the bolus relative to its axial center, which moves with the cross-sectional average velocity. At the trailing edge, diffusion is inward, again reducing the variance of the velocity of the bolus. In particular, a cylindrical frame of reference (r, z) is considered with the z-axis along the tube's axis of symmetry (Fig. 19.2.1). Assuming that the particulates constituting the solute have the same velocity as that of the fluid they are displacing (this implies that the particles are sufficiently small), the governing advection–diffusion equation is given by (see Eq. (3.3.9) and Example 1.5-5)

$$\frac{\partial c}{\partial t} + v(r)\frac{\partial c}{\partial z} = \frac{D_m}{r}\frac{\partial}{\partial r}\left(r\frac{\partial c}{\partial r}\right) + D_m\frac{\partial^2 c}{\partial z^2} \quad (19.2.1)$$

where $v(r)$ is the non-uniform axial velocity and $c(r, z; t)$ is the nanovector concentration; D_m is the diffusion coefficient considered constant along the capillary. Taylor's first approximation was that of neglecting the Brownian diffusion in favour of pure convective diffusion along z, thus cancelling out the term $D_m \cdot \partial^2 c/\partial z^2$ in (19.2.1). Furthermore, considering an auxiliary frame of reference (ζ, r) moving at the mean velocity V along z, so that $z = \zeta + V \times t$, Equation (19.2.1) results in

$$\check{v}(r)\frac{\partial c}{\partial \zeta} = \frac{D_m}{r}\frac{\partial}{\partial r}\left(r\frac{\partial c}{\partial r}\right) \quad (19.2.2)$$

being $\check{v}(r) = v(r) - V$ the velocity deviation from the mean value V. Similarly, Taylor introduced $\check{c}(r, z; t) = c(r, z; t) - C_m(z; t)$ as the deviation of concentration from the mean value C_m, and assumed that transverse variations in concentration, lumped in the term

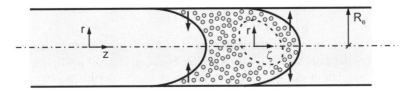

Fig. 19.2.1 Shear augmented dispersion mechanisms in a straight capillary

$\check{c}(r, z; t)$, were much smaller than the longitudinal variations, lumped in the term $C_m(z; t)$. This second Taylor's approximation leads to the fundamental equation

$$\check{v}(r)\frac{\partial C_m}{\partial \zeta} = \frac{D_m}{r}\frac{\partial}{\partial r}\left(r\frac{\partial \check{c}}{\partial r}\right) \quad (19.2.3)$$

stating that molecular diffusion along the radial direction (right-hand-side term) induces longitudinal diffusion (left-hand-side term) owing to the nonuniform radial velocity profile $\check{v}(r)$ (convective diffusion). The nonuniform flow stretches the species concentration profile along the capillary generating transverse variations in concentration, which are destroyed by transverse molecular diffusion reshaping the concentration profile.

The rate of transfer J_ζ (the mass flux q_m in (3.2.5)) of particulates across the plane $\zeta = const$ owing to pure convection can be expressed, recalling Fick's law (3.2.5), as

$$J_\zeta = -AD_{app}\frac{\partial C_m}{\partial \zeta} \quad (19.2.4)$$

or alternatively, based on the definition of J_ζ, as

$$J_\zeta = \int_A c(r, z; t)\,\check{v}(r)\,dS = \int_A \check{c}(r, z; t)\,\check{v}(r)\,dS \quad (19.2.5)$$

being

$$\int_A C_m(r, z; t)\,\check{v}(r)\,dS = 0 \quad (19.2.6)$$

by definition; where A is the cross-sectional area. Thus, the apparent diffusion coefficient D_{app} can be introduced as

$$D_{app} = -\int_A \check{c}(r, z; t)\,\check{v}(r)\,dS \Big/ A\frac{\partial C_m}{\partial \zeta} \quad (19.2.7)$$

As a consequence, the whole problem of determining the convective contribution to the effective longitudinal diffusion is reduced to evaluating the concentration deviation from the mean value. The effective diffusion is obtained introducing the Brownian contribution to diffusion $D_{eff} = D_m + D_{app}$. For a laminar flow in a circular pipe of radius R_e with nonpermeable walls, an explicit expression for D_{eff} is derived as:

$$D_{eff} = D_m + \frac{VR_e}{48D_m} \quad (19.2.8)$$

Dispersion is maximized as the molecular diffusivity goes to zero since any lateral diffusion reduces the axial spread of the material: Equation (19.2.8) predicts that D_{eff} goes to zero as $D_m \sim 0$, but in this limit the assumption of small lateral concentration gradients breaks down and the result is no longer valid. In contrast to the basis for the Taylor results, the transport becomes purely convective.

Assumptions implicit in this analysis are: (i) the dispersion is quasi-steady, thereby eliminating the temporal term from the species transport equation; (ii) an assumption of unidirectional, usually fully developed, flow eliminates all convective terms except the axial one; (iii) axial convection is dominant over axial diffusion; and (iv) lateral variations in concentration are small compared with those in the longitudinal direction. Considerable effort has been expended in attempts to relax Taylor's assumptions. The first assumption can be particularly troublesome, since in a problem involving the spread of a tracer introduced into the flow stream in some arbitrary configuration, the analysis is restricted to the limit of large time. Specifically, the Taylor (1953) and Aris (1956) analysis is valid for $t \gg 1/2\, R_e^2/D_m$ that, given particular values of diffusivity and outer diameter describing common physiological conditions, can be also significantly big (considering the dispersion of submicrometric particles, with a molecular diffusivity D_m typically ranging between 10^{-11} and $10^{-9}\ m^2/s$, it follows that the Taylor–Aris asymptotic solution is strictly valid in large vessels (arteries) with $R_e \sim 10^{-2}\ m$ at times larger than 10^5–$10^7\ s$, whereas in small capillaries with $R_e \sim 10^{-6}\ m$ at times larger than 10^{-3} to $10^{-1}\ s$). Observing that blood in large vessels has a mean velocity V of about $10^2\ mm/s$, the Taylor–Aris regime would be fully developed in arteries only after $10^4\ m$. In small capillaries with V of about 1 mm/s, the asymptotic solution would hold true after 10^{-3} to $10^{-1}\ mm$, which is smaller than the characteristic length of normal capillaries typically ranging between 100 μm and few millimetres). The analysis carried so far is incomplete to the extent that: (i) the permeability of the vessels where dispersion takes place is not considered; (ii) blood is treated as a Newtonian fluid, whereas more sophisticated and reliable (and realistic) rheological models exist (for instance, the Casson model, see Section 13.1.2) that may describe blood more accurately; and (iii) the transient time of dispersion is disregarded, and the solution is given in terms of mean concentration (the radial distribution of concentration cannot be deducted from the mono-dimensional analysis of Taylor/Aris).

Ananthakrishnan et al. (1965) solved numerically the complete convective–diffusion equation describing the dispersion of the solute within a cylindrical steady laminar flow and observed a perfect agreement with the approximate results of the Taylor and Aris theory in the limit of sufficiently large times t ($t \gg 1/2\, R_e^2/D_m$), as widely reported above. Gill (1967) extended Taylor's framework to obtain the local concentration distribution, by means of a series expansion about the mean concentration, while Gill and Sankarasubramanian (1970) established that the above-mentioned method of series solution (known as the generalized dispersion model) could exactly reproduce the centroid and the width of the concentration for all times, by solving the following simplified convective–diffusive equation

$$\frac{\partial C_m}{\partial t} = \sum_{i=1}^{\infty} K_i(t) \frac{\partial^i C_m}{\partial z^i} \qquad (19.2.9)$$

provided that the coefficients of the models $K_i(t)$ are chosen as suitable functions of time. Also, they showed that the series in (19.2.9) may be truncated to the second order as

$$\frac{\partial C_m}{\partial t} = K_1(t) \frac{\partial C_m}{\partial z} + K_2(t) \frac{\partial^2 C_m}{\partial z^2} \qquad (19.2.10)$$

where

$$K_3(\tau) = -\frac{1}{23040} \qquad (19.2.11)$$

Sankarasubramanian and Gill (1973) elaborated the generalized dispersion model (GDM) by including the effects of interphase mass transfer (i.e. by removing the hypothesis of impermeability of the walls to the solute; in such a circumstance, summation in (19.2.9) would start from $i = 0$).

Biomedical Applications

In biomedical applications, macromolecules and nanoparticles are systemically administered and transported within capillaries with different radii, lengths and properties. Depending on the organ, the capillary walls can be impermeable, as for the blood–brain endothelium, or can be highly permeable, as for the capillary of the kidney or those of developing tumor masses. In addition to this, the velocity profile in capillaries can be significantly different from parabolic (Poiseuille flow), because of the presence of red blood cells, which tend to accumulate in a central 'core' region of the capillary leaving a marginal 'cell free layer'. In arterioles and venules, the blood velocity profile follows quite accurately the Casson law with a central plug region (zero radial velocity gradient) of radius r_c (plug radius) and an outer region with a parabolic velocity profile.

The velocity profile as well as the wall permeability have a significant effect on the convective transport of a solute. In 1993, Sharp derived explicit expressions for D_{eff} considering non-Newtonian fluids with different rheological laws, namely for a Casson, Bingham plastic, and power-law fluid. In particular, for a Casson fluid, it was determined

$$D_{eff} = D_m \left[1 + \frac{P_e^2}{48} \frac{E(\xi_c)}{A^2(\xi_c)} \right] \quad (19.2.12)$$

with $A(\xi_c)$ and $E(\xi_c)A^2(\xi_c)$ given as

$$A(\xi_c) = 1 - \frac{16}{7}\sqrt{\xi_c} + \frac{4}{3}\xi_c - \frac{1}{21}\xi_c^4, \quad (19.2.13)$$

$$E(\xi_c)A^2(\xi_c) = 1 - \frac{5888}{1555}\xi_c^{1/2} + \frac{558368}{56595}\xi_c - \frac{6144}{715}\xi_c^{3/2} + \frac{128}{45}\xi_c^2 + \frac{244}{21}\xi_c^4$$
$$- \frac{272128}{3773}\xi_c^{9/2} + \frac{385312}{2205}\xi_c^5 - \frac{4096}{21}\xi_c^{11/2} + \frac{11464}{165}\xi_c^6 +$$
$$+ \frac{55808}{1155}\xi_c^{13/2} + - \frac{6976}{165}\xi_c^7 + \frac{430331}{66885}\xi_c^8 - \frac{512}{147}\xi_c^{17/2} + \frac{64}{21}\xi_c^9 +$$
$$- \frac{872}{1155}\xi_c^{10} + \frac{4}{147}\xi_c^{12} - \frac{8}{147}\xi_c^8 \ln(\xi_c) \quad (19.2.14)$$

and depending on the rheological parameter $\xi_c = r_c/R_e$, the ratio between the plug radius r_c and the capillary radius R_e.

This solution asymptotes to the Taylor's results as ξ_c goes to zero, differently approaches 0 as ξ_c approaches 1: in these limits, the flow becomes more and more plug-like and the area over which shear may augment dispersion disappears. The factor E/A^2 (shown in Fig. 19.2.2) gives the reduction in dispersion due to non-Newtonian rheology at equivalent flow. It is worth revisiting an assumption implicit in these solutions since it may have a great impact on the application of the results to blood. While it has been assumed that the medium is homogenous, blood is actually a concentrated mixture of elastic particles (red blood cells). The tumbling and deformation of individual cells and aggregates will obviously affect the

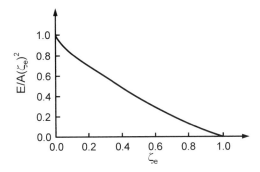

Fig. 19.2.2 The factor E/A^2 giving the reduction in dispersion due to non-Newtonian rheology (Sharp's model of dispersion, equation (19.2.12))

lateral mixing of any solute introduced into blood and will, therefore, modify the rate of axial dispersion.

Dash et al. (2000) and Nagarani et al. (2004) used the Sharp model in conjunction with the GDM to investigate the effects of yield stress (or equivalently the plug radius) and of the irreversible solute-reaction mechanism at the flow boundaries on the dispersion in a Casson fluid through a conduit. In this scenario, the entire phenomenon of solute propagation was described in terms of three effective transport coefficients: exchange (K_0, which arises due to adsorption mechanisms at the walls, that is null in Dash et al. (2000) in that no adsorption reactions are therein considered); convection (K_1, due to the velocity of the solute); and dispersion (K_2, which can be related to the Taylor's effective diffusivity as $D_{eff} = R_e^2 w_0^2 K_2 / D_m$). While in Dash et al. (2000) a closed form solution was provided, in Nagarani et al. (2004), due to the complexity of the equations involved, an exact solution was found for K_0 solely, whereas for K_1 and K_2 the asymptotic values were derived. It was seen that the asymptotic dispersion coefficient decreases with increase in the wall solute-absorption parameter β, and yield stress of the fluid.

The discussed methods and solutions strongly depend on the assumption that the transverse concentration distribution can be expanded in terms of eigenfunctions (Bessel functions for a circular pipe), which is properly verified only in the limit of complete transverse mixing. When the solute has not yet strongly interacted with the boundary, a free space expansion would be more suitable in describing the problem. Lighthill (1966) first studied this transient and anomalous regime and found a solution for the concentration (which accounts for the transverse diffusion, but neglects longitudinal diffusion and interactions with the pipe's boundary) in terms of a Fourier transform, and showed that the tracer distribution, for small times, spreads longitudinally proportional to t (which is properly a superdiffusive behavior). Latini and Bernoff (2001), more recently, have revisited the problem of dispersion of a point discharge of tracer in laminar pipe Poiseuille flow. Assuming a δ-function initial condition at the center of the pipe, and by means of a Fourier transform of the advection–diffusion equation, they fully modelled the three initial stages of dispersion, that is: (i) at small times, when diffusion dominates advection yielding a spherically symmetric Gaussian dispersion cloud; (ii) at large times, in correspondence of which the flow is in the classical Taylor regime; and (iii) at an intermediate regime, where the longitudinal mean concentration profile is either asymmetric and anomalous.

Most recently, Decuzzi et al. (2006a,b; 2007a,b) have extended the Taylor and Aris theory including the permeability of the walls to the sole solvent and leading to a new and more general expression for D_{eff} being

$$D_{eff} = D_m + \frac{v_0^2 R_e^2}{192 D_m} \left[\frac{(1+e^{2\zeta\Pi}) - e^{\Pi}(1+e^{2(\zeta-1)\Pi})\Omega}{2 - e^{\Pi}(1+e^{-2\Pi})\Omega} e^{-\Pi\zeta} \right]^2 \quad (19.2.15)$$

with ζ the dimensionless longitudinal coordinate $(= D_m z/(R_e^2 v_0))$, where v_0 is the initial centerline velocity), Π the permeability parameter

$$\Pi = \zeta_1 \sqrt{2 L_p / \pi} \quad (19.2.16)$$

related to L_p – the vascular hydraulic conductivity, and Ω – the pressure parameter

$$\Omega = (p_0/\pi_i - 1)/(p_1/\pi_i - 1) \quad (19.2.17)$$

related to the inlet p_0, outlet p_1 and the interstitial fluid π_i pressures, respectively. Whereas the model proposed from Sharp has been subsequently refined by Gentile et al. (2007a,b,c), to introduce the effect of permeability of the channel to the solvent, inducing a reduction of velocity along the longitudinal coordinate as broadly discussed in Decuzzi et al. (2006a,b).

In the following, the generalized dispersion model re-proposed by Dash et al. (2000) and Nagarani et al. (2004) is combined with the steady-state solution given in Decuzzi et al. (2006a,b) to analyze the unsteady dispersion of nanoparticles in permeable capillaries (Gentile et al. 2007).

19.3 The mathematical model

A straight circular capillary with radius R_e and length l is considered (Fig. 19.3.1), the flow being described by a Newtonian fluid law. The capillary walls may be permeable or impermeable to the fluid, but are impermeable and not adsorbent for the solute.

In the following the generalized dispersion model is recalled and revised to consider the effective perfusion of the solvent through the walls. The dimensionless coefficients constituting the model are deducted and given in terms of the time and spatial variables, and of the permeability and pressure parameters Π and Ω, respectively. The relationship between the above cited coefficients and the effective diffusion coefficient D_{eff} is shown.

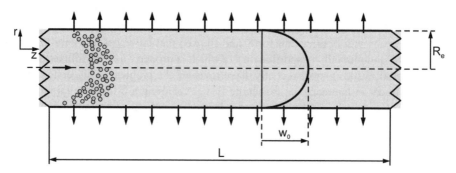

Fig. 19.3.1 The geometry of the channel where the nanoparticles are dislodged

19.3.1 The governing equations

Following an approach firstly proposed by Sankarasubramanian and Gill (1973), and more recently employed also by Dash *et al.* (2000) and Nagarani *et al.* (2004), the dispersion of a passive tracer or particles in a Poiseuille flow may be described in a dimensionless form by the advection–diffusion equation

$$\frac{\partial \psi}{\partial t} + \bar{v}\frac{\partial \psi}{\partial z} = \left(\frac{1}{r}\frac{\partial}{\partial r}\left(r\frac{\partial}{\partial r}\right) + \frac{1}{P_e^2}\frac{\partial^2}{\partial z^2}\right)\psi \quad (19.3.1)$$

expressed in terms of nondimensional physical quantities defined as

$$\psi = \frac{C}{C_0}, \quad \bar{v} = \frac{v}{v_0}, \quad \rho = \frac{r}{R_e}, \quad \zeta = \frac{D_m z}{R_e^2 v_0}, \quad \tau = \frac{D_m t}{R_e^2} \quad (19.3.2)$$

where C and Ψ are the dimensional and nondimensional concentration of the passive species respectively, C_0 is a concentration of reference, v_0 is the initial centerline velocity, and v is the velocity distribution within the pipe given explicitly in the following paragraph, R_e is the radius of the capillary, D_m is the molecular diffusivity, r and z are the radial and longitudinal coordinates as from the frame of reference in Fig. 19.3, and τ is the dimensional time variable. In (19.3.1) we used $P_e (= R_e v_0 / D_m)$ as the characteristic Peclet number. It is assumed that the particles are sufficiently small to have the same velocity of the dislodging fluid so that the diffusion/advection problem and the fluid-dynamic problem may be treated separately.

The solution for Ψ may be derived as (Gill 1967, Dash et al. 2000)

$$\psi = \sum_{i=0}^{\infty} f_i(\rho, \zeta; \tau) \frac{\partial^i \psi_m}{\partial \zeta^i} \quad (19.3.3)$$

where the parameters f_i are weight functions relating the local concentration Ψ to the derivative of order i of the mean concentration Ψ_m with respect to the spatial variable ζ (Notice that for $i = 0$, (19.3.3) gives Ψ_m, with $f_0 = 1$). The mean concentration Ψ_m is averaged over the cross-section as

$$\psi_m = 2\int_0^1 \psi \rho d\rho \quad (19.3.4)$$

and must obey the condition

$$\frac{\partial \psi_m}{\partial \tau} = \sum_{i=0}^{\infty} K_i(\zeta, \tau) \frac{\partial^i \psi_m}{\partial \zeta^i} \quad (19.3.5)$$

where the auxiliary functions $K_i(\zeta, \tau)$ are given by

$$K_i(\zeta, \tau) = \frac{\delta_{i2}}{P_e^2} + 2\frac{\partial f_i}{\partial r}(1, \zeta; \tau) - 2\int_0^1 f_{i-1}(\rho, \zeta; \tau) \bar{v}(\rho, \zeta; \tau)\rho d\rho,$$
$$f_{-1} = 0 \quad (19.3.6)$$

Here, δ_{ij} denotes the Kronecker delta symbol.

Solving (19.3.5), with the appropriate initial and boundary conditions, the mean concentration $\Psi_m(\zeta, \tau)$ is derived and the local concentration $\Psi(\zeta, \rho; \tau)$ is eventually obtained using (19.3.3). Thus, the problem is basically reduced to estimating $f_i(\zeta, \rho; \tau)$ and $K_i(\zeta, \tau)$ for each i. For the weight functions $f_i(\zeta, \rho; \tau)$ a set of differential equations may be derived in a general form

$$\frac{\partial f_n}{\partial \tau} = \frac{1}{\rho}\frac{\partial}{\partial \rho}\left(\rho \frac{\partial f_n}{\partial \rho}\right) - \bar{v}(\rho, \zeta) f_{n-1} + \frac{1}{P_e^2} f_{n-2} - \sum_{i=0}^{n} K_i \, f_{n-i}, \qquad (19.3.7)$$
$$f_{-1} = f_{-2} = 0$$

which relates $K_i(\zeta, \tau)$ and $f_i(\zeta, \rho; \tau)$. The equations (19.3.6) and (19.3.7) together with the initial and boundary conditions completely define the dispersion problem under analysis. It has also been shown by Sankarasubramanian and Gill that sufficiently accurate results are obtained by limiting the summation in (19.22) to the first three terms ($i = 2$), that is

$$\frac{\partial \psi_m}{\partial \tau} = K_0 \Psi_m + K_1 \frac{\partial \psi_m}{\partial \zeta} + K_2 \frac{\partial^2 \psi_m}{\partial \zeta^2} + O\left(\frac{\partial^3 \psi_m}{\partial \zeta^3}\right) \qquad (19.3.8)$$

and in the sequel higher order terms are neglected.

19.3.2 The initial and boundary conditions

It is assumed that a bolus of particles at the time $\tau = 0$ is introduced instantaneously and uniformly along the radius ρ, thus to satisfy the initial condition

$$\psi_m(\zeta; 0) = \psi_{m0}(\zeta) \qquad (19.3.9)$$

Note that no particular restrictions apply to the initial distribution profile of Ψ_m, i.e. the mean concentration at the initial time $\tau = 0$ may be the most general. On the other hand, since for $\tau = 0$ the solute is uniformly spread along every cross-section of the channel, the local concentration has to satisfy the condition:

$$\psi(\rho, \zeta; 0) \equiv \psi_m(\zeta; 0) \qquad (19.3.10)$$

It is further assumed that the pipe walls are impermeable to the particles constituting the solute,

$$\frac{\partial \psi}{\partial \rho}(1, \zeta; \tau) = 0 \qquad (19.3.11)$$

while, due to the conservation of mass of the species diffusing in the channel, infinitely far away from the inlet section, the concentration as well as the derivatives of concentrations up to a generic order i go to zero

$$\psi(\rho, \infty; \tau) = \frac{\partial^i \psi}{\partial \rho^i}(\rho, \infty; \tau) = 0, \; \psi_m(\infty; \tau) = \frac{\partial^i \psi}{\partial \rho^i}(\infty; \tau) = 0 \qquad (19.3.12)$$

and on the center line the symmetry condition imposes that

$$\psi(0, \zeta; \tau) = \text{finite and} \quad \frac{\partial \psi}{\partial \rho}(0, \zeta; \tau) = 0 \qquad (19.3.13)$$

MODELING IN CANCER NANOTECHNOLOGY

Considering (19.3.5) and (19.3.9)–(19.3.13), the initial and boundary conditions on Ψ and Ψ_m can be rephrased in terms of the weight functions f_i, leading to a new set of conditions which may be effective within the governing equations to derive explicit relations for K_i and f_i. In particular, from the definition of the average concentration (19.3.4), the solvability condition is straightforwardly derived as

$$\int_0^1 f_n(\rho, \zeta; \tau)\rho\, d\rho = \frac{\delta_{0n}}{2}, \quad n \geq 0 \tag{19.3.14}$$

where δ_{0n} is the Kronecker delta.

The initial condition of uniformity on Ψ can be analytically expressed through

$$\left.\frac{\partial \psi}{\partial \rho}\right|_{\tau=0} = 0 \tag{19.3.15}$$

and, substituting (19.3.3) into (19.3.15):

$$f_n(\rho, \zeta; 0) \equiv f_n(\zeta; 0), \quad n \geq 0 \tag{19.3.16}$$

From (19.3.16), (19.3.14) and (19.3.10) the initial conditions on the f_i are deducted as

$$f_n(\zeta; 0) = \delta_{0n}, \tag{19.3.17}$$

while the boundary conditions are derived from (19.3.4), (19.3.11) and (19.3.13) as

$$\frac{\partial f_n}{\partial \rho}(1, \zeta; \tau) = 0, \quad n \geq 0, \tag{19.3.18}$$

and

$$f_n(0, \zeta; \tau) = \text{finite and } \frac{\partial f_n}{\partial \rho}(0, \zeta; \tau) = 0, \quad n \geq 0. \tag{19.3.19}$$

19.3.3 Solution for K_0 and f_0

The function f_0 and the exchange coefficient K_0 do not depend on the velocity field and can be solved directly. For $n = 0$, equation (19.3.6) reduces to

$$K_0(\zeta, \tau) = 2\frac{\partial f_0}{\partial \rho}(1, \zeta; \tau), \tag{19.3.20}$$

that, through (19.3.18) allows for the determination of the exchange coefficient K_0 as

$$K_0(\zeta, \tau) = 0. \tag{19.3.21}$$

Note that the coefficient K_0 is zero, therefore there are no absorption effects at the walls.

The function f_0 dictates the deviation of the local concentration Ψ from the mean concentration Ψ_m, due to solute absorption at the walls mechanisms. When there are no depletion effects of solute at the border, f_0 is set to one

$$f_0 = 1 \tag{19.3.22}$$

which is the sole solution of (19.3.7), which satisfies both the boundary conditions (19.3.18) and (19.3.19) and the initial condition (19.3.17). This may be proved as follows: from (19.3.7), imposing $n=0$, the expression

$$\frac{\partial f_0}{\partial \tau} = \frac{1}{\rho}\frac{\partial}{\partial \rho}\left(\rho\frac{\partial f_0}{\partial \rho}\right) \quad (19.3.23)$$

is obtained. Using the solvability condition (19.3.14) in (19.3.23), the expressions

$$\frac{\partial f_0}{\partial \rho} = 0, \quad \forall \tau, \quad (19.3.24)$$

$$\frac{\partial f_0}{\partial \tau} = 0, \quad \forall \rho \quad (19.3.25)$$

may be derived, which, together with (19.3.17), allow for the deconvolution of f_0 as $f_0 = 1$.[1]

19.3.4 Solution for K_1 and f_1

For $n=1$, Equation (19.3.7) becomes

$$\frac{\partial f_1}{\partial \tau} = \frac{1}{\rho}\frac{\partial}{\partial \rho}\left(\rho\frac{\partial f_1}{\partial \rho}\right) - v(\rho, \zeta) - K_1, \quad (19.3.26)$$

recalling that $f_1 = 0$. Multiplying (19.3.26) by ρ and integrating from 0 to 1 with respect to ρ, along with the solvability condition (19.3.14), it follows that

$$K_1(\zeta) = -2\int_0^1 \bar{v}(\rho, \zeta)\rho d\rho \equiv -\bar{v}_m(\zeta). \quad (19.3.27)$$

If the conduit is impermeable then the velocity profile depends on the radius solely and it may be described by the classical Poiseuille parabolic velocity distribution. It follows that $K_1(\zeta) = -\bar{v}_m(\zeta) \equiv -0.5$ (as obtained in Dash et al. 2000).

The distribution function f_1 is a solution for the partial differential equation (19.3.26); that can be decomposed (Dash et al. 2000; see also Gill & d Sankarasabrumanian 1970, Nagarani et al. 2004) as

$$f_1(\rho, \zeta; \tau) = f_{1s}(\rho, \zeta) + f_{1t}(\rho, \zeta; \tau), \quad (19.3.28)$$

---

[1] Notice that in Dash et al. (2000), K_0 is null and $f_0 = 1$ in that no reaction mechanisms at the walls are considered. In Sankarasubramanian and Gill (1973), and Nagarani et al. (2004), on the other hand, given that the heterogenous reaction mechanism occurs at the wall described by $\frac{\partial \psi}{\partial \rho}(\zeta, 1; \tau) = -\beta\psi(\zeta, 1; \tau)$ where β is the nondimensional wall absorption parameter, f_0 and K_0 are in general not constant and are derived as

$$f_0(\tau, \rho) = \frac{\sum_0^\infty A_n J_0(\mu_n \rho) e^{-\mu_n^2 \tau}}{2\sum_0^\infty \left(\frac{A_n}{\mu_n}\right) J_1(\mu_n) e^{-\mu_n^2 \tau}}; \quad K_0(\tau, \rho) = -\frac{\sum_0^\infty A_n \mu_n J_1(\mu_n) e^{-\mu_n^2 \tau}}{\sum_0^\infty \left(\frac{A_n}{\mu_n}\right) J_1(\mu_n) e^{-\mu_n^2 \tau}}.$$

where $f_{1s}(\rho, \zeta)$ is the steady-state solution, whereas $f_{1t}(\rho, \zeta; \tau)$ is the transient, time-dependent solution. Following an approach as in Dash et al. (2000), the expressions for f_{1s} and f_{1t} may be derived as

$$f_{1s}(\zeta, \rho) = \bar{v}_0(\zeta) \left(\frac{1}{8}\rho^2 - \frac{1}{16}\rho^4 - \frac{1}{24} \right), \tag{19.3.29}$$

where $\bar{v}_0(\zeta)$ is the nondimensional center-line velocity of the flow and

$$f_{1t}(\rho, \zeta; \tau) = \sum_{n=0}^{\infty} -\frac{2 \int_0^1 J_0(\lambda_n \rho) f_{1s}(\rho, \zeta) d\rho}{[J_0(\lambda_n)]^2} e^{-\lambda_n^2 \tau} J_0(\lambda_n \rho) =$$

$$= \sum_{n=0}^{\infty} -\frac{2 \int_0^1 J_0(\lambda_n \rho) \bar{v}_0(\zeta) \left(\frac{1}{8}\rho^2 - \frac{1}{16}\rho^4 - \frac{1}{24} \right) \rho d\rho}{[J_0(\lambda_n)]^2} e^{-\lambda_n^2 \tau} J_0(\lambda_n \rho), \tag{19.3.30}$$

are the eigenvalues λ_n as the roots of the equation $J_1(\rho) = 0$. Note that for impermeable channels (where the nondimensional centerline velocity is constant and given by $\bar{v}_0(\zeta) \equiv 1$ equations (19.3.29) and (19.3.30) are deducted as

$$f_{1_s}(\rho) = \frac{1}{8}\rho^2 - \frac{1}{16}\rho^4 - \frac{1}{24}, \tag{19.3.31}$$

$$f_{1t}(\rho; \tau) = \sum_{n=0}^{\infty} -\frac{2 \int_0^1 J_0(\lambda_n \rho) \left(\frac{1}{8}\rho^2 - \frac{1}{16}\rho^4 - \frac{1}{24} \right) \rho d\rho}{[J_0(\lambda_n)]^2} e^{-\lambda_n^2 \tau} J_0(\lambda_n \rho) \tag{19.3.32}$$

These results coincide with those derived by Dash et al. (2000), provided that the plug radius r_p is null (meaning that the Casson fluid degenerates into Newtonian).

19.3.5 Solution for K_2

To derive the expression for K_2, the same approach as for K_0 and K_1 is used. Imposing $n = 2$ within (19.3.7), multiplying by ρ and integrating from 0 to 1, after some algebra the expression for K_2 is obtained as

$$K_2(\zeta, \tau) = \frac{1}{P_e^2} - 2 \int_0^1 f_1 \bar{v}(\rho, \zeta) \rho d\rho, \tag{19.3.33}$$

which may be simplified in

$$K_2(\zeta, \tau) = \frac{1}{P_e^2} + \frac{\bar{v}_0(\zeta)}{192} - 2 \int_0^1 f_{1t} \bar{v}(\rho, \zeta) \rho d\rho. \tag{19.3.34}$$

Note that in the limit of $\tau \to \infty$ ($\tau > 0.5 \times R_e^2/D_m$) the classical solution by Taylor and Aris can be recalled, where the effective longitudinal diffusion coefficient D_{eff} is given as

$$D_{eff} = \frac{R_e^2 \bar{v}_0^2(\zeta)}{D_m} K_2 \tag{19.3.35}$$

whereas in general D_{eff} would depend also on time τ.

It is important to note that, in the original formulation by Gill and Sankarasubramanian (1970), where for the first time the idea of a time-dependent effective diffusion was introduced, the auxiliary functions K_i were only depending on time τ. In the present formulation, the fluid velocity is no longer constant along the capillary because of its lateral permeability which induces a continuous reduction in flow velocity with ζ. Consequently, the auxiliary functions K_i would in general depend on ζ too. And, in particular, the problem would be determined if the velocity field in the capillary is known.

19.3.6 The velocity distribution (effect of boundary depletion of the solvent)

Recalling the dimensionless variables (19.3.2) and introducing the nondimensional pressure

$$p = \underline{p} D_m / (4\mu v_0^2) \qquad (19.3.36)$$

the classical governing equation for the laminar flow in a circular pipe of radius R_e is given by

$$\rho \frac{\partial p}{\partial \zeta} = \frac{1}{4} \frac{\partial}{\partial \rho} \left(\rho \frac{\partial \bar{v}}{\partial \rho} \right), \qquad (19.3.37)$$

with μ being the dynamic viscosity of the fluid and $\underline{p}$ the dimensional pressure within the capillary. Imposing the no-slip condition at the wall ($\bar{v}(1, \zeta) = 0$ and the symmetry condition at the center line ($\partial w(0, \rho)/\partial \rho = 0$), with the assumption that the gradient of pressure along the longitudinal direction is constant, the classical Poiseuille parabolic velocity distribution is readily recovered as

$$\bar{v}(\rho, \zeta) = -(1 - \rho^2) \frac{dp}{d\zeta}, \qquad (19.3.38)$$

from which the nondimensional centerline velocity $v_0(\zeta)$ is derived:

$$\bar{v}_0(\zeta) = \bar{v}(0, \zeta) = -\frac{dp}{d\zeta}; \qquad (19.3.39)$$

while the dimensional mean velocity v_m is given by

$$V = \frac{1}{\pi R_e^2} \int_0^{R_e} v(r, z) \, 2\pi r dr = \frac{2}{R_e^2} \int_0^{R_e} v(r, z) r dr \qquad (19.3.40)$$

and the nondimensional mean velocity $\bar{v}_m$

$$\bar{v}_m = 2 \int_0^1 \bar{v}(\rho, \zeta) \rho d\rho = 2 \int_0^1 -(1 - \rho^2) \frac{dp}{d\zeta} \rho d\rho = \frac{1}{2} \frac{dp}{d\zeta}. \qquad (19.3.41)$$

If the walls of the capillary are permeable to the solvent, there would be fluid leaking across the walls leading to a continuous reduction of the flow rate along the channel. Still assuming that the fluid lateral flux does not modify the velocity profile within the channel which still obeys to the Poiseuille parabolic distribution, i.e. the hypothesis of monodimensional flow still holds true. Mass continuity for an incompressible flow imposes that

$$\frac{\partial Q}{\partial z} + v_p \lambda_p = 0, \tag{19.3.42}$$

where Q is the volume flow rate, defined as

$$Q = \int_0^{R_e} v\, 2\pi r\, dr = V\pi R_e^2, \tag{19.3.43}$$

that, in nondimensional terms, has the form:

$$\Theta = \int_0^1 \bar{v}\, 2\pi \rho\, d\rho = \frac{\pi}{2}\frac{dp}{d\zeta}, \tag{19.3.44}$$

while $\lambda_p = 2\pi R_e$ is the lateral profile of the wall, and v_p the perfusing velocity derived from Darcy's law as

$$v_p = -L_p(\pi_i - p); \quad L_p = \frac{k}{\mu\delta}, \tag{19.3.45}$$

where L_p is the vascular hydraulic conductivity expressed as a function of the lateral thickness δ and the permeability k of the capillary wall; π_i is the interstitial fluid pressure (IFP). The mass continuity can be then rephrased in nondimensional terms as

$$\frac{\partial \Theta}{\partial \zeta} - L_p(\pi_i - p) = 0 \tag{19.3.46}$$

which, through (19.3.44), obtains the partial differential equation that dictates the change in pressure along the channel (whose length is l)

$$\frac{\partial^2 p}{\partial \zeta^2} + \left(\frac{\Pi}{\zeta_l}\right)^2 (1-p) = 0; \quad p = \frac{p}{\pi_i}, \quad \zeta_l = \frac{l \times D_m}{R_e^2 v_0}, \quad \Pi = \zeta_l \sqrt{\frac{2\hat{L}_p}{\pi}}; \tag{19.3.47}$$

provided that the following holds true

$$\Theta = \frac{Q}{v_0 R_e^2}, \tag{19.3.48}$$

$$\hat{L}_p = 8\pi R_e^2 \frac{L_p}{D_m^2} v_0^2 \mu. \tag{19.3.49}$$

Solving with the boundary conditions

$$\begin{aligned} p(0) &= p_0 \text{ inlet pressure} \\ p(\zeta_l) &= p_1 \text{ outlet pressure} \end{aligned} \tag{19.3.50}$$

the pressure distribution along the channel is finally derived as

$$p(\zeta) = \frac{1}{2}\Big[\left(-e^{(\zeta_l-\zeta)\Pi/\zeta_l} + e^{(\zeta+\zeta_l)\Pi/\zeta_l}\right)(p_1 - 1) + \\ + \left(e^{2(\zeta_l-\zeta)\Pi/\zeta_l} - e^{\zeta\Pi/\zeta_l}\right)(p_0 - 1) + e^{2\Pi} - 1\Big](\coth[\Pi] - 1). \tag{19.3.51}$$

The effective velocity distribution may be finally obtained as

$$\bar{v}(\rho, \zeta) = \frac{-\pi_i \frac{dp}{d\zeta}(1-\rho^2)}{-\pi_i \frac{dp}{d\zeta}\big|_{\zeta=0}}$$

$$= e^{-\zeta\Pi/\zeta_l} \times$$

$$\times \frac{\left[e^{\Pi}(p_1-1) + e^{(\zeta_l+2\zeta)\Pi/\zeta_l}(p_1-1) - e^{2\Pi}(p_0-1) - e^{2\zeta\Pi/\zeta_l}(p_0-1)\right]}{2e^{\Pi}(p_1-1)-(p_0-1)(1+e^{2\Pi})} \quad (19.3.52)$$

$$\times (1-\rho^2)$$

$$= e^{-\zeta\Pi/\zeta_l} \frac{\left[1 + e^{2\zeta\Pi/\zeta_l} - \Omega\left(1 + e^{2(\zeta-\zeta_l)\Pi/\zeta_l}\right)e^{\Pi}\right]}{2 - e^{\Pi}(e^{-2\Pi}+1)\Omega}(1-\rho^2),$$

where $\Omega = (p_0 - \pi_i)/(p_1 - \pi_i)$ is the pressure parameter, while Π is the permeability parameter as defined in (19.3.47). From (19.3.52) it appears that the permeability of the walls does not modify the Poiseuille characteristic velocity profile along the cross-section of the capillary; nevertheless it induces a reduction in velocity along ζ. In consideration of these results, the variables of the model of diffusion can be determined, and are shown in Section 19.4.

19.4 The concentration profile

As shown in (19.3.35), the diffusive term K_2 is proportional to the effective diffusion coefficient. In an impermeable capillary, the diffusive term K_2 grows with time along the capillary as shown by Gill and Sankarasubramanian (1970). This is shown in Fig. 19.4.1, which gives the contour plot of K_2 as a function of time $\tau(0, 0.5)$ and position along the capillary $\zeta(0, 1)$. Note that as time increases, the solution for K_2 tends to a constant asymptotic value coinciding with that of Taylor and Aris.

In Fig. 19.4.2 the same contour plot is shown for a nonzero capillary wall permeability ($\Pi = 2$). As predicted in Decuzzi et al. (2006a,b), the effective diffusion coefficient D_{eff}, and thus K_2 is not uniform along the capillary: it reduces from the inlet of the capillary,

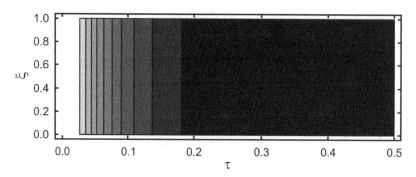

Fig. 19.4.1 K_2 contour plot ($\zeta \in [0, 1]$, $\tau \in [0, 0.5]$, $\Pi = 0$, $\Omega = -2$)

MODELING IN CANCER NANOTECHNOLOGY

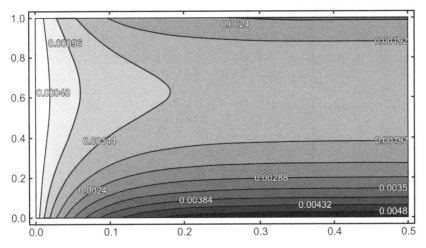

Fig. 19.4.2 K_2 contour Plot ($\zeta \in [0, 1]$, $\tau \in [0, 0.5]$, $\Pi = 2$, $\Omega = -2$)

reaches a minimum value and then increases again as the outlet of the capillary is approached (see also (19.2.15) and the discussion in Section 19.1). And this same behavior is shown at each time interval. Again it is verified that the asymptotic solution, in this case coinciding with that derived in Decuzzi et al. (2006a,b), is reached after a sufficiently large time $\tau > 0.5$. As Π increases, the variation of K_2 along the capillary becomes steeper and a central area of the capillary can be identified where the beneficial effect of the convection on the longitudinal dispersion of the solute particles is null.

19.4.1 The mean dimensionless concentration Ψ_m

In Fig. 19.4.3 the mean concentration profile is shown versus the nondimensional longitudinal coordinate ζ and for different values of the nondimensional time, namely $\tau = 0, 0.1, 0.5, 1$, while the permeability and pressure parameters are held constant as $\Pi = 1$ and $\Omega = -2$. As time increases, the centroid of the distribution gradually moves to the right, while the

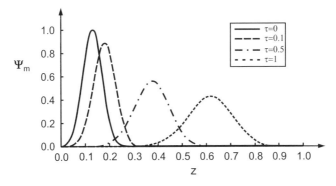

Fig. 19.4.3 Mean concentration profile at different time steps ($\zeta \in [0, 1]$ $\tau = \{0, 0.1, 0.5, 1\}$ $\Pi = 1$, $\Omega = -2$)

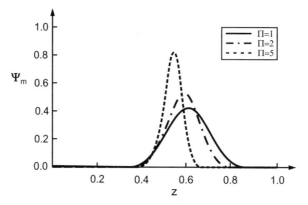

Fig. 19.4.4 Mean concentration profile at different values of the permeability parameter Π ($\zeta \in [0, 1]$, $\tau = 1$, $\Omega = -2$)

concentration flattens more and more, meaning that the bolus of solute is progressively transported downstream with time, experiencing a dispersion along ζ due to the nonuniform velocity profile of the flow, and described in nondimensional terms by K_2. Notably, the peak of concentration at the release ($\tau = 0$) is located at $\zeta_1 = 0.127$, whereas after one unit time it moves to $\zeta_2 = 0.610$, with the difference $\zeta_2 - \zeta_1 \sim 0.483$ less than 0.5, that would instead have been predicted using the Taylor and Aris model (in that, according to the mentioned model, the peak of mean concentration would move, in the time $\tau = 1$, of $\Delta \zeta = \tau \, v_m = 0.5$). Evidently, this is due to permeability of the capillary, that induces an overall delay on the convection–diffusion of the solute.

To illustrate the effect of the permeability of the capillary on the overall dispersion properties, the mean concentration Ψ_m is shown in Fig. 19.4.4 at different values of Π ($\Pi = 1, 2, 5$) and at fixed $\Omega = -2$. Although the traces are derived at the same time $\tau = 1$, they clearly do not coincide: the larger the value of permeability, the more the dispersion is retarded or, in other words, the concentration is closer to the point of release, and less spread out. The same effects, but far less pronounced, can be ascribed to the pressure parameter Ω, as shown in Fig. 19.4.5 ($\Omega = -2, -12$; $\Pi = 1$; $\tau = 1$).

19.4.2 The local dimensionless concentration Ψ

The local concentration Ψ is derived according to (19.3.3) truncated at the first order and is shown in Fig. 19.4.6 (see color plate) as a function of the dimensionless radius ρ and longitudinal coordinate ζ, at the time $\tau = 0.4$. The permeability parameter Π and the pressure parameter Ω are hold constant as $\Pi = 1$ and $\Omega = -2$. To some extent the concentration resembles a wave with its front traveling faster downstream along the centerline of the capillary, and its tails following the peak of concentration with some delay. This is due to the nonuniform velocity profile along ρ showing a maximum at $\rho = 0$ (center of the capillary) and being instead zero at $\rho = 1$ (boundary of the capillary). As a consequence, the bolus of nanoparticles either (i) cluster around the centerline or (ii) aggregate near the borders of the channel, depending on the particular cross-section under study (downstream with respect to the peak of concentration ($\Psi_{m,max}$) the first behavior is observed, 19.4.6). Notice that this

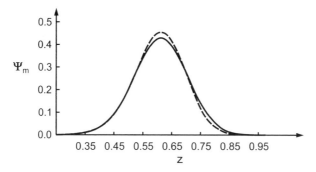

Fig. 19.4.5 Mean concentration profile at different values of the pressure parameter Ω ($\zeta \in [0, 1]$, $\tau = 1$, $\Omega = \{-2 = \text{solid line}; -12 = \text{dashed line}\}$, $\Pi = 1$)

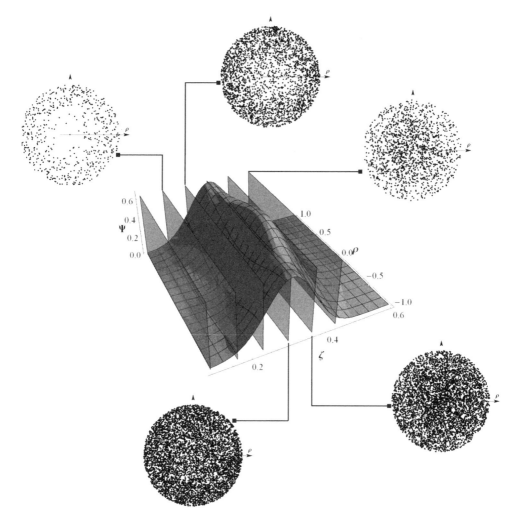

Fig. 19.4.6 The local concentration Ψ ($\tau = 0.4$; $\Pi = 1$, $\Omega = -2$) (see Plate 33)

mechanism is mathematically described within (19.20) by the term f_1 times $\partial \Psi_m/\partial \zeta$: when the mean concentration has a maximum, that is $\partial \Psi_m/\partial \zeta = 0$, then $\Psi \equiv \Psi_m$ regardless of the transverse coordinate ρ; elsewhere the function f_1 dominates, and Ψ would be in general different from Ψ_m.

The function f_1 depends, among others, on the centerline velocity v_0 along ζ, which means that a change in Π or Ω would locally affect the concentration as well. In Fig. 19.4.7 a comparison is made between two different conditions of permeability, namely (i) $\Pi = 1$ and $\Omega = -2$, and (ii) $\Pi = 5$ and $\Omega = -2$. In the first case, the local concentration Ψ_1 is derived at the time $\tau_1 = 0.8$ and is shown in Fig. 19.4.7 versus the radial coordinate ρ (solid line); in the second case, the concentration Ψ_2 (dashed line) is determined for $\tau_2 = 0.91$, and the time at which the centroids of the distributions Ψ_{m1} and Ψ_{m2} along ζ coincide (notice that $\tau_1 \neq \tau_2$ necessarily, and in particular $\tau_2 > \tau_1$ – due to the increased values of permeability, Ψ_{m2} experiences a delay in dispersion). In both cases the longitudinal coordinate is chosen as $\zeta = 0.47$; that is the section where $\partial \Psi_{m2}/\partial \zeta$ attains a maximum ($\partial^2 \Psi_{m2}/\partial \zeta^2|_{\zeta=0.47} = 0$ and $\partial^3 \Psi_{m2}/\partial \zeta^3|_{\zeta=0.47} > 0$) meaning that, according to (19.3.3), within that section the local concentration Ψ_2 would most deviate from the mean. Although Ψ_2 is higher than Ψ_1 everywhere within the section (and this well agrees with the above derived results for the diffusion coefficient K_2, which decreases as Π increases), the concentration Ψ_2 is more uniformly distributed along ρ than Ψ_1, and the deviations from the mean are less important and significant. This is evidently due to the increased effects of permeability – the perfusion of the solvent through the walls redistribute the concentration, reshaping the distribution of solute in the capillary and reducing the gradients of concentration along the radius ρ. This effect is also clear from Fig. 19.4.8 that illustrates the function f_1 versus ρ at different values of permeability ($\Pi = 0, 2, 5$; $\Omega = -2$), and different cross-sections ($\zeta = 0, 0.2, 0.5, 0.8$), while the time being held constant as $\tau = 0.5$. The higher the values of permeability, the lower the modulus of f_1 everywhere in the channel, and less sensible the differences between the local and average concentration. Notice that at the center of the channel (where the fluid is most likely stagnant) these effects are more dramatic, whereas in the close proximity of the inlet they are negligible, up to be null at the limit $\zeta \to 0$ (at the entrance of the channel the function f_1 is invariant whatever the value of permeability Π).

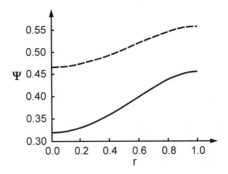

Fig. 19.4.7 The local concentration profile versus the radius ρ (solid line: $\tau = 0.8$; $\Pi = 1$; dashed line: $\tau = 0.91$; $\Pi = 5$; both curves: $\Omega = -2$ and $\zeta = 0.47$)

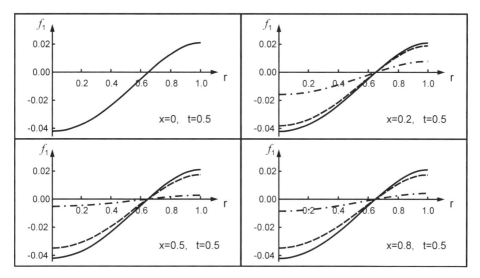

Fig. 19.4.8 The shape function f_1 versus the radius ρ at different cross-sections ($\tau = 0.5$; $\Omega = -2$; solid lines: $\Pi = 0$; dashed lines: $\Pi = 2$; dashed-dotted lines: $\Pi = 5$)

19.5 Comments and discussions of the analytical models and solutions

Stemming from the generalized dispersion model (Gill 1967, Dash et al. 2000, Nagarani et al. 2004), the unsteady dispersion of a solute in a permeable channel was derived in terms of the dimensionless effective diffusion coefficient K_2. It was found that for a given set of permeability parameters different from zero, K_2 increases with time up to a value that depends on the position ζ within the channel, and that can never be higher than the theoretical limit $K_{2,max} = D_{eff0} \times D_m/R_e^2 v_0^2$ where D_{eff0} is the Taylor and Aris diffusion coefficient derived at the entrance of the channel. In general, K_2 would be lower in the central regions of the capillary, where the velocity of the fluid dramatically reduces, and the higher the permeability, the smaller the dimensionless diffusion coefficient. Nevertheless, whatever the longitudinal coordinate ζ, or the permeability parameters Π and Ω, the time employed to reach the steady-state regime is the same ($\tau_{steady} = 0.5 \times R_e^2/D_m$), meaning that an increased leakage would not modify the coefficients K_1 and K_2 in time (but through K_1 and K_2 the bolus of solute would experience different histories of dispersion). Most important, it is found that the perfusion of the solvent at the walls would uniformly redistribute the concentration along the radius of the channel.

As discussed in Decuzzi et al. (2006a,b), and Gentile et al. (2007a,b,c), in a network of capillaries a solute would most likely follow the path presenting the largest effective diffusivity. Based on these theoretical findings it may be concluded that a bolus of nanovectors would preferentially move in larger vessels (where high Reynolds number flows occur, $V = O(1\,mm/s)$) rather than in small, leaky capillaries of the tumor districts (with small blood velocities, $V = O(100\,\mu m/s)$ or less). Also, the quasi-uniform radial distribution of the solute in permeable capillaries represents a novel biological barrier. Margination and extravasation of nanovectors is in fact hindered (or, at least, it is not favorable), and only a

428   COMPUTER MODELING IN BIOENGINEERING

small amount of such nanocarriers (those in close proximity of the walls) would be candidate to sediment on the surface and extravasate.

However, the analysis presented so far is approximate in that it relies on the strong assumption that the particles are sufficiently small to have the same velocity of water molecules: as it stands, this approach completely disregards the physical, chemical and geometrical properties of the nanovectors constituting the solute. Considering the above-mentioned properties is equivalent to introducing into the problem more degrees of freedom that may be suitably tailored to enhance the performance of these nanocarriers. For instance, in recent experiments (Gentile *et al.*, 2007a,b,c) it is shown that in a flow chamber system and under the influence of a gravitational field, the number of marginating particles over time increases if discoidal or quasi-hemispherical particles are used in place of silica spheres. And that would demonstrate that nonspherical inertial particles would perform better than classical spherical particles, pertaining to drug delivery and bio-imaging.

19.6 Numerical modeling of particle motion within capillary

In this section we present computer modeling of particle motion of circular and elliptical particles in microchannels. A brief description of the loose coupling procedure is given, assuming the finite element (FE) method (Section 7.6) or element-free Galerkin method (Section 8.5). Trajectories of these two particle types are shown as examples.

19.6.1 Computational procedure

The motion of a solid particle within a fluid is considered. It is assumed that fluid flow and particle motion occur within a plane $x-y$, that the solid has high rigidity (practically undeformable under the action of the fluid) and that the fluid is Newtonian and incompressible. Fluid flow within a 2D channel with rigid walls is driven by a difference in pressure between the inlet and outlet of the channel, Fig. 19.6.1a. A particle moves due to fluid mechanical forces acting on its surface.

The fluid and solid (particle) domains are discretized by generating two independent meshes for either FE or EFG models, as symbolically shown in Fig. 19.6.1a. Then, according to the loose coupling concept, we solve fluid flow for the current time step, i.e. calculate fluid velocities and pressures at the end of the time step using the position and velocities on the particle surface at the start of the time step. With the forces at the fluid FE nodes (or EFG free points) along the polygon surrounding the particle, the forces acting in the particle surface are evaluated. As shown in Fig. 19.6.1b, for each segment of the fluid polygon we calculate the distributed load. For example, for fluid nodes 1 and 2 and the segment L_{F12} we have the continuous load $\mathbf{q}_{F12}$:

$$\mathbf{q}_{F12} = (\mathbf{F}_{F1} + \mathbf{F}_{F2})/L_{F12} \qquad (19.6.1)$$

where $\mathbf{F}_{F1}$ and $\mathbf{F}_{F2}$ are nodal point fluid forces at nodes 1 and 2. On the other hand, we have the polygon at the particle surface and for node 2 of the solid let the closest two nodes at the fluid polygon be nodes 1 and 2. Then, the force $\mathbf{F}_{s2}$ at node 2 of the solid is:

$$\mathbf{F}_{s2} = \mathbf{q}_{F12}(l_{s1} + l_{s2})/2 \qquad (19.6.2)$$

MODELING IN CANCER NANOTECHNOLOGY

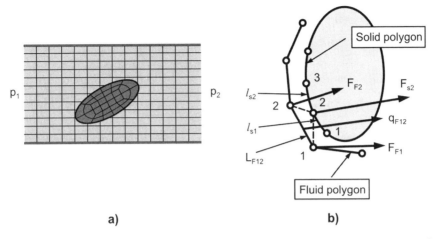

Fig. 19.6.1 Fluid flow within a 2D channel with motion of a stiff particle due to action of fluid. (a) FE model; (b) Forces of fluid acting on the particle

where l_{s1} and l_{s2} are lengths of the segments belonging to node 2 of the solid. With nodal forces acting on the particle nodes we determine increments of displacements of the particle by solving a system of differential equations of motion of the form (5.2.7) (see Section 5.3).

The new position of the particle and velocities at the particle surface are used as the boundary conditions for the fluid and a new solution for the fluid is calculated for the current time step. The process of transferring boundary data from fluid to solid and then from solid to fluid continues until the difference between the two solutions in the solid–fluid loop is small.

19.6.2 Example – trajectories of spherical and elliptical particles

Using the above described procedure we calculated trajectories of spherical (i.e. circular in 2D) and elliptical particles for steady-state conditions at the inlet and outlet boundaries of a channel. It is assumed that initially the fluid flow field is uniform and corresponds to a steady solution. Initial velocities of the particle are equal to zero.

The data used in this example are: channel diameter $D = 10\,\mu m$, fluid density $\rho = 1050\,kg/m^3$, fluid viscosity $\mu = 3.675 \times 10^{-3}\,Pa\,s$, maximal fluid velocity at inlet $v_0 = 150\,\mu m/s$, circle diameter $d = 2\,\mu m$, ellipse diameters $a = 2.39\,\mu m$ and $b = 1.33\,\mu m$.

The shape of the trajectory for the circular particle corresponds to the analytical solutions (e.g. Goldman *et al.* 1967a,b). The circular particle rotates and its center moves along the channel. In this example the rotation is in the clockwise direction since velocities of points on the surface closer to the channel axis have larger velocities. The trajectory is ultimately parallel to the channel axis with an offset which does not depend on the initial particle position (see Fig. 19.6.2a).

On the other hand, the elliptical particle has an oscillatory trajectory and crosses the central line of the channel, with rotation which changes the direction during motion (Fig. 19.6.2b). This shape of the particle trajectory is of particular interest in the process of particle margination, since it can provide good conditions for drug delivery within tumor blood vessels (see discussion in Section 19.5; also Decuzzi *et al.* 2005).

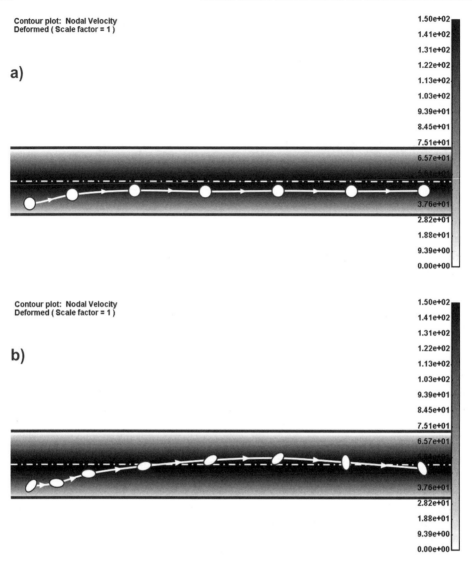

Fig. 19.6.2 Trajectories of particles in a microchannel (and fluid velocity field, velocity in $\mu m/s$). (a) Circular particle; (b) Elliptical particle

References

Ananthakrishnan, V., Gill, W.N. & Barduhn, A.J. (1965). Laminar dispersion in capillaries: Part I. Mathematical analysis, *AIChE J.*, **11**, 1063–72.

Aris, R. (1956). On the dispersion of a solute in a fluid flowing through a tube, *Proc. Roy. Soc. Lon. A*, **235**(1200), 67–77.

Dash, R.K. Jayaraman, G. & Mehta, K.N. (2000). Shear-augmented dispersion of a solute in a Casson fluid flowing in a conduit, *Ann. Biomed. Eng.*, **28**, 373–85.

Decuzzi, P., Causa, F., Ferrari, M. & Netti, P.A. (2006b). The effective dispersion of nanovectors within the tumor microvasculature, *Ann. Biomed. Eng.*, **34**(4), 633–41.

Decuzzi, P. & Ferrari, M. (2006a). The adhesive strength of non-spherical particles mediated by specific interactions, *Biomaterials*, 27(30), 5307–34.

Decuzzi, P. & Ferrari, M. (2007a). The role of specific and non-specific interactions in receptor-mediated endocytosis of nanoparticles, *Biomaterials*, **28**(18), 2915–22.

Decuzzi, P., Gentile, F. *et al.* (2007b). Flow chamber analysis of size effects in the adhesion of spherical particles, *Int. J. NanoMed.*, **2**(4), 1–8.

Decuzzi, P., Lee, S., Bhushan, B. & Ferrari, M. (2005). A theoretical model for the margination of particles within blood vessels, *Ann. Biomed. Eng.*, **33**(2), 179–90.

Ferrari, M. (2005a). Cancer nanotechnology: opportunities and challenges, *Nature Rev. Cancer*, **5**, 161–71.

Ferrari, M. (2005b). Nanovector therapeutics, *Curr. Opin. Chem. Biol.*, **9**, 343–6.

Gentile, F., Chiappini, C., Fine, D. *et al.* (2007a). The margination dynamics of nonspherical inertial particles in a microchannel, submitted to *BioMechanics*.

Gentile, F., Curcio, A., Indolfi, C., Decuzzi, P. & Ferrari, M. (2007b). Margination dynamics of nanoparticles in a microfluidic chamber and the efficiency of delivery, submitted to *BioMaterials*.

Gentile, F., Ferrari, M. & Decuzzi, P. (2007c). The longitudinal transport of nanoparticles within a casson fluid in a permeable capillary, submitted to *Ann. Biomed. Eng.*

Gentile, F., Ferrari, M. & Decuzzi, P. (2007). Transient diffusion of nanovectors in permeable capillaries, *J. Serbian Soc. Comp. Mech.*, **1**(1), 1–19.

Gill, W.N. (1967). A note on the solution of transient dispersion problems, *Proc. Roy. Soc. Lon. A*, **298**, 335–9.

Gill, W.N. & Sankarasubramanian, R. (1970). Exact analysis of unsteady convective diffusion, *Proc. Roy. Soc. Lon. A*, **316**, 341–50.

Goldman, A.J., Cox, R.G. & Brenner, H. (1967a). Slow viscous motion of a sphere parallel to a plane wall. I. Motion through a quiescent fluid, *Chem. Eng. Sci.*, **22**, 637–51.

Goldman, A.J., Cox, R.G. & Brenner, H. (1967b). Slow viscous motion of a sphere parallel to a plane wall. II. Couette flow, *Chem. Eng. Sci.*, **22**, 653–60.

Heldin, C.-H., Rubin, K., Pietras, K. & Ostman, A. (2004). High interstitial fluid pressure – an obstacle in cancer therapy, *Nature Rev. Cancer*, **4**, 806–13.

Imai, K. & Takaoka, K. (2006). A comparing antibody and small-molecules therapies for cancer, *Nature Rev. Cancer*, **6**, 714–27.

Latini, M. & Bernoff, A.J. (2001). Transient anomalous diffusion in Poiseuille flow, *J. Fluid Mech.*, **441**, 399–411.

Li, K.C.P., Pandit, S.D., Guccione, S. & Bednarski, M.D. (2004). Molecular imaging applications in nanomedicine, *Biomed. Microdevices*, **6**, 113–16.

Lighthill, M.J. (1966). Initial development of diffusion in Poiseuille flow, *J. Inst. Math. Appl.*, **2**, 97–108.

Minchinton, A.I. & Tannock, I.F. (2006). Drug penetration in solid tumors, *Nature Rev. Cancer*, **6**, 583–92.

Nagarani, P., Sarojamma, G. & Jayaraman, G. (2004). Effect of boundary absorption in dispersion in a casson fluid flow in a tube, *Ann. Biomed. Eng.*, **32**, 706–19.

Neri, D. & Bicknell, R. (2005). Tumor vascular targeting, *Nature Rev. Cancer*, **5**, 436–46.

Sankarasubramanian, R. & Gill, W.N. (1973). Unsteady convective diffusion with interphase mass transfer, *Proc. Roy. Soc. Lon. A*, **333**, 115–32.

Sharp, M.K. (1993). Shear-augmented dispersion in non-Newtonian fluids, *Ann. Biomed. Eng.*, **21**, 407–15.

Taylor, G.I. (1953). The dispersion of soluble matter in a solvent flowing through a tube, *Proc. Roy. Soc. Lon. A*, **219**, 186–203.

Trepel, M., Arap, W. & Pasqualini, R. (2002). *In vivo* phage display and vascular heterogeneity: implications for targeted medicine, *Curr. Opin. Chem. Biol.*, **6**, 399–404.

Index

A-bands, 228
abdominal aorta aneurism (AAA), 265
acceleration; vector, 18, 100, 103
accuracy of the solution, 117
acetylcholine, 230
actin; filament, 228–229, 334
action of surfactant, 218–219
action potential, 230, 289
activation; function, 237, 281, 291, 305, 341, 342, 347
activation threshold, 242
adductor canal, 279
adherent cell, 333, 346
ADP, 229
advancements in bioengineering, 174
advancements in computer modeling, 177
advection–diffusion equation, 413, 415
adventia, 255
agonist; transport, 302, 303
airway epithelium, 350
airway remodeling, 352
albumin, 297
albumin concentration, 300
albumin transport, 299, 301
Almansi strain, 43
alpha motoneurons, 230
alveolar tissue, 202, 203
amplitude vector, 104
analytical solution, 57, 63, 168
angular frequency, 103
annulus, 315, 328

aorta arch, 262
apparent (equivalent) viscosity, 256
apparent density, 185, 186
apparent diffusion coefficient, 412
arbitrary Lagrangian–Eulerian (ALE) formulation, 135, 136
area ratio, 207, 213, 218
arterial system, 252
arterial wall, 256, 278, 280
arterioles, 252
artery; bifurcation, 38, 296
arthritis, 315
articular cartilage, 313
assemblage of finite elements, 77
asthma, 350, 352
asthmatic airway, 352
atherosclerosis, 273, 295
ATP, 228
atrium, 288
attachment, 346
attractive force, 309, 310
autocrine ligand concentration, 376
autocrine loop, 351, 352, 361
average velocity, 409
axial extension, 343
axial force, 110, 190
axial stress, 33, 34, 282
axial stretch, 222
axial symmetry, 28
axial velocity, 139, 275
axisymmetric conditions, 400
axisymmetric element, 87, 89, 139, 291, 299, 304

434 INDEX

balance equation of diffusion, 354
balance of linear momentum, 18, 58, 163, 319, 340
balance of mass, 56, 163, 304
bar; structure, 73, 110
barrier to diffusion, 375
basal boundary, 363
base vectors, 166, 208
basis size, 166
bead; indentation, 344
bending; of cantilever, 85, 118
bending stiffness, 190
biaxial curve, 204, 210, 211
biaxial loading, 204
biaxial model, 203, 209, 218, 219
biaxial stress–strain state, 204
biaxial test, 203, 204
biconcave RBC, 344
biconcave shape, 343, 344
binding force, 308
binding sites, 229
bioengineering models, 175
bioinformatics, 175
bioinstrumentation, 175
biological barriers, 178
biological fluids, 58
biological membrane, 46, 203, 207, 209, 220
biological pathway, 376
biological soft tissue, 201
biomaterials, 175
biomechanics, 175, 176
biomedical applications, 412
bionics, 176
Biot number, 381, 390, 392, 395, 400
biphasic FE model, 343
biphasic finite element, 343
biphasic medium, 345
biphasic model, 339, 342
bisection method, 386
bladder, 201
blood, 251, 253
blood density, 263, 300
blood flow, 254
blood pressure, 38, 251, 279
blood velocity; field, 258, 305
blood vessel, 222, 255
blood vessel deformation, 260
blood viscosity, 253
blood volume, 251
body force, 18, 64
Boltzmann constant, 153

bolus, 409
bone: lamellar; fiborous; cortical; trabecular; tubular; woven, 182, 183, 185
bone density, 184, 185
bone fracture, 187
bone mechanical properties, 185
bone structure, 181
bone tissue, 181, 185, 190
bounce-back reflections boundary conditions, 153
boundary conditions, 37, 79, 159, 418
boundary conditions in DPD, 153
bovine pericardium, 204
bridging scale (BS) method, 155
bronchoconstriction, 352
B-spline function, 163
bulk modulus, 29, 65, 342, 346
bulk modulus of solid; of fluid, 317, 325, 340
bulk modulus of solid skeleton, 340
Burger's viscous equation, 63

cable, 336, 338
calcium ions, 228
cancellous bone, 315, 328
cancer cells, 178
cancer nanotechnology, 407
cantilever, 118
capillary; straight circular, 257, 409, 414, 422
capillary radius, 412
capillary system, 252
capillary wall, 421
carbohydrate, 228
cardiac cycle, 263
cardiovascular system, 250
carotid artery bifurcation, 272, 273
carotid artery model, 272
Cartesian coordinate system, 6, 14
cartilage; hyaline, 314, 315
cartilage deformation, 320
cartilage model, 318, 320
cartilage permeability, 315
cartilage stiffness, 322
cartilage structure, 314, 315
cartilaginous end plate, 329
Casson fluid; relation; law, 255, 412, 413, 419
cat mesentery, 204
Cauchy formula, 17, 259
Cauchy stress; true stress, 17, 45, 114, 216
Cauchy–Green deformation tensor; left; right; modified, 42, 43, 118, 209, 216, 233
cavity flow, 160

INDEX

cell adhesion, 332
cell crawling, 345
cell cytoplasm, 345
cell cytoskeleton, 341
cell mechanics, 331
cell membrane, 342
centerline velocity, 419, 420, 426
channel section, 118, 119
characteristic equation, 48, 104
characteristic length, 381
chemical energy, 228
circular frequency, 324
circular pipe, 410, 420
circular stress, 33
circulatory system, 249, 266
circumferential strain, 89
clamped plate, 96
closed cylinder, 219
coarse-scale; velocity, 155, 156
coarse graining, 151
coefficient of chemical contraction, 317
coefficient of convection, 52
coefficients in EFG, 166
collagen, 182, 186, 202, 306, 308, 310
collagen fiber; bundles, 202
collagen wall, 306
collapse of tube, 142
column; settlement, 66, 67
comminuted fracture, 190
common boundary, 159
common carotid artery, 263
common coordinate system, 118
compliance matrix, 29
component form, 11
compressibility modulus, 66
compression therapy, 282, 285
compressive stress, 350, 352, 353, 355, 369, 372
computed hysteresis, 206
computer models, 356
concentrated solution, 56
concentration at node, 365
concentration average; mean; initial, 57, 373, 413, 417
concentration dimennsional; dimensionless; nondimensional, 415, 423, 424
concentration distribution, 57, 135
concentration gradient, 410
concentration local, 411, 415, 416, 424, 425, 426
concentration of mass total, 400
concentration of polymer, 382

concentration of protein, 400
concentration of solvent, 382
concentration profile; profiles, 359, 362, 366, 370, 371, 395, 398, 410, 422, 423, 426
concentration strain, 317
conductivity matrix, 125
configuration initial; current; deformed; reference; stress-free, 41, 45, 47, 64, 114, 118, 119, 212, 223
conjugate stress, 45
connective tissue; biological, 181, 202, 205, 227, 235
conservation of mass, 364
conservative force, 310
consistent mass matrix, 100
consistent tangent constitutive matrix, 117, 261
constitutive coefficient, 215
constitutive curve; uniaxial; biaxial, 46, 218
constitutive equation, 35, 60
constitutive law; uniaxial, 233, 345
constitutive matrix, 114, 186, 211, 212, 216, 217
constitutive relation; relations; relationships; tensor, 26, 59, 65, 114, 116, 185
contact traction, 206
continuity equation, 58, 59, 65, 131, 143, 319, 340
continuity of mass, 393, 421
continuous load, 428
continuum model, 296, 306
continuum-based method; methods, 258, 302
contour plot, 422
contractile element, 235
contractile unit, 228
control volume, 58
convection, 60, 413
convection and shedding vector, 365
convection external, 390
convection flux, 52, 125
convection matrix, 299
convection resistance, 390
convection–diffusion equation, 63, 134, 299, 424
convection–diffusion–reaction equations, 303
convective derivative, 11
convective mass transfer coefficient, 381, 390
convective transport, 412
convergence check, 211
convergence criteria, 141
convergence rate, 112, 117
conversion rate, 304
coordinate transformations, 18

coronary artery, 264
cortical bone, 182, 185, 186, 315, 329
conservative (repulsive) force, 152
Coulomb friction coefficient, 206
coupling DPD-FE, 155
coupling EFG and FE, 168
creep; deformation; response, 38, 64, 145, 201, 324
cross-bridges, 229, 234
CSK prestress, 334
curl of vector field, 9
current area, 384
current density, 318, 319, 324
cycle period, 272
cyclic displacement; internal pressure;loading; stretching, 217, 219, 221
cycling rate, 229
cylindrical biological membrane, 219
cylindrical coordinate system, 13, 14, 19, 20
cytoplasm, 334
cytoskeletal mechanics; actin; prestress; stiffness, 332, 335, 336
cytoskeleton (CSK), 228, 332, 334
cytosol, 342

damping; coefficient; matrix, 35, 100, 101, 260
Darcy's law; velocity, 64, 66, 318, 319, 339, 421
daughter branches, 38
decomposition of deformation; of kinetic energy, 44, 157
deflection, 118
deformable body, 232
deformable wall, 266, 275
deformation gradient; modified, 41, 47, 118, 209, 215
deformed shape, 96
Delfino tissue model, 216
density of material; of substance; of solid; of mixture; total, 55, 56, 64, 65, 66, 100, 145, 322, 342; 384
depletion; effects, 417, 420
detachment, 345, 346
deviatoric stress; strains, 18, 29
dextran molecules, 368
diastole, 287
diastolic; flow; pressure, 251, 273, 275
discretized domain, 168
differential equation; of motion, 99, 121, 157, 321
differential operator; 'nabla', 9, 13, 64

diffusion; and convection; axially symmetric, 51, 55, 127, 362, 363, 392
diffusion Brownian, 409
diffusion coefficient, 56, 57, 302, 351, 361, 363, 368, 369, 371, 374, 375, 410
diffusion coefficient constant; variable, 388
diffusion coefficient of water, 380, 398, 40
diffusion convective, 411
diffusion equation; equations, 60, 127, 365
diffusion internal; lateral ; with convection, 60, 380, 392, 410
diffusion resistance, 390
diffusion–transport equation, 298
diffusive regime; -convective, 370
diluted species; solution; mixture, 56, 60, 304
discrete particle method, 147
discrete system, 113
discretization; of fluid, 72, 156, 165
dispersion; longitudinal, 411, 414, 423
displacement field, 23
displacement increment, 109, 110, 112, 114, 115
displacement of solid; of fluid; of mixture, 64, 320, 328
displacement vector, 72, 77, 113
dissipative force, 152, 308
dissipative particle dynamics (DPD), 151
distortion of material, 20
distribution function, 418
divergence of vector field; of tensor field; of velocity, 9, 59
divergence theorem, 10
domain of influence; in DPD, 152, 165, 166
DPD method; equations; particle, 158, 296, 308, 310
drug delivery, 429
dummy index, 5, 53
duration of collapse, 361
dyadic notation; multiplication; product, 5, 9, 12
dynamic analysis, 99
dynamic diffusion; equation, 356, 357, 362
dynamic equations of motion, 18
dynamic hip; device; implant, 194, 195, 197, 199
dynamic response, 106, 328
dynamic stiffness, 322, 323
dynamic viscosity, 59, 259, 300, 420

effective diffusion; coefficient; diffusivity, 304, 305, 414, 419, 422, 427
effective longitudinal diffusion, 410
effective spring constant, 308

effective stress, 65, 191, 196, 345
EFG cell; point, 168
EGF receptor, 351, 352
eigenvalue analysis, 104
eigenvalues; eigenvector, 8, 12, 104
Einstein summation convention, 5
ejection phase, 289
elastic constitutive matrix, 26, 317, 340
elastic matrix, 27, 39, 215
elastic modulus, 27, 186, 324, 332, 397
elastin; bundles; fiber, 202
electrical conductivity, 318
electrical potential, 318, 320, 325
electrical-to-mechanical transduction, 324
electrokinetic coupling, 316, 318, 319, 320, 327, 328
electrokinetic coupling coefficient, 318
electrokinetic transduction, 322
element balance equation for diffusion, 127
element forces, 78
element geometry interpolation, 81
element resistance force, 76
element stiffness matrix; mass matrix, 76, 88, 100, 101
element surface flux, 128
element-free Galerkin (EFG) method, 164
element-free points, 165
elongational flow, 379, 389, 390, 396, 398
embolization, 270
endocardium, 201
endomysium, 227
endoplasmic reticulum, 228
endoprosthesis, 277
endurance, 240
energy loss within cycle, 219
energy source, 228
engineering shear strains, 20
epimysium, 227
equation of balance for heat conduction, 124
equation of motion, 61, 186, 260
equilibrium equation; of finite element, 18, 64, 77, 78, 232, 338
equilibrium iteration, 117, 258
error function, 385
Euclidean norm, 7
Eulerian description, 55, 60
evaporation; experiment, 380, 381, 391
exchange coefficient, 417
excitation, 230
expanding pipe, 138
expansion in plane, 23

extended Hill's model, 243
extensible protein, 202
extension force, 206, 343
external fixator, 188
external force; body force; force increment; loading, 38, 73, 101, 109, 110, 113, 143, 157, 231
external nodal force, 186, 260, 261
external pressure, 89
external skeletal fixation, 188
external virtual work, 38, 76, 113
extracellular fibrous proteins, 202
extracellular matrix, 314, 332, 408
extracellular mechanism, 376
extracellular mechanotransduction, 351, 368, 374, 376
extravasation, 408

Fåhraeus–Lindquist effect, 254
fast-twitch, 231, 242
fatigue curve; factor; rate, 240, 241
FE equations of balance, 320, 359
FE equilibrium equations, 392
FE matrices, 383, 392
FE model of diffusion and convection, 362
FE model of dynamic diffusion, 357
FE model of pan experiment, 382
FE volume, 393
femoral artery, 276
femoral neck, 195
femur; comminuted fracture, 189, 190, 192
FES, 240
fiber diameter, 395
fiber direction; unit vector, 214, 340
fiber spinning, 379, 388, 389, 392, 394, 397, 402, 404
fiber strain; stretch, 233, 339
fiber–fiber kinetics model, 206
fibrocartilagenous cartilage, 314
fibrous bone, 182
Fick's law, 55, 56, 317, 363, 382, 410
fictitious elastic configuration, 118
field problems, 121
filaments overlap, 230
filtration velocity, 300
fine scale, 157
Finger deformation tensor, 43
finite element; 3D, 71, 278, 342
finite element description, 41
finite element equations, 340
finite element node, 72

finite element stiffness matrix, 78, 88
fitness function; level, 240
flat bone, 183, 184
flow resistance, 254
flow through porous deformable media, 143
fluid compressibility, 317
fluid density, 58, 164
fluid domain, 140
fluid flow; with heat transfer; with mass transfer, 51, 129, 132
fluid polygon, 428
fluid pressure, 64, 318, 326, 339
fluid–structure interaction, 263, 268
fold-mean concentration, 372, 374
force development, 230
force function, 36
force generation, 229
force–displacement loop, 222
Fourier's law, 52, 56
four-node shell elements, 118
fracture treatment, 188
free point, 165, 166, 167
free-diffusion, 373
free-swelling, 325
frequency, 324
friction coefficient, 152
frictional force, 206
full Newton iteration, 112
Fung tissue model, 217

Galerkin method; procedure, 121, 136, 143, 258, 320, 357, 383
Gauss quadrature, 167
Gauss' theorem, 10, 123, 143, 320, 358
Gauss–Ostrogradskii theorem, 10
generalized Darcy's law, 64
generalized dispersion model, 414, 427
geometric nonlinearity, 40, 116
geometrically nonlinear only, 40, 116
geometrically nonlinear stiffness matrix, 208
global coordinate system, 78, 209
global domain, 159
gluteal muscles, 195
glycocalyx, 351, 354, 355, 357, 360, 368
Gordon curve, 234
gradient of function; of vector field, 9
gradient to mid-surface, 92
Green–Lagrange strain, 43, 45, 114, 204, 216, 217

hardening behavior, 203
Haversian canal, 182
HB-EGF, 353, 360, 368, 371, 373, 375
heart chambers, 286
heart cycle, 279
heart model, 286
heat and mass transfer, 129
heat conduction, 51, 124
heat conduction coefficient, 53, 54, 127
heat conduction governing equation, 127
heat conduction matrix, 53
heat flux; through surface, 52, 125
heat source, 52
heat transfer, 51, 60
hematocrit, 251, 253, 255
Hencky (logarithmic) strain, 118
Hill's muscle model, 234, 235
hindered diffusion; HB-EGF, 371, 373, 374
hip fracture, 194, 195
homogenous deformation, 23
Hooke's law, 26, 32, 39
hoop strain, 34
hoop stress, 39, 90, 280
human aorta, 262, 263
human bronchial epithelial cells, 351, 356, 367, 368
hydraulic permeability, 318
hyperelastic models, 45
hysteresis; of ring; resulting, 202, 203, 206, 219, 220
hysteretic action of surfactant, 222
hysteretic behavior; character; characteristic, 203, 205, 212, 215
hysteretic curve; constitutive, 205, 219
hysteretic model, 205

I-bands, 228
identity matrix, 5
imaging agents, 178
impermeable; wall, 358, 412, 416, 418
implant, 194
implicit method, 103
in vitro, 350, 352, 353, 362, 363, 370, 371
in vivo, 355, 363, 370, 371
incompressibility, 60, 364, 394
incompressible fluid; flow, 56, 59, 257, 421
incremental: analysis; equations; form; solution, 109, 116, 124, 359
incremental-iterative equations, 124, 261, 298, 383, 365

INDEX

incremental-iterative procedure; scheme; solution, 113, 208, 232
inelastic large strain, 118
inertial force, 99, 100, 117
initial conditions, 416, 417
in-plane stresses, 28
integral theorems, 8
integrated bioengineering science, 174
integration of differential equations, 101
integration of DPD equations, 154
integration parameter, 102
interaction force, 148, 151, 152, 156, 308
intermediate filament, 332, 333
internal energy, 52
internal excitation, 231
internal finite element forces, 158
internal fixation, 189
internal force, 15, 110, 111, 157, 208
internal nodal force, 76, 116, 158
internal pressure, 33, 89, 222
internal rim, 220
internal virtual work, 38, 76, 113, 114
interphase mass transfer, 412
interpolation by EFG, 165
interpolation function, 71, 74, 75, 82, 94, 123, 129, 155, 167, 357, 383
interpolation functions for pressure; for velocity, 301
interpolation matrix, 75, 82, 143, 320
interpolation of displacement field, 81
interstitial: fluid; fluid pressure, 408, 414, 421
intervertebral disc, 314, 328
intima, 255, 256
intracapsular fracture, 195, 197
intracellular substance, 181
intramedullary nail, 189, 190, 192, 193
intraventricular pressure, 290
invariants of deformation tensors, 216
inverse deformation gradient, 42, 233
inverse matrix, 5
ion concentration, 317, 322, 325
ionic diffusion; coefficient, 325, 326
Irving–Kirkwood model, 158
isochoric deformation, 215
isometric contraction, 229, 234
isoparametric element (3D), 117
isoparametric finite element, 190
isoparametric formulation, 75, 357
isoparametric function, 74
isotonic contraction, 229
isotropic: material, medium, 26, 53

isotropic membrane, 204
isotropy, 30
isovolumetric contraction, 287
iteration counter, 383
iteration loop on cycles, 215
iteration scheme; procedure, 112, 385
iterations, 140
iterative scheme, 131, 215

Jacobian matrix, 95
Jacobian of transformation, 75, 83, 212

Kelvin model; Kelvin-Voigt model, 29, 35
kernel approximation, 161
kernel support domain, 162
kinematic viscosity, 164, 391
kinematically admissible, 39
kinematics of deformation, 19, 20
kinetic energy, 157
kinetics of geometric changes, 375
Kronecker symbols, 5

Lagrangian description, 15, 58
laminar flow, 410, 411, 420
Laplacian of: scalar field; vector field, 9
Laplacian operator, 9
large blood vessel, 250, 253
large displacements, 118, 231
large strain deformation, 41
large strains, 47, 231
lateral heat conduction, 53
lateral intercellular space, 349, 362, 363
LDL concentration, 298, 301
LDL transport, 295, 298, 300, 301
leapfrog method, 150
Lees–Edwards method, 153
left basis, 43, 47, 48, 118, 216
length of finite element, 365
Lennard-Jones (LJ) potential, 148
leukocyte, 252
ligaments, 202
ligand, 178
ligand accumulation, 375
ligand concentration; concentration profile, 349, 350, 358, 365, 371, 375
ligand diffusivity, 363
ligand–receptor affinities, 374
ligand–receptor specific interactions, 408
line integrals, 384
linear base, 166, 168
linear collapse, 370

linear elastic; constitutive law; model, 26, 185, 223
linear finite element, 82, 86
linear interpolation, 210
linear stiffness matrices, 208
linear strain–displacement matrix, 115, 144, 208
linear stress–strain relationship, 202
linear viscoelasticity, 29
linearization, 137
linearized form of virtual work, 114
LIS collapse, 354, 359, 369, 375
LIS concentration; dynamics, 360
LIS domain, 363, 364
LIS model; finite element model, 355, 358, 364
LIS width, 354, 359, 363, 364, 366, 369, 371, 375
living cell, 331
loading; curve; part; regime, 203, 215, 219
local buckling, 118, 119
local coordinate system, 77
local derivative, 11, 59
local domain, 159
local membrane coordinate system, 209, 211
local shell plane, 93
logarithmic strain, 43, 117, 209, 316
long bone, 183, 187
long column, 53, 125
longitudinal diffusion; coefficient, 413, 421
loop direction, 222
loose coupling; method, 139, 140, 261, 428
low density lipoprotein (LDL), 295
lumped mass matrix, 100
lung tissue, 46

macromolecule, 412
macromolecule concentration, 297
magnetic twisting cytometry, 344
main artery, 38
margination, 429
marrow core, 184
mass concentration, 55
mass curves, 388
mass density; initial; current, 18, 45, 52, 54, 127
mass flux, 58, 383, 393, 410
mass matrix, 117, 186, 260, 299
mass transport, 297
material body, 15
material coordinates, 41
material derivative; of volume integral, 10, 59, 60, 136
material description, 58

material element, 19, 47, 92, 203
material ellipsoid;sphere; line; plane, 25, 26
material model with hysteresis, 219
material models, 26
material nonlinearity, 40, 116
material particle, 10
material point, 15, 165, 166, 167
material rotation, 43
material volume, 52
materially nonlinear only (MNO), 116
materials with memory, 29
mathematical model, 414
matrix; two-dimensional; square; diagonal, 3, 4, 13
matrix algebra; addition; multiplication; subtraction; summation, 4, 5
matrix determinant; cofactor; inverse, 5, 11
maximum force, 229
maximum rate of ligand accumulation, 371
maximum shear strain rate, 302
maximum shear stresses, 17
Maxwell model, 29, 35
Maxwellian reflection method, 153
MD algorithm, 149
MD limitation, 149, 151
MD particle, 148, 152
mean LIS concentration; molar, 360, 373
mean strain, 29
mean stress, 18
mean velocity, 279, 284, 300, 420
mean wall shear stress, 259
mechanical model, 175
mechanical power, 45
mechanical properties, 379, 389, 394, 395, 397, 398
mechanical response, 231
mechanically superior fiber, 404
mechanical-to-electrical transduction, 322, 323
mechanobiology, 176
membrane conditions, 224
membrane covered with surfactant, 212
membrane stress, 215
membrane tangential plane, 209
membrane with a hole, 220
mesh fixed in space, 129
mesh-referential time derivative, 136, 137
mesoscale particle, 151, 156
mesoscopic bridging scale method (MBS), 155
microchannel, 308, 428, 430
microtubules, 334, 335
microvelocity, 316

INDEX 441

microvessel, 254
mid-surface, 92, 94
Mindlin's plate theory, 92
mineral salt, 182
mitochondria, 228
mixture formulation, 339
modal shape, 99, 103, 104, 105
modeling field problems, 121
modified Newton iteration, 112
Mohr circle, 17, 21
molar concentration, 373
molecular diffusion, 410
molecular diffusivity, 410, 415
molecular dynamics (MD), 147
molecular size; small, 373; 407
molecular weight, 355, 360, 367, 368, 373, 391
monoclonal antibody, 407
motor unit, 230
mucous membranes, 201
multi-fiber model, 243
multi-file flow, 254
multilinear curve, 385
multiphasic material, 313
multiscale biomechanics, 177
multiscale MBS method, 160
multiscale method, 308
multiscale modeling, 155
multistage nanodevice, 178
multi-targeted therapy, 408
muscle; skeletal; striated, 227; 228, 231
muscle activation, 237, 279
muscle cell, 203, 228
muscle contraction; shortening; extension, 228, 234, 279, 282
muscle fatigue, 239
muscle fiber; bundle, 227, 242, 278
muscle material; tissue; structure, 227, 231, 279
muscle model, 290
muscle modeling, 231, 234
muscle physiology, 227
muscle recovery; relaxation, 239; 285
muscle stress; tension; force, 230, 232, 234, 235, 239
muscle tone, 231
myofibrils, 228
myosin; crossbridge, 229, 333

nanoparticle, 408, 412, 414
nanoparticle delivery, 178
nanovector concentration, 409

natural coordinates; coordinate space, 41, 82, 83, 86, 156
natural frequencies, 104, 105
Navier–Stokes equations; for a finite element, 60, 131, 135, 159
nebulin, 228
neighbor-list method, 154
Nephila clavipes spider, 379, 389, 397–398
Nephila spinning: canal, process, 402
neurotransmitter, 230
neutralization plate, 189, 190, 191–194
Newmark method, 99, 101, 103, 321
Newtonian fluid, 263, 290, 299, 409, 414428
Newton's method, 237
Nitinol; stent, 277, 280
nodal acceleration vector, 100–101, 117, 186
nodal blood pressure; velocity, 258
nodal concentration, 128
nodal coordinate vector, 82
nodal displacement; vector, 73, 74, 82, 94, 103, 144, 186
nodal flux, 128
nodal force; vector, 74, 84, 101, 130, 141, 429
nodal force due to surfactant, 209
nodal point, 72
nodal pressure vector, 129
nodal relative velocity vector, 143
nodal rotation vector, 94
nodal surface flux vector, 123
nodal temperatures, 125
nodal velocity vector, 117, 129, 156, 158
nodal volume flux vector, 125
node numbering, 80
nonlamellar bone, 182
nonlinear constitutive law; constitutive relations, 45, 116, 201, 233
nonlinear continuum mechanics, 40
nonlinear elastic material model, 45, 215
nonlinear finite element analysis, 109
nonlinear incremental analysis, 113
nonlinear strain–displacement matrix, 115, 117, 144
nonlinear strains, 114
non-Newtonian, 258, 412
normal stress; components, 17
no-slip boundary condition, 153
nucleus, 315, 328, 345, 346
number density of particles, 158
numerical tolerance, 393

one-dimensional diffusion, 57, 129
one-dimensional matrix, 18
one-dimensional model, 327
one-dimensional relationship, 30
optical tweezers, 342
orthogonal; bases; matrix, 5, 12, 13
orthogonality; condition; property; relation, 12, 13, 33, 48, 155
orthonormal, 13
orthotropic medium, 53
oscillatory shear index, 259
osteoporosis; assessment, 184, 185, 315
outlet pressure, 289
out-of-balance vector, 365
oxygen transport, 297

PAK; PAK-T, 107, 386
pan experiment, 382, 399
parabolic finite element, 86
parallel elastic element, 235
parallel screws, 194, 195
parameter space, 356, 359, 363, 366
partial density, 55, 382
particle circular; elliptical, 428, 429
particle in SPH, 162
particle mass, 158
particle motion, 428
particle surface, 428
particulate formulation, 407
partition coefficient, 381
passive state, 231
Peclet number, 371, 415
penalty method; parameter, 131
perfusing velocity, 421
perfussion chamber, 309
pericardium, 201
perimysium, 227
periodic boundary conditions, 149, 154
permeability, 325, 342, 346, 411, 414, 420, 421, 423, 426
permeability coefficient, 66, 298
permeability matrix, 64
permeability parameter, 414, 422, 424
permeable, 412, 414
permutation symbol, 5, 93
persistence length, 332
phase, 103
phosphorylation, 352, 353, 355
physical quantity, 11
physical space, 82, 86
Piola–Kirchhoff stress, 45, 204, 216

plane strain; element, 27, 28, 87
plane stress conditions; element, 30, 31, 88, 211
plaque; rupture, 295
plasma, 251, 252
plaster cast, 188
plate with hole, 89, 168
platelet accumulation; rate, 295, 303, 304, 305
platelet; activated; resting; deposited, 252, 295, 303, 307, 310
platelet activation; adhesion; aggregation, 296, 302, 306, 310
platelet concentration; deposition, 304, 306, 310
platelet flux, 303, 305
platelet mediated thrombosis, 303
plates and screws fixation, 189
plug radius, 412
Poiseuille flow, 61, 150, 154, 409, 412, 415
Poiseuille parabolic distribution, 420
Poisson's ratio; coefficient, 27, 39, 96, 325
polar decomposition theorem, 43
pole of the Mohr circle, 21, 22
polymer solution, 380, 381, 384, 389, 392
pore fluid pressure, 64
poroelastic material, 66
porosity, 64, 66, 145, 316, 321, 325, 346
porous deformable media, 51, 63
position vector, 92, 93
post-collapse steady-state solution, 366
potential function, 148
pre-collapse solution, 367
prescribed displacement, 118
prescribed temperature, 52
pressure; dimensional, 59, 420
pressure distribution, 142, 269, 421
pressure drop, 254
pressure field, 133
pressure inlet; outlet, 421
pressure parameter, 414, 422, 424
pressure profile, 269
pressure total, 317, 319, 326
pre-strain, 341
prestress; cytoskeletal, 334
prestressed structure, 334, 336
principal basis; vectors; right, 8, 12, 43, 47, 48, 118
principal direction; directions, 8, 21, 47, 49, 209, 211, 316
principal plane; planes, 17, 22
principal strain directions, 217
principal stresses, 17, 21, 118, 209, 217

principal stretch; stretches, 43, 47, 49, 118, 208, 209, 211, 316
principal value; values, 8, 48
principle of virtual work; for continuum, 37, 76, 113, 114, 143
profile of the spinning canal, 400
projection operator, 155, 157
protrusion, 345
proximal femur, 195, 197
pulmonary circulation, 249
pulsatile flow, 259, 268
pure shear; in plane, 23, 47
pure stretch, 43
pure-diffusion model, 362

quadratic base, 166
quadratic convergence rate, 112
quasi-static problem, 208
quintic spline, 163

radial displacement, 106, 218, 219
radial domain, 363, 364
radial stress, 90
radiation flux, 125
random force, 152
rate of change, 10
rate of deformation tensor, 49
rate of LIS collapse, 373
rate of shedding, 354
rate sensitivity, 368
ratio of the stresses, 204
receptor; receptors, 230, 408
receptor activation; trafficking, 376
recovery curve, 242
recovery factor, 240
recruitment, 230
red blood cell (RBC) or erythrocyte, 251, 342
relative viscosity; apparent, 254, 255
relative fluid velocity, 143, 328
relaxation; function, 29, 203, 346
renal arteries, 262
resistance force, 117, 319
resistance matrix, 117
resonant regime, 104
retrograde flow, 283
Reynolds number, 300, 390
rheological parameter, 412
rigid body displacement, 79
rigid strut, 338
rigid walls, 273, 283

rings, 219, 220
rotated coordinate system, 30
rotation displacement, 92
rotation tensor, 7, 11, 12, 43, 48
rupture risk, 266

sarcolemma, 227
sarcomere, 242
sarcoplasm, 228
sarcoplasmic reticulum, 228
scalar, 3
scaling procedure, 213
scope of bioengineering, 173
second invariant of the strain rate, 253
second moment of inertia, 190
second Piola–Kirchhoff stress, 114
semi-infinite medium, 54
semipermeable wall, 301
serial elastic element, 235
settlement, 145
shape function, 427
shear modulus, 29, 190
shear rate, 253, 305, 306, 310
shear strain, 20
shear stress, 17, 259, 263, 275
shear-augmented dispersion, 409
shedding, 357
shedding rate, 349, 356, 359, 363, 366, 367, 369, 373, 374, 37
shell conditions, 30, 31
shell constitutive matrix, 28, 32, 95
shell finite element, 91, 263, 268, 278
shell in-plane terms, 211
shell surface, 212
shell tangential plane, 92–93, 208
Sherwood number, 298, 390
short bone, 183, 184
signaling properties, 374
single file flow, 254
singular stiffness matrix, 79
six-strut tensegrity model; structure, 336, 337, 341
skew-symmetric, 21
slow-twitch, 231, 242
smooth muscle tissue, 205
smoothed particle hydrodynamics (SPH), 161
smoothing kernel function, 161
solid matrix, 314, 316
solid skeleton, 340
solid tumor, 408

solid–fluid interaction; algorithm, 139, 140, 141, 339
solid–fluid mixture, 64, 316, 346
solute, 412
solute-absorption parameter, 413
solvability condition, 417, 418
solvent, 56, 414, 42
solvent concentration; distribution; mean, 379, 380, 384, 387, 388, 391, 393, 395
solvent diffusion coefficient, 379, 380, 381, 384, 387, 390, 391
solvent removal, 379, 380, 388, 389, 390, 392, 397
solvent removal modeling, 388, 397
source term, 357
space cell, 165
space membrane, 28
space volume, 10
spatial description, 58
spatial field, 11, 58
spatial gradient, 56
spatial point, 58
species concentration, 56
specific heat, 52, 127
specific interactions, 408
spectral decomposition, 13
specular boundary conditions, 153
SPH interpolation, 162
SPH method, 308
spherical biological membrane, 217, 218
spider spinning process, 398
spin line, 390, 392, 393, 394, 395, 397
spin tensor, 21, 49
spinal cord, 315
spinal motion segment (SMS), 66, 145, 315, 327, 328
spine anatomy, 315
spinning canal, 379, 389, 397, 398, 399, 400, 402, 403
spinning dope, 389, 397
spring constant, 35
squared membrane, 221
static condensation, 27, 32
steady flow, 60
steady heat conduction, 125
steady state solution, 53, 150, 354, 355
steady-state, 357, 363, 370, 414
stenosis, 270, 300, 304, 305, 308
stent, 276, 278, 280, 281
step pressure, 106
stiffness matrix, 73, 84, 103, 113, 186, 260

straight aorta, 268
strain; small (linear), 19, 20, 114
strain energy function, 45, 201, 215
strain field, 20
strain measures, 43
strain rate, 19, 21, 30, 59, 186, 187
strains; strain tensor, 20, 95
strain-displacement matrix, 83
strains-axial components; circumferential, 21
streaming potential, 322, 323
streamline contours, 305
stress; tensor; vector; components, 15, 16, 17, 208
stress active; passive, 339, 346
stress concentration, 90, 196
stress distribution, 168, 270
stress effective; total, 316, 344
stress fiber, 332
stress in the solid phase, 64
stress increment, 211
stress integration, 116, 117, 235
stress matrix, 117
stress measures, 45
stress ratio, 210
stress recovery; relaxation, 203
stress–strain curves; relationship, 202, 397
stress–strain hysteretic loops, 206
stress–stretch curves; relations; relationships, 46, 205, 209, 210
stress–stretch loop, 214
stretch, 118, 210, 233
stretch ratio, 234
stretch tensor left; right, 43, 48
stretched biaxially, 220
strong coupling method, 139
structural axial strain, 341
structural stiffness, 79, 110, 342
structure of biological tissue, 201
strut, 338, 340
subdomain, 167
surface flux; heat, 54, 123, 358
surface integral, 10
surface tension, 207, 2012, 213, 218
surface traction vector, 259
surfactant; with hysteresis, 201, 207, 211, 218, 219
surfactant area, 207
surrounding material, 214
surrounding temperature, 52
swelling of column, 325

INDEX

swelling pressure, 316, 317, 318, 321, 322, 326, 327, 334
symmetric AAA, 268
symmetric matrix; tensor, 4, 12
symmetry conditions, 224, 416
symmetry of structure, 80
synthetic fiber, 379, 388, 389, 394, 397
system configuration, 113
system frequencies, 99, 103
system matrix, 365
system stiffness matrix, 79, 110
systemic circulation, 250, 252
systole, 287
systolic; flow; phase; pressure, 269, 266, 276, 292

tangent constitutive matrix, 208, 211, 232, 238
tangent modulus, 210, 211
tangential stress; components, 16
tangential velocity, 259
Taylor series, 110
Taylor's approximation, 410
temperature distribution; field, 53, 126
temperature gradient, 51
temperature profile, 54
tendon, 202, 227
tensile stiffness, 186
tensegrity architecture, 334, 341
tensegrity model, 334, 337, 343
tensile element, 334
tensile force, 332
tension–length relationship, 235
tension–stretch relationship, 234
tension–velocity relation, 234
tensor, 6
tensor - cross product; dot product; scalar product, 7
tensorial shear strains, 20
tensorial transformation, 6, 17
tetanic contraction, 230, 234
tetanic stress, 237
tetanized condition; state, 234, 240
TGF-alpha, 372, 373, 374, 375
therapeutic antibody, 408
thermal diffusivity, 302, 305
thick filaments, 227
thick-walled cylinder, 89
thin filaments, 227
three-dimensional (3D) finite element, 81, 126, 127, 327
thrombosis, 295

tight junction, 365
time integration, 359
time step, 124
tissue; soft, 201
tissue density, 186
tissue histeresis, 205
titin, 229
torsional force, 187
total derivative, 59
total mass of polymer solution, 384
total strain; stress, 65
total water content, 325
traction normal; tangential, 345
trajectory, 429, 430
transformation matrix, 6, 12, 32, 209
transformation of constitutive: matrix; relations, 30, 95
transformation of stiffness matrix, 75
transformation of : strains; stresses, 18, 31, 32
transformation rule, 13
transitional domain, 363, 364
transport equation, 363, 411
transposed matrix, 4
transverse concentration, 413
transverse diffusion, 413
traveling cross section, 403
tropomyosin, 228
troponin, 230
truss finite element, 73, 74
T-tubules, 230
tumor, 408
tumor cell, 178, 408
tumor vasculature, 408
twitch, 230
two-dimensional finite element, 85
two-dimensional steady flow, 133

unbalanced force, 112
undeformed configuration, 212
uniaxial curve, 204, 210, 211
uniaxial deformation; extension, 23, 343
uniaxial loading; load, 203, 204, 343
uniaxial model, 203
uniaxial tension, 89, 168
uniaxial test, 203
uniqueness theorem, 37
unit normal, 92
unit step force, 36
unit vectors, 13
unloading, 205, 221
unloading curve, 205

unloading regime, 214
unsteady diffusion, 57, 128, 129
unsteady fluid flow, 62, 134, 138, 164
unsteady heat conduction, 54, 126, 127
urinary bladder, 208, 223
uterus, 201

valve aortic, 287
valvular plane, 289
valvular pressure, 289
vascular hydraulic conductivity, 414, 421
vector; components, 3, 6
vein; deformable, 285, 286
vein valve, 283
velocity deviation, 409
velocity distribution, 134, 422
velocity field, 269, 290, 300, 420
velocity fluctuation, 155
velocity gradient, 259
velocity of contraction, 234
velocity profile, 61, 133, 155, 161, 164, 269, 284, 285, 301, 309, 403, 410, 412
velocity vector; field, 61, 103
velocity–load relationship, 229
velocity–pressure interpolation, 129
venous system, 282
ventricle left; right, 287, 289
ventricle valve, 287
ventricular pressure, 290
venula, 250
Verlet algorithm, 154
vertebrae, 315
vessel wall, 271
virtual displacements, 37
virtual strain, 37, 38, 40, 76
virtual vascular surgery, 177
virtual work, 37
viscoelastic; constitutive law, 29, 186, 313
viscoelastic materials, 29
viscoelastic models, 206
viscoelastic response, 36

viscoelasticity, 29, 201, 339
viscosity, 259
viscosity (viscous) coefficient, 35, 59
viscous constitutive law, 30
viscous friction coefficients, 206
viscous nodal force, 159
viscous resistance, 260
viscous stress, 30, 59, 130, 158, 254, 259
virtual displacement, 76
Voigt model, 35
volume flow rate, 421
volume integral, 10, 123
volumetric concentration, 55, 128
volumetric deformation, 215
volumetric force; nodal, 60, 84
volumetric fraction, 215
volumetric source, 383
volumetric strain; strain rate, 29, 30
von Mises stress, 118, 119, 270, 274, 328
Voronoi cell, 151

wall permeability, 422
wall reactions, 34
water concentration ratio, 404
water content, 321, 325
water diffusion coefficient, 398, 399
water diffusion internal, 400
water diffusion modeling, 400
wave equation, 62
waveform, 268
weak form, 122, 131, 357
weighted (weight, weighting) function, 122, 152, 165, 166, 167, 357, 415, 417
weighted quadratic form, 166
weighting method, 122

Young's modulus, 27, 39, 66, 202, 325, 338, 346

zero traction stress, 300
Z-line, 228